中国植物
大化石记录

Record of Megafossil Plants
from China

Ⅸ

Record of Megafossil Plant
New Generic Names
Established for
Chinese Specimens
(1865－2010)

中国植物大化石新属名记录(1865－2010)

吴向午 / 编著

科学技术部科技基础性工作专项
(2013FY113000) 资助

中国科学技术大学出版社

内 容 简 介

本书是"中国植物大化石记录"丛书的一个分册,由内容基本相同的中文和英文两部分组成,共收录 1865—2010 年间正式发表的中国植物大化石新属(或新属名)380 个(其中古生代的新属名 195 个,中生代的新属名 177 个,新生代的新属名 8 个)。书中对每个属的属名(含汉译名)、属的创建者、创建年代、异名表、模式种、分类位置、分布时代、模式标本及标本的存放单位等资料做了详细记录。各部分附有属、种名索引,书末附有参考文献。

本书在广泛查阅国内外古植物学文献和系统采集数据的基础上编写而成,是一份资料收集较齐全、查阅较方便的文献,可供国内外古植物学、生命科学和地球科学的科研、教育及数据库等有关人员参阅。

图书在版编目(CIP)数据

中国植物大化石新属名记录:1865—2010/吴向午编著.—合肥:中国科学技术大学出版社,2023.7

(中国植物大化石记录)

ISBN 978-7-312-04422-9

Ⅰ.中… Ⅱ.吴… Ⅲ.植物化石—中国—1865-2010 Ⅳ.Q914.2

中国国家版本馆 CIP 数据核字(2023)第 047062 号

中国植物大化石新属名记录(1865—2010)

ZHONGGUO ZHIWU DA HUASHI XIN SHUMING JILU (1865—2010)

出版	中国科学技术大学出版社
	安徽省合肥市金寨路 96 号
	http://press.ustc.edu.cn
	https://zgkxjsdxcbs.tmall.com
印刷	合肥华苑印刷包装有限公司
发行	中国科学技术大学出版社
开本	787 mm×1092 mm 1/16
印张	24
插页	1
字数	754 千
版次	2023 年 7 月第 1 版
印次	2023 年 7 月第 1 次印刷
定价	258.00 元

总序

　　古生物学作为一门研究地质时期生物化石的学科，历来十分重视和依赖化石的记录，古植物学作为古生物学的一个分支，亦是如此。对古植物化石名称的收录和编纂，早在19世纪就已经开始了。在K. M. von Sternberg 于1820年开始在古植物研究中采用林奈双名法不久后，F. Unger 就注意收集和整理植物化石的分类单元名称，并于1845年和1850年分别出版了 *Synopsis Plantarum Fossilium* 和 *Genera et Species Plantarium Fossilium* 两部著作，对古植物学科的发展起了历史性的作用。在这以后，多国古植物学家和相关的机构相继编著了古植物化石记录的相关著作，其中影响较大的先后有：由大英博物馆主持，A. C. Seward 等著名学者在19世纪末20世纪初编著的该馆地质分部收藏的标本目录；荷兰 W. J. Jongmans 和他的后继者 S. J. Dijkstra 等用多年时间编著的 *Fossilium Catalogus II : Plantae*；英国 W. B. Harland 等和 M. J. Benton 先后主编的 *The Fossil Record (Volume 1)* 和 *The Fossil Record (Volume 2)*；美国地质调查所出版的由 H. N. Andrews Jr. 及其继任者 A. D. Watt 和 A. M. Blazer 等编著的 *Index of Generic Names of Fossil Plants*，以及后来由隶属于国际生物科学联合会的国际植物分类学会和美国史密森研究院以这一索引作为基础建立的"Index Nominum Genericorum (ING)"电子版数据库等。这些记录尽管详略不一，但各有特色，都早已成为各国古植物学工作者的共同资源，是他们进行科学研究十分有用的工具。至于地区性、断代的化石记录和单位库存标本的编目等更是不胜枚举：早年 F. H. Knowlton 和 L. F. Ward 以及后来的 R. S. La Motte 等对北美白垩纪和第三纪植物化石的记录，S. Ash 编写的美国西部晚三叠世植物化石名录，荷兰 M. Boersma 和 L. M. Broekmeyer 所编的石炭纪、二叠纪和侏罗纪大化石索引，R. N. Lakhanpal 等编写的印度植物化石目录，S. V. Meyen 的植物化石编录以及 V. A. Vachrameev 的有关苏联中生代孢子植物和裸子植物的索引等。这些资料也都对古植物学成果的交流和学科的发展起到了积极的作用。从上述目录和索引不难看出，编著者分布在一些古植物学比较发达、有关研究论著和专业人员众多的国家或地区。显然，目录和索引的编纂，是学科发展到一定阶段的需要和必然的产物，因而代表了这些国家或地区古植物学研究的学术水平和学科发展的程度。

虽然我国地域广大，植物化石资源十分丰富，但古植物学的发展较晚，直到20世纪50年代以后，才逐渐有较多的人员从事研究和出版论著。随着改革开放的深化，国家对科学日益重视，从20世纪80年代开始，我国古植物学各个方面都发展到了一个新的阶段。研究水平不断提高，研究成果日益增多，不仅迎合了国内有关科研、教学和生产部门的需求，也越来越多地得到了国际同行的重视和引用。一些具有我国特色的研究材料和成果已成为国际同行开展相关研究的重要参考资料。在这样的背景下，我国也开始了植物化石记录的收集和整理工作，同时和国际古植物学协会开展的"Plant Fossil Record (PFR)"项目相互配合，编撰有关著作并筹建了自己的数据库。吴向午研究员在这方面是我国起步最早、做得最多的。早在1993年，他就发表了文章《中国中、新生代大植物化石新属索引(1865—1990)》，出版了专著《中国中生代大植物化石属名记录(1865—1990)》。2006年，他又整理发表了1990年以后的属名记录。刘裕生等(1996)则编制了《中国新生代植物大化石目录》。这些都对学科的交流起到了有益的作用。

由于古植物学内容丰富、资料繁多，要对其进行全面、综合和详细的记录，显然是不可能在短时间内完成的。经过多年的艰苦奋斗，现终能根据资料收集的情况，将中国植物化石记录按照银杏植物、真蕨植物、苏铁植物、松柏植物、被子植物等门类，结合地质时代分别编纂出版。与此同时，还要将收集和编录的资料数据化，不断地充实已经初步建立起来的"中国古生物和地层学专业数据库"和"地球生物多样性数据库(GBDB)"。

"中国植物大化石记录"丛书的编纂和出版是我国古植物学科发展的一件大事，无疑将为学科的进一步发展提供良好的基础信息，同时也有利于国际交流和信息的综合利用。作为一个长期从事古植物学研究的工作者，我热切期盼该丛书的出版。

前言

在我国,对植物化石的研究有着悠久的历史。最早的文献记载,可追溯到北宋学者沈括(1031—1095)的《梦溪笔谈》。在该书第 21 卷中,详细记述了陕西延州永宁关(今陕西省延安市延川县延水关)的"竹笋"化石[据邓龙华(1976)考辨,可能为似木贼或新芦木髓模]。此文也对古地理、古气候等问题做了阐述。

和现代植物一样,对植物化石的认识、命名和研究离不开双名法。双名法系瑞典探险家和植物学家 Carl von Linné 于 1753 年在其巨著《植物种志》(*Species Plantarum*)中创立的用于现代植物的命名法。捷克矿物学家和古植物学家 K. M. von Sternberg 在 1820 年开始发表其系列著作《史前植物群》(*Flora der Vorwelt*)时率先把双名法用于化石植物,确定了化石植物名称合格发表的起始点(McNeill et al.,2006)。因此收录于本丛书的现生属、种名以 1753 年后(包括 1753 年)创立的为准,化石属、种名则采用 1820 年后(包括 1820 年)创立的名称。用双名法命名中国的植物化石是从美国史密森研究院(Smithsonian Institute)的 J. S. Newberry [1865(1867)]撰写的《中国含煤地层植物化石的描述》(*Description of Fossil Plants from the Chinese Coal-bearing Rocks*)一文开始的,本丛书对数据的采集时限也以这篇文章的发表时间作为起始点。

我国幅员辽阔,各地质时代地层发育齐全,蕴藏着丰富的植物化石资源。新中国成立后,特别是改革开放以来,随着国家建设的需要,尤其是地质勘探、找矿事业以及相关科学研究工作的不断深入,我国古植物学的研究发展到了一个新的阶段,积累了大量的古植物学资料。据不完全统计,1865(1867)—2000 年间正式发表的中国古植物大化石文献有 2000 多篇[周志炎、吴向午(主编),2002];1865(1867)—1990 年间发表的用于中国中生代植物大化石的属名有 525 个(吴向午,1993a);至 1993 年止,用于中国新生代植物大化石的属名有 281 个(刘裕生等,1996);至 2000 年,根据中国中、新生代植物大化石建立的属名有 154 个(吴向午,1993b,2006)。但这些化石资料零散地刊载于浩瀚的国内外文献之中,使古植物学工作者的查找、统计和引用极为不便,而且有许多文献仅以中文或其他文字发表,不利于国内外同行的引用与交流。

为了便于检索、引用和增进学术交流,编者从 20 世纪 80 年代开始,在广泛查阅文献和系统采集数据的基础上,把这些分散的资料做了系统编录,并进行了系列出版。如先后出版了《中国中生代大植物化石属名记录(1865—1990)》(吴向午,1993a)、《中国中、新生代大植物化石新属索引(1865—1990)》(吴向午,1993b)和《中国中、新生代大植物化石新属记录(1991—2000)》(吴向午,2006)。这些著作仅涉及属名记录,未收录种名信息,因此编写一部包括属、种名记录的中国植物大化石记录显得非常必要。本丛书主要编录1865—2005年间正式发表的中国中生代植物大化石信息。由于篇幅较大,我们按苔藓植物、石松植物、有节植物、真蕨植物、苏铁植物、银杏植物、松柏植物、被子植物等门类分别编写和出版。

本丛书以种和属为编写的基本单位。科、目等不立专门的记录条目,仅在属的"分类位置"栏中注明。为了便于读者全面地了解植物大化石的有关资料,对模式种(模式标本)并非产自中国的属(种),我们也尽可能做了收录。

属的记录:按拉丁文属名的词序排列。记述内容包括属(属名)的创建者、创建年代、异名表、模式种[现生属不要求,但在"模式种"栏以"(现生属)"形式注明]及分类位置等。

种的记录:在每一个属中首先列出模式种,然后按种名的拉丁文词序排列。记录种(种名)的创建者、创建年代等信息。某些附有"aff.""Cf.""cf.""ex gr.""?"等符号的种名,作为一个独立的分类单元记述,排列在没有此种符号的种名之后。每个属内的未定种(sp.)排列在该属的最后。如果一个属内包含两个或两个以上未定种,则将这些未定种罗列在该属的未定多种(spp.)的名称之下,以发表年代先后为序排列。

种内的每一条记录(或每一块中国标本的记录)均以正式发表的为准;仅有名单,既未描述又未提供图像的,一般不做记录。所记录的内容包括发表年代、作者(或鉴定者)的姓名,文献页码、图版、插图、器官名称,产地、时代、层位等。已发表的同一种内的多个记录(或标本),以文献发表年代先后为序排列;年代相同的则按作者的姓名拼音升序排列。如果同一作者同一年内发表了两篇或两篇以上文献,则在年代后加"a""b"等以示区别。

在属名或种名前标有"△"者,表示此属名或种名是根据中国标本建立的分类单元。凡涉及模式标本信息的记录,均根据原文做了尽可能详细的记述。

为了全面客观地反映我国古植物学研究的基本面貌,本丛书一律按原始文献收录所有属、种和标本的数据,一般不做删舍,不做修改,也不做评论,但尽可能全面地引证和记录后来发表的不同见解和修订意见,尤其对于那些存在较大问题的,包括某些不合格发表的属、种名等做了注释。

《国际植物命名法规》(《维也纳法规》)第36.3条规定:自1996年1月1日起,植物(包括孢粉型)化石名称的合格发表,要求提供拉丁文或英文的特征集要和描述。如果仅用中文发表,属不合格发表[McNeill et al.,2006;周志炎,2007;周志炎、梅盛吴(编译),1996,《古植物学简讯》第38期]。为便于读者查证,本记录在收录根据中国标本建立的分类单元时,从1996年起注明原文的发表语种。

为了增进和扩大学术交流,促使国际学术界更好地了解我国古植物学研究现状,所有属、种的记录均分为内容基本相同的中文和英文两个部分。参考文献用英文(或其他西文)列出,其中原文未提供英文(或其他西文)题目的,参考周志炎、吴向午(2002)主编的《中国古植物学(大化石)文献目录(1865－2000)》的翻译格式。

"中国植物大化石记录"丛书的出版,不仅是古植物学科积累和发展的需要,而且将为进一步了解中国不同类群植物化石在地史时期的多样性演化与辐射以及相关研究提供参考,同时对促进国内外学者在古植物学方面的学术交流也会有诸多益处。

本书是"中国植物大化石记录"丛书的一个分册,是在《中国中生代大植物化石属名记录(1865－1990)》、《中国中、新生代大植物化石新属索引(1865－1990)》、《中国中、新生代大植物化石新属记录(1991－2000)》和《中国中、新生代大植物化石新属记录(2001－2010)》(吴向午,1993a、1993b、2006、2013)等文献的基础之上进行编写的。本书除了对中国中生代、新生代大植物化石新属的资料做必要的查核、校正和补遗外,还收录了中国古生代大植物化石新属的有关资料。书中记录数据的采集时间始于1865(1867)年。终止的时间和《中国中、新生代大植物化石新属记录(2001－2010)》(吴向午,2013)一致,延续到2010年12月底。全书共收录1865－2010年间正式发表的中国植物大化石新属名380个(其中依据古生代植物大化石建立的新属名195个、依据中生代植物大化石建立的新属名177个、依据新生代植物大化石建立的新属名8个)。书中对每个属的属名(含汉译名)、属的创建者、年代、异名表、模式种、分类位置及产地层位等作系统编录。对模式种的模式标本的存放单位等资料也做了编撰。分散保存的孢子、花粉不属于当前记录的范畴,未做收录。本记录在文献收录和数据采集中存在的不足、错误和遗漏,请读者多提宝贵意见。

本项工作得到了国家科学技术部科技基础性工作专项(2013FY113000)及国家基础研究发展计划项目(2012CB822003,2006CB700401)、国家自然科学基金项目(No. 41272010)、现代古生物学和地层学国家重点实验室项目(No. 103115)、中国科学院知识创新工程重要方向性项目(ZKZCX2-YW-154)、信息化建设专项(INF105-

SDB-1-42)以及中国科学院科技创新交叉团队项目等的联合资助。

  本书在编写过程中得到了中国科学院南京地质古生物研究所古植物学与孢粉学研究室有关专家和同行的关心与支持,尤其是周志炎院士给予了多方面帮助和鼓励,并撰写总序;南京地质古生物研究所图书馆冯曼、褚存英和袁道俊等协助借阅图书文献;本书的顺利编写和出版与詹仁斌所长、王军所长、现代古生物学和地层学国家重点实验室戎嘉余院士、袁训来主任的关心和帮助是分不开的,编者在此一并致以衷心的感谢。

编　者

目 录

总序 | i

前言 | iii

系统记录 | 1

附录 | 133
  附录1 属名索引 | 133
  附录2 种名索引 | 144

GENERAL FOREWORD | 169

INTRODUCTION | 171

SYSTEMATIC RECORDS | 175

APPENDIXES | 320
  Appendix 1 Index of Generic Names | 320
  Appendix 2 Index of Specific Names | 331

REFERENCES | 352

系 统 记 录

华脉蕨属 Genus *Abropteris* Lee et Tsao,1976
1976　李佩娟、曹正尧,见李佩娟等,100 页。
模式种:*Abropteris virginiensis*(Fontaine)Lee et Tsao,1976
分类位置:真蕨纲紫萁科(Osmundaceae,Filicopsida)
分布与时代:中国四川,美国弗吉尼亚;晚三叠世。

弗吉尼亚华脉蕨 *Abropteris virginiensis* (Fontaine) Lee et Tsao,1976
1883　*Lonchopteris virginiensis* Fontaine,Fontaine W M,53 页,图版 28,图 1,2;图版 29,图 1-4;蕨叶;美国弗吉尼亚;晚三叠世。
1976　李佩娟、曹正尧,见李佩娟等,100 页;蕨叶;美国弗吉尼亚;晚三叠世。

此属创建时同时报道 2 种,除模式种外尚有:

永仁华脉蕨 *Abropteris yongrenensis* Lee et Tsao,1976
1976　李佩娟、曹正尧,见李佩娟等,102 页,图版 12,图 1-3;图版 13,图 6,10,11;插图 3-1;裸羽片;登记号:PB5215-PB5217,PB5220-PB5222;正模:PB5215(图版 12,图 1);标本保存在中国科学院南京地质古生物研究所;四川渡口摩沙河;晚三叠世纳拉箐组大荞地段。

华丽羊齿属 Genus *Abrotopteris* Mo,1980
1980　莫壮观,见赵修祜等,84 页。
模式种:*Abrotopteris guizhouensis*(Gu et Zhi)Mo,1980
分类位置:分类位置不明植物(Plantae incertae sedis)
分布与时代:中国贵州;晚二叠世早期。

贵州华丽羊齿 *Abrotopteris guizhouensis* (Gu et Zhi) Mo,1980
1974　*Glossopteris guizhouensis* Gu et Zhi,中国科学院南京地质古生物研究所、植物研究所,见《中国古生代植物》,137 页,图版 110,图 3,4;叶;登记号:PB4983,PB4984;合模①:PB4983,PB4984(图版 110,图 3,4);标本保存在中国科学院南京地质古生物研究所;贵州盘县;晚二叠世早期宣威组下段。
1980　莫壮观,见赵修祜等,84 页,图版 19,图 1-7,8a(?);叶;登记号:PB4983,PB4984,PB7075-PB7080;标本保存在中国科学院南京地质古生物研究所;贵州盘县;晚二叠世早期宣威组下段。

---

① 注:依据《国际植物命名法规》《维也纳法规》第 37.2 条,1958 年起,模式标本只能是 1 块标本。

刺枝属 Genus *Acanthocladus* Yang,2006（中文和英文发表）

2006 杨关秀等,165,321页。

模式种:*Acanthocladus xyloides* Yang,2006

分类位置:分类位置不明植物(Plantae incertae sedis)

分布与时代:中国河南禹州;晚二叠世。

木质刺枝 *Acanthocladus xyloides* Yang,2006（中文和英文发表）

2006 杨关秀等,165,321页,图版38,图7;茎或枝;正模:HEP0948(图版38,图7);标本保存在中国地质大学(北京);河南禹州大风口、云盖山;晚二叠世云盖山组6段。

刺蕨属 Genus *Acanthopteris* Sze,1931

1931 斯行健,53页。

模式种:*Acanthopteris gothani* Sze,1931

分类位置:真蕨纲蚌壳蕨科(Dicksoniaceae,Filicopsida)

分布与时代:中国辽宁阜新;早侏罗世(Lias)。

高腾刺蕨 *Acanthopteris gothani* Sze,1931

1931 斯行健,53页,图版7,图2—4;蕨叶;辽宁阜新孙家沟;早侏罗世(Lias)。

似槭树属 Genus *Acerites* Pan,1983（nom. nud.）

1983 潘广,1520页。(中文)

1984 潘广,959页。(英文)

模式种:(没有种名)

分类位置:"原始被子植物类群"("primitive angiosperms")

分布与时代:华北燕辽地区;中侏罗世。

似槭树(sp. indet.) *Acerites* sp. indet.

(注:原文仅有属名,没有种名)

1983 *Acerites* sp. indet.,潘广,1520页;华北燕辽地区东段(45°58′N,120°21′E);中侏罗世海房沟组。(中文)

1984 *Acerites* sp. indet.,潘广,959页;华北燕辽地区东段(45°58′N,120°21′E);中侏罗世海房沟组。(英文)

似乌头属 Genus *Aconititis* Pan,1983（nom. nud.）

1983 潘广,1520页。(中文)

1984 潘广,959页。(英文)

模式种:(没有种名)

分类位置:"原始被子植物类群"("primitive angiosperms")

分布与时代:华北燕辽地区;中侏罗世。

似乌头(sp. indet.) *Aconititis* sp. indet.

(注:原文仅有属名,没有种名)

1983　*Aconititis* sp. indet.,潘广,1520页;华北燕辽地区东段(45°58′N,120°21′E);中侏罗世海房沟组。(中文)

1984　*Aconititis* sp. indet.,潘广,959页;华北燕辽地区东段(45°58′N,120°21′E);中侏罗世海房沟组。(英文)

奇叶属 Genus *Acthephyllum* Duan et Chen,1982

1982　段淑英、陈晔,510页。

模式种:*Acthephyllum kaixianense* Duan et Chen,1982

分类位置:裸子植物类(Gymnospermae incertae sedis)

分布与时代:中国四川开县;晚三叠世。

开县奇叶 *Acthephyllum kaixianense* Duan et Chen,1982

1982　段淑英、陈晔,510页,图版11,图1—5;蕨叶;登记号:No. 7173—No. 7176,No. 7219;正模:No. 7219(图版11,图3);标本保存在中国科学院植物研究所;四川开县桐树坝;晚三叠世须家河组。

辐射叶属 Genus *Actinophyllus* Xiao,1985 (non *Actinophyllum* Phillips,1848)

[注:*Actinophyllum* 属的模式种*Actinophyllum plicatum* Phillips,1848(见 Phillips J 和 Salter J W,1848,386页,图版30,图4;藻类?;苏格兰;泥盆纪)]

1985　肖素珍,见肖素珍、张恩鹏,585页。

模式种:*Actinophyllus cordaioides* Xiao,1985

分类位置:分类位置不明植物(Plantae incertae sedis)

分布与时代:中国山西太原;早二叠世。

科达状辐射叶 *Actinophyllus cordaioides* Xiao,1985

1985　肖素珍,见肖素珍、张恩鹏,585页,图版205,图1—3;插图39;叶;登记号:SH721,SH722,SH725;正模:SH722(图版205,图1);副模:SH721(图版205,图2),SH725(图版205,图3);山西太原西山;早二叠世下石盒子组。

剌羊齿属 Genus *Aculeovinea* Li et Taylor,1998(英文发表)

1998　李洪起、Taylor D W,1024页。

模式种:*Aculeovinea yunguiensis* Li et Taylor,1998

分类位置:大羽羊齿目大羽羊齿科(Gigantopteridaceae,Gigantopteridales)

分布与时代:中国贵州盘县;晚二叠世。

云贵刺羊齿 *Aculeovinea yunguiensis* Li et Taylor,1998(英文发表)

1998 李洪起、Taylor D W,1024 页,图 1－3;图 4.25;茎干;正模:PLY02(岩石标本),薄片 Slides PLY02♯C10(L)-1,PLY02♯C10(L)-2,PLY02♯C10-4S1,PLY02♯C10-4S2(图 1.1－1.3);副模:PLY02(岩石标本),薄片 Slides PLY02♯A,PLY02♯A1-1-1-3,PLY02♯A1-1-2-2,PLY02♯A1-2-1-2,PLY02♯C7-6T,PLY02♯C10(L)-2,PLY02♯C15(R-A)-2,PLY02♯A2-3-3-2(图 1.4,2.10－3.23);标本保存在中国科学院植物研究所;贵州盘县;晚二叠世宣威组。

奇异叶属 Genus *Adoketophyllum* Li et Edwards,1992

1992 李承森、Edwards D,259 页。

模式种:*Adoketophyllum subverticillatum* (Li et Cai) Li et Edwards,1992

分类位置:裸蕨类(Psilophytes incertae sedis)

分布与时代:中国云南文山;早泥盆世早期。

亚轮生奇异叶 *Adoketophyllum subverticillatum* (Li et Cai) Li et Edwards,1992

1977 *Zosterophyllum subverticillatum* Li et Cai,李星学、蔡重阳,24 页,图版 3,图 1－3,3a;插图 8;孢子囊穗;采集号:ACE187(图版 3,图 1,2),ACE188(图版 3,图 3,3a);登记号:PB6464－PB6466;正模:PB6466(图版 3,图 3,3a);标本保存在中国科学院南京地质古生物研究所;云南文山古木;早泥盆世坡松冲组。

1992 李承森、Edwards D,259 页,图版 1－4;插图 2;孢子囊穗;云南文山古木;早泥盆世坡松冲组。

奇羊齿属 Genus *Aetheopteris* Chen G X et Meng,1984

1984 陈公信、孟繁松,见陈公信,587 页。

模式种:*Aetheopteris rigida* Chen G X et Meng,1984

分类位置:裸子植物类(Gymnospermae incertae sedis)

分布与时代:中国湖北荆门;晚三叠世。

坚直奇羊齿 *Aetheopteris rigida* Chen G X et Meng,1984

1984 陈公信、孟繁松,见陈公信,587 页,图版 261,图 3,4;图版 262,图 3;插图 133;蕨叶;登记号:EP685;正模:EP685(图版 262,图 3);标本保存在湖北省地质局;副模:图版 261,图 3,4;标本保存在宜昌地质矿产研究所;湖北荆门分水岭;晚三叠世九里岗组。

准爱河羊齿属 Genus *Aipteridium* Li et Yao,1983

1983 李星学、姚兆奇,322 页。

模式种:*Aipteridium pinnatum* (Sixtel) Li et Yao,1983

分类位置:种子蕨纲(Pteridospermopsida)

分布与时代:南费尔干纳;晚三叠世。中国;古生代。

羽状准爱河羊齿 *Aipteridium pinnatum* (Sixtel) Li et Yao,1983
1961　*Aipteris pinnatum* Sixtel,Sixtel T A,153 页,图版 3;南费尔干纳;晚三叠世。
1983　李星学、姚兆奇,322 页。

异麻黄属 Genus *Alloephedra* Tao et Yang,2003(中文和英文发表)
2003　陶君容、杨永,209,212 页。
模式种:*Alloephedra xingxuei* Tao et Yang,2003
分类位置:买麻藤目麻黄科(Ephedraceae,Gnetales)
分布与时代:中国吉林延边地区;早白垩世。

星学异麻黄 *Alloephedra xingxuei* Tao et Yang,2003(中文和英文发表)
2003　陶君容、杨永,209,212 页,图版 1,2;草本状小灌木,带雌球花的枝;标本号:No. 54018a,No. 54018b;模式标本:No. 54018a,No. 54018b(图版 1,图 1);标本保存在中国科学院植物研究所;吉林延边地区;早白垩世大拉子组。

奇异木属 Genus *Allophyton* Wu,1982
1982　吴向午,53 页。
模式种:*Allophyton dengqenensis* Wu,1982
分类位置:真蕨纲?(Filicopsida?)
分布与时代:中国西藏丁青;晚三叠世?。

丁青奇异木 *Allophyton dengqenensis* Wu,1982
1982　吴向午,53 页,图版 6,图 1;图版 7,图 1,2;茎干;采集号:RN0038,RN0040,RN0045;登记号:PB7263—PB7265;正模:PB7263(图版 6,图 1);标本保存在中国科学院南京地质古生物研究所;西藏丁青八达松多;中生代含煤地层(晚三叠世?)。

拟安杜鲁普蕨属 Genus *Amdrupiopsis* Sze et Lee,1952
1952　斯行健、李星学,6,24 页。
模式种:*Amdrupiopsis sphenopteroides* Sze et Lee,1952
分类位置:裸子植物类(Gymnospermae incertae sedis)
分布与时代:中国四川威远;早侏罗世。

楔羊齿型拟安杜鲁普蕨 *Amdrupiopsis sphenopteroides* Sze et Lee,1952
1952　斯行健、李星学,6,24 页,图版 3,图 7—7b;插图 1;蕨叶;标本保存在中国科学院南京地质古生物研究所;四川威远矮山子;早侏罗世。[注:此标本曾改定为 *Amdrupia stenodonta* Harris(徐仁,1954)和 *Amdrupia sphenopteroides* (Sze et Lee) Lee(斯行健、李星学等,1963)]

花穗杉果属 Genus *Amentostrobus* Pan,1983（nom. nud.）
1983　潘广,1520 页。（中文）
1984　潘广,958 页。（英文）
模式种：（仅有属名）
分类位置：松柏纲（Coniferopsida）
分布与时代：华北燕辽地区；中侏罗世。

花穗杉果（sp. indet.） *Amentostrobus* sp. indet.
（注：原文仅有属名,没有种名）
1983　*Amentostrobus* sp. indet.,潘广,1520 页；华北燕辽地区东段（40°58′N,120°21′E）；中侏罗世海房沟组。（中文）
1984　*Amentostrobus* sp. indet.,潘广,958 页；华北燕辽地区东段（40°58′N,120°21′E）；中侏罗世海房沟组。（英文）

疑麻黄属 Genus *Amphiephedra* Miki,1964
1964　Miki S,19,21 页。
模式种：*Amphiephedra rhamnoides* Miki,1964
分类位置：买麻藤纲麻黄科（Ephedraceae,Gnetopsida）
分布与时代：中国辽宁凌源；晚侏罗世。

鼠李型疑麻黄 *Amphiephedra rhamnoides* Miki,1964
1964　Miki S,19,21 页,图版 1,图 F；带叶枝；辽宁凌源；晚侏罗世狼鳍鱼层。

抱囊蕨属 Genus *Amplectosporangium* Geng,1992
1992b　耿宝印,451 页。
模式种：*Amplectosporangium jiagyouense* Geng,1992
分类位置：裸蕨类（Psilophytes incertae sedis）
分布与时代：中国四川江油；早泥盆世。

江油抱囊蕨 *Amplectosporangium jiagyouense* Geng,1992
1992b　耿宝印,451 页,图版 1,图 1－7；图版 2,图 1－8；插图 1；孢子囊穗；标本号：8255－8258,8356－8360；正型：8356（图版 1,图 1）；副型：8357－8360,8255－8258；标本保存在中国科学院植物研究所；四川江油雁门坝；早泥盆世平驿铺组。

窄叶属 Genus *Angustiphyllum* Huang,1983
1983　黄其胜,33 页。

模式种：*Angustiphyllum yaobuense* Huang，1983
分类位置：种子蕨纲（Pteridospermopsida）
分布与时代：中国安徽怀宁；早侏罗世。

腰埠窄叶 *Angustiphyllum yaobuense* Huang，1983
1983　黄其胜，33 页，图版 4，图 1—7；叶；登记号：Ahe8132，Ahe8134—Ahe8138，Ahe8140；正模：Ahe8132，Ahe8134（图版 4，图 1，2）；标本保存在武汉地质学院古生物教研室；安徽怀宁拉犁尖；早侏罗世象山群下部。

似轮叶属 Genus *Annularites* Halle，1927
1927　Halle T G，19 页。
模式种：*Annularites ensilolius* Halle，1927
分类位置：木贼目（Equisetales）
分布与时代：中国山西中部（Ch'en-chia-yu）；晚二叠世早期。

剑瓣似轮叶 *Annularites ensilolius* Halle，1927
1927　Halle T G，19 页，图版 1，图 1—5；图版 2，图 1，2；图版 3，图 1—4；图版 4，图 1—3；枝和叶轮；山西中部（Ch'en-chia-yu）；早二叠世晚期上石盒子系。

此属创建时同时报道 4 种，除模式种外尚有：

舌形似轮叶 *Annularites lingulatus* Halle，1927
1927　Halle T G，26 页，图版 1，图 6；图版 2，图 3，4；图版 6，图 4；枝和叶轮；山西中部（Ch'en-chia-yu）；早二叠世晚期上石盒子系。

平安似轮叶 *Annularites heianensis*（Kodaira）Halle，1927
1924　*Schizoneura heianensis* Kodaira，Kodaira R，163 页，图版 23。
1927　Halle T G，27 页，图版 5，图 13，图 13，14；枝和叶轮；山西中部（Ch'en-chia-yu）；早二叠世晚期上石盒子系。

中国似轮叶 *Annularites sinensis* Halle，1927
1927　Halle T G，28 页，图版 5，图 6—11，12(?)；图版 6(?)，图 8，9；枝和叶轮；山西中部（Ch'en-chia-yu）；早二叠世晚期上石盒子系。

古果属 Genus *Archaefructus* Sun，Dilcher，Zheng et Zhou，1998（英文发表）
1998　孙革、Dilcher D L、郑少林、周浙昆，1692 页。
模式种：*Archaefructus liaoningensis* Sun，Dilcher，Zheng et Zhou，1998
分类位置：双子叶植物纲（Dicotyledoneae）
分布与时代：中国辽西北票；晚侏罗世。

辽宁古果 *Archaefructus liaoningensis* Sun，Dilcher，Zheng et Zhou，1998（英文发表）
1998　孙革、Dilcher D L、郑少林、周浙昆，1692 页，图 2A—2C；被子植物果枝和角质层；标本号：SZ0916；正模：SZ0916（图 2A）；辽西北票上园黄半吉沟；晚侏罗世义县组下部尖山沟层。

始木兰属 Genus *Archimagnolia* Tao et Zhang,1992
1992　陶君容、张川波,423,424 页。
模式种:*Archimagnolia rostrato-stylosa* Tao et Zhang,1992
分类位置:双子叶植物纲(Dicotyledoneae)
分布与时代:中国吉林延吉;早白垩世。

喙柱始木兰 *Archimagnolia rostrato-stylosa* Tao et Zhang,1992
1992　陶君容、张川波,423,424 页,图版 1,图 1-6;着生雌蕊的花托;标本号:503882;正模:503882(图版 1,图 1-6);标本保存在中国科学院植物研究所;吉林延吉;早白垩世大拉子组。

始苹果属 Genus *Archimalus* Tao,1992
1992　陶君容,240,241 页。
模式种:*Archimalus calycina* Tao,1992
分类位置:双子叶植物纲蔷薇科苹果亚科(Maloidea,Rosaceae,Dicotyledoneae)
分布与时代:中国山东临朐山旺;中中新世。

大萼始苹果 *Archimalus calycina* Tao,1992
1992　陶君容,240,241 页,图版 1,图 6-8;两性花;标本号:053880;标本保存在中国科学院植物研究所;山东临朐山旺;中中新世。

华网蕨属 Genus *Areolatophyllum* Li et He,1979
1979　李佩娟、何元良,见何元良等,137 页。
模式种:*Areolatophyllum qinghaiense* Li et He,1979
分类位置:真蕨纲双扇蕨科(Dipteridaceae,Filicopsida)
分布与时代:中国青海都兰;晚三叠世。

青海华网蕨 *Areolatophyllum qinghaiense* Li et He,1979
1979　李佩娟、何元良,见何元良等,137 页,图版 62,图 1,1a,2,2a;蕨叶;采集号:58-7a-12;登记号:PB6327,PB6328;正模:PB6328(图版 62,图 1,1a);副模:PB6327(图版 62,图 2,2a);标本保存在中国科学院南京地质古生物研究所;青海都兰八宝山;晚三叠世八宝山群。

亚洲叶属 Genus *Asiatifolium* Sun,Guo et Zheng,1992
1992　孙革、郭双兴、郑少林,见孙革等,546 页。(中文)
1993　孙革、郭双兴、郑少林,见孙革等,254 页。(英文)

模式种：*Asiatifolium elegans* Sun,Guo et Zheng,1992

分类位置：双子叶植物纲(Dicotyledoneae)

分布与时代：中国黑龙江鸡西；早白垩世。

雅致亚洲叶 *Asiatifolium elegans* Sun,Guo et Zheng,1992

1992　孙革、郭双兴、郑少林，见孙革等，546 页，图版 1，图 1—3；叶；登记号：PB16766，PB16767；正模：PB16766(图版 1，图 1)；标本保存在中国科学院南京地质古生物研究所；黑龙江鸡西城子河；早白垩世城子河组上部。(中文)

1993　孙革、郭双兴、郑少林，见孙革等，253 页，图版 1，图 1—3；叶；登记号：PB16766，PB16767；正模：PB16766(图版 1，图 1)；标本保存在中国科学院南京地质古生物研究所；黑龙江鸡西城子河；早白垩世城子河组上部。(英文)

亚洲羊齿属 Genus *Asiopteris* Zhang,1987

1987　张泓，203 页。

模式种：*Asiopteris huairenensis* Zhang,1987

分类位置：真蕨纲或种子蕨纲(Filices or Pteridospermopsida)

分布与时代：中国山西怀仁；早二叠世。

怀仁亚洲羊齿 *Asiopteris huairenensis* Zhang,1987

1987　张泓，203 页，图版 17，图 1—5；蕨叶；标本号：Hs-No.5；登记号：Mp-85152—Mp-85154；山西怀仁小峪；早二叠世。

星壳斗属 Genus *Astrocupulites* Halle,1927

(注：又名星托属)

1927　Halle T G,219 页。

模式种：*Astrocupulites acuminatus* Halle,1927

分类位置：分类位置不明植物(Plantae incertae sedis)

分布与时代：中国山西中部(Ch'en-chia-yu)；晚二叠世早期。

渐尖星壳斗 *Astrocupulites acuminatus* Halle,1927

(注：又名渐尖星托)

1927　Halle T G,219 页，图版 48，图 10,11；壳斗化石；山西中部(Ch'en-chia-yu)；早二叠世晚期下石盒子系。

白果叶属 Genus *Baiguophyllum* Duan,1987

1987　段淑英，52 页。

模式种：*Baiguophyllum lijianum* Duan,1987

分类位置：茨康目(Czekanowskiales)

分布与时代：中国北京西山；中侏罗世。

利剑白果叶 *Baiguophyllum lijianum* Duan,1987
1987　段淑英,52 页,图版 16,图 4,4a;图版 17,图 1;插图 14;长、短枝及叶;标本号:S-PA-86-680(1),S-PA-86-680(2);正模:S-PA-86-680(2)(图版 17,图 1);标本保存在瑞典历史自然博物馆古植物室;北京西山斋堂;中侏罗世门头沟煤系。

荆棘果属 Genus *Batenburgia* Hilton et Geng,1998(英文发表)
1998　Hilton J、耿宝印,265 页。
模式种:*Batenburgia sakmarica* Hilton et Geng,1998
分类位置:松柏类(Coniferales)
分布与时代:中国河南;早二叠世早期(萨克马尔阶)。

萨克马尔荆棘果 *Batenburgia sakmarica* Hilton et Geng,1998(英文发表)
1998　Hilton J、耿宝印,265 页,图版 1—6;插图 3,4;球果;标本号:CBP9199—CPB9200;正模:CBP9199a(图版 1,图 1),CBP9199b(反面)(图版 1,图 2);标本保存在中国科学院植物研究所;河南观音滩;早二叠世早期(萨克马尔阶)山西组。

北票果属 Genus *Beipiaoa* Dilcher,Sun et Zheng,2001(中文和英文发表)
2001　Dilcher D L、孙革、郑少林,见孙革等,25,151 页。
模式种:*Beipiaoa spinosa* Dilcher,Sun et Zheng,2001
分类位置:被子植物门?(Angiospermae?)
分布与时代:中国辽宁北票;晚侏罗世。

强刺北票果 *Beipiaoa spinosa* Dilcher,Sun et Zheng,2001(中文和英文发表)
2001　Dilcher D L、孙革、郑少林,见孙革等,26,152 页,图版 5,图 1—4,5(?);图版 33,图 11—19;插图 4.7G;果实;登记号:PB18959—PB18962,PB18966,PB18967,ZY3004—ZY3006;正模:PB18959(图版 5,图 1);辽宁北票上园黄半吉沟;晚侏罗世尖山沟组。

此属创建时同时报道 3 种,除模式种外尚有:

小北票果 *Beipiaoa parva* Dilcher,Sun et Zheng,2001(中文和英文发表)
1999　*Trapa*? sp.,吴舜卿,22 页,图版 16,图 1—2a,6(?),6a(?),8(?);果实;辽西北票上园黄半吉沟;晚侏罗世义县组下部尖山沟层。
2001　Dilcher D L、孙革、郑少林,见孙革等,25,151 页,图版 5,图 7;图版 33,图 1—8,21;插图 4.7A;果实;登记号:PB18953,ZY3001—ZY3003;正模:PB18953(图版 5,图 7);辽宁北票上园黄半吉沟;晚侏罗世尖山沟组。

圆形北票果 *Beipiaoa rotunda* Dilcher,Sun et Zheng,2001(中文和英文发表)
2001　Dilcher D L、孙革、郑少林,见孙革等,25,151 页,图版 5,图 8,6(?);图版 33,图 10,9(?);插图 4.7B;果实;登记号:PB18958,ZY3001—ZY3003;正模:PB18958(图版 5,图 8);辽宁北票上园黄半吉沟;晚侏罗世尖山沟组。

本内缘蕨属 Genus *Bennetdicotis* Pan,1983（nom. nud.）

1983　潘广,1520页。（中文）
1984　潘广,958页。（英文）
模式种:（没有种名）
分类位置:"半被子植物类群"（"hemiangiosperms"）
分布与时代:华北燕辽地区;中侏罗世。

本内缘蕨（sp. indet.）*Bennetdicotis* sp. indet.

（注:原文仅有属名,没有种名）
1983　*Bennetdicotis* sp. indet.,潘广,1520页;华北燕辽地区东段（45°58′N,120°21′E）;中侏罗世海房沟组。（中文）
1984　*Bennetdicotis* sp. indet.,潘广,958页;华北燕辽地区东段（45°58′N,120°21′E）;中侏罗世海房沟组。（英文）

本溪羊齿属 Genus *Benxipteris* Zhang et Zheng,1980

1980　张武、郑少林,见张武等,263页。
模式种:*Benxipteris acuta* Zhang et Zheng,1980[注:此属创建时同时报道4种。原文未指定模式种,吴向午（1993）将列在第一的种 *Benxipteris acuta* Zhang et Zheng 选作本属的代表种]
分类位置:种子蕨纲（Pteridospermopsida）
分布与时代:中国辽宁本溪;中三叠世。

尖叶本溪羊齿 *Benxipteris acuta* Zhang et Zheng,1980

1980　张武、郑少林,见张武等,263页,图版108,图1-13;插图193;营养蕨叶和生殖器官;登记号:D323-D335;标本保存在沈阳地质矿产研究所;辽宁本溪林家崴子;中三叠世林家组。

此属创建时同时报道4种,除模式种外尚有:

密脉本溪羊齿 *Benxipteris densinervis* Zhang et Zheng,1980

1980　张武、郑少林,见张武等,264页,图版107,图3-6;插图194;营养蕨叶和生殖器官;登记号:D319-D322;标本保存在沈阳地质矿产研究所;辽宁本溪林家崴子;中三叠世林家组。

裂缺本溪羊齿 *Benxipteris partita* Zhang et Zheng,1980

1980　张武、郑少林,见张武等,265页,图版107,图7-9;图版109,图6-7;蕨叶;登记号:D344-D346,D336-D337;标本保存在沈阳地质矿产研究所;辽宁本溪林家崴子;中三叠世林家组。

多态本溪羊齿 *Benxipteris polymorpha* Zhang et Zheng,1980

1980　张武、郑少林,见张武等,265页,图版109,图1-5;蕨叶;登记号:D338-D342;辽宁本溪林家崴子；中三叠世林家组。

双网羊齿属 Genus *Bicoemplectopteris* Asama, 1959

1959　Asama K,57 页。

模式种：*Bicoemplectopteris hallei* Asama,1959

分类位置：大羽羊齿类（Gigantopterides）

分布与时代：中国辽宁本溪、山西太原、福建和朝鲜；晚古生代。

赫勒双网羊齿 *Bicoemplectopteris hallei* Asama, 1959

1927　*Gigantopteris nicotianaefolia* Halle,Halle T G,162 页,图版 43－44,图 1－13；图版 45,图 1－5；图版 46,图 1；图版 47,图 10；图版 48,图 1(?)－6(?),7。

1959　Asama K,57 页,图版 5,图 1,6；图版 6,图 1－4；图版 7,图 1－3；蕨叶和种子；中国辽宁本溪、山西太原、福建和朝鲜；晚古生代。

双列囊蕨属 Genus *Bifariusotheca* Zhao, 1980

1980　赵修祜等,80 页。

模式种：*Bifariusotheca qinglongensis* Zhao,1980

分类位置：栉羊齿类（Pecopterides）

分布与时代：中国贵州；晚二叠世早期。

晴隆双列囊蕨 *Bifariusotheca qinglongensis* Zhao, 1980

1980　赵修祜等,81 页,图版 12,图 5,5a,6；生殖羽片；采集号：H11-7；登记号：PB7032,PB7033；标本保存在中国科学院南京地质古生物研究所；贵州晴隆；晚二叠世早期。

两瓣叶属 Genus *Bilobphyllum* He, Liang et Shen, 1996（中文和英文发表）

1996　何锡麟、梁敦士、沈树忠,78,165 页。

模式种：*Bilobphyllum fengchengensis* He,Liang et Shen,1996

分类位置：真蕨纲或种子蕨纲（Filices or Pteridospermopsida）

分布与时代：中国江西；晚二叠世。

丰城两瓣叶 *Bilobphyllum fengchengensis* He, Liang et Shen, 1996（中文和英文发表）

1996　何锡麟、梁敦士、沈树忠,78,165 页,图版 79,图 1,2；图版 80,图 1－5；叶；登记号：X88324－X88330；合模：X88324－X88330（图版 79,图 1,2；图版 80,图 1－5）；标本保存在中国矿业大学地质系古生物实验室；江西丰城；晚二叠世乐平组下老山亚段。

鲍斯木属 Genus *Boseoxylon* Zheng et Zhang, 2005（中文和英文发表）

2005　郑少林、张武,见郑少林等,209,212 页。

模式种：*Boseoxylon andrewii*（Bose et Sah）Zheng et Zhang，2005

分类位置：苏铁类（Cycadophytes）

分布与时代：印度拉杰马哈尔；侏罗纪。

安德鲁斯鲍斯木 *Boseoxylon andrewii*（Bose et Sah）Zheng et Zhang，2005（中文和英文发表）

1954　*Sahnioxylon andrewii* Bose et Sah，Bose M N 和 Sah S C D，4 页，图版 2，图 11—18；苏铁类木化石；印度拉杰马哈尔；侏罗纪。

2005　郑少林、张武，见郑少林等，209，212 页；印度拉杰马哈尔；侏罗纪。

似阴地蕨属 Genus *Botrychites* Wu S，1999（中文发表）

1999　吴舜卿，13 页。

模式种：*Botrychites reheensis* Wu S，1999

分类位置：真蕨纲阴地蕨科？（Botrychiaceae?，Filicopsida）

分布与时代：中国辽宁北票；晚侏罗世。

热河似阴地蕨 *Botrychites reheensis* Wu S，1999（中文发表）

1999a　吴舜卿，13 页，图版 4，图 8—10A，10a；图版 6，图 1—3a；营养叶和生殖叶；采集号：AE-233，AE-65，AE-66，AE-233a，AE-233，AE-117，AE-119；登记号：PB18248—PB18253；正模：PB18257（图版 6，图 2）；标本保存在中国科学院南京地质古生物研究所；辽宁北票上园黄半吉沟；晚侏罗世义县组下部尖山沟层。

苞片蕨属 Genus *Bracteophyton* Wang et Hao，2004（英文发表）

2004　王德明、郝守刚，337 页。

模式种：*Bracteophyton variatum* Wang et Hao，2004

分类位置：裸蕨类（Psilophytes incertae sedis）

分布与时代：中国云南曲靖；早泥盆世。

变异苞片蕨 *Bracteophyton variatum* Wang et Hao，2004（英文发表）

2004　王德明、郝守刚，337 页，图 1—5；生殖茎轴；正模：WH9901A（图 1a），WH9901B2（图 3a）；副模：WH9902（图 2）；标本保存在北京大学地质系；云南曲靖徐家冲；早泥盆世徐家冲组。

丽花属 Genus *Callianthus* Wang et Zheng，2009（英文发表）

2009　王鑫、郑少林，800 页。

模式种：*Callianthus dilae* Wang et Zheng，2009

分类位置：被子植物（Angiosperms）

分布与时代：中国辽宁北票；早白垩世。

迪拉丽花 *Callianthus dilae* Wang et Zheng, 2009（英文发表）

2009　王鑫、郑少林, 801 页, 图 1－5；被子植物两性花；登记号：PB21047a, PB21047b, PB18320, PB21091a, PB21091b, PB21092；正模：PB21047a（图 1-A, C, D）, PB21047b（图 1-B-E, G-L）；标本保存在中国科学院南京地质古生物研究所；辽宁北票上园黄半吉沟（41°12′N, 119°22′E）；早白垩世（Barremian）义县组。

似木麻黄属 Genus *Casuarinites* Pan, 1983（nom. nud.）

1983　潘广, 1520 页。（中文）
1984　潘广, 959 页。（英文）
模式种：（没有种名）
分类位置："原始被子植物类群"（"primitive angiosperms"）
分布与时代：华北燕辽地区；中侏罗世。

似木麻黄（sp. indet.）*Casuarinites* sp. indet.

（注：原文仅有属名，没有种名）

1983　*Casuarinites* sp. indet., 潘广, 1520 页；华北燕辽地区东段（45°58′N, 120°21′E）；中侏罗世海房沟组。（中文）
1984　*Casuarinites* sp. indet., 潘广, 959 页；华北燕辽地区东段（45°58′N, 120°21′E）；中侏罗世海房沟组。（英文）

掌裂蕨属 Genus *Catenalis* Hao et Beck, 1991

1991a　郝守刚、Beck C B, 874 页。
模式种：*Catenalis dichotoma* Hao et Beck, 1991
分类位置：裸蕨类（Psilophytes incertae sedis）
分布与时代：中国云南文山；早泥盆世早期。

指状掌裂蕨 *Catenalis digitata* Hao et Beck, 1991

1991a　郝守刚、Beck C B, 874 页, 图 1－24；植物体（具营养枝和生殖枝）；正模：PUH-Ch 801（图 1）；副模：PUH-Ch 803（图 6）, PUH-Ch 807（图 7）, PUH-Ch 806（图 8）；标本保存在北京大学地质系；云南文山；早泥盆世坡松冲组。

中国羊齿属 Genus *Cathaiopteridium* Obrhel, 1966

1966　Obrhel Jiri, 442 页。
模式种：*Cathaiopteridium minutum*（Halle）Obrhel, 1966
分类位置：原始蕨目（Primofilices）
分布与时代：中国云南；早泥盆世。

细小中国羊齿 *Cathaiopteridium minutum*（Halle）Obrhel, 1966

1936　*Protopteridium minutum* Halle, Halle T G, 16 页, 图版 4, 5；插图 2；具有孢子囊的草本

植物;云南;早泥盆世。
1966　Obrhel Jiri,442 页。

华夏穗属 Genus *Cathayanthus* Wang,Tian et Galtier,2003(英文发表)
2003　王士俊、田宝霖、Galtier J,98 页。

模式种:*Cathayanthus ramentrarus*(Wang et Tian)Wang,Tian et Galtier,2003

分类位置:科达植物(Cordaitalean)

分布与时代:中国太原西山;早二叠世早期。

少鳞华夏穗 *Cathayanthus ramentrarus*(Wang et Tian)Wang,Tian et Galtier,2003(英文发表)
1991　*Cardaianthus ramentrarus* Wang et Tian,王士俊、田宝霖,743 页,图版 1—3;插图 1—4;科达植物雄性繁殖器官;山西太原西山煤田;早二叠世早期太原组 7 号煤层。
2003　王士俊、田宝霖、Galtier J,98 页,图 7—9;繁殖器官;正模:Coal balls TN8 Slide W113,T7-70B W133;副模:Coal ball T7-90C,Slide W137;标本保存在中国矿业大学北京研究生部地质研究室;山西太原西山煤田;早二叠世早期太原组 7 号煤层。

此属创建时同时报道 2 种,除模式种外尚有:

中国华夏穗 *Cathayanthus sinensis*(Wang et Tian)Wang,Tian et Galtier,2003(英文发表)
1993　*Cardaianthus sinensis* Wang et Tian,王士俊、田宝霖,760 页,图版 1,图 1—3;图版 2,图 1—4;科达植物雌性繁殖器官;山西太原西山煤田;早二叠世早期太原组 7 号煤层。
2003　王士俊、田宝霖、Galtier J,101 页,图 10d—10h,11a—11h;繁殖器官;正模:Coal ball N61B,Slide W402—W406(图 10g,11a);标本保存在中国矿业大学北京研究生部地质研究室;山西太原西山煤田;早二叠世早期太原组 7 号煤层。

华夏苏铁属 Genus *Cathaysiocycas* Yang,1990
1990　杨关秀,38,41 页。

模式种:*Cathaysiocycas rectanervis* Yang,1990

分类位置:苏铁类(Cycadophytes)

分布与时代:中国河南禹县;早二叠世。

直脉华夏苏铁 *Cathaysiocycas rectanervis* Yang,1990
1990　杨关秀,39,41 页,图版 1,图 1—5;插图 1,2;羽叶;标本号:HEP890—HEP894;合模:HEP890(图版 1,图 1),HEP891(图版 1,图 2),HEP892(图版 1,图 3),HEP893(图版 1,图 4),HEP894(图版 1,图 5);标本保存在中国地质大学(北京)古生物学及地层学研究室;河南禹县;早二叠世早期神后组("山西组")。

华夏木属 Genus *Cathaysiodendron* Lee,1963
1963　李星学,22,126 页。

模式种:*Cathaysiodendron incertum* (Sze et Lee) Lee,1963
分类位置:石松目(Lycopodiales)
分布与时代:中国宁夏、山西、河北、辽宁、内蒙古;石炭纪—二叠纪。

不定华夏木 *Cathaysiodendron incertum* (Sze et Lee) Lee,1963
1945　*Lepidodendron? incertum* Sze et Lee,斯行健、李星学,243 页,图版 1,图 1－3;具叶座的茎干化石;宁夏贺兰山、葫芦斯台;石炭纪太原组下部(?)。
1963　李星学,23,127 页,图版 19,图 6(?);图版 20,图 8(?);图版 21,图 1－6(?);具叶座的茎干化石;登记号:PB3089(图版 19,图 6),PB3096(图版 20,图 8),PB3097,PB855－PB857(图版 21,图 1－6);标本保存在中国科学院南京地质古生物研究所;河北磁县峰峰北大岭;二叠纪山西组下部(?);山西太原西山;石炭纪太原组下部(?);宁夏贺兰山、葫芦斯台;石炭纪太原组(?)。

此属创建时同时报道 3 种,除模式种外尚有:

朱森华夏木 *Cathaysiodendron chuseni* Lee,1963
1963　李星学,24,128 页,图版 21,图 7,8;具叶座的茎干化石;登记号:PB3098(图版 21,图 7,8);标本保存在中国科学院南京地质古生物研究所;辽宁锦西南票苇子沟;二叠纪山西组上部(虹螺岘群上煤组)。

南票华夏木 *Cathaysiodendron nanpiaoense* Lee,1963
1963　李星学,24,129 页,图版 17,图 5;图版 21,图 9,10;具叶座的茎干化石;登记号:PB3082(图版 17,图 5),PB3099(图版 21,图 9,10);标本保存在中国科学院南京地质古生物研究所;辽宁锦西南票三家子;石炭纪太原组下部(虹螺岘群下煤组);内蒙古鄂尔多斯十里铺;石炭纪太原组。

华夏叶属 Genus *Cathaysiophyllum* Lan,Li H et Wang,1982
1982　蓝善先、李汉民、王国平,见李汉民等,367 页。
模式种:*Cathaysiophyllum lobifolium* (Yang et Chen) Lan,Li H et Wang,1982
分类位置:分类位置不明植物(Plantae incertae sedis)
分布与时代:中国江苏、福建、广东;晚二叠世。

裂瓣华夏叶 *Cathaysiophyllum lobifolium* (Yang et Chen) Lan,Li H et Wang,1982
1979　*Adiantites? lobifolius* Yang et Chen,杨关秀、陈芬,110 页,图版 19,图 5,5a;蕨叶;广东;晚二叠世。
1982　蓝善先、李汉民、王国平,见李汉民等,367 页,图版 147,图 4－12;插图 92;蕨叶;江苏镇江;晚二叠世龙潭组;福建龙岩;晚二叠世翠屏山组。

中华羊齿属 Genus *Cathaysiopteridium* Li (MS) ex Mei et Li,1989
1989　梅美棠、李生盛,见黄联盟等,46 页。
模式种:*Cathaysiopteridium fasciculatum* Li (MS) ex Mei et Li,1989
分类位置:种子蕨纲或大羽羊齿类(Pteridospermopsida or Gigantopterides)

分布与时代：中国福建安溪、永定；早二叠世。

束脉中华羊齿 *Cathaysiopteridium fasciculatum* Li (MS) ex Mei et Li,1989
1989　梅美棠、李生盛,见黄联盟等,47 页,图版 26,图 1—4；图版 27,图 15；蕨叶；福建安溪剑斗、永定坎市；早二叠世童子岩组 3 段。

华夏羊齿属 Genus *Cathaysiopteris* Koidzumi,1934
1934　Koidzumi G,113 页。
模式种：*Cathaysiopteris whitei* (Halle) Koidzumi,1934
分类位置：种子蕨纲或大羽羊齿类 (Pteridospermopsida or Gigantopterides)
分布与时代：中国山西中部；早二叠世晚期。

怀特华夏羊齿 *Cathaysiopteris whitei* (Halle) Koidzumi,1934
(注：又名华夏羊齿)
1927　*Gigantopteris whitei* Halle, Halle T G,162 页,图版 47,图 10；蕨叶；山西中部；早二叠世晚期上石盒子系。
1934　Koidzumi G,113 页。

隐囊蕨属 Genus *Celathega* Hao et Gensel,1995
1995　郝守刚、Gensel P G,897 页。
模式种：*Celathega beckii* Hao et Gensel,1995
分类位置：裸蕨类 (Psilophytes incertae sedis)
分布与时代：中国云南文山；早泥盆世。

贝氏隐囊蕨 *Celathega beckii* Hao et Gensel,1995
1995　郝守刚、Gensel P G,897 页,图 1—35；茎轴；正模：BUHG-110 (图 10)；副模：BUHG-101,102,105,116,116',119,126 (图 2—6,13)；标本保存在北京大学地质系；云南文山；早泥盆世坡松冲组。

鹿角蕨属 Genus *Cervicornus* Li et Hueber,2000 (英文发表)
2000　李承森、Hueber F M,116 页。
模式种：*Cervicornus wenshanensis* Li et Hueber,2000
分类位置：石松植物门原始鳞木目 (Protolepidodendrales, Lycophyta)
分布与时代：中国云南文山；早泥盆世。

文山鹿角蕨 *Cervicornus wenshanensis* Li et Hueber,2000 (英文发表)
2000　李承森、Hueber F M,116 页,图版 1,图 1—8；插图 1；草本植物；正模：CBYn8802001 (图版 1,图 1)；标本保存在中国科学院植物研究所；云南文山纸厂峪；早泥盆世坡松冲组。

纤木属 Genus *Chamaedendron* Schweitzer et Li,1996（英文发表）

1996　Schweitzer H J、李承森,45,50 页。

模式种:*Chamaedendron multisporangiatum* Schweitzer et Li,1996

分类位置:石松植物门原始鳞木目（Protolepidodendrales,Lycophyta）

分布与时代:中国湖北武汉;晚泥盆世。

异囊纤木 *Chamaedendron multisporangiatum* Schweitzer et Li,1996（英文发表）

1996　Schweitzer H J、李承森,45,50 页,图版 1—4;插图 4—13;小型树状植物;标本号:CBHb 8912001a, CBHb 8912002a, CBHb 8912003a, CBHb 8912003b, CBHb 8912004a, CBHb 8912006b, CBHb 8912007, CBHb 8912009b, CBHb 8912010a, CBHb 8912011, CBHb 8912013, CBHb 8912014, CBHb 8912015, CBHb 8912016, CBHb 8912017, CBHb 8912018, CBHb 8912019, CBHb 8912020, CBHb 8912021, CBHb 8912022, CBHb 8912024;标本保存在中国科学院植物研究所;湖北武汉;晚泥盆世。

钱耐果属 Genus *Chaneya* Wang et Manchester,2000（英文发表）

2000　王宇飞、Manchester R,169 页。

模式种:*Chaneya tenuis*（Lesquereux）Wang et Manchester,2000

分类位置:被子植物门（Dicotyledoneae）

分布与时代:北美、朝鲜、中国山东和黑龙江;渐新世（或晚始新世）、中新世。

细小钱耐果 *Chaneya tenuis*（Lesquereux）Wang et Manchester,2000（英文发表）

1883　*Porana tenuis* Lesquereux,Lesquereux L,173 页。

2000　王宇飞、Manchester R,169 页,图 2,3;具翼果实;北美;渐新世或晚始新世;黑龙江宜兰煤田;始新世。

此属创建时同时报道 2 种,除模式种外尚有:

科干钱耐果 *Chaneya kokangensis*（Endo）Wang et Manchester,2000（英文发表）

1939　*Porana kokangensis* Endo,Endo S,346 页,图版 23,图 6。

2000　王宇飞、Manchester R,173 页,图 4,5;具翼果实;朝鲜（Tyosen）、中国山东山旺;中新世。

长武蕨属 Genus *Changwuia* Hilton et Li,2000（英文发表）

2000　Hilton J、李承森,10 页。

模式种:*Changwuia schweitzeri* Hilton et Li,2000

分类位置:分类位置不明植物（Plantae incertae sedis）

分布与时代:中国广西长武;早泥盆世中部（Siegenian）。

施魏策尔长武蕨 *Changwuia schweitzeri* Hilton et Li,2000（英文发表）

2000　Hilton J、李承森,10 页,图版 1,图 1—6;插图 1;草本植物;正模:CBG9805001a（正面）,

CBG9805001b(反面)(图版1,图1-6;插图1);标本保存在中国科学院植物研究所;广西长武石桥;早泥盆世(Siegenian)石桥群。

长阳木属 Genus *Changyanophyton* Sze,1952
1952a 斯行健,22 页。(中文)
1952b 斯行健,185 页。(英文)
模式种:*Changyanophyton hupeiense* Sze,1952
分类位置:原始鳞木目?(Protolepidodendrales?)
分布与时代:中国湖北长阳;晚泥盆世。

湖北长阳木 *Changyanophyton hupeiense* Sze,1952
1952a 斯行健,22 页,图版5,图1-3;图版6,图1;茎干;湖北长阳马鞍山;晚泥盆世。(中文)
1952b 斯行健,185 页,图版4,图1-3a;茎干;湖北长阳马鞍山;晚泥盆世。(英文)

朝阳序属 Genus *Chaoyangia* Duan,1997 (1998)(中文和英文发表)
1997 段淑英,519 页。(中文)
1998 段淑英,15 页。(英文)
模式种:*Chaoyangia liangii* Duan,1997 (1998)
分类位置:被子叶植物门(Angiosperm)[注:此属后改归于买麻藤类(Chlamydospermopsida) 或买麻藤目(Gnetales)(郭双兴、吴向午,2000;吴舜卿,1999)]
分布与时代:中国辽宁朝阳;晚侏罗世。

梁氏朝阳序 *Chaoyangia liangii* Duan,1997 (1998)(中文和英文发表)
1997 段淑英,519 页,图1-4;雌性生殖器官;正模:9341(图1,图2);辽宁朝阳;晚侏罗世义县组。(中文)
1998 段淑英,15 页,图1-4;雌性生殖器官;正模:9341(图1,图2);辽宁朝阳;晚侏罗世义县组。(英文)

城子河叶属 Genus *Chengzihella* Guo et Sun,1992
1992 郭双兴、孙革,见孙革等,546 页。(中文)
1993 郭双兴、孙革,见孙革等,254 页。(英文)
模式种:*Chengzihella obovata* Guo et Sun,1992
分类位置:双子叶植物纲(Dicotyledoneae)
分布与时代:中国黑龙江鸡西;早白垩世。

倒卵城子河叶 *Chengzihella obovata* Guo et Sun,1992
1992 郭双兴、孙革,见孙革等,546 页,图版1,图4-9;叶部化石;登记号:PB16768-PB16772;正模:PB16768(图版1,图4);标本保存在中国科学院南京地质古生物研究所;黑龙江鸡西城子河;早白垩世城子河组上部。(中文)

1993　郭双兴、孙革,见孙革等,254页,图版1,图4—9;叶部化石;登记号:PB16768—16772;正模:PB16768(图版1,图4);标本保存在中国科学院南京地质古生物研究所;黑龙江鸡西城子河;早白垩世城子河组上部。(英文)

小蛟河蕨属　Genus *Chiaohoella* Li et Ye,1980
1978　*Chiaohoella* Lee et Yeh,见杨学林等,图版3,图2—4。(裸名)
1980　李星学、叶美娜,7页。
模式种:*Chiaohoella mirabilis* Li et Ye,1980
分类位置:真蕨纲铁线蕨科(Adiantaceae,Filicopsida)
分布与时代:中国吉林蛟河;早白垩世。

奇异小蛟河蕨　*Chiaohoella mirabilis* Li et Ye,1980
1978　*Chiaohoella mirabilis* Lee et Yeh,见杨学林等,图版3,图2—4;蕨叶;吉林蛟河盆地杉松剖面;早白垩世磨石砬子组。(裸名)
1980　李星学、叶美娜,7页,图版2,图7;图版4,图1—3;蕨叶;登记号:PB8970,PB4606,PB4608;正模PB4606(图版4,图1);标本保存在中国科学院南京地质古生物研究所;吉林蛟河杉松;早白垩世中—晚期杉松组。

此属创建时同时报道2种,除模式种外尚有:

新查米叶型小蛟河蕨　*Chiahooella neozamioide* Li et Ye,1980
1980　李星学、叶美娜,8页,图版3,图1;蕨叶;登记号:PB8971;正模PB8971(图版3,图1);标本保存在中国科学院南京地质古生物研究所;吉林蛟河杉松;早白垩世中—晚期杉松组。

吉林羽叶属　Genus *Chilinia* Li et Ye,1980
1980　李星学、叶美娜,7页。
模式种:*Chilinia ctenioides* Li et Ye,1980
分类位置:苏铁纲苏铁目(Cycadales,Cycadopsida)
分布与时代:中国吉林蛟河;早白垩世。

篦羽叶型吉林羽叶　*Chilinia ctenioides* Li et Ye,1980
1980　张武等,273页,图版171,图3;图版172,图4,4a;羽叶;吉林蛟河杉松;早白垩世磨石砬子组。(裸名)
1980　李星学、叶美娜,7页,图版2,图1—6;羽叶和角质层;登记号:PB8966—PB8969;正模:PB8966(图版2,图1);标本保存在中国科学院南京地质古生物研究所;吉林蛟河杉松;早白垩世中—晚期杉松组。

细毛蕨属　Genus *Ciliatopteris* Wu X W,1979
1979　吴向午,见何元良等,139页。

模式种:*Ciliatopteris pecotinata* Wu X W,1979

分类位置:真蕨纲蚌壳蕨科?(Dicksoniaceae?,Filicopsida)

分布与时代:中国青海刚察;早—中侏罗世。

栉齿细毛蕨 *Ciliatopteris pecotinata* Wu X W,1979

1979 吴向午,见何元良等,139 页,图版 63,图 3—6;插图 9;裸羽片和实羽片;采集号:002,003;登记号:PB6339—PB6342;正模:PB6340(图版 63,图 4);副模:PB6339(图版 63,图 4),PB6342(图版 63,图 6);标本保存在中国科学院南京地质古生物研究所;青海刚察海德尔;早—中侏罗世木里群江仓组。

准枝脉蕨属 Genus *Cladophlebidium* Sze,1931

1931 斯行健,4 页。

模式种:*Cladophlebidium wongi* Sze,1931

分类位置:真蕨纲(Filicopsida)

分布与时代:中国江西萍乡;早侏罗世。

翁氏准枝脉蕨 *Cladophlebidium wongi* Sze,1931

1931 斯行健,4 页,图版 2,图 4;蕨叶;江西萍乡;早侏罗世(Lias)。

枝带羊齿属 Genus *Cladotaeniopteris* Zhang et Mo,1981

1981 张善桢、莫壮观,240 页。

模式种:*Cladotaeniopteris shaanxiensis* Zhang et Mo,1981

分类位置:苏铁类植物(Cycadophytes)

分布与时代:中国陕西韩城;早二叠世。

陕西枝带羊齿 *Cladotaeniopteris shaanxiensis* Zhang et Mo,1981

1981 张善桢、莫壮观,240 页,图版 3,图 1—4;插图 2;带状叶(呈串状排列在茎上);正模:PB8773(图版 3,图 1);副模:S51-19(图版 3,图 4);标本保存在中国科学院南京地质古生物研究所;陕西韩城;早二叠世下石盒子组。

似铁线莲叶属 Genus *Clematites* ex Tao et Zhang,1990,emend Wu,1993

[注:此属名为陶君容、张川波(1990)首次使用,但未注明是新属名(吴向午,1993a)]

1990 见陶君容、张川波,221,226 页。

1993a 吴向午,12,217 页。

1993b 吴向午,508,511 页。

模式种:*Clematites lanceolatus* Tao et Zhang,1990

分类位置:双子叶植物毛茛科?(Ranunculaceae?,Dicotyledoneae)

分布与时代:中国吉林延吉;早白垩世。

披针似铁线莲叶 *Clematites lanceolatus* Tao et Zhang,1990
1990　陶君容、张川波,221,226 页,图版 1,图 9;插图 4;叶;标本号:K_1d_{41-3};标本保存在中国科学院植物研究所;吉林延吉;早白垩世大拉子组。
1993a　吴向午,12,217 页。
1993b　吴向午,508,511 页。

普通蕨属 Genus *Coenosophyton* Wang et Xu,2003(英文发表)
2003　王怿、徐洪河,78 页。
模式种:*Coenosophyton tristichus* Wang et Xu,2003
分类位置:种子蕨门(Pteridophyta incertae sedis)
分布与时代:中国安徽巢湖;早石炭世。

三叉普通蕨 *Coenosophyton tristichus* Wang et Xu,2003(英文发表)
2003　王怿、徐洪河,78 页,图 2－7;真蕨植物;正模:PB19285(图 2d);副模:PB19291,PB19292b,PB19294(图 4i,4k,6a);标本保存在中国科学院南京地质古生物研究所;安徽巢湖北山;早石炭世梧桐组。

粘合囊蕨属 Genus *Cohaerensitheca* Liu et Yao,2006(英文发表)
2006　刘陆军、姚兆奇,69 页。
模式种:*Cohaerensitheca sahnii*(Hsu)Liu et Yao,2006
分类位置:观音座莲目合囊蕨科(Marattiaceae,Marattiales)
分布与时代:中国江苏镇江;中二叠世。

沙尼粘合囊蕨 *Cohaerensitheca sahnii* (Hsu) Liu et Yao,2006(英文发表)
1952　*Pecopteris sahnii* Hsu,徐仁,250 页,图版 3,图 30,33。
2006　刘陆军、姚兆奇,70 页,图 1－4;蕨叶;江苏镇江;中二叠世龙潭组。

贝叶属 Genus *Conchophyllum* Schenk,1883
1883　Schenk A,223 页。
模式种:*Conchophyllum richthofenii* Schenk,1883
分类位置:瓢叶目或科达目?(Noeggerathiales or Cordaitales?)
分布与时代:中国河北开平;石炭纪。

李氏贝叶 *Conchophyllum richthofenii* Schenk,1883
(注:又名贝叶)
1883　Schenk A,223 页,图版 42,图 21－26;枝叶;河北开平;石炭纪。

隐羊齿属 Genus *Cryptonoclea* Li,1992

1992 李中明,162 页。

模式种:*Cryptonoclea primitiva* Li,1992

分类位置:大羽羊齿目(Gigantopteridales)

分布与时代:中国贵州水城;晚二叠世。

原始隐羊齿 *Cryptonoclea primitiva* Li,1992

1992 李中明,162 页,图 1-57;整体化石植物,包括种子和孢子;正模和合模标本号:GP2.309-4-3,GP2.312-6-2,GP2.713-2;标本保存在中国科学院植物研究所;贵州水城汪家寨煤矿;晚二叠世汪家寨组。

拟苏铁籽属 Genus *Cycadeoidispermum* Hu et Zhu,1982

1982 胡雨帆、朱家楠,见朱家楠等,80 页。

模式种:*Cycadeoidispermum petiolatum* Hu et Zhu,1982

分类位置:苏铁类植物(Cycadophytes)

分布与时代:中国山西太原;晚二叠世。

具柄拟苏铁籽 *Cycadeoidispermum petiolatum* Hu et Zhu,1982

1982 胡雨帆、朱家楠,见朱家楠等,80 页,图版 2,图 8;插图 3;种子;标本号:B69a;正模:B69a(图版 2,图 8);标本保存在中国科学院植物研究所;山西太原;晚二叠世早期上石盒子组。

苏铁缘蕨属 Genus *Cycadicotis* Pan,1983 (nom. nud.)

[注:此属名后被 Kimura 等(1994)改定为 *Pankuangia*,模式种 *Cycadicotis nissonervis* 改定为 *Pankuangia haifanggouensis* Kimura,Ohana,Zhao et Geng]

1983 潘广,1520 页。(中文)

1983 潘广,见李杰儒,22 页。

1984 潘广,958 页。(英文)

模式种:*Cycadicotis nissonervis* Pan (MS) ex Li,1983

分类位置:"半被子植物类群"中华缘蕨科(Sinodicotiaceae)或苏铁类(Cycadopsida?)

分布与时代:中国辽宁南票;中侏罗世。

蕉羽叶脉苏铁缘蕨 *Cycadicotis nissonervis* Pan (MS) ex Li,1983 (nom. nud.)

1983 潘广,见李杰儒,22 页,图版 2,图 3;叶和雌性生殖器官;标本号:Jp1h2-30;标本保存在辽宁省地质矿产局区域调查地质队;辽宁南票后富隆山盘道沟;中侏罗世海房沟组 3 段。

此属创建时同时报道2种,除模式种外尚有:
苏铁缘蕨(sp. indet.) *Cycadicotis* sp. indet.
(注:原文仅有属名,没有种名)
1983 *Cycadicotis* sp. indet.,潘广,1520页;华北燕辽地区东段(45°58′N,120°21′E);中侏罗世海房沟组。(中文)
1984 *Cycadicotis* sp. indet.,潘广,958页;华北燕辽地区东段(45°58′N,120°21′E);中侏罗世海房沟组。(英文)

苏铁鳞叶属 Genus *Cycadolepophyllum* Yang,1978
1978 杨贤河,510页。
模式种:*Cycadolepophyllum minor* Yang,1978
分类位置:苏铁纲本内苏铁目(Bennettiales,Cycadopsida)
分布与时代:中国四川长宁、广东乐昌;晚三叠世。

较小苏铁鳞叶 *Cycadolepophyllum minor* Yang,1978
1978 杨贤河,510页,图版163,图11;图版175,图4;羽叶;标本号:Sp0041;正模:Sp0041(图版163,图11);标本保存在地质部成都地质矿产研究所;四川长宁双河;晚三叠世须家河组。

此属创建时同时报道2种,除模式种外尚有:
等形苏铁鳞叶 *Cycadolepophyllum aequale* (Brongniart) Yang,1978
1942 *Pterophyllum aequale* (Brongniart) Nathorst,斯行健,189页,图版1,图1—4;羽叶;广东乐昌;晚三叠世—早侏罗世。
1978 杨贤河,510页;羽叶;广东乐昌;晚三叠世。

铁花属 Genus *Cycadostrobilus* Zhu,1994
1994 朱家楠等,341页。
模式种:*Cycadostrobilus paleozoicus* Zhu,1994
分类位置:苏铁类(Cycadophytes)
分布与时代:中国山西太原;晚二叠世。

古生铁花 *Cycadostrobilus paleozoicus* Zhu,1994
1994 朱家楠等,342页,图版1,图1—6;大孢子叶;标本号:PZ4006;正模:PZ4006(图版1,图1);标本保存在中国科学院植物研究所;山西太原东山;早二叠世下石盒子组。

道虎沟叶状体属 Genus *Daohugouthallus* Wang,Krings et Taylor,2010(英文发表)
2010 王鑫、Krings M 和 Taylor T N,592页。
模式种:*Daohugouthallus ciliiferus* Wang,Krings et Taylor,2010

分类位置：地衣植物门（Lichenes）
分布与时代：中国内蒙古宁城；中侏罗世。

细毛道虎沟叶状体 *Daohugouthallus ciiliferus* Wang, Krings et Taylor, 2010（英文发表）
2010　王鑫、Krings M 和 Taylor T N，592 页，图版 1，图 1—5；图版 2，图 1—6；图版 3，图 1—4；叶状体；登记号：PB21398—PB21400；正模：PB21398（图版 1，图 1）；副模：PB21399（图版 1，图 5），PB21400（图版 3，图 1）；标本保存在中国科学院南京地质古生物研究所；内蒙古宁城道虎沟村（119°14.318′E，41°18.979′N）；中侏罗世九龙山组。

大同叶属 Genus *Datongophyllum* Wang, 1984
1984　王自强，281 页。
模式种：*Datongophyllum longipetiolatum* Wang, 1984
分类位置：分类位置未定之银杏目植物（Ginkgoales incertae sedis）
分布与时代：中国山西怀仁；早侏罗世。

长柄大同叶 *Datongophyllum longipetiolatum* Wang, 1984
1984　王自强，281 页，图版 130，图 5—13；带叶的营养枝和生殖枝；标本 7 块；登记号：P0174，P0175（合模），P0176，P0177（合模），P0182，P0179，P0180（合模）；标本保存在中国科学院南京地质古生物研究所；山西怀仁；早侏罗世永定庄组。

此属创建时同时报道 2 种，除模式种外尚有：

大同叶（未定种）*Datongophyllum* sp.
1984　*Datongophyllum* sp.，王自强，282 页，图版 130，图 14；叶；山西怀仁；早侏罗世永定庄组。

华美木属 Genus *Decoroxylon* Zhang et Zheng, 2006（2008）（中文和英文发表）
2006　张武、郑少林，见张武等，94 页。（中文）
2008　张武、郑少林，见张武等，97 页。（英文）
模式种：*Decoroxylon chaoyangense* Zhang et Zheng, 2006（中文），2008（英文）
分类位置：分类位置不明植物（Plantae incertae sedis）
分布与时代：中国辽宁；早二叠世太原组。

朝阳华美木 *Decoroxylon chaoyangense* Zhang et Zheng, 2006（2008）（中文和英文发表）
2006　张武、郑少林，见张武等，97 页，图版 3-46；图版 3-47；图版 3-48；木化石；标本号：Xt-8；正模：Xt-8（图版 3-46；图版 3-47；图版 3-48）；标本保存在沈阳地质矿产研究所；辽宁朝阳薛台子煤矿；早二叠世太原组。（中文）
2008　张武、郑少林，见张武等，97 页，图版 3-46；图版 3-47；图版 3-48；木化石；标本号：Xt-8；正模：Xt-8（图版 3-46；图版 3-47；图版 3-48）；标本保存在沈阳地质矿产研究所；辽宁朝阳薛台子煤矿；早二叠世太原组。（英文）

正三角籽属 Genus *Deltoispermum* Yang, 2006 (中文和英文发表)

2006　杨关秀等, 175, 330 页。
模式种: *Deltoispermum henanense* Yang, 2006
分类位置: 裸子植物种子 (Gymnospermarum)
分布与时代: 中国河南临汝; 晚二叠世。

河南正三角籽 *Deltoispermum henanense* Yang, 2006 (中文和英文发表)

2006　杨关秀等, 175, 330 页, 图版 57, 图 3; 种子; 正模: HEP3441 (图版 57, 图 3); 标本保存在中国地质大学 (北京); 河南临汝坡池; 晚二叠世云盖山组 8 段。

扁囊蕨属 Genus *Demersatheca* Li et Edwards, 1996 (英文发表)

1996　李承森、Edwards D, 79 页。
模式种: *Demersatheca contigua* (Li et Cai) Li et Edwards, 1996
分类位置: 工蕨类 (Zosterophyllophytes)
分布与时代: 中国云南文山; 早泥盆世早期。

紧贴扁囊蕨 *Demersatheca contigua* (Li et Cai) Li et Edwards, 1996 (英文发表)

1977　*Zosterophyllum contiguum* Li et Cai, 李星学、蔡重阳, 24 页, 图版 3, 图 4, 5, 7, 9; 插图 7; 孢子囊穗和孢子囊; 采集号: ACE194 (图版 3, 图 4), ACE186 (图版 3, 图 5), ACE192 (图版 3, 图 7, 9); 登记号: PB6467—PB6470; 标本保存在中国科学院南京地质古生物研究所; 云南文山古木; 早泥盆世坡松冲组。

1996　李承森、Edwards D, 79 页, 图版 1—4; 插图 1, 2; 孢子囊穗和孢子囊; 云南文山古木; 早泥盆世坡松冲组。

登封籽属 Genus *Dengfengia* Yang, 2006 (中文和英文发表)

2006　杨关秀等, 172, 328 页。
模式种: *Dengfengia bifurcata* Yang, 2006
分类位置: 裸子植物种子 (Gymnospermarum)
分布与时代: 中国河南登封; 中二叠世。

双翅登封籽 *Dengfengia bifurcata* Yang, 2006 (中文和英文发表)

2006　杨关秀等, 172, 328 页, 图版 51, 图 8, 9; 图版 74, 图 4, 5; 苞鳞-种鳞复合体; 正模: HEP3068 (图版 51, 图 8); 副模: HEP3069 (图版 51, 图 9); 标本保存在中国地质大学 (北京); 河南登封磴槽; 中二叠世小风口组 4 煤段。

灯笼蕨属 Genus *Denglongia* Xue et Hao, 2008 (英文发表)

2008　薛进庄、郝守刚, 1315 页。

模式种：*Denglongia hubeiensis* Xue et Hao,2008
分类位置：真蕨植物门枝木蕨目（Cladoxylates,Pteridophyta）
分布与时代：中国湖北；晚泥盆世（Frasnian）。

湖北灯笼蕨 *Denglongia hubeiensis* Xue et Hao,2008（英文发表）
2008　薛进庄、郝守刚,1315 页,图 1－14；蕨类植物；正模：PKU-XH120（图 8a,8b）；副模：PKU-XH110,113a,122b,140,144（图 2a,2b;4a,4c,4f;11a）；标本保存在北京大学地质系；湖北长阳；晚泥盆世（Frasnian）黄家蹬组。

牙羊齿属 Genus *Dentopteris* Huang,1992
1992　黄其胜,179 页。
模式种：*Dentopteris stenophylla* Huang,1992
分类位置：裸子植物类（Gymnospermae incertae sedis）
分布与时代：中国四川达县铁山；晚三叠世。

窄叶牙羊齿 *Dentopteris stenophylla* Huang,1992
1992　黄其胜,179 页,图版 18,图 1,1a；蕨叶；登记号：SD87001；标本保存在中国地质大学（武汉）古生物教研室；四川达县铁山；晚三叠世须家河组 7 段。

此属创建时同时报道 2 种,除模式种外尚有：

宽叶牙羊齿 *Dentopteris platyphylla* Huang,1992
1992　黄其胜,179 页,图版 19,图 3,5,7；图版 20,图 13；蕨叶；采集号：SD5；登记号：SD87003－SD87005；正模：SD87003（图版 19,图 7）；标本保存在中国地质大学（武汉）古生物教研室；四川达县铁山；晚三叠世须家河组 3 段。

苏铁籽属 Genus *Dioonocarpus* Hu et Zhu,1982
1982　胡雨帆、朱家楠,见朱家楠等,79 页。
模式种：*Dioonocarpus ovatus* Hu et Zhu,1982
分类位置：苏铁类（Cycadophytes）
分布与时代：中国山西太原；晚二叠世。

卵形苏铁籽 *Dioonocarpus ovatus* Hu et Zhu,1982
1982　胡雨帆、朱家楠,见朱家楠等,79 页,图版 2,图 7；插图 1；种子；标本号：B69；正模：B69（图版 2,图 7）；标本保存在中国科学院植物研究所；山西太原；晚二叠世早期上石盒子组。

盘囊蕨属 Genus *Discalis* Hao,1989
1989a　郝守刚,158 页。
模式种：*Discalis longistipa* Hao,1989

分类位置:工蕨类(Zosterophyllophytes)

分布与时代:中国云南;早泥盆世(Siegenian)。

长柄盘囊蕨 *Discalis longistipa* Hao,1989

1989a 郝守刚,158 页,图版 1,图 1－4;图版 2,图 1－6;图版 3,图 1－11;图版 4,图 1－16;插图 1－5;营养枝和生殖枝;标本号:PUH301－PUH311;正模:PUH301(图版 1,图 1);副模:PUH302－PUH306;标本保存在北京大学地质系;云南文山古木;早泥盆世(Siegenian)坡松冲组。

两列羊齿属 Genus *Distichopteris* Yabe et Shimakura,1940

1940b Yabe H 和 Shimakura M,179 页。

模式种:*Distichopteris heteropinna* Yabe et Shimakura,1940

分类位置:蕨类植物门(Pteridopsida incertae sedis)

分布与时代:中国江苏江宁;中二叠世。

异常两列羊齿 *Distichopteris heteropinna* Yabe et Shimakura,1940

1940b Yabe H 和 Shimakura M,179 页,图版 1,图 1－7;蕨叶(或羽片?);江苏江宁龙潭煤矿;中二叠世龙潭组。〔注:此标本后改定为 *Pecopteris heteropinna* (Yabe et Shimakura) Gu et Zhi(中国科学院南京地质古生物研究所、植物研究所,1974,95 页)〕

缨囊属 Genus *Distichotheca* Gu et Zhi,1974

1974 中国科学院南京地质古生物研究所、中国科学院植物研究所,见《中国古生代植物》,167 页。

模式种:*Distichotheca crossothecoides* Gu et Zhi,1974

分类位置:分类位置不明植物(Plantae incertae sedis)

分布与时代:中国江苏江宁;晚二叠世早期。

具边缨囊 *Distichotheca crossothecoides* Gu et Zhi,1974

(注:又名缨囊)

1974 中国科学院南京地质古生物研究所、植物研究所,见《中国古生代植物》,167 页,图版 129,图 1－4;生殖羽叶;登记号:PB3803(图版 129,图 1),PB3800(图版 129,图 3);正模:PB3800;标本保存在中国科学院南京地质古生物研究所;江苏江宁;晚二叠世早期龙潭组。

对枝柏属 Genus *Ditaxocladus* Guo et Sun,1984

1984 郭双兴、孙喆华,见郭双兴等,126 页。

模式种:*Ditaxocladus planiphyllus* Guo et Sun,1984

分类位置:松柏纲柏科(Cupressaceae,Coniferopsida)

分布与时代:中国新疆阿尔泰;古新世。

扁叶对枝柏 *Ditaxocladus planiphyllus* Guo et Sun,1984
1984　郭双兴、孙喆华,见郭双兴等,128 页,图版 1,图 5,5a,6,6a,8,8a;图版 6,图 8;枝叶;采集号:790HK-3-4,790HK-3-3,790HK-3-42,790HK-3-40;登记号:PB9863,PB9864,PB9866,PB9867;正模:PB9863(图版 1,图 5);副模:PB9864(图版 1,图 6),PB9866(图版 1,图 8),PB9867(图版 6,图 8);标本保存在中国科学院南京地质古生物研究所;新疆阿尔泰;古新世。

龙蕨属 Genus *Dracopteris* Deng,1994
1994　邓胜徽,18 页。
模式种:*Dracopteris liaoningensis* Deng,1994
分类位置:真蕨纲(Filicopsida)
分布与时代:中国辽宁阜新、铁法;早白垩世。

辽宁龙蕨 *Dracopteris liaoningensis* Deng,1994
1994　邓胜徽,18 页,图版 1,图 1—8;图版 2,图 1—15;图版 3,图 1—9;图版 4,图 1—9;插图 2;蕨叶、生殖羽片、囊群和孢子;标本号:Fxt5-086 — Fxt5-090,TDMe622;正模:Fxt5-087(图版 1,图 6);标本保存在石油勘探开发科学研究院;辽宁阜新盆地和铁法盆地;早白垩世阜新组和小明安碑组。

渡口叶属 Genus *Dukouphyllum* Yang,1978
1978　杨贤河,525 页。
模式种:*Dukouphyllum noeggerathioides* Yang,1978
分类位置:苏铁纲(Cycadopsida)[注:杨贤河(1982)后把此属归于银杏目楔拜拉科(Sphenobaieraceae,Ginkgoales)]
分布与时代:中国四川渡口;晚三叠世。

诺格拉齐蕨型渡口叶 *Dukouphyllum noeggerathioides* Yang,1978
1978　杨贤河,525 页,图版 186,图 1—3;图版 175,图 3;叶;标本号:Sp0134 — Sp0137;合模:Sp0134 — Sp0137;标本保存在地质部成都地质矿产研究所;四川渡口摩沙河;晚三叠世大荞地组。

渡口痕木属 Genus *Dukouphyton* Yang,1978
1978　杨贤河,518 页。
模式种:*Dukouphyton minor* Yang,1978
分类位置:苏铁纲本内苏铁目(Bennettiales,Cycadopsida)
分布与时代:中国四川渡口;晚三叠世。

较小渡口痕木 *Dukouphyton minor* Yang,1978
1978　杨贤河,518 页,图版 160,图 2;羽叶;标本号:Sp0021;正模:Sp0021(图版 160,图 2);标

本保存在地质部成都地质矿产研究所；四川渡口摩沙河；晚三叠世大荞地组。

拟爱博拉契蕨属 Genus *Eboraciopsis* Yang,1978

1978　杨贤河,495 页。

模式种：*Eboraciopsis trilobifolia* Yang,1978

分类位置：真蕨纲(Filicopsida)

分布与时代：中国四川渡口；晚三叠世。

三裂叶拟爱博拉契蕨 *Eboraciopsis trilobifolia* Yang,1978

1978　杨贤河,495 页,图版 163,图 6；图版 175,图 5；蕨叶；标本号：Sp0036；正模：Sp0036（图版 163,图 6）；标本保存在地质部成都地质矿产研究所；四川渡口太平场；晚三叠世大荞地组。

织羊齿属 Genus *Emplectopteris* Halle,1927

1927　Halle T G,119 页。

模式种：*Emplectopteris trangularis* Halle,1927

分类位置：种子蕨纲(Pteridospermopsida)

分布与时代：中国山西中部；早二叠世晚期。

三角织羊齿 *Emplectopteris trangularis* Halle,1927

1927　Halle T G,122 页,图版 31；蕨叶；山西中部；早二叠世晚期下石盒子系。

始水松属 Genus *Eoglyptostrobus* Miki,1964

1964　Miki S,14,21 页。

模式种：*Eoglyptostrobus sabioides* Miki,1964

分类位置：松柏纲松柏目(Coniferales,Coniferopsida)

分布与时代：中国辽西凌源；晚侏罗世。

清风藤型始水松 *Eoglyptostrobus sabioides* Miki,1964

1964　Miki S,14,21 页,图版 1,图 E；枝叶；辽西凌源；晚侏罗世狼鳍鱼层。

始团扇蕨属 Genus *Eogonocormus* Deng,1995（non Deng,1997）

1995　邓胜徽,14,108 页。

模式种：*Eogonocormus cretaceum* Deng,1995

分类位置：真蕨纲膜蕨科(Hymenophyllaceae,Filicopsida)

分布与时代：中国内蒙古霍林河盆地；早白垩世。

白垩始团扇蕨 *Eogonocormus cretaceum* Deng,1995（non Deng,1997）

1995　邓胜徽,14,108 页,图版 3,图 1-2；图版 4,图 1-2,6-8；图版 5,图 1-6；插图 4；营

养叶和生殖叶;标本号:H17-431;标本保存在石油勘探开发科学研究院;内蒙古霍林河盆地;早白垩世霍林河组。

此属创建时同时报道 2 种,除模式种外尚有:
线形始团扇蕨 *Eogonocormus linearifolium* (Deng) Deng,1995
1993 *Hymenophyllites linearifolius* Deng,邓胜徽,256 页,图版 1,图 5—7;插图 d—f;蕨叶和生殖羽片;内蒙古霍林河盆地;早白垩世霍林河组。
1995 邓胜徽,17,108 页,图版 3,图 3—4;营养叶和生殖叶;标本号:H14-509,H14-510;标本保存在石油勘探开发科学研究院;内蒙古霍林河盆地;早白垩世霍林河组。

始团扇蕨属 Genus *Eogonocormus* Deng,1997 (non Deng,1995)(英文发表)
(注:此属名为 *Eogonocormus* Deng,1995 的晚出等同名)
1997 邓胜徽,60 页。
模式种:*Eogonocormus cretaceum* Deng,1997
分类位置:真蕨纲膜蕨科(Hymenophyllaceae,Filicopsida)
分布与时代:中国内蒙古霍林河盆地;早白垩世。

白垩始团扇蕨 *Eogonocormus cretaceum* Deng,1997 (non Deng,1995)(英文发表)
(注:此种名为白垩始团扇蕨 *Eogonocormus cretaceum* Deng,1995 的晚出等同名)
1997 邓胜徽,60 页,图 2—5;营养叶和生殖叶;标本号:H17-431;正模:H17-431(图 3a);标本保存在石油勘探开发科学研究院;内蒙古霍林河盆地;早白垩世霍林河组。

始羽蕨属 Genus *Eogymnocarpium* Li,Ye et Zhou,1986
1986 李星学、叶美娜、周志炎,14 页。
模式种:*Eogymnocarpium sinense* (Li et Ye) Li,Ye et Zhou,1986
分类位置:真蕨纲蹄盖蕨科(Athyriaceae,Filicopsida)
分布与时代:中国吉林蛟河;早白垩世。

中国始羽蕨 *Eogymnocarpium sinense* (Li et Ye) Li,Ye et Zhou,1986
1978 *Dryopterites sinense* Lee et Yeh,见杨学林等,图版 2,图 3,4;图版 3,图 7;蕨叶;吉林蛟河盆地杉松剖面;早白垩世磨石砬子组。(裸名)
1980 *Dryopterites sinense* Li et Ye,李星学、叶美娜,6 页,图版 1,图 1—5;生殖羽片;吉林蛟河杉松;早白垩世中—晚期杉松组。
1986 李星学、叶美娜、周志炎,14 页,图版 12;图版 13;图版 14,图 1—6;图版 15,图 5—7a;图版 16,图 3;图版 40,图 4;图版 45,图 1—3;插图 4A,4B;生殖蕨叶;吉林蛟河杉松(127°15′E,43°30′N);早白垩世蛟河群。

始鳞木属 Genus *Eolepidodendron* Wu et Zhao,1981
1981 吴秀元、赵修祜,54 页。

模式种：*Eolepidodendron jurongense* Wu et Zhao,1981

分类位置：石松植物门原始鳞木目（Protolepidodendrales,Lycophyta）

分布与时代：中国江苏句容、无锡；早石炭世。

句容始鳞木 *Eolepidodendron jurongense* Wu et Zhao,1981

1981　吴秀元、赵修祜,54页,图版1,图5,5a；茎干；采集号：MWS-66；登记号：PB7494；标本保存在中国科学院南京地质古生物研究所；江苏句容高资；早石炭世高骊山组。

此属创建时同时报道3种,除模式种外尚有：

无锡始鳞木 *Eolepidodendron wusihense* (Sze) Wu et Zhao,1981

1936　*Lepidodendron* aff. *leeianum* Gothan et Sze (? n. sp.),斯行健,141页,图版2,图1—6；图版3,图1,2；图版5,图1,2；茎干；江苏无锡；早石炭世乌桐石英岩。

1943　*Lepidodendron wusihense* Sze,斯行健,63页。

1956c　*Sublepidodendron wusihense* (Sze) Sze,斯行健,49页。

1981　吴秀元、赵修祜,54页。

无锡始鳞木（比较种）*Eolepidodendron* cf. *wusihense* (Sze) Wu et Zhao,1981

1981　吴秀元、赵修祜,54页,图版1,图7；茎干；江苏句容高资；早石炭世高骊山组。

始鳞木（未定种）*Eolepidodendron* sp.

1981　*Eolepidodendron* sp.,吴秀元、赵修祜,54页,图版1,图6；茎干；江苏句容高资镇；早石炭世高骊山组。

始叶羊齿属 Genus *Eophyllogonium* Mei,Dilcher et Wan,1992

1992　梅美棠、Dilcher D L、万志辉,99页。

模式种：*Eophyllogonium cathayense* Mei,Dilcher et Wan,1992

分类位置：带羊齿类（Taeniopterides）

分布与时代：中国江西乐平；二叠纪。

华夏始叶羊齿 *Eophyllogonium cathayense* Mei,Dilcher et Wan,1992

1992　梅美棠、Dilcher D L、万志辉,99页,图版1,图1—5；图版2,图1—4；图版3,图2；图版4,图1—5；图版5,图1—4；插图1；带种子的叶；正模：X 9-005（图版1,图3）；副模：X 9-147（图版2,图3），X 9-282（图版2,图1）；标本保存在中国矿业大学北京研究生部地质研究室；江西乐平；二叠纪乐平组关山段。

始叶蕨属 Genus *Eophyllophyton* Hao,1988

1988　郝守刚,442页。

模式种：*Eophyllophyton bellum* Hao,1988

分类位置：裸蕨类（Psilophytes incertae sedis）

分布与时代：中国云南文山；早泥盆世。

优美始叶蕨 *Eophyllophyton bellum* Hao, 1988
1988 郝守刚,442 页,图版 1—3;插图 1,2;植物体;正模:BUPb101(图版 1,图 1);副模:BUPb113,152,131(图版 1,图 2,6,7),BUPb110,127(图版 2,图 1,2,11) BUPb137,116,121,112(图版 3,图 1,2,4,6);标本保存在北京大学地质系;云南文山古木乡;早泥盆世坡松冲组。

似画眉草属 Genus *Eragrosites* Cao et Wu S Q, 1997 (1998) (中文和英文发表)
[注:此属模式种后被郭双兴、吴向午(2000)改归于买麻藤类(Chlamydospermopsida)或买麻藤目(Gnetales)的 *Ephedrites* 属,定名为 *Ephedrites chenii* (Cao et Wu S Q) Guo et Wu X W;被吴舜卿改归于买麻藤目(Gnetales),定名为 *Liaoxia chenii* (Cao et Wu S Q) Wu S Q(吴舜卿,1999)]
1997 曹正尧、吴舜卿,见曹正尧等,1765 页。(中文)
1998 曹正尧、吴舜卿,见曹正尧等,231 页。(英文)
模式种:*Eragrosites changii* Cao et Wu S Q, 1997 (1998)
分类位置:单子叶植物纲禾本科(Gramineae,Monocotyledoneae)
分布与时代:中国辽西北票;晚侏罗世。

常氏似画眉草 *Eragrosites changii* Cao et Wu S Q, 1997 (1998) (中文和英文发表)
1997 曹正尧、吴舜卿,见曹正尧等,1765 页,图版 2,图 1—3;插图 1;草本植物、花枝;登记号:PB17801—PB17802;正模:PB17803(图版 2,图 2);标本保存在中国科学院南京地质古生物研究所;辽西北票上园炒米店附近;晚侏罗世义县组尖山沟层。(中文)
1998 曹正尧、吴舜卿,见曹正尧等,231 页,图版 2,图 1—3;插图 1;草本植物、花枝;登记号:PB17801—PB17802;正模:PB17803(图版 2,图 2);标本保存在中国科学院南京地质古生物研究所;辽西北票上园炒米店附近;晚侏罗世义县组尖山沟层。(英文)

似杜仲属 Genus *Eucommioites* ex Tao et Zhang, 1992
[注:此属名为陶君容、张川波(1992)首次使用,见于 *Eucommioites orientalis* Tao et Zhang,1992]
1992 见陶君容、张川波,423,425 页。
模式种:*Eucommioites orientalis* Tao et Zhang, 1992
分类位置:双子叶植物纲(Dicotyledoneae)
分布与时代:中国吉林延吉;早白垩世。

东方似杜仲 *Eucommioites orientalis* Tao et Zhang, 1992
1992 陶君容、张川波,423,425 页,图版 1,图 7—9;翅果;标本号:503883;正模:503883(图版 1,图 7—9);标本保存在中国科学院植物研究所;吉林延吉;早白垩世大拉子组。

准束羊齿属 Genus *Fascipteridium* Zhang et Mo, 1985
1985 张善桢、莫壮观,175 页。

模式种：*Fascipteridium ellipticum* Zhang et Mo,1985

分类位置：种子蕨纲(Pteridospermopsida)

分布与时代：中国河南；晚二叠世。

椭圆准束羊齿 *Fascipteridium ellipticum* Zhang et Mo,1985

1985 张善桢、莫壮观,175页,图版3,图1—5；着生种子的蕨叶；登记号：PB8114—PB8117；合模：PB8114(图版3,图1),PB8117(图版3,图5)；标本保存在中国科学院南京地质古生物研究所；河南；晚二叠世上石盒子系。

束羊齿属 Genus *Fascipteris* Gu et Zhi,1974

1974 中国科学院南京地质古生物研究所、植物研究所,见《中国古生代植物》,99页。

模式种：*Fascipteris hallei* (Kawasaki) Gu et Zhi,1974

分类位置：真蕨纲或种子蕨纲(Filices or Pteridospermopsida)

分布与时代：中国河北、山西、江苏；二叠纪。

赫勒束羊齿 *Fascipteris hallei* (Kawasaki) Gu et Zhi,1974

(注：又名弧束羊齿)

1939 *Validopteris hallei* (Kawasaki) Stockmans et Mathieu, Stockmans F 和 Mathieu F F,75页,图版34,图1；河北开平；晚二叠世早期。

1974 中国科学院南京地质古生物研究所、植物研究所,见《中国古生代植物》,99页,图版68,图8—12；插图83.1；羽叶；山西太原；晚二叠世早期上石盒子组；河北开平；晚二叠世早期古冶组。

此属创建时同时报道5种,除模式种外尚有：

垂束羊齿 *Fascipteris recta* Gu et Zhi,1974

1927 *Pecopteris* sp. a, Halle T H,100页,图版23,图4—13；羽片；山西中部；晚二叠世上石盒子组。

1974 中国科学院南京地质古生物研究所、植物研究所,见《中国古生代植物》,100页,图版69,图1—4；羽片；山西太原；晚二叠世早期上石盒子组。

中国束羊齿 *Fascipteris sinensis* (Stockmans et Mathieu) Gu et Zhi,1974

1957 *Validopteris sinensis* Stockmans et Mathieu, Stockmans F 和 Mathieu F F,22页,图版13,图2,3；河北开平；晚二叠世。

1974 中国科学院南京地质古生物研究所、植物研究所,见《中国古生代植物》,100页,图版69,图5—7；插图84.1；小羽片；河北开平，晚二叠世早期古冶组。

密囊束羊齿(皱囊蕨) *Fascipteris* (*Ptychocarpus*) *densata* Gu et Zhi,1974

1974 中国科学院南京地质古生物研究所、植物研究所,见《中国古生代植物》,100页,图版69,图8—14；插图85—86；羽叶；登记号：PB3686,PB3688,PB3690；合模：PB3686(图版69,图8),PB3688(图版69,图10),PB3690(图版69,图12)；标本保存在中国科学院南京地质古生物研究所；江苏江宁；晚二叠世早期龙潭组。

狭束羊齿 *Fascipteris stena* Gu et Zhi,1974

1974 中国科学院南京地质古生物研究所、植物研究所,见《中国古生代植物》,101页,图版

69,图 15—17;插图 84.2;小羽片;登记号:PB3697,PB3699;正模:PB3699(图版 69,图 15);标本保存在中国科学院南京地质古生物研究所;江苏江宁;晚二叠世早期龙潭组。

羊齿缘蕨属 Genus *Filicidicotis* Pan,1983 (nom. nud.)
1983　潘广,1520 页。(中文)
1984　潘广,958 页。(英文)
模式种:(没有种名)
分类位置:"半被子植物类群"("hemiangiosperms")
分布与时代:华北燕辽地区;中侏罗世。

羊齿缘蕨(sp. indet.) *Filicidicotis* sp. indet.
(注:原文仅有属名,没有种名)
1983　*Filicidicotis* sp. indet.,潘广,1520 页;华北燕辽地区东段(45°58′N,120°21′E);中侏罗世海房沟组。(中文)
1984　*Filicidicotis* sp. indet.,潘广,958 页;华北燕辽地区东段(45°58′N,120°21′E);中侏罗世海房沟组。(英文)

纤细蕨属 Genus *Filiformorama* Wang,Hao et Cai,2006(英文发表)
2006　王怿、郝守刚、蔡重阳,见王怿等,25 页。
模式种:*Filiformorama simplexa* Wang,Hao et Cai,2006
分类位置:瑞尼蕨类?(?Rhynophytoid incertae sedis)
分布与时代:中国新疆准噶尔盆地;晚志留世(Late Pridoli)。

简单纤细蕨 *Filiformorama simplexa* Wang,Hao et Cai,2006(英文发表)
2006　王怿、郝守刚、蔡重阳,见王怿等,25 页,图 1—4;植物体光滑、二分叉;正模:PB20338(图 1C,1D);副模:PB20336,PB20337,PB20339—PB20346(图 1A,F—K;图 3A—J);标本保存在中国科学院南京地质古生物研究所;新疆准噶尔盆地;晚志留世(Late Pridoli)乌图布拉克组中部。

睫囊蕨属 Genus *Fimbriotheca* Zhu et Chen,1981
1981　朱家楠、陈公信,488 页。
模式种:*Fimbriotheca tomentosa* Zhu et Chen,1981
分类位置:真蕨纲观音座莲目(Marattiales,Filicopsida)
分布与时代:中国湖北;晚二叠世。

毛状睫囊蕨 *Fimbriotheca tomentosa* Zhu et Chen,1981
1981　朱家楠、陈公信,488 页,图版 1,图 1—7;插图 1;生殖羽片;标本号:2004/55-13,2004/55-14;湖北阳新大王殿;晚二叠世早期。

似茎状地衣属 Genus *Foliosites* Ren,1989
［注：此属原归于地衣植物门，但有人认为有属于苔藓植物门的可能（吴向午、厉宝贤,1992）］
1989　任守勤,见任守勤、陈芬,634,639 页。
1992　吴向午、厉宝贤,272 页。
模式种：*Foliosites formosus* Ren,1989
分类位置：地衣植物门？（Lichenes?）或苔藓植物门？（Bryophytes?）
分布与时代：中国内蒙古海拉尔；早白垩世。

美丽似茎状地衣 *Foliosites formosus* Ren,1989
1989　任守勤,见任守勤、陈芬,634,639 页,图版 1,图 1－4;插图 1;叶状体;登记号：HW043, HW044,HWS012;正模：HW043（图版 1,图 1）;标本保存在中国地质大学（北京）;内蒙古海拉尔五九煤盆地；早白垩世大磨拐河组。
1992　吴向午、厉宝贤,272 页。

福建羊齿属 Genus *Fujianopteris* Liu et Yao,2004（中文和英文发表）
2004　刘陆军、姚兆奇,474,480 页。
模式种：*Fujianopteris fukianensis* (Yabe et Ôishi) Liu et Yao,2004
分类位置：种子蕨纲或大羽羊齿类（Pteridospermopsida or Gigantopterides）
分布与时代：中国福建、广东；二叠纪。

闽福建羊齿 *Fujianopteris fukianensis* (Yabe et Ôishi) Liu et Yao,2004（中文和英文发表）
1938　*Gigantopteris fukianensis* Yabe et Ôishi,Yabe H 和 Ôishi S,231 页,图版 32,图 8,9, 12C;插图 9（非插图 10）;福建龙岩；二叠纪。
2004　刘陆军、姚兆奇,478,484 页,图版 1,图 2,79;图版 2,图 1;插图 6;蕨叶;福建龙岩苏邦、南靖长塔;中二叠世（Guadalupian）童子岩组上段。

此属创建时同时报道 4 种,除模式种外尚有：

狭角福建羊齿 *Fujianopteris angustiangla* (Yang et Chen) Liu et Yao,2004（中文和英文发表）
1979　*Gigantonoclea angustiangla* Yang et Chen,杨关秀、陈芬,127 页,图版 37,图 2;插图 40;蕨叶;广东仁化；晚二叠世龙潭组。
2004　刘陆军、姚兆奇,475,482 页;插图 2;蕨叶;广东仁化;"晚二叠世龙潭组"（层位与中二叠世童子岩组上段相当）。

枝脉福建羊齿 *Fujianopteris cladonervis* (S Li) Liu et Yao,2004（中文和英文发表）
1989　*Gigantonoclea cladonervis* S Li,李生盛,见黄联盟等,49 页,图版 29,图 1－3;图版 30, 图 1－3;蕨叶;福建安溪剑斗；早二叠世。
2004　刘陆军、姚兆奇,476,482 页;插图 3;蕨叶;福建南靖长塔、安溪剑斗;中二叠世（Guadalupian）童子岩组上段。

中间福建羊齿 *Fujianopteris intermedia* Liu et Yao,2004（中文和英文发表）
2004　刘陆军、姚兆奇,477,483页,图版1,图1,3—6;图版2,图2—5;插图4,5;蕨叶;登记号:PB9154,PB9155,PB9156,PB9157,PB9158,PB9160,PB9161;正模:PB9158(图版1,图5);副模:PB9154(图版1,图1),PB9155(图版2,图2),PB9156(图版1,图3),PB9157(PB9156标本负面),PB9160(图版1,图1),PB9161(图版2,图3);标本保存在中国科学院南京地质古生物研究所;福建龙岩、永定;中二叠世(Guadalupian)童子岩组上段。

甘肃芦木属 Genus *Gansuphyllite* Xu et Shen,1982
1982　徐福祥、沈光隆,见刘子进,118页。
模式种:*Gansuphyllite multivervis* Xu et Shen,1982
分类位置:楔叶纲木贼目(Equisetales,Sphenopsida)
分布与时代:中国甘肃武都;中侏罗世。

多脉甘肃芦木 *Gansuphyllite multivervis* Xu et Shen,1982
1982　徐福祥、沈光隆,见刘子进,118页,图版58,图5;茎和轮生叶;标本号:LP00013-3;甘肃武都大岭沟;中侏罗世龙家沟组上部。

双生叶属 Genus *Geminofoliolum* Zeng,Shen et Fan,1995
1995　曾勇、沈树忠、范炳恒,49,76页。
模式种:*Geminofoliolum gracilis* Zeng,Shen et Fan,1995
分类位置:楔叶纲芦木科(Calamariaceae,Sphenopsida)
分布与时代:中国河南义马;中侏罗世。

纤细双生叶 *Geminofoliolum gracilis* Zeng,Shen et Fan,1995
1995　曾勇、沈树忠、范炳恒,49,76页,图版7,图1—2;插图9;茎干;采集号:No. 117146,No. 117144;登记号:YM94031,YM94032;正模:YM94032(图版7,图2);副模:YM94031(图版7,图1);标本保存在中国矿业大学地质系;河南义马;中侏罗世义马组。

单网羊齿属 Genus *Gigantonoclea* Koidzumi,1936
1936　Koidzumi G,138页。
模式种:*Gigantonoclea lagrelii*（Halle）Koidzumi,1938
分类位置:大羽羊齿类种子蕨纲(Pteridospermopsida,Gigantopterides)
分布与时代:中国山西中部;早二叠世晚期。

波缘单网羊齿 *Gigantonoclea lagrelii*（Halle）Koidzumi,1936
（注:又名华夏羊齿）
1927　*Gigantopteris lagrelii* Halle,Halle T G,170页,图版46;蕨叶;山西中部;早二叠世晚

期上石盒子系。
1936　Koidzumi G,138 页。

带囊蕨属 Genus *Gigantonomia* Li et Yao,1983

1983　李星学、姚兆奇,14 页。
模式种:*Gigantonomia*（*Gigatonoclea*）*fukienensis*（Yabe et Ôishi）Li et Yao,1983
分类位置:大羽羊齿类种子蕨纲（Pteridospermopsida,Gigantopterides）
分布与时代:华南;二叠纪。

福建带囊蕨（单网羊齿）*Gigantonomia*（*Gigatonoclea*）*fukienensis*（Yabe et Ôishi）Li et Yao,1983

1938　*Gigatonoclea fukienensis* Yabe et Ôishi,Yabe H 和 Ôishi S,231 页,图版 1,图 8,9,12C;插图 9,12;福建龙岩;二叠纪。
1983　李星学、姚兆奇,14 页,图版 1,图 1－3;图版 2,图 1－4;图版 3,图 1－5;插图 1－4;大羽羊齿类生殖器官;华南;二叠纪。

大羽羊齿属 Genus *Gigantopteris* Schenk ex Whit D,1912

1883　*Megalopteris* Schenk,Schenk A,238 页。
1912　见 Whit D,494 页。
模式种:*Gigantopteris nicotianaefolia* Schenk ex Whit D,1912
分类位置:大羽羊齿类种子蕨纲（Pteridospermopsida,Gigantopterides）
分布与时代:中国湖南、江苏、云南、河南;晚石炭世（晚二叠世）。

烟叶大羽羊齿 *Gigantopteris nicotianaefolia* Schenk ex Whit D,1912

1883　*Megalopteris nicotianaefolia* Schenk,Schenk A,238 页,图版 32,图 6－8;图版 33,图 1－3;图版 35,图 6;蕨叶;湖南耒阳;晚石炭世。
1912　见 Whit D,494 页。
1974　中国科学院南京地质古生物研究所、植物研究所,见《中国古生代植物》,130 页,图版 100,图 2－4;图版 101,图 1;图版 102,图 7;插图 103－105,108;蕨叶;中国湖南、江苏、云南、河南;晚二叠世。

大囊蕨属 Genus *Gigantotheca* Li et Yao,1983

1983　李星学、姚兆奇,19 页。
模式种:*Gigantotheca paradoxa* Li et Yao,1983
分类位置:大羽羊齿类种子蕨纲（Pteridospermopsida,Gigantopterides）
分布与时代:华南;二叠纪。

奇异大囊蕨 *Gigantotheca paradoxa* Li et Yao,1983

1983　李星学、姚兆奇,20 页,图版 4,图 1－3;图版 5,图 1－4;图版 6,图 1－5;插图 5,6;大

羽羊齿类生殖器官;登记号:PB9059(图版4,图1),PB9077(图版4,图2),PB9067(图版4,图3),PB9073(图版5,图1),PB9060(图版5,图2),PB9078(图版5,图3),PB9074(图版5,图4),PB9062(图版6,图1),PB9063(图版6,图2),PB9076(图版6,图3),PB9071(图版6,图4),PB9066(图版6,图5);华南;二叠纪。

广南蕨属 Genus *Guangnania* Wang et Hao, 2002（英文发表）

2002　王德明、郝守刚,14 页。

模式种:*Guangnania cuneata* Wang et Hao,2002

分类位置:工蕨类(Zosterophyllophytes)

分布与时代:中国云南;早泥盆世。

楔形广南蕨 *Guangnania cuneata* Wang et Hao, 2002（英文发表）

2002　王德明、郝守刚,15 页,图版 1－3;插图 1－3;囊穗;标本号:WH-D01－WH-D13,WH-G,WH-X01－WH-X03;正型:WH-D01(图版1,图1);标本保存在北京大学地质系;云南广南;早泥盆世坡松冲组;云南曲靖;早泥盆世徐家冲组。

广西叶属 Genus *Guangxiophyllum* Feng, 1977

1977　冯少南等,247 页。

模式种:*Guangxiophyllum shangsiense* Feng,1977

分类位置:裸子植物类(Gymnospermae incertae sedis)

分布与时代:中国广西上思;晚三叠世。

上思广西叶 *Guangxiophyllum shangsiense* Feng, 1977

1977　冯少南等,247 页,图版 95,图 1;羽叶;标本号:P25281;正模:P25281(图版95,图1);标本保存在湖北地质科学研究所;广西上思那汤汪门;晚三叠世。

黔囊属 Genus *Guizhoua* Zhao, 1980

1980　赵修祜等,89 页。

模式种:*Guizhoua gregalis* Zhao,1980

分类位置:分类位置不明植物(Plantae incertae sedis)

分布与时代:中国贵州;晚二叠世早期。

堆黔囊 *Guizhoua gregalis* Zhao, 1980

1980　赵修祜等,89 页,图版 23,图 1,1a,1b,2－4;生殖叶和孢子囊群;采集号:PZ2-7;登记号:PB7107－PB7110;标本保存在中国科学院南京地质古生物研究所;贵州盘县;晚二叠世早期宣威组下段。

贵州木属 Genus *Guizhouoxylon* Tian et Li,1992

1992　田宝霖、李洪起,336,343页。

模式种:*Guizhouoxylon dahebianense* Tian et Li,1992

分类位置:分类位置不明植物(Plantae incertae sedis)

分布与时代:中国贵州水城;晚二叠世。

大河边贵州木 *Guizhouoxylon dahebianense* Tian et Li,1992

1992　田宝霖、李洪起,337,343页,图版1,图1—9;图版2,图1—10;图版3,图1—10;图版4,图1—10;茎化石;标本保存在中国矿业大学北京研究生部地质研究室;贵州水城大河边煤矿;晚二叠世龙潭组。

古木蕨属 Genus *Gumuia* Hao,1989

1989b　郝守刚,954,955页。

模式种:*Gumuia zyzzata* Hao,1989

分类位置:工蕨类(Zosterophyllophytes)

分布与时代:中国云南;早泥盆世。

曲轴古木蕨 *Gumuia zyzzata* Hao,1989

1989b　郝守刚,955页,图版1,图1—3;图版2,图1—11;插图1—3;标本号:Bupb601,602,603,604,604',605,606,607,608,608',609;正型:Bupb602(图版2,图1);标本保存在北京大学地质系;云南文山古木;早泥盆世坡松冲组。

似雨蕨属 Genus *Gymnogrammitites* Sun et Zheng,2001(中文和英文发表)

2001　孙革、郑少林,见孙革等,75,185页。

模式种:*Gymnogrammitites ruffordioides* Sun et Zheng,2001

分类位置:真蕨纲(Filicopsida)

分布与时代:中国辽宁北票;晚侏罗世。

鲁福德似雨蕨 *Gymnogrammitites ruffordioides* Sun et Zheng,2001(中文和英文发表)

2001　孙革、郑少林,见孙革等,75,185页,图版7,图6;图版9,图1—2;图版40,图5—8;蕨叶;标本号:PB19020,PB19020A(正、反模);正模:PB19020(图版7,图6);标本保存在中国科学院南京地质古生物研究所;辽宁北票上园黄半吉沟;晚侏罗世尖山沟组。

哈勒角籽属 Genus *Hallea* Mathews,1947—1948(non Yang et Wu,2006)

1947—1948　Mathews G B,241页。

模式种:*Hallea pekinensis* Mathews,1947—1948

分类位置:不明(Incertae sedis)
分布与时代:中国北京西山;二叠纪(?)、三叠纪(?)。

北京哈勒角籽 *Hallea pekinensis* Mathews,1947－1948
1947－1948　Mathews G B,241 页,图 4;种子;北京西山;二叠纪(?)、三叠纪(?)双泉群。

赫勒单网羊齿属 Genus *Hallea* Yang et Wu,2006（non Mathews,1947－1948）(中文和英文发表)
(注:此属名为 *Hallea* Mathews,1947－1948 的晚出同名)
2006　杨关秀、吴跃辉,见杨关秀等,194,314 页。
模式种:*Hallea dengfengensis* Yang et Wu,2006
分类位置:大羽羊齿目(Gigantopteridales)
分布与时代:中国河南登封;晚二叠世。

登封赫勒单网羊齿 *Hallea dengfengensis* Yang et Wu,2006(中文和英文发表)
2006　杨关秀、吴跃辉,见杨关秀等,194,314 页,图版 56,图 3,3a;羽状复叶(?);正模:HEP3150(图版 56,图 3,3a);标本保存在中国地质大学(北京);河南登封磴槽;晚二叠世云盖山组 8 煤段。

哈氏蕨属 Genus *Halleophyton* Li et Edwards,1997(英文发表)
1997　李承森、Edwards D,1447 页。
模式种:*Halleophyton zhichangense* Li et Edwards,1997
分类位置:石松植物门镰蕨目(Drepanophycales,Lycophyta)
分布与时代:中国云南文山;早泥盆世。

纸厂哈氏蕨 *Halleophyton zhichangense* Li et Edwards,1997(英文发表)
1997　李承森、Edwards D,1448 页,图 1－29;枝和孢子囊;标本号:CBYn9003001a,CBYn9003001b,CBYn9003002a,CBYn9003002b,CBYn9003003,CBYn9003004b,CBYn9003005a,CBYn9003008,CBYn9003012a,CBYn9003014a,CBYn9003015a,CBYn9003015b,CBYn9003018,CBYn9003021,CBYn9003023b,CBYn9003030;正模:CBYn9003001a,CBYn9003001b(图 17,18);标本保存在中国科学院植物研究所;云南文山纸厂村;早泥盆世坡松冲组。

钩蕨属 Genus *Hamatophyton* Gu et Zhi,1974
1974　中国科学院南京地质古生物研究所、植物研究所,见《中国古生代植物》,38 页。
模式种:*Hamatophyton verticillatum* Gu et Zhi,1974
分类位置:楔叶纲歧叶目(Hyeniales,Sphenopsida)
分布与时代:中国江苏、浙江、江西;晚泥盆世。

轮生钩蕨 *Hamatophyton verticillatum* Gu et Zhi,1974

（注：又名钩蕨）

1974　中国科学院南京地质古生物研究所、植物研究所,见《中国古生代植物》,38页,图版19,图3—5;图版20,图1—6;显示钩叶的枝;登记号:PB4893—PB4897;合模:PB4893,PB4894（图版19,图3,4）,PB4895—PB4897（图版20,图1,3,4）;标本保存在中国科学院南京地质古生物研究所;江苏江阴秦皇山、龙潭、句容、震泽、浙江长兴;晚泥盆世五通群;江西莲花、永新;中—晚泥盆世峡山群上部;浙江杭州;晚泥盆世千里岗群。

和丰孢穗属 Genus *Hefengistrobus* Xu et Wang,2002（中文和英文发表）

2002　徐洪河、王怿,251,256页。

模式种:*Hefengistrobus bifurcus* Xu et Wang,2002

分类位置:石松纲（Lycopsida）

分布与时代:中国新疆准噶尔盆地;晚泥盆世。

二歧和丰孢穗 *Hefengistrobus bifurcus* Xu et Wang,2002（中文和英文发表）

2002　徐洪河、王怿,252,256页,图版1,图1—9;图版2,图11;插图2;石松类植物;登记号:PB19243—PB19250;正模:PB19245（图版1,图6）;副模:PB19244A,PB19244B（图版1,图2,3）,PB19247A,PB19247B（图版2,图1,2）;标本保存在中国科学院南京地质古生物研究所;新疆准噶尔盆地;晚泥盆世洪古勒楞组。

缠绕蕨属 Genus *Helicophyton* Wang et Xu,2002（英文发表）

2002　王怿、徐洪河,475页。

模式种:*Helicophyton dichotomum* Wang et Xu,2002

分类位置:蕨类植物门（Pteridopsida incertae sedis）

分布与时代:中国江苏南京孔山;早石炭世。

二叉缠绕蕨 *Helicophyton dichotomum* Wang et Xu,2002（英文发表）

2002　王怿、徐洪河,475页,图1—6;植物体具主轴和次级枝;正模:PB19210（图3a）;副模:PB19206,PB19208（图1a,1c）;标本保存在中国科学院南京地质古生物研究所;江苏南京孔山;早石炭世五通组上段上部。

河南叶属 Genus *Henanophyllum* Xi et Feng,1977

1977　席运宏、冯少南,见冯少南等,673页。

模式种:*Henanophyllum palamifolium* Xi et Feng,1977

分类位置:分类位置不明植物（Plantae incertae sedis）

分布与时代:中国河南;晚二叠世。

掌河南叶 *Henanophyllum palamifolium* Xi et Feng,1977

1977 席运宏、冯少南,见冯少南等,674 页,图版 248,图 6,7;掌状叶;标本号:P0021,P0022;合模:P0021(图版 248,图 6),P0022(图版 248,图 7);河南禹县;晚二叠世上石盒子组。

河南羊齿属 Genus *Henanopteris* Yang,1987

1987a 杨关秀,52 页。

模式种:*Henanopteris lanceolatus* Yang,1987

分类位置:种子蕨纲(Pteridospermopsida)

分布与时代:中国河南;晚二叠世。

披针河南羊齿 *Henanopteris lanceolatus* Yang,1987

1987a 杨关秀,52 页,图版 15,图 2—5;蕨叶;登记号:HEP728,HEP729,HEP730,HEP731;合模 1:HEP728(图版 15,图 2);合模 2:HEP729(图版 15,图 3);合模 3:HEP730(图版 15,图 4);合模 4:HEP731(图版 15,图 5);标本保存在中国地质大学(北京);河南禹县大风口;晚二叠世上石盒子组。

豫囊蕨属 Genus *Henanotheca* Yang,2006(中文和英文发表)

2006 杨关秀等,107,260 页。

模式种:*Henanotheca*(*Sphenopteris*)*ovata* Yang,2006

分类位置:真蕨目里白科(Gleicheniaceae,Filicales)

分布与时代:中国河南禹州;中二叠世。

卵豫囊蕨(楔羊齿) *Henanotheca*(*Sphenopteris*)*ovata* Yang,2006(中文和英文发表)

2006 杨关秀等,107,260 页,图版 16,图 10,10a,10b;图版 24,图 5;蕨叶;正模:HEP0908(图版 16,图 10);标本保存在中国地质大学(北京);河南禹州大风口;中二叠世小风口组。

六叶属 Genus *Hexaphyllum* Ngo,1956

1956 敖振宽,25 页。

模式种:*Hexaphyllum sinense* Ngo,1956

分类位置:不明或木贼目?(Plantae incertae sedis or Equisetales?)

分布与时代:中国广东广州小坪;晚三叠世。

中国六叶 *Hexaphyllum sinense* Ngo,1956

1956 敖振宽,25 页,图版 1,图 2;图版 6,图 1,2;插图 3;轮叶;标本号:A4;登记号:0015;标本保存在中南矿冶学院地质系古生物地史教研组;广东广州小坪;晚三叠世小坪煤系。[注:此标本后改定为 *Annulariopsis*? *sinensis* (Ngo) Lee(斯行健、李星学等,1963)]

全泽米属 Genus *Holozamites* Wang X,Li,Wang Y et Zheng,2009（nom. nud.）

2009a 王鑫、李楠、王永栋、郑少林,1937 页。(中文)(仅有属名)
2009b 王鑫、李楠、王永栋、郑少林,3116 页。(英文)(仅有属名)
模式种：*Holozamites hongtaoi* Wang X,Li,Wang Y et Zheng,2009
分类位置 被子植物？(Angiospermae?)
分布与时代：中国辽宁北票；晚侏罗世。

洪涛全查米亚 *Holozamites hongtaoi* Wang X,Li,Wang Y et Zheng,2009（nom. nud.）

2009a 王鑫、李楠、王永栋、郑少林,1937 页；辽宁北票上园黄半吉沟；晚侏罗世。(中文)(仅有种名)
2009b 王鑫、李楠、王永栋、郑少林,3116 页；辽宁北票上园黄半吉沟；晚侏罗世。(英文)(仅有种名)

香溪叶属 Genus *Hsiangchiphyllum* Sze,1949

1949 斯行健,28 页。
模式种：*Hsiangchiphyllum trinerve* Sze,1949
分类位置：苏铁纲(Cycadopsida)
分布与时代：中国湖北秭归香溪；早侏罗世。

三脉香溪叶 *Hsiangchiphyllum trinerve* Sze,1949

1949 斯行健,28 页,图版 7,图 6；图版 8,图 1；羽叶；湖北秭归香溪；早侏罗世香溪煤系。

徐氏蕨属 Genus *Hsuea* Li,1982

1982 李承森,331,341 页。
模式种：*Hsuea robusta*（Li et Cai）Li,1982
分类位置：裸蕨植物门鹿角蕨目顶囊蕨科(Cooksoniaceae,Rhynales,Psilophyta)
分布与时代：中国云南曲靖；早泥盆世。

粗壮徐氏蕨 *Hsuea robusta*（Li et Cai）Li,1982

1978 *Cooksonia zhanyiensis* Li et Cai,李星学、蔡重阳,10 页,图版 2,图 6；植物体；云南曲靖；早泥盆世。
1978 *Taeniocrada robusta* Li et Cai,李星学、蔡重阳,10 页,图版 2,图 7—14；植物体；云南曲靖；早泥盆世。
1982 李承森,331,341 页,图版 3—10；植物体；标本号：7771—7775,7787,7788,7791,7793,7796,7799,7800,7802,7853,7863,7870,7874,7880,7888,7895 等；正模：7771(图版 3,图 1)；标本保存在中国科学院植物研究所；云南曲靖龙华山、翠峰山；早泥盆世徐家冲组。

汲清羊齿属 Genus *Huangia* Si, 1989

1989 斯行健, 55, 196 页。

模式种: *Huangia elliptica* Si, 1989

分类位置: 真蕨纲或种子蕨纲 (Filices or Pteridospermopsida)

分布与时代: 中国内蒙古准格尔旗; 早二叠世。

椭圆汲清羊齿 *Huangia elliptica* Si, 1989

1989 斯行健, 55, 196 页, 图版 63, 图 1—4; 图版 64, 图 1; 蕨叶; 标本号: YF545, YF546, YF549; 登记号: PB4219, PB4218, PB4217, PB4221 (图版 63, 图 1—4), PB4220 (图版 64, 图 1); 正模: PB4221 (图版 63, 图 4); 标本保存在中国科学院南京地质古生物研究所; 内蒙古准格尔旗黑带沟; 早二叠世山西组? (或太原组?)。

湖北蕨属 Genus *Hubeiia* Xue, Hao, Wang et Liu, 2005 (英文发表)

2005 薛进庄、郝守刚、王德明、刘振峰, 520 页。

模式种: *Hubeiia dicrofollia* Xue, Hao, Wang et Liu, 2005

分类位置: 石松植物门原始鳞木目原始鳞木科 (Protolepidodendraceae, Protolepidodendrales, Lycophyta)

分布与时代: 中国湖北长阳; 晚泥盆世 (Famennian)。

叉叶湖北蕨 *Hubeiia dicrofollia* Xue, Hao, Wang et Liu, 2005 (英文发表)

2005 薛进庄、郝守刚、王德明、刘振峰, 520 页, 图 2—6; 草本石松植物; 正模: PKU-X-11a (图 2a); 副模: PKU-X-5b, 7, 19, PKU-X-16-3, PKU-X-17-3, 17-4a (图 2h—k, 图 4a, 图 5a, 5d); 标本保存在北京大学地质系; 湖北长阳; 晚泥盆世 (Famennian) 写经寺组。

湖北叶属 Genus *Hubeiophyllum* Feng, 1977

1977 冯少南等, 247 页。

模式种: *Hubeiophyllum cuneifolium* Feng, 1977

分类位置: 裸子植物类 (Gymnospermae incertae sedis)

分布与时代: 中国湖北远安铁炉湾; 晚三叠世。

楔形湖北叶 *Hubeiophyllum cuneifolium* Feng, 1977

1977 冯少南等, 247 页, 图版 100, 图 1—4; 叶; 标本号: P25298—P25301; 合模: P25298—P25301 (图版 100, 图 1—4); 标本保存在湖北地质科学研究所; 湖北远安铁炉湾; 晚三叠世香溪群下煤组。

此属创建时同时报道 2 种, 除模式种外尚有:

狭细湖北叶 *Hubeiophyllum angustum* Feng, 1977

1977 冯少南等, 247 页, 图版 100, 图 5—7; 叶; 标本号: P25302—P25304; 合模: P25302—

P25304(图版100,图5－7);标本保存在湖北地质科学研究所;湖北远安铁炉湾;晚三叠世香溪群下煤组。

先骕蕨属 Genus *Huia* Geng,1985
1985　耿宝印,419,425 页。
模式种:*Huia recurvata* Geng,1985
分类位置:裸蕨植物门鹿角蕨类带蕨科(Taeniocradaceae,Rhyniophytina,Psilophyta)
分布与时代:中国云南文山;早泥盆世。

回弯先骕蕨 *Huia recurvata* Geng,1985
1985　耿宝印,420,425 页,图版 1,2;植物体;标本号:8122－8133;正模:8122(图版 1,图 10);副模:8123－8133;标本保存在中国科学院植物研究所;云南文山古木公社;早泥盆世坡松冲组。

湖南木贼属 Genus *Hunanoequisetum* Zhang,1986
1986　张采繁,191 页。
模式种:*Hunanoequisetum liuyangense* Zhang,1986
分类位置:楔叶纲木贼目(Equisetales,Sphenopsida)
分布与时代:中国湖南浏阳跃龙;早侏罗世。

浏阳湖南木贼 *Hunanoequisetum liuyangense* Zhang,1986
1986　张采繁,191 页,图版 4,图 4－4a,5;插图 1;木贼类茎干;登记号:PH472,PH473;正模:PH472(图版 4,图 4);标本保存在湖南省地质博物馆;湖南浏阳跃龙;早侏罗世跃龙组。

似八角属 Genus *Illicites* Pan,1983(nom. nud.)
1983　潘广,1520 页。(中文)
1984　潘广,959 页。(英文)
模式种:(没有种名)
分类位置:"原始被子植物类群"("primitive angiosperms")
分布与时代:华北燕辽地区;中侏罗世。

似八角(sp. indet.) *Illicites* sp. indet.
(注:原文仅有属名,没有种名)
1983　*Illicites* sp. indet. ,潘广,1520 页;华北燕辽地区东段(45°58′N,120°21′E);中侏罗世海房沟组。(中文)
1984　*Illicites* sp. indet. ,潘广,959 页;华北燕辽地区东段(45°58′N,120°21′E);中侏罗世海房沟组。(英文)

耶氏蕨属 Genus *Jaenschea* Mathews,1947－1948

1947－1948　Mathews G B,239 页。

模式种：*Jaenschea sinensis* Mathews,1947－1948

分类位置：真蕨纲紫萁科？（Osmundaceae,Filicopsida?）

分布与时代：中国北京西山；二叠纪(?)、三叠纪(?)。

中国耶氏蕨 *Jaenschea sinensis* Mathews,1947－1948

1947－1948　Mathews G B,239 页,图 2；实羽片印痕；北京西山；二叠纪(?)、三叠纪(?)双泉群。

江西叶属 Genus *Jiangxifolium* Zhou,1988

1988　周贤定,126 页。

模式种：*Jiangxifolium mucronatum* Zhou,1988

分类位置：真蕨纲（Filicopsida）

分布与时代：中国江西丰城攸洛；晚三叠世。

短尖头江西叶 *Jiangxifolium mucronatum* Zhou,1988

1988　周贤定,126 页,图版 1,图 1,2,5,6；插图 1；蕨叶；登记号：No. 2228, No. 1862, No. 1348, No. 2867；正模：No. 2228（图版 1,图 1）；标本保存在江西省 195 地质队；江西丰城攸洛；晚三叠世安源组。

此属创建时同时报道 2 种,除模式种外尚有：

细齿江西叶 *Jiangxifolium denticulatum* Zhou,1988

1988　周贤定,127 页,图版 1,图 3,4；蕨叶；登记号：No. 2135, No. 2867；正模：No. 2135（图版 1,图 3）；标本保存在江西省 195 地质队；江西丰城攸洛；晚三叠世安源组。

赣囊蕨属 Genus *Jiangxitheca* He,Liang et Shen,1996（中文和英文发表）

1996　何锡麟、梁敦士、沈树忠,50,160 页。

模式种：*Jiangxitheca xinanensis* He,Liang et Shen,1996

分类位置：真蕨纲或种子蕨纲（Filices or Pteridospermopsida）

分布与时代：中国江西；晚二叠世。

新安赣囊蕨 *Jiangxitheca xinanensis* He,Liang et Shen,1996（中文和英文发表）

1996　何锡麟、梁敦士、沈树忠,50,160 页,图版 36,图 5；图版 37,图 1－4；蕨叶；登记号：X88163－X88167；合模：X88163（图版 36,图 5）,X88164－X88167（图版 37,图 1－4）；标本保存在中国矿业大学地质系古生物实验室；江西铅山新安五都煤矿；晚二叠世上饶组童家段。

荆门叶属 Genus *Jingmenophyllum* Feng,1977

1977　冯少南等,250 页。

模式种:*Jingmenophyllum xiheense* Feng,1977

分类位置:裸子植物类(Gymnospermae incertae sedis)

分布与时代:中国湖北荆门西河;晚三叠世。

西河荆门叶 *Jingmenophyllum xiheense* Feng,1977

1977　冯少南等,250 页,图版 94,图 9;羽叶;标本号:P25280;正模:P25280(图版 94,图 9);标本保存在湖北地质科学研究所;湖北荆门西河;晚三叠世香溪群下煤组。[注:此标本后改定为 *Compsopteris xiheensis* (Feng) Zhu,Hu et Meng(朱家楠等,1984)]

鸡西叶属 Genus *Jixia* Guo et Sun,1992

1992　郭双兴、孙革,见孙革等,547 页。(中文)

1993　郭双兴、孙革,见孙革等,254 页。(英文)

模式种:*Jixia pinnatipartita* Guo et Sun,1992

分类位置:双子叶植物纲(Dicotyledoneae)

分布与时代:中国黑龙江鸡西城子河;早白垩世。

羽裂鸡西叶 *Jixia pinnatipartita* Guo et Sun,1992

1993　郭双兴、孙革,见孙革等,547 页,图版 1,图 10－12;图版 2,图 7;叶部化石;登记号:PB16773－PB16775,PB16773A;正模:PB16774(图版 1,图 10);标本保存在中国科学院南京地质古生物研究所;黑龙江鸡西城子河;早白垩世城子河组上部。(中文)

1993　郭双兴、孙革,见孙革等,254 页,图版 1,图 10－12;图版 2,图 7;叶部化石;登记号:PB16773－PB16775,PB16773A;正模:PB16774(图版 1,图 10);标本保存在中国科学院南京地质古生物研究所;黑龙江鸡西城子河;早白垩世城子河组上部。(英文)

准噶尔蕨属 Genus *Junggaria* Dou,1983

1983　窦亚伟等,562 页。

模式种:*Junggaria spinosa* Dou,1983

分类位置:裸蕨类(Psilophytes incertae sedis)

分布与时代:中国新疆准噶尔盆地;早泥盆世。

刺状准噶尔蕨 *Junggaria spinosa* Dou,1983

1983　窦亚伟等,562 页,图版 189,图 1－4;植物体;采集号:730H-1-44;登记号:XPA001－XPA003,XPA004;合模:XPA001－XPA003(图版 189,图 1－3);新疆准噶尔盆地和布克赛尔芒克鲁;早泥盆世乌图布拉克组。

侏罗缘蕨属 Genus *Juradicotis* Pan,1983（nom. nud.）

1983　潘广,1520 页。（中文）
1984　潘广,958 页。（英文）
模式种:（没有种名）
分类位置:"半被子植物类群"（"hemiangiosperms"）
分布与时代:华北燕辽地区;中侏罗世。

侏罗缘蕨(sp. indet.) *Juradicotis* sp. indet.

（注:原文仅有属名,没有种名）

1983　*Juradicotis* sp. indet. ,潘广,1520 页;华北燕辽地区东段（45°58′N,120°21′E）;中侏罗世海房沟组。（中文）
1984　*Juradicotis* sp. indet. ,潘广,958 页;华北燕辽地区东段（45°58′N,120°21′E）;中侏罗世海房沟组。（英文）

侏罗木兰属 Genus *Juramagnolia* Pan,1983（nom. nud.）

1983　潘广,1520 页。（中文）
1984　潘广,959 页。（英文）
模式种:（没有种名）
分类位置:"原始被子植物类群"（"primitive angiosperms"）
分布与时代:华北燕辽地区;中侏罗世。

侏罗木兰(sp. indet.) *Juramagnolia* sp. indet.

（注:原文仅有属名,没有种名）

1983　*Juramagnolia* sp. indet. ,潘广,1520 页;华北燕辽地区东段（45°58′N,120°21′E）;中侏罗世海房沟组。（中文）
1984　*Juramagnolia* sp. indet. ,潘广,959 页;华北燕辽地区东段（45°58′N,120°21′E）;中侏罗世海房沟组。（英文）

侏罗球果属 Genus *Jurastrobus* Wang,Li et Cui,2006（英文发表）

2006　王鑫、李楠、崔金钟,214 页。
模式种:*Jurastrobus chenii* Wang,Li et Cui,2006
分类位置:苏铁目（Cycadales）
分布与时代:中国内蒙古锡林浩特;早侏罗世。

陈氏侏罗球果 *Jurastrobus chenii* Wang,Li et Cui,2006（英文发表）

2006　王鑫、李楠、崔金钟,214 页,图 2-6;带花粉球的植物体;标本号:No. 9221;正模:No. 9221（图 2）;标本保存在中国科学院植物研究所;内蒙古锡林浩特 795 煤井;早侏罗世晚期（? Toarcian）马尼特庙群 K-1 层。

似南五味子属 Genus *Kadsurrites* Pan,1983（nom. nud.）

1983　潘广,1520 页。（中文）
1984　潘广,959 页。（英文）
模式种:（没有种名）
分类位置:"原始被子植物类群"（"primitive angiosperms"）
分布与时代:华北燕辽地区;中侏罗世。

似南五味子（sp. indet.）*Kadsurrites* sp. indet.

（注:原文仅有属名,没有种名）
1983　*Kadsurrites* sp. indet.,潘广,1520 页;华北燕辽地区东段（45°58′N,120°21′E）;中侏罗世海房沟组。（中文）
1984　*Kadsurrites* sp. indet.,潘广,959 页;华北燕辽地区东段（45°58′N,120°21′E）;中侏罗世海房沟组。（英文）

开平木属 Genus *Kaipingia* Stockmans et Mathieu,1957

1957　Stockmans F 和 Mathieu F F,62 页。
模式种:*Kaipingia sinica* Stockmans et Mathieu,1957
分类位置:石松目（Lycopodiales）
分布与时代:中国河北开平;晚石炭世。

中国开平木 *Kaipingia sinica* Stockmans et Mathieu,1957

1957　Stockmans F 和 Mathieu F F,62 页,图版 1,图 1,1a;具叶座的茎干化石;河北开平;晚石炭世。

契丹穗属 Genus *Khitania* Guo,Sha,Bian et Qiu,2009（英文发表）

2009　郭双兴、沙金庚、边力增、仇寅龙,94 页。
模式种:*Khitania columnispicata* Guo,Sha,Bian et Qiu,2009
分类位置:买麻藤纲买麻藤目买麻藤科（Gnetaceae,Gnetales,Gnetopsida）
分布与时代:中国辽宁北票;早白垩世。

柱状契丹穗 *Khitania columnispicata* Guo,Sha,Bian et Qiu,2009（英文发表）

2009　郭双兴、沙金庚、边力增、仇寅龙,94 页,图 1（左）,2A—2F;雄性果穗;登记号:PB20189;正模:PB20189[图 1（左）];标本保存在中国科学院南京地质古生物研究所;辽宁北票黄半吉沟（41°12′N,119°22′E）;早白垩世（Barremian）热河群义县组尖山沟层。

似克鲁克蕨属 Genus *Klukiopsis* Deng et Wang,1999（2000）（中文和英文发表）

1999　邓胜徽、王士俊,552页。（中文）
2000　邓胜徽、王士俊,356页。（英文）
模式种：*Klukiopsis jurassica* Deng et Wang,1999（2000）
分类位置：真蕨纲海金沙科（Schzaeaceae,Filicopsida）
分布与时代：中国河南义马；中侏罗世。

侏罗似克鲁克蕨 *Klukiopsis jurassica* Deng et Wang,1999（2000）（中文和英文发表）

1999　邓胜徽、王士俊,552页,图版1；蕨叶、生殖羽片、孢子囊和孢子；标本号：YM98-303；正模：YM98-303（图1a）；河南义马；中侏罗世。（中文）
2000　邓胜徽、王士俊,356页,图版1；蕨叶、生殖羽片、孢子囊和孢子；标本号：YM98-303；正模：YM98-303（图1a）；河南义马；中侏罗世。（英文）

凹尖枝属 Genus *Koilosphenus* Bohlin,1971

1971　Bohlin B,47页。
模式种：*Koilosphenus cuneifolius* Bohlin,1971
分类位置：松柏类（Conifers）
分布与时代：中国甘肃；晚古生代。

楔裂凹尖枝 *Koilosphenus cuneifolius* Bohlin,1971

1971　Bohlin B,47页,图版7,图6,7；插图91A-D；茎和枝叶；甘肃"鱼儿红"；晚古生代。

此属创建时同时报道2种,除模式种外尚有：

? 凹尖枝（未定种）? *Koilosphenus* sp.

1971　? *Koilosphenus* sp.,Bohlin B,48页,图版7,图4；插图90A-D；枝叶碎片；甘肃"鱼儿红"；晚古生代。

孔山羊齿属 Genus *Kongshania* Wang,2000（英文发表）

2000　王怿,47页。
模式种：*Kongshania synangioides* Wang,2000
分类位置：楔羊齿类（Sphenopterides）
分布与时代：中国江苏江宁；晚泥盆世。

类连生孔山羊齿 *Kongshania synangioides* Wang,2000（英文发表）

2000　王怿,47页,图版1—4；插图1—6；植物体羽状分裂；登记号：PB17201—PB17204,PB17205a,PB17205b,PB17206—PB17210,PB17211a,PB17211b,PB17212,PB17213,PB17214a,PB17214b；正模：PB17203（图版1,图3）；副模：PB17205b（图版1,图10）,PB17211a（图版2,图1）,PB17211b（图版2,图2）,PB17212（图版2,图3）；标本保存在

中国科学院南京地质古生物研究所；江苏江宁孔山；晚泥盆世五通组。

今野羊齿属 Genus *Konnoa* Asama,1959

1959　Asama K,63页。

模式种：*Konnoa koraiensis*（Tokunaga）Asama,1959

分类位置：美羊齿类（Callipterides）

分布与时代：中国辽宁本溪、山西太原和朝鲜；晚古生代。

高丽今野羊齿 *Konnoa koraiensis*（Tokunaga）Asama,1959

1951　*Alethopteris koraiensis* Tokunaga,Tokunaga S,52页；插图。

1959　Asama K,64页,图版14,图1—3；图版15,图1—3；图版16,图1—4；图版17,图5；蕨叶；朝鲜（Jido Series）和中国辽宁（Huangchi and Lintang Series）、山西（Taiyuan and Shansi Series）；晚古生代。

此属创建时同时报道2种,除模式种外尚有：

本溪今野羊齿 *Konnoa penchihuensis* Asama,1959

1959　Asama K,65页,图版17,图1—4；图版18,图1—3；图版19,图1,2；蕨叶；辽宁本溪；晚古生代（Huangchi Series）。

宽甸叶属 Genus *Kuandiania* Zheng et Zhang,1980

1980　郑少林、张武,见张武等,279页。

模式种：*Kuandiania crassicaulis* Zheng et Zhang,1980

分类位置：苏铁纲（Cycadopsida）

分布与时代：中国辽宁本溪宽甸；中侏罗世。

粗茎宽甸叶 *Kuandiania crassicaulis* Zheng et Zhang,1980

1980　郑少林、张武,见张武等,279页,图版144,图5；羽叶；登记号：D423；标本保存在沈阳地质矿产研究所；辽宁本溪宽甸；中侏罗世转山子组。

李氏穗属 Genus *Leeites* Zodrow et Gao,1991

1991　Zodrow E L、高志峰,63页。

模式种：*Leeites oblongifolis* Zodrow et Gao,1991

分类位置：楔叶纲楔叶目（Sphenophyllales,Sphenopsida）

分布与时代：加拿大；晚石炭世。

椭圆李氏穗 *Leeites oblongifolis* Zodrow et Gao,1991

1991　Zodrow E L、高志峰,64页,图版1—8；插图1B—1F；插图2—7；插图8A,8B；插图9；孢子叶穗；正模：982FG-315″5″（图版1,图1；插图3D）；副模：982FG-315″1″（图版2,图1）；加拿大（Cape Breton Island,Nova Scotia,Canada）；晚石炭世。

乐平苏铁属 Genus *Lepingia* Liu et Yao,2002（英文发表）

2002　刘陆军、姚兆奇,177 页。

模式种:*Lepingia emarginata* Liu et Yao,2002

分类位置:苏铁类?（Cycadalean?）

分布与时代:中国江西乐平;晚二叠世。

缺顶乐平苏铁 *Lepingia emarginata* Liu et Yao,2002（英文发表）

2002　刘陆军、姚兆奇,177 页,图 3,5—7;带状叶;正模:PB18843（图 5A）;标本保存在中国科学院南京地质古生物研究所;江西乐平;晚二叠世早期乐平组崂山段下部。

拉萨木属 Genus *Lhassoxylon* Vozenin-Serra et Pons,1990

1990　Vozenin-Serra C 和 Pons D,110 页。

模式种:*Lhassoxylon aptianum* Vozenin-Serra et Pons,1990

分类位置:松柏纲?（Coniferopsida?）

分布与时代:中国西藏林周;早白垩世（Aptian）。

阿普特拉萨木 *Lhassoxylon aptianum* Vozenin-Serra et Pons,1990

1990　Vozenin-Serra C 和 Pons D,110 页,图版 1,图 1—7;图版 2,图 1—8;图版 3,图 1—7;图版 4,图 1—3;插图 2,3;木化石;采集号:X/2 Pj/2（J. J. Jaeger 采集）;登记号:n°10468;模式标本:n°10468;标本保存在巴黎居里夫人大学古植物和孢粉实验室;西藏林周附近（Lamba）;早白垩世（Aptian）。

连山草属 Genus *Lianshanus* Pan,1983（nom. nud.）

1983　潘广,1520 页。（中文）

1984　潘广,959 页。（英文）

模式种:（没有种名）

分类位置:"原始被子植物类群"（"primitive angiosperms"）

分布与时代:华北燕辽地区;中侏罗世。

连山草(sp. indet.) *Lianshanus* sp. indet.

（注:原文仅有属名,没有种名）

1983　*Lianshanus* sp. indet.,潘广,1520 页;华北燕辽地区东段（45°58′N,120°21′E）;中侏罗世海房沟组。（中文）

1984　*Lianshanus* sp. indet.,潘广,959 页;华北燕辽地区东段（45°58′N,120°21′E）;中侏罗世海房沟组。（英文）

辽宁缘蕨属 Genus *Liaoningdicotis* Pan,1983（nom. nud.）

1983　潘广,1520 页。（中文）
1984　潘广,958 页。（英文）
模式种:（没有种名）
分类位置:"半被子植物类群"（"hemiangiosperms"）
分布与时代:华北燕辽地区;中侏罗世。

辽宁缘蕨（sp. indet.）*Liaoningdicotis* sp. indet.

（注:原文仅有属名,没有种名）
1983　*Liaoningdicotis* sp. indet.,潘广,1520 页;华北燕辽地区东段（45°58′N,120°21′E）;中侏罗世海房沟组。（中文）
1984　*Liaoningdicotis* sp. indet.,潘广,958 页;华北燕辽地区东段（45°58′N,120°21′E）;中侏罗世海房沟组。（英文）

辽宁枝属 Genus *Liaoningocladus* Sun,Zheng et Mei,2000（英文发表）

2000　孙革、郑少林、梅盛吴,202 页。
模式种:*Liaoningocladus boii* Sun,Zheng et Mei,2000
分类位置:松柏类（Conifers）
分布与时代:中国辽宁北票;晚侏罗世。

薄氏辽宁枝 *Liaoningocladus boii* Sun,Zheng et Mei,2000（英文发表）

2000　孙革、郑少林、梅盛吴,202 页,图版 1,图 1—5;图版 2,图 1—7;图版 3,图 1—5;图版 4,图 1—5;长短枝、叶和角质层;正模:YB001（图版 1,图 1）;标本保存在中国科学院南京地质古生物研究所;辽宁北票;晚侏罗世义县组上部。

辽宁木属 Genus *Liaoningoxylon* Zhang et Zheng,2006（2008）（中文和英文发表）

2006　张武、郑少林,见张武等,110 页。（中文）
2008　张武、郑少林,见张武等,110 页。（英文）
模式种:*Liaoningoxylon chaoyangehse* Zhang et Zheng,2006（2008）
分类位置:松柏类（Conifers incertae sedis）
分布与时代:中国辽宁朝阳;早三叠世。

朝阳辽宁木 *Liaoningoxylon chaoyangehse* Zhang et Zheng,2006（2008）（中文和英文发表）

2006　张武、郑少林,见张武等,110 页,图版 4-8 — 4-10;木化石;正模:GJ6-49;副模:GJ11-1;标本保存在沈阳地质矿产研究所;辽宁朝阳段木头沟;早三叠世红砬组。（中文）
2008　张武、郑少林,见张武等,110 页,图版 4-8 — 4-10;木化石;正模:GJ6-49;副模:GJ11-1;标本保存在沈阳地质矿产研究所;辽宁朝阳段木头沟;早三叠世红砬组。（英文）

辽西草属 Genus *Liaoxia* Cao et Wu S Q,1997（1998）（中文和英文发表）

［注：此属模式种后被郭双兴、吴向午归于盖子植物纲（Chlamydospermopsida）或买麻藤目（Gnetales）的 *Ephedrites* 属，定名为 *Ephedrites chenii* (Cao et Wu S Q) Guo et Wu X W（郭双兴、吴向午，2000）；此属名也被吴舜卿（1999）改归于买麻藤目（Gnetales）］

1997　曹正尧、吴舜卿，见曹正尧等，1765页。（中文）
1998　曹正尧、吴舜卿，见曹正尧等，231页。（英文）

模式种：*Liaoxia chenii* Cao et Wu S Q,1997（1998）

分类位置：单子叶植物纲莎草科（Cyperaceae,Monocotyledoneae）

分布与时代：中国辽宁北票；晚侏罗世。

陈氏辽西草 *Liaoxia chenii* Cao et Wu S Q,1997（1998）（中文和英文发表）

1997　曹正尧、吴舜卿，见曹正尧等，1765页，图版1，图1－2,2a,2b,2c；草本植物、花枝；登记号：PB17800,PB17801；正模：PB17800（图版I，图1）；标本保存在中国科学院南京地质古生物研究所；辽西北票上园炒米店附近；晚侏罗世义县组下部尖山沟层。（中文）

1998　曹正尧、吴舜卿，见曹正尧等，231页，图版1，图1－2,2a,2b,2c；草本植物、花枝；登记号：PB17800,PB17801；正模：PB17800（图版I，图1）；标本保存在中国科学院南京地质古生物研究所；辽西北票上园炒米店附近；晚侏罗世义县组下部尖山沟层。（英文）

李氏苏铁属 Genus *Liella* Yang et Zhao,2006（中文和英文发表）

2006　杨关秀、赵继明，见杨关秀等，144,288页。

模式种：*Liella mirabilis* Yang et Zhao,2006

分类位置：苏铁植物门苏铁目（Cycadales,Cycadophyta）

分布与时代：中国河南临汝、宜阳；中二叠世。

奇异李氏苏铁 *Liella mirabilis* Yang et Zhao,2006（中文和英文发表）

2006　杨关秀、赵继明，见杨关秀等，144,288页，图版40，图3；*Taeniopteris* 型的叶；正模：HEP3931（图版40，图3）；标本保存在中国地质大学（北京）；河南临汝坡池、宜阳李沟；中二叠世小风口组。

似百合属 Genus *Lilites* Wu S Q,1999（中文发表）

1999　吴舜卿，23页。

模式种：*Lilites reheensis* Wu S Q,1999

分类位置：单子叶植物纲百合科（Liliaceae,Monocotyledoneae）

分布与时代：中国辽宁北票；晚侏罗世。

热河似百合 *Lilites reheensis* Wu S Q,1999（中文发表）

1999　吴舜卿，23页，图版18，图1,1a,2,4,5,7,7a,8A；枝叶和果实；采集号：AEO-11,219,245,134,158,246；登记号：PB18327－PB18332；合模1：PB18327（图版18，图1）；合模

2：PB18330（图版18，图5）；标本保存在中国科学院南京地质古生物研究所；辽西北票上园黄半吉沟；晚侏罗世义县组下部尖山沟层。[注：此属模式种后被孙革等（2001）归于松柏类的 *Podocarpites* 属，名为 *Podocarpites reheensis*（Wu）Sun et Zheng]

灵乡叶属 Genus *Lingxiangphyllum* Meng，1981

1981　孟繁松，100页。

模式种：*Lingxiangphyllum princeps* Meng，1981

分类位置：分类位置不明植物（Plantae incertae sedis）

分布与时代：中国湖北大冶；早白垩世。

首要灵乡叶 *Lingxiangphyllum princeps* Meng，1981

1981　孟繁松，100页，图版1，图12－13；插图1；叶；登记号：CHP7901，CHP7902；正模：CHP7901（图版1，图12）；标本保存在宜昌地质矿产研究所；湖北大冶灵乡长坪湖；早白垩世灵乡群。

网叶属 Genus *Linophyllum* Zhao，1980

1980　赵修祜等，83页。

模式种：*Linophyllum xuanweiense* Zhao，1980

分类位置：大羽羊齿类（Gigantopterides）

分布与时代：中国云南宣威；晚二叠世。

宣威网叶 *Linophyllum xuanweiense* Zhao，1980

1980　赵修祜等，83页，图版22，图5，5a；叶；采集号：XL-16；登记号：PB7103；标本保存在中国科学院南京地质古生物研究所；云南宣威；晚二叠世早期宣威组下段。

李氏木属 Genus *Lioxylon* Zhang，Wang，Saiki，Li and Zheng，2006（英文发表）

2006　张武、王永栋、Ken'ichi Saiki、李楠、郑少林，237页。

2006　见王永栋等，125页。（中文）

2008　见王永栋等，125页。（英文）

模式种：*Lioxylon liaoningense* Zhang，Wang，Saiki，Li et Zheng，2006

分类位置：苏铁目（Cycadales）

分布时代：中国辽宁北票；中侏罗世（Bathonian－Callovian）。

辽宁李氏木 *Lioxylon liaoningense* Zhang，Wang，Saiki，Li et Zheng，2006（英文发表）

2006　张武、王永栋、Ken'ichi Saiki、李楠、郑少林，见张武等，237页，图1－6；保存苏铁类构造的茎干；标本号：DMG-1，DMG-2，DMG-3，DMG-6，LMY-133，LMY-138，XCG-19；正模：DMG-1；副模：DMG-2，DMG-3，DMG-6，LMY-133，LMY-138，XCG-19；标本保存在沈阳地质矿产研究所；辽宁北票长皋；中侏罗世（Bathonian－Callovian）髫髻山组。

2006　见王永栋等，125页，图版5-1－5-6；茎干化石；辽宁北票长皋；中侏罗世（Bathonian－

Callovian)髫髻山组。（中文）
2008　见王永栋等，125 页，图版 5-1 — 5-6；茎干化石；辽宁北票长皋；中侏罗世（Bathonian — Callovian)髫髻山组。（英文）

柳林果属 Genus *Liulinia* Wang, 1986
1986　王自强，611，615 页。
模式种：*Liulinia lacinulata* Wang, 1986
分类位置：苏铁纲苏铁目（Cycadales, Cycaopsida）
分布与时代：中国山西柳林；晚二叠世。

条裂柳林果 *Liulinia lacinulata* Wang, 1986
1986　王自强，611，615 页，图版 1，图 1 — 7；图版 2，图 1 — 11；插图 1；雄性球果；采集号：8402-208，8402-209，8309-157，8309-158；正模：8402-209（图版 1，图 1）；标本保存在中国科学院南京地质古生物研究所；山西柳林；晚二叠世孙家沟组中段。

李氏蕨属 Genus *Lixotheca* Yao, Liu et Zhang, 1993
1993　姚兆奇、刘陆军、张士，526，535 页。
模式种：*Lixotheca* (*Cladophlebis*) *permica* (Lee et Wang) Yao, Liu et Zhang, 1993
分类位置：真蕨纲膜蕨科（Hymenophyllaceae, Filicopsida）
分布与时代：中国山西、江苏、河南、福建；二叠纪。

二叠李氏蕨（枝脉蕨）*Lixotheca* (*Cladophlebis*) *permica* (Lee et Wang) Yao, Liu et Zhang, 1993
1956　*Cladophlebis permica* Lee et Wang，李星学、王水，346 页，图版 1，2；图版 3，图 1（非图 2 — 4)；蕨叶；登记号：PB2538 — PB2545；标本保存在中国科学院南京地质古生物研究所；山西武乡；晚二叠世上盒子组；江苏南京龙潭；晚二叠世。
1993　姚兆奇、刘陆军、张士，526，535 页，图版 1 — 3；插图 1，2；营养羽片和生殖羽片；河南登封、禹县，山西武乡、安泽；晚二叠世上盒子组；福建永春、龙岩；晚二叠世童子岩组。

拟瓣轮叶属 Genus *Lobatannulariopsis* Yang, 1978
1978　杨贤河，472 页。
模式种：*Lobatannulariopsis yunnanensis* Yang, 1978
分类位置：楔叶纲木贼目（Equisetales, Sphenopsida）
分布与时代：中国云南广通；晚三叠世。

云南拟瓣轮叶 *Lobatannulariopsis yunnanensis* Yang, 1978
1978　杨贤河，472 页，图版 158，图 6；枝叶；标本号：Sp0009；正模：Sp0009（图版 158，图 6）；标本保存在成都地质矿产研究所；云南广通一平浪；晚三叠世干海子组。

龙井叶属 Genus *Longjingia* Sun et Zheng, 2000（nom. nud.）
2000　孙革、郑少林,见孙革等,图版 4,图 5,6。
模式种:*Longjingia gracilifolia* Sun et Zheng, 2000
分类位置:双子叶植物纲(Dicotyledoneae)
分布与时代:中国吉林龙井;早白垩世。

细叶龙井叶 *Longjingia gracilifolia* Sun et Zheng, 2000（nom. nud.）
2000　孙革、郑少林,见孙革等,图版 4,图 5,6;叶;吉林龙井智新(大拉子);早白垩世大拉子组。

长穗属 Genus *Longostachys* Zhu, Hu et Feng, 1983
1983　朱家楠、胡雨帆、冯少南,79 页。
模式种:*Longostachys latisporophyllus* Zhu, Hu et Feng, 1983
分类位置:石松植物类(Lycopods incertae sedis)
分布与时代:中国湖南澧县;中泥盆世。

宽叶长穗 *Longostachys latisporophyllus* Zhu, Hu et Feng, 1983
1983　朱家楠、胡雨帆、冯少南,79 页,图版 1,图 1a,1b;生殖枝(孢子叶穗);标本号:PB25604;正模:PB25604(图版 1,图 1a,1b);湖南澧县山门水库;中泥盆世云台观组。

碟囊属 Genus *Lopadiangium* Zhao, 1980
1980　赵修祜等,90 页。
模式种:*Lopadiangium acmodontum* Zhao, 1980
分类位置:分类位置不明植物(Plantae incertae sedis)
分布与时代:中国贵州;晚二叠世早期。

齿缘碟囊 *Lopadiangium acmodontum* Zhao, 1980
1980　赵修祜等,90 页,图版 23,图 9,9a,10,10a,11－14;生殖叶和孢子囊群;采集号:H11-6;登记号:PB7115－PB7119;标本保存在中国科学院南京地质古生物研究所;贵州盘县;晚二叠世早期宣威组下段;贵州晴隆;晚二叠世早期。

梳囊属 Genus *Lophotheca* Zhao, 1980
1980　赵修祜等,89 页。
模式种:*Lophotheca panxianensis* Zhao, 1980
分类位置:分类位置不明植物(Plantae incertae sedis)
分布与时代:中国贵州;晚二叠世早期。

盘县梳囊 *Lophotheca panxianensis* Zhao,1980
1980　赵修祜等,89 页,图版 23,图 5,5a,6—8;生殖叶和孢子囊群;采集号:PL18-15;登记号:PB7111—PB7114;标本保存在中国科学院南京地质古生物研究所;贵州盘县;晚二叠世早期宣威组下段。

乐平蕨属 Genus *Lopinopteris* Sze,1958
(注:又名乐平羊齿属)
1958　斯行健,383 页。(中文)
1959　斯行健,322 页。(英文)
模式种:*Lopinopteris intercalata* Sze,1958
分类位置:真蕨纲或种子蕨纲(Filices or Pteridospermopsida)
分布与时代:中国江西乐平;中石炭世(Westphalian)。

插入乐平蕨 *Lopinopteris intercalata* Sze,1958
(注:又名乐平羊齿)
1958　斯行健,383 页,图版 2,图 1—4;图版 3,图 4—6;蕨叶;登记号:PB2626—PB2629,PB2632,PB2633;标本保存在中国科学院南京地质古生物研究所;江西乐平;中石炭世(Westphalian)榐山煤系。(中文)
1959　斯行健,322 页,图版 2,图 1—4;图版 3,图 4,5,5a;蕨叶;登记号:PB2626—PB2629,PB2632,PB2633;标本保存在中国科学院南京地质古生物研究所;江西乐平;中石炭世(Westphalian)榐山煤系。(英文)

带状鳞穗属 Genus *Loroderma* Geng et Hilton,1999 (英文发表)
1999　耿宝印、Hilton J,127 页。
模式种:*Loroderma henania* Geng et Hilton,1999
分类位置:松柏类植物(Coniferophytes)
分布与时代:中国河南义马;早二叠世(Sakmarian)。

河南带状鳞穗 *Loroderma henania* Geng et Hilton,1999 (英文发表)
1999　耿宝印、Hilton J,127 页,图 6—22;松柏类植物的胚珠;标本号:CBP9191—CBP9197;正模:CBP9191(图 14);标本保存在中国科学院植物研究所;河南义马;早二叠世(Sakmarian)山西组。

条形蕨属 Genus *Lorophyton* Fairon-Demaret et Li,1993
1993　Fairon-Demaret M、李承森,15 页。
模式种:*Lorophyton goense* Fairon-Demaret et Li,1993
分类位置:枝木蕨纲(Cladoxylopsida)
分布与时代:比利时;中泥盆世(Lower Givetian)。

高氏条形蕨 *Lorophyton goense* Fairon-Demaret et Li,1993

1993　Fairon-Demaret M、李承森,15 页,图版 1—6;插图 1—6;正模:ULg2056a,ULg2056b(图版 1,图 3;图版 4,图 1—5;图版 6,图 1,4—7);副模:ULg2057a(图版 1,图 1;图版 2,图 2),ULg2057b(图版 1,图 2),ULg2065(图版 3,图 1),ULg2066(图版 2,图 1),ULg2066(图版 5,图 2,3);标本保存在 Paleobotanical collections of the University of Liege(Belgium);比利时;中泥盆世(Lower Givetian)。

吕蕨属 Genus *Luereticopteris* Hsu et Chu,1974

1974　徐仁、朱家柟,见徐仁等,270 页。
模式种:*Luereticopteris megaphylla* Hsu et Chu,1974
分类位置:真蕨纲(Filicopsida)
分布与时代:中国云南永仁;晚三叠世。

大叶吕蕨 *Luereticopteris megaphylla* Hsu et Chu,1974

1974　徐仁、朱家柟,见徐仁等,270 页,图版 2,图 5—11;图版 3,图 2—3;插图 2;蕨叶;编号:No.742a-c,2515;合模:No.742a-c,2515 图版 2,图 5—11;图版 3,图 2—3);标本保存在中国科学院植物研究所;云南永仁花山;晚三叠世大荞地组中部。

大舌羊齿属 Genus *Macroglossopteris* Sze,1931

1931　斯行健,5 页。
模式种:*Macroglossopteris leeiana* Sze,1931
分类位置:种子蕨纲(Pteridospermopsida)
分布与时代:中国江西萍乡;早侏罗世(Lias)。

李氏大舌羊齿 *Macroglossopteris leeiana* Sze,1931

1931　斯行健,5 页,图版 3,图 1;图版 4,图 1;蕨叶;江西萍乡;早侏罗世(Lias)。[注:此属模式种后改定为 *Anthrophyopsis leeiana* (Sze) Florin (Florin,1933)]

东北穗属 Genus *Manchurostachys* Kon'no,1960

1960　Kon'no E,164 页。
模式种:*Manchurostachys manchuriensis* Kon'no,1960
分类位置:木贼目(Equisetales)
分布与时代:中国辽宁本溪;二叠纪。

裂鞘叶东北穗 *Manchurostachys manchuriensis* Kon'no,1960

1960　Kon'no E,164 页,图版 20;东北裂鞘叶(*Schizoneura manchuriensis*)的生殖器官;辽宁本溪煤田;二叠纪(Tsaichia Formation)。

袖套杉属 Genus *Manica* Watson,1974

1974　Watson J,428 页。

模式种:*Manica parceramosa*(Fontaine)Watson,1974

分类位置:松柏纲掌鳞杉科(Cheirolepidiaceae,Coniferopsida)

分布与时代:美国弗吉尼亚;早白垩世。

希枝袖套杉 *Manica parceramosa* (Fontaine) Watson,1974

1889　*Frenilopsis parceramosa* Fontaine,218 页,图版 111,112,158;枝叶;美国弗吉尼亚;早白垩世。

1974　Watson J,428 页;美国弗吉尼亚;早白垩世。

袖套杉(长岭杉)亚属 Subgenus *Manica* (*Chanlingia*) Chow et Tsao,1977

1977　周志炎、曹正尧,172 页。

模式种:*Manica*(*Chanlingia*)*tholistoma* Chow et Tsao,1977

分类位置:松柏纲伏脂杉科希默杉亚科(Cheirolepidiaceae,Coniferopsida)

分布与时代:中国吉林长岭、扶余,浙江兰溪;白垩纪。

穹孔袖套杉(长岭杉) *Manica* (*Chanlingia*) *tholistoma* Chow et Tsao,1977

[注:此种后改定为 *Pseudofrenelopsis tholistoma*(Chow et Tsao)Cao,1989]

1977　周志炎、曹正尧,172 页,图版 2,图 16,17;图版 5,图 1－10;插图 4;枝叶、角质层;登记号:PB6265,PB6272;正型:PB6272(图版 5,图 1,2);标本保存在中国科学院南京地质古生物研究所;吉林长岭孙文屯;早白垩世青山口组;吉林扶余五家屯;早白垩世泉头组;浙江兰溪沈店;晚白垩世衢江群。

袖套杉(袖套杉)亚属 Subgenus *Manica* (*Manica*) Chow et Tsao,1977

1977　周志炎、曹正尧,169 页。

模式种:*Manica*(*Manica*)*parceramosa*(Fontaine)Chow et Tsao,1977

分类位置:松柏纲伏脂杉科希默杉亚科(Cheirolepidiaceae,Coniferopsida)

分布与时代:美国弗吉尼亚和中国吉林、宁夏、浙江;早白垩世。

希枝袖套杉(袖套杉) *Manica* (*Manica*) *parceramosa* (Fontaine) Chow et Tsao,1977

1889　*Frenilopsis parceramosa* Fontaine,Fontaine O,218 页,图版 111,112,158;枝叶;美国弗吉尼亚;早白垩世。

1977　周志炎、曹正尧,169 页。

此亚属创建时同时报道 4 种,除模式种外尚有:

大拉子袖套杉(袖套杉) *Manica* (*Manica*) *dalatzensis* Chow et Tsao,1977

[注:此种后改定为 *Pseudofrenelopsis dalatzensis*(Chow et Tsao)Cao ex Zhou(周志炎,1995)]

1977　周志炎、曹正尧,171 页,图版 3,图 5－11;图版 4,图 13;插图 3;枝叶、角质层;登记号:

PB6267,PB6268；正型：PB6267（图版 3，图 5）；标本保存在中国科学院南京地质古生物研究所；吉林延吉智新（大拉子）；早白垩世大拉子组。

窝穴袖套杉（袖套杉）*Manica* (*Manica*) *foveolata* Chow et Tsao,1977

［注：此种后改定为 *Pseudofrenelopsis foveolata* (Chow et Tsao)（曹正尧,1989）,*Pseudofrenelopsis papillosa* (Chow et Tsao) Cao ex Zhou（周志炎,1995）］

1977　周志炎、曹正尧,171 页,图版 4,图 1—7,14；枝叶、角质层；登记号：PB6269,PB6270；正型：PB6269（图版 4,图 1,2）；标本保存在中国科学院南京地质古生物研究所；宁夏固原蒿店、西吉牵羊河；早白垩世六盘山群。

乳突袖套杉（袖套杉）*Manica* (*Manica*) *papillosa* Chow et Tsao,1977

［注：此种后改定为 *Pseudofrenelopsis papillosa* (Chow et Tsao) Cao ex Zhou（周志炎,1995）］

1977　周志炎、曹正尧,169 页,图版 2,图 15；图版 3,图 1—4；图版 4,图 12；插图 2；枝叶、角质层和球果；登记号：PB6264,PB6266；正型：PB6266（图版 3,图 1）；标本保存在中国科学院南京地质古生物研究所；浙江新昌苏秦；早白垩世馆头组；宁夏固原青石咀；早白垩世六盘山群。

中间苏铁属 Genus *Mediocycas* Li et Zheng,2005（中文和英文发表）

2005　李楠、郑少林,见李楠等,425,433 页。

模式种：*Mediocycas kazuoensis* Li et Zheng,2005

分类位置：苏铁纲苏铁目（Cycadales,Cycadopsida）

分布与时代：中国辽宁西部；早三叠世。

喀左中间苏铁 *Mediocycas kazuoensis* Li et Zheng,2005（中文和英文发表）

1986　Problematicum 1,郑少林、张武,181 页,图版 1,图 10—11；辽宁西部喀左杨树沟；早三叠世红砬组。

1986　*Carpolithus* sp.,郑少林、张武,图版 3,图 11；种子；辽宁西部喀左杨树沟；早三叠世红砬组。

1986　*Carpolithus*? sp.,郑少林、张武,14 页,图版 3,图 12—14；种子；辽宁西部喀左杨树沟；早三叠世红砬组。

2005　李楠、郑少林,见李楠等,425,433 页；插图 3A—3F,5E；大孢子叶；标本号：SG110280—SG110283（正反印痕）,SG11026—SG11028；正模：SG110280—SG110283（插图 3A）；副模：SG110280—SG110283（插图 3B）；标本保存在沈阳地质矿产研究所；辽宁西部喀左杨树沟；早三叠世红砬组。

大叶羊齿属 Genus *Megalopteris* Schenk,1883 (non Andrews E B,1875)

［注：此属名为 *Megalopteris* Andrews E B,1875 的晚出同名,其模式种为 *Megalopteris dawsoni* (Hartt) Andrews 1875 (Andrews E B,1875,415 页；蕨叶；加拿大；泥盆纪？。即 *Neuropteris dawsoni* Hartt,见 Dawson,1868,551 页,图 193）］

1883　Schenk A,238 页。

模式种：*Megalopteris nicotianaefolia* Schenk，1883

分类位置：种子蕨纲大羽羊齿类（Pteridospermopsida，Gigantopterides）

分布与时代：中国湖南；晚石炭世。

烟叶大叶羊齿 *Megalopteris nicotianaefolia* Schenk，1883

［注：此种名后改定为 *Gigantopteris nicotianaefolia* Schenk（Whit D，1912；中国科学院南京地质古生物研究所、植物研究所，见《中国古生代植物》，1974）］

1883　Schenk A，238 页，图版 32，图 6—8；图版 33，图 1—3；图版 35，图 6；蕨叶；湖南耒阳；晚石炭世。

梅氏叶属 Genus *Meia* He，Liang et Shen，1996（中文和英文发表）

1996　何锡麟、梁敦士、沈树忠，77，163 页。

模式种：*Meia mingshanensis* He，Liang et Shen，1996

分类位置：真蕨纲或种子蕨纲（Filices or Pteridospermopsida）

分布与时代：中国江西、贵州；晚二叠世。

鸣山梅氏叶 *Meia mingshanensis* He，Liang et Shen，1996（中文和英文发表）

1996　何锡麟、梁敦士、沈树忠，77，164 页，图版 75；图版 76，图 1；图版 77，图 1，2；图版 78，图 1；图版 79，图 3；图版 92；枝叶和角质层；登记号：X88313—X88317；合模：X88313—X88315（图版 75，图 1—3），X88316（图版 78，图 1），X88317（图版 79，图 3）；标本保存在中国矿业大学地质系古生物实验室；江西乐平鸣山煤矿；晚二叠世乐平组老山下亚段。

此属创建时同时报道 2 种，除模式种外尚有：

大叶梅氏叶 *Meia magnifolia* He，Liang et Shen，1996（中文和英文发表）

1996　何锡麟、梁敦士、沈树忠，78，164 页，图版 78，图 2—4；图版 90，91；枝叶和角质层；登记号：X88318—X88320；合模：X88318—X88320（图版 78，图 2—4）；标本保存在中国矿业大学地质系古生物实验室；江西乐平鸣山煤矿；晚二叠世乐平组老山下亚段。

膜质叶属 Genus *Membranifolia* Sun et Zheng，2001（中文和英文发表）

2001　孙革、郑少林，见孙革等，108，208 页。

模式种：*Membranifolia admirabilis* Sun et Zheng，2001

分类位置：分类位置不明植物（Plantae incertae sedis）

分布与时代：中国辽宁凌源；晚侏罗世。

奇异膜质叶 *Membranifolia admirabilis* Sun et Zheng，2001（中文和英文发表）

2001　孙革、郑少林，见孙革等，108，208 页，图版 26，图 1—2；图版 67，图 3—6；膜质叶；标本号：PB19184，PB19185，PB19187，PB19196；正模：PB19184（图版 26，图 1）；标本保存在中国科学院南京地质古生物研究所；辽宁凌源；晚侏罗世尖山沟组。

异枝蕨属 Genus *Metacladophyton* Wang et Geng,1997(英文发表)
1997　王忠、耿宝印,93 页。
模式种:*Metacladophyton tetraxylum* Wang et Geng,1997
分类位置:真蕨植物门枝木蕨目(Cladoxylates,Pteridophyta)
分布与时代:中国湖北;中泥盆世晚期。

四木质柱异枝蕨 *Metacladophyton tetraxylum* Wang et Geng,1997(英文发表)
1997　王忠、耿宝印,93 页,图版 1—11;插图 1—7;蕨类植物;正模:No. 9167(图版 3,图 1);副模:No. 9168—No. 9190;标本保存在中国科学院植物研究所;湖北长阳;晚中泥盆世云台观组。

变态鳞木属 Genus *Metalepidodendron* Shen (MS) ex Wang X F,1984(nom. nud.)
1984　沈光隆,见王喜富,297 页。
模式种:*Metalepidodendron sinensis* Shen(MS)ex Wang X F,1984
分类位置:石松纲石松目(Lycopodiales,Lycoposida)
分布与时代:中国甘肃、河北承德;早一中三叠世。

中国变态鳞木 *Metalepidodendron sinensis* Shen (MS) ex Wang X F,1984(nom. nud.)
1984　沈光隆,见王喜富,297 页;甘肃;中三叠世。

此属创建时同时报道 2 种,除模式种外尚有:
下板城变态鳞木 *Metalepidodendron xiabanchengensis* Wang X F et Cui,1984
1984　王喜富,297 页,图版 175,图 8—11;茎干;登记号:HB-57,HB-58;河北承德下板城;早三叠世和尚沟组上部。

水杉属 Genus *Metasequoia* Miki,1941 ex Hu et Cheng,1948
1941　Miki S,262 页。(化石种)
1948　胡先骕、郑万钧,153 页。(现生种)
模式种:*Metasequoia disticha* Miki,1941(化石种)
　　　Metasequoia glyptostroboides Hu et Cheng,1948(现生种)
分类位置:松柏纲杉科(Taxodiaceae,Coniferopsida)
分布与时代:北半球;白垩纪—第三纪(化石种)。

水松型水杉 *Metasequoia glyptostroboides* Hu et Cheng,1948(现生种)
1948　胡先骕、郑万钧,153 页;插图 1—2;四川万县磨刀溪;现生种。

二列水杉 *Metasequoia disticha* Miki,1941(化石种)
1876　*Sequoia disticha* Heer,Heer O,63 页,图版 12,图 2a;图版 13,图 9—11;小枝和球果;北半球;白垩纪、古新世和始新世。

1941　Miki S,262 页,图版 5,图 A－Ca;插图 8,A－G;小枝和球果;北半球;白垩纪、古新世和始新世。

似叉苔属　Genus *Metzgerites* Wu et Li,1992

1992　吴向午、厉宝贤,268,276 页。

模式种:*Metzgerites yuxinanensis* Wu et Li,1992

分类位置:苔纲(Hepaticae)

分布与时代:中国河北蔚县;中侏罗世。

蔚县似叉苔　*Metzgerites yuxinanensis* Wu et Li,1992

1992　吴向午、厉宝贤,268,276 页,图版 3,图 3－5a;图版 6,图 1,2;插图 6;叶状体;采集号:ADN41-01,ADN41-02;登记号:PB15480－PB15483;正模:PB15481(图版 3,图 4);标本保存在中国科学院南京地质古生物研究所;河北蔚县涌泉庄附近;中侏罗世乔儿涧组。

塔状木属　Genus *Minarodendron* Li,1990

1990　李承森,105 页。

模式种:*Minarodendron cathaysiense* (Schweitzer et Cai) Li,1990

分类位置:石松植物门原始鳞木目原始鳞木科(Protolepidodendraceae, Protolepidophytales, Lycophyta)

分布与时代:中国云南、湖南、广西;中泥盆世晚期(Givetium)。

华夏塔状木　*Minarodendron cathaysiense* (Schweitzer et Cai) Li,1990

1987　*Protolepidodendron cathaysiense* Schweitzer et Cai,Schweitzer H J、蔡重阳,33(5)页,图版 1,2;插图 4;石松植物;云南沾益龙华山地区;中泥盆世晚期(Givetium)海口组(西充组);湖南浏阳;中泥盆世晚期跳马涧组;广西鹿寨、罗城;中泥盆世晚期东岗岭组。

1990　李承森,105 页,图版 1－11;插图 1－7;草本石松植物;云南沾益龙华山地区;中泥盆世晚期海口组;湖南浏阳;中泥盆世晚期跳马涧组;广西鹿寨、罗城;中泥盆世晚期东岗岭组。

奇异羊齿属　Genus *Mirabopteris* Mi et Liu,1993

1993　米家榕、刘茂强,见米家榕等,102 页。

模式种:*Mirabopteris hunjiangensis* (Mi et Liu) Mi et Liu,1993

分类位置:种子蕨纲(Pteridospermopsida)

分布与时代:中国吉林浑江;晚三叠世。

浑江奇异羊齿　*Mirabopteris hunjiangensis* (Mi et Liu) Mi et Liu,1993

1977　*Paradoxopteris hunjiangensis* Mi et Liu,米家榕、刘茂强,见长春地质学院地勘系、吉林省地质局区测大队、吉林省煤田地质勘探公司 102 队调查队,8 页,图版 3,图 1;插图

1;蕨叶;登记号:X-008;标本保存在长春地质学院地史古生物教研室;吉林浑江石人北山;晚三叠世"北山组"。
1993 米家榕、刘茂强,见米家榕等,102 页,图版 18,图 3;图版 53,图 1,2,6;插图 21;蕨叶和角质层;吉林浑江石人北山;晚三叠世北山组(小河口组)。

奇脉叶属 Genus *Mironeura* Zhou,1978
1978 周统顺,114 页。
模式种:*Mironeura dakengensis* Zhou,1978
分类位置:苏铁纲蕉羽叶目或苏铁目(Nilssoniales or Cycadales,Cycadopsida)
分布与时代:中国福建漳平;晚三叠世。

大坑奇脉叶 *Mironeura dakengensis* Zhou,1978
1978 周统顺,114 页,图版 25,图 1,2,2a;插图 4;羽叶;采集号:$WFT_3W_1^1$-9;登记号:FKP135;标本保存在地质科学研究院地质矿产所;福建漳平大坑(文宾山);晚三叠世文宾山组下段。

间羽叶属 Genus *Mixophylum* Meng,1983
1983 孟繁松,228 页。
模式种:*Mixophylum simplex* Meng,1983
分类位置:分类位置不明植物(Plantae incertae sedis)
分布与时代:中国湖北南漳;晚三叠世。

简单间羽叶 *Mixophylum simplex* Meng,1983
1983 孟繁松,228 页,图版 3,图 1;匙叶;登记号:D76018;正模:D76018(图版 3,图 1);标本保存在湖北地质矿产研究所;湖北南漳东巩;晚三叠九里岗组。

间羽蕨属 Genus *Mixopteris* Hsu et Chu C N,1974
1974 徐仁、朱家楠,见徐仁等,271 页。
模式种:*Mixopteris intercalaris* Hsu et Chu C N,1974
分类位置:真蕨纲?(Filicopsida?)
分布与时代:中国云南永仁;晚三叠世。

插入间羽蕨 *Mixopteris intercalaris* Hsu et Chu C N,1974
1974 徐仁、朱家楠,见徐仁等,271 页,图版 3,图 4—7;插图 4;蕨叶;标本号:No. 2610;标本保存在中国科学院植物研究所;云南永仁纳拉箐;晚三叠世大荞地组底部。

似提灯藓属 Genus *Mnioites* Wu X W,Wu X Y et Wang,2000(英文发表)
2000 吴向午、吴秀元、王永栋,170 页。

模式种:*Mnioites brachyphylloides* Wu X W,Wu X Y et Wang,2000
分类位置:真藓类(Bryiidae)
分布与时代:中国新疆克拉玛依;中侏罗世。

短叶杉型似提灯藓 *Mnioites brachyphylloides* Wu X W,Wu X Y et Wang,2000(英文发表)
2000　吴向午、吴秀元、王永栋,170页,图版2,图5;图版3,图1—2d;茎叶体;采集号:92-T-61;登记号:PB17797—PB17799;正模:PB17798(图版3,图1—1c);副模:PB17797(图版3,图2—2d),PB17797(图版2,图5);标本保存在中国科学院南京地质古生物研究所;新疆克拉玛依吐孜阿克内沟;中侏罗世西山窑组。

单叶单网羊齿属 Genus *Monogigantonoclea* Yang,2006(中文和英文发表)
2006　杨关秀等,189,311页。
模式种:*Monogigantonoclea colocasifolia* (Yang) Yang,2006
分类位置:大羽羊齿目(Gigantopteridales)
分布与时代:中国河南禹州;二叠纪。

芋叶单叶单网羊齿 *Monogigantonoclea colocasifolia* (Yang) Yang,2006(中文和英文发表)
1987b　*Gigantonoclea colocasifolia* Yang,杨关秀等,185页,图版2,图1—6。
2006　杨关秀等,190,311页,图版65,图1—3;图版66,图1—4;图版67,图6;插图7-10;单叶;合模:HEP0152,0149,0124(图版65,图1—3),HEP0147,0150,0153,3621(图版66,图1—4),HEP0148(图版67,图6);标本保存在中国地质大学(北京);河南禹州大风口、云盖山、方山滴水潭;晚二叠世云盖山组6煤段。

此属创建时同时报道5种,除模式种外尚有:

圆形单叶单网羊齿 *Monogigantonoclea rotundifolia* (Yang) Yang,2006(中文和英文发表)
1987b　*Gigantonoclea rotundifolia* Yang,杨关秀等,186页,图版2,图8,9。
2006　杨关秀等,191,312页,图版67,图1—5;插图7-11;单叶;合模:HEP0172(图版67,图1),HEP0168(图版67,图2),HEP0170(图版67,图3),HEP0171(图版67,图4),HEP0169(图版67,图5);标本保存在中国地质大学(北京);河南禹州大风口、云盖山;晚二叠世云盖山组6煤段。

阔卵单叶单网羊齿 *Monogigantonoclea latiovata* Yang et Wu,2006(中文和英文发表)
2006　杨关秀、吴跃辉,见杨关秀等,192,313页,图版68,图1—3;插图7-12;单叶;合模:HEP0157(图版68,图1),HEP0159(图版68,图2),HEP3149(图版68,图3);标本保存在中国地质大学(北京);河南登封、禹州云盖山;晚二叠世云盖山组7、8煤段。

巨齿单叶单网羊齿 *Monogigantonoclea grandidenia* Yang et Sheng,2006(中文和英文发表)
2006　杨关秀、盛阿兴,见杨关秀等,193,313页,图版55,图3,4;单叶;合模:HEP5106(图版55,图3),HEP0203(图版55,图4);标本保存在中国地质大学(北京);河南禹州大风口、平顶山矿;中二叠世小风口组。

似槭单叶单网羊齿 *Monogigantonoclea aceroides* Yang,2006(中文和英文发表)
2006　杨关秀等,194,314页,图版65,图4;单叶;正模:HEP0154(图版65,图4);标本保存在中国地质大学（北京）；河南禹州云盖山；晚二叠世云盖山组。

单叶大羽羊齿属 Genus *Monogigantopteris* Yang，2006（中文和英文发表）
2006　杨关秀等，199，319 页。
模式种：*Monogigantopteris clathroreticulatus* Yang，2006
分类位置：大羽羊齿目（Gigantopteridales）
分布与时代：中国河南禹州、登封；晚二叠世。

格网单叶大羽羊齿 *Monogigantopteris clathroreticulatus* Yang，2006（中文和英文发表）
2006　杨关秀等，199，319 页，图版 70，图 1－7；单叶；合模：HEP0207（图版 70，图 1），HEP0205（图版 70，图 2），HEP0209（图版 70，图 3），HEP0208（图版 70，图 4），HEP0206（图版 70，图 5），HEP0210（图版 70，图 6），HEP0211（图版 70，图 7）；标本保存在中国地质大学（北京）；河南禹州大风口；中二叠世小风口组 4 煤段。

此属创建时同时报道 2 种，除模式种外尚有：

密网单叶大羽羊齿 *Monogigantopteris densireticulatus* Yang，2006（中文和英文发表）
2006　杨关秀等，200，320 页，图版 72，图 1－6；插图 7-15；单叶；合模：HEP0123（图版 72，图 1），HEP0124（图版 72，图 3），HEP0125（图版 72，图 4），HEP0127（图版 72，图 2），HEP3121（图版 72，图 6），HEP3140（图版 72，图 5）；标本保存在中国地质大学（北京）；河南禹州大风口、云盖山，登封磴槽；晚二叠世云盖山组 8 煤段。

尖籽属 Genus *Muricosperma* Seyfullah et Hilton，2010（英文发表）
2010　Seyfullah L J 和 Hilton J，见 Seyfullah L J 等，99 页。
模式种：*Muricosperma guizhouensis* Seyfullah et Hilton，2010
分类位置：种子植物门（Spermatophyta）
分布与时代：中国贵州盘县；晚二叠世（Wuchiapigian）。

贵州尖籽 *Muricosperma guizhouensis* Seyfullah et Hilton，2010（英文发表）
2010　Seyfullah L J 和 Hilton J，见 Seyfullah L J 等，99 页，图 1－20；胚珠化石；正模：GPP 2-001（图 1－7）；标本保存在中国科学院植物研究所；中国贵州盘县；晚二叠世（Wuchiapigian）宣威组。

密叶属 Genus *Myriophyllum* Xiao，1985
1985　肖素珍、张恩鹏，584 页。
模式种：*Myriophyllum shanxiense* Xiao，1985
分类位置：分类位置不明植物（Plantae incertae sedis）
分布与时代：中国山西山阴；晚石炭世晚期。

山西密叶 *Myriophyllum shanxiense* Xiao，1985
1985　肖素珍、张恩鹏，584 页，图版 203，图 13；图版 204，图 1－3；羽叶；登记号：SH430，

SH390,SH388,SH389；正模：SH390（图版204，图1）；副模：SH430（图版203，图13），SH388（图版204，图2），SH389（图版204，图3）；山西山阴；晚石炭世晚期山西组。

南票叶属 Genus *Nanpiaophyllum* Zhang et Zheng, 1984
1984　张武、郑少林，389页。
模式种：*Nanpiaophyllum cordatum* Zhang et Zheng, 1984
分类位置：分类位置不明植物（Plantae incertae sedis）
分布与时代：中国辽宁西部南票；晚三叠世。

心形南票叶 *Nanpiaophyllum cordatum* Zhang et Zheng, 1984
1984　张武、郑少林，389页，图版3，图4—9；插图8；蕨叶；登记号：J005-1—J005-6；标本保存在沈阳地质矿产研究所；辽宁西部南票沙锅屯；晚三叠世老虎沟组。

南漳叶属 Genus *Nanzhangophyllum* Chen, 1977
1977　陈公信，见冯少南等，246页。
模式种：*Nanzhangophyllum donggongense* Chen, 1977
分类位置：裸子植物类（Gymnospermae incertae sedis）
分布与时代：中国湖北南漳；晚三叠世。

东巩南漳叶 *Nanzhangophyllum donggongense* Chen, 1977
1977　陈公信，见冯少南等，246页，图版99，图6—7；插图82；叶；标本号：P5014，P5015；合模1：P5014（图版99，图6）；合模2：P5015（图版99，图7）；标本保存在湖北省地质局；湖北南漳东巩大道场；晚三叠世香溪群下煤组。

新轮叶属 Genus *Neoannularia* Wang, 1977
1977　王喜富，186页。
模式种：*Neoannularia shanxiensis* Wang, 1977
分类位置：楔叶纲木贼目（Equisetales, Sphenopsida）
分布与时代：中国陕西宜君、四川渡口；晚三叠世。

陕西新轮叶 *Neoannularia shanxiensis* Wang, 1977
1977　王喜富，186页，图版1，图1—9；带轮叶枝；采集号：JP672001—JP672009；登记号：76003—76011；陕西宜君焦坪；晚三叠世延长群上部。

此属创建时同时报道2种，除模式种外尚有：

川滇新轮叶 *Neoannularia chuandianensis* Wang, 1977
1977　王喜富，187页，图版1，图10；插图1；带轮叶枝；采集号：DK70502；登记号：76002；四川渡口摩沙河；晚三叠世大青组。

新科达属 Genus *Neocordaites* Yang et Wang,2006（中文和英文发表）
2006　杨关秀、王洪山,见杨关秀等,159,297 页。
模式种:*Neocordaites lanceolatus* Yang et Wang,2006
分类位置:科达纲(Cordaitopsida)
分布与时代:中国河南平顶山煤矿;中二叠世。

披针新科达 *Neocordaites lanceolatus* Yang et Wang,2006（中文和英文发表）
2006　杨关秀、王洪山,见杨关秀等,159,297 页,图版 48,图 9,9a;叶;主模:HEP4362(图版 48,图 9);标本保存在中国地质大学(北京);河南平顶山煤矿;中二叠世神垕组。

新准大羽羊齿属 Genus *Neogigantopteridium* Yang,1987
1987b　杨关秀,188,194 页。
模式种:*Neogigantopteridium spiniferum* Yang,1987
分类位置:大羽羊齿类(Gigantopterides)
分布与时代:中国河南禹县;晚二叠世。

具刺新准大羽羊齿 *Neogigantopteridium spiniferum* Yang,1987
1987b　杨关秀,188,194 页,图版 3,图 3－5;插图 9;蕨叶;登记号:HEP 223－HEP 225;合模 1:HEP 223(图版 2,图 4,4a);合模 2:HEP 224(图版 2,图 5);合模 3:HEP 225;标本保存在中国地质大学(北京);豫西禹县陈庄南李家门;晚二叠世上部云盖山组 7 煤段顶部。

新孢穗属 Genus *Neostachya* Wang,1977
1977　王喜富,188 页。
模式种:*Neostachya shanxiensis* Wang,1977
分类位置:楔叶纲木贼目(Equisetales,Sphenopsida)
分布与时代:中国陕西宜君;晚三叠世。

陕西新孢穗 *Neostachya shanxiensis* Wang,1977
1977　王喜富,188 页,图版 2,图 1－10;有节类生殖枝;采集号:JP672010－JP672017;登记号:76012－76019;陕西宜君焦坪;晚三叠世延长群上部。

翅叶属 Genus *Neurophyllites* Zhang,1980
1980　张善桢,见赵修祜等,85 页。
模式种:*Neurophyllites pecopteroides* Zhang,1980
分类位置:分类位置不明植物(Plantae incertae sedis)

分布与时代：中国云南富源；晚二叠世。

栉状翅叶 *Neurophyllites pecopteroides* Zhang,1980
1980　张善桢,见赵修祜等,85页,图版14,图1,1a,2,2a,2b；蕨叶；采集号：PQ-118；登记号：PB7045,PB7046；标本保存在中国科学院南京地质古生物研究所；云南富源；晚二叠世早期宣威组下段。

宁夏叶属 Genus *Ningxiaphyllum* Zhao,Wu et Gu,1986
1986　赵修祜、吴秀元、顾其昌,554,558页。
模式种：*Ningxiaphyllum trilobatum* Zhao,Wu et Gu,1986
分类位置：分类位置不明植物（Plantae incertae sedis）
分布与时代：中国宁夏青铜峡县；晚泥盆世。

三裂宁夏叶 *Ningxiaphyllum trilobatum* Zhao,Wu et Gu,1986
1986　赵修祜、吴秀元、顾其昌,555,558页,图版4,图3；图版5,图1,2；插图9；叶片；采集号：QD-H-60,QD-H-84,QD-H-85；登记号：PB11493,PB11495,PB11496；正模：PB11495（图版5,图1）；标本保存在中国科学院南京地质古生物研究所；宁夏青铜峡广武应声台；晚泥盆世中宁组。

那琳壳斗属 Genus *Norinia* Halle,1927
（注：又名帚叶属）
1927　Halle T G,218页。
模式种：*Norinia cucullata* Halle,1927
分类位置：分类位置不明植物（Plantae incertae sedis）
分布与时代：中国山西中部（Ch'en-chia-yu）；晚二叠世早期。

僧帽状那琳壳斗 *Norinia cucullata* Halle,1927
（注：又名帚叶）
1927　Halle T G,218页,图版56,图8-12；叶部化石；山西中部（Ch'en-chia-yu）；晚二叠世早期上石盒子系。

裸囊穗属 Genus *Nudasporestrobus* Feng,Wang et Bek,2008（英文发表）
2008　冯卓、王军、Bek J,152页。
模式种：*Nudasporestrobus ningxicus* Feng,Wang et Bek,2008
分类位置：鳞木目封印木科（Sigillariaceae,Lepidodendrales）
分布与时代：中国宁夏中卫；晚石炭世。

宁夏裸囊穗 *Nudasporestrobus ningxicus* Feng,Wang et Bek,2008（英文发表）
2008　冯卓、王军、Bek J,152页,图版1-4；孢子叶穗；登记号：PB20826-PB20832；正模：

PB20826(图版1,图1－4);标本保存在中国科学院南京地质古生物研究所;中国宁夏中卫;晚石炭世(早宾夕法尼亚纪)杨虎沟组。

新常富籽属 Genus *Nystroemia* Halle,1927

(注:又名髻籽羊齿属)

1927　　Halle T G,221 页。

模式种:*Nystroemia pectiniformis* Halle,1927

分类位置:种子蕨纲?(Pteridospermopsida?)

分布与时代:中国山西中部(Ch'en-chia-yu);晚二叠世早期。

篦形新常富籽 *Nystroemia pectiniformis* Halle,1927

(注:又名髻籽羊齿)

1927　　Halle T G,221 页,图版59;着生种子的羽叶;山西中部(Ch'en-chia-yu);晚二叠世早期上石盒子系。

似齿囊蕨属 Genus *Odontosorites* Kobayashi et Yosida,1944

1944　　Kobayashi T 和 Yosida T,257 页。

模式种:*Odontosorites heerianus* (Yokoyama) Kobayashi et Yosida,1944

分类位置:真蕨纲(Filicopsida)

分布与时代:中国黑龙江黑河附近(Rykusin);侏罗纪。日本;早白垩世。

海尔似齿囊蕨 *Odontosorites heerianus* (Yokoyama) Kobayashi et Yosida,1944

1889　　*Adiantites heerianus* Yokoyama,Yokoyama M,28 页,图版12,图1,1a,1b,2;日本;早白垩世(Tetori Series)。

1944　　Kobayashi T 和 Yosida T,257 页,图版28,图6－7;插图 a－c;实羽片和裸羽片;黑龙江黑河附近(Rykusin);侏罗纪。[注:此标本后改定为? *Coniopteris burejensis* (Zalessky) Seward(斯行健、李星学等,1963)]

似兰属 Genus *Orchidites* Wu S Q,1999(中文发表)

1999　　吴舜卿,23 页。

模式种:*Orchidites linearifolius* Wu S Q,1999(注:原文未指定模式种,本文暂把原文列在第一的种作为模式种编录)

分类位置:单子叶植物纲兰科(Orchidaceae,Monocotyledoneae)

分布与时代:中国辽西北票;晚侏罗世。

线叶似兰 *Orchidites linearifolius* Wu S Q,1999(中文发表)

1999　　吴舜卿,23 页,图版16,图7;图版17,图1－3;营养枝;采集号:AEO-123,104,29;登记号:PB18321,PB18324,PB18325;标本保存在中国科学院南京地质古生物研究所;辽西北票上园黄半吉沟;晚侏罗世义县组下部尖山沟层。

此属创建时同时报道 2 种,除模式种外尚有:
披针叶似兰 *Orchidites lancifolius* Wu S Q,1999(中文发表)
1999 吴舜卿,23 页,图版 17,图 4,4a;草本、枝叶;采集号:AEO196;登记号:PB18326;标本保存在中国科学院南京地质古生物研究所;辽西北票上园黄半吉沟;晚侏罗世义县组下部尖山沟层。

耳叶属 Genus *Otofolium* Gu et Zhi,1974
1974 中国科学院南京地质古生物研究所、植物研究所,见《中国古生代植物》,164 页。
模式种:*Otofolium polymorphum* Gu et Zhi,1974
分类位置:分类位置不明植物(Plantae incertae sedis)
分布与时代:中国江苏、安徽、广东、江西、贵州;晚二叠世。

多形耳叶 *Otofolium polymorphum* Gu et Zhi,1974
1974 中国科学院南京地质古生物研究所、植物研究所,见《中国古生代植物》,165 页,图版 127,图 2-6;插图 134;羽状复叶;登记号:PB3772(图版 127,图 2),PB3771(图版 127,图 3),PB3715(图版 127,图 5),PB3713;正模:PB3771;标本保存在中国科学院南京地质古生物研究所;江苏江宁、安徽泾县、广东曲江;晚二叠世早期龙潭组;江西进贤;晚二叠世乐平组。

此属创建时同时报道 2 种,除模式种外尚有:
卵耳叶 *Otofolium ovatum* Gu et Zhi,1974
1974 中国科学院南京地质古生物研究所、植物研究所,见《中国古生代植物》,165 页,图版 127,图 7-9;羽状复叶;登记号:PB4993(图版 127,图 7),PB4994(图版 127,图 8);正模:PB4994;标本保存在中国科学院南京地质古生物研究所;贵州盘县;晚二叠世早期宣威组。

古银杏型木属 Genus *Palaeoginkgoxylon* Feng,Wang et Roessler,2010(英文发表)
2010 冯卓、王军、Roessler R,149 页。
模式种:*Palaeoginkgoxylon zhoui* Feng,Wang et Roessler,2010
分类位置:银杏类(Ginkgophytes)
分布与时代:中国内蒙古阿拉善左旗;中二叠世(Guadalupian)。

周氏古银杏型木 *Palaeoginkgoxylon zhoui* Feng,Wang et Roessler,2010(英文发表)
2010 冯卓、王军、Roessler R,149 页,图版 1-4;木化石;登记号:YKLP20006;正模:YKLP20006(图版 1-4);标本保存在云南大学云南古生物学重点实验室;内蒙古阿拉善左旗呼鲁斯太;中二叠世(Guadalupian)下石盒子组。

古买麻藤属 Genus *Palaeognetaleaana* Wang,2004(英文发表)
2004 王自强,282 页。

模式种：*Palaeognetaleaana auspicia* Wang，2004

分类位置：买麻藤目（Gnetales）

分布与时代：中国山西；晚二叠世。

吉祥古买麻藤 *Palaeognetaleaana auspicia* Wang，2004（英文发表）
2004　王自强，282 页，图 2—6；球果；标本号：9107-1，9107-8，9705-0，9705-37，9705-39，9705-39a，9705-41，9705-45，9805-045，9805-048；正模：9107-1（图 2—3）；标本保存在中国科学院南京地质古生物研究所；山西西北；晚二叠世。

古舟藤属 Genus *Palaeoskapha* Jacques et Guo，2007（英文发表）
2007　Jacques Frédérrie M B，郭双兴，578 页。

模式种：*Palaeoskapha sichuanensis* Jacques et Guo，2007

分类位置：双子叶植物纲防己科（Menispermaceae，Dicotyledoneae）

分布与时代：中国四川理塘；始新世。

四川古舟藤 *Palaeoskapha sichuanensis* Jacques et Guo，2007（英文发表）
2007　Jacques Frédérrie M B，郭双兴，578 页，图 2A，2B；果实；登记号：PB12702，PB12703；合模：PB12703（图 2A），PB12702（图 2B）；标本保存在中国科学院南京地质古生物研究所；四川理塘热鲁村（约 30°N，100°32′E）；始新世热鲁组。

潘氏果属 Genus *Pania* Yang，2006（中文和英文发表）
2006　杨关秀等，144，289 页。

模式种：*Pania cycadina* Yang，2006

分类位置：苏铁植物门苏铁目（Cycadales，Cycadophyta）

分布与时代：中国河南禹州；晚二叠世。

似苏铁潘氏果 *Pania cycadina* Yang，2006（中文和英文发表）
2006　杨关秀等，144，289 页，图版 40，图 4—6；雄性球果（小孢子叶球）；合模：HEP0885（图版 40，图 4），HEP0888（图版 40，图 5），HEP0887（图版 40，图 6）；标本保存在中国地质大学（北京）；河南禹州大风口；晚二叠世云盖山组 7 段。

潘广叶属 Genus *Pankuangia* Kimura，Ohana，Zhao et Geng，1994
1994　Kimura T、Ohana T、赵立明、耿宝印，256 页。

模式种：*Pankuangia haifanggouensis* Kimura，Ohana，Zhao et Geng，1994

分类位置：苏铁纲苏铁目（Cycadales，Cycadopsida）

分布与时代：中国辽宁锦西；中侏罗世。

海房沟潘广叶 *Pankuangia haifanggouensis* Kimura，Ohana，Zhao et Geng，1994
1994　Kimura T、Ohana T、赵立明、耿宝印，257 页，图 2—4，8；叶部化石；标本号：LJS-8690，

8555,8554,8807,L0407A[潘广定名为 *Juradicotis elrecta* Pan（MS）]；正模：LJS-8690（图2A）；标本保存在中国科学院植物研究所；辽宁锦西（约 40°58′N,120°21′E）；中侏罗世海房沟组。[注：这些标本后改定为 *Anomozamites haifanggouensis*（Kimura,Ohana,Zhao et Geng）Zheng et Zhang(郑少林等,2003)]

蝶叶属 Genus *Papilionifolium* Cao,1999（中文和英文发表）
1999　曹正尧,102,160 页。
模式种：*Papilionifolium hsui* Cao,1999
分类位置：分类位置不明植物（Plantae incertae sedis）
分布与时代：中国浙江文成；早白垩世。

徐氏蝶叶 *Papilionifolium hsui* Cao,1999（中文和英文发表）
1999　曹正尧,102,160 页,图版 21,图 12 — 15；插图 35；茎和叶；采集号：Zh301；登记号：PB14467 — PB14470；正模：PB14469（图版 21,图 14）；标本保存在中国科学院南京地质古生物研究所；浙江文成孔龙；早白垩世馆头组。

副开通尼亚属 Genus *Paracaytonia* Wang,2010（英文发表）
2010　王鑫,208 页。
模式种：*Paracaytonia hongtaoi* Wang,2010
分类位置：开通目（Caytoniales）
分布与时代：中国辽宁建昌；早白垩世（Barremian）。

洪涛副开通尼亚 *Paracaytonia hongtaoi* Wang,2010（英文发表）
2010　王鑫,208 页,图 1 — 3；雌性生殖器官；标本号：GBM1A,B；正模：GBM1（图 1）；标本保存在深圳仙湖植物园古生物博物馆；辽宁建昌郎家沟、要路沟；早白垩世（Barremian）义县组。

副球果属 Genus *Paraconites* Hu,1984（nom. nud.）
1984　胡雨帆,571 页。
模式种：*Paraconites longifolius* Hu,1984
分类位置：松柏纲杉科（Taxodiaceae,Coniferopsida）
分布与时代：中国山西大同；早侏罗世。

伸长副球果 *Paraconites longifolius* Hu,1984（nom. nud.）
1984　胡雨帆,571 页；松柏类球果；山西大同煤峪口；早侏罗世永定庄组。

奇异羊齿属 Genus *Paradoxopteris* Mi et Liu,1977（non Hirmer,1927）
[注：此属名为埃及晚白垩世 *Paradoxopteris* Hirmer,1927 的晚出同名（吴向午,1993a,

1993b),其模式种为 *Paradoxopteris strumeri* Hirmer,1927(Hirmer,1927,609 页;插图 733—736;埃及;晚白垩世)。此属名后改定为 *Mirabopteris*（Mi et Liu）Mi et Liu(米家榕等,1993)]

1977 米家榕、刘茂强,见长春地质学院勘探系调查组等,8 页。

模式种:*Paradoxopteris hunjiangensis* Mi et Liu,1977

分类位置:种子蕨纲(Pteridospermopsida)

分布与时代:中国吉林浑江;晚三叠世。

浑江奇异羊齿 *Paradoxopteris hunjiangensis* Mi et Liu,1977

1977 米家榕、刘茂强,见长春地质学院勘探系调查组等,8 页,图版 3,图 1;插图 1;蕨叶;标本号:X-08;标本保存在长春地质学院勘探系;吉林浑江石人;晚三叠世小河口组。〔注:此种后改定为 *Mirabopteris hunjiangensis*（Mi et Liu）Mi et Liu(米家榕等,1993)]

副镰羽叶属 Genus *Paradrepanozamites* Chen,1977

1977 陈公信,见冯少南等,236 页。

模式种:*Paradrepanozamites dadaochangensis* Chen,1977

分类位置:苏铁纲(Cycadopsida)

分布与时代:中国湖北南漳;晚三叠世。

大道场副镰羽叶 *Paradrepanozamites dadaochangensis* Chen,1977

1977 陈公信,见冯少南等,236 页,图版 99,图 1—2;插图 81;羽叶;标本号:P5107,P25269;合模 1:P5107(图版 99,图 1);标本保存在湖北省地质局;合模 2:P25269(图版 99,图 2);标本保存在湖北地质科学研究所;湖北南漳东巩;晚三叠世香溪群下煤组。

拟楔叶属 Genus *Parasphenophyllum* Asama,1970

1970 Asama K,301 页。

模式种:*Parasphenophyllum shansiense*（Asama）Asama,1970

分类位置:楔叶目(Sphenophyllales)

分布与时代:中国和朝鲜;二叠纪。

山西拟楔叶 *Parasphenophyllum shansiense*（Asama）Asama,1970

1970 Asama K,301 页,图版 3,图 1;楔叶类的叶;中国和朝鲜;二叠纪。

拟楔羊齿属 Genus *Parasphenopteris* Sun et Deng,2006(英文发表)

2006 孙克勤、邓胜徽,161 页。

模式种:*Parasphenopteris orientalis* Sun et Deng,2006

分类位置:真蕨纲楔羊齿类或种子蕨纲(Sphenopterides of Filicopsida or Pteridospermopsia)

分布与时代:中国内蒙古乌达地区;早二叠世。

东方拟楔羊齿 *Parasphenopteris orientalis* Sun et Deng, 2006（英文发表）
2006　孙克勤、邓胜徽, 162 页, 图 1; 蕨叶; 正模: WD20026（图 1）; 标本保存在中国地质大学（北京）; 内蒙古乌达地区; 早二叠世山西组。

拟斯托加枝属 Genus *Parastorgaardis* Zeng, Shen et Fan, 1995
1995　曾勇、沈树忠、范炳恒, 67 页。
模式种: *Parastorgaardis mentoukouensis*（Stockmans et Mathieu）Zeng, Shen et Fan, 1995
分类位置: 松柏纲杉科（Taxodiaceae, Coniferopsida）
分布与时代: 中国北京门头沟、河南义马; 侏罗纪。

门头沟拟斯托加枝 *Parastorgaardis mentoukouensis*（Stockmans et Mathieu）Zeng, Shen et Fan, 1995
1941　*Podocarpites mentoukouensis* Stockmans et Mathieu, Stockmans F 和 Mathieu F F, 53 页, 图版 7, 图 5, 6; 枝叶; 北京门头沟; 侏罗纪。
1995　曾勇、沈树忠、范炳恒, 67 页, 图版 20, 图 3; 图版 23, 图 1; 图版 29, 图 6－8; 枝叶和角质层; 河南义马; 中侏罗世义马组下含煤段。

类紫杉籽属 Genus *Parataxospermum* Li, 1993
1993b　李中明, 66 页。
模式种: *Parataxospermum taiyuanesis* Li, 1993
分类位置: 科达目（Cardiocarpales）
分布与时代: 中国山西太原; 晚石炭世。

太原类紫杉籽 *Parataxospermum taiyuanesis* Li, 1993
1993b　李中明, 66 页, 图版 1－4; 插图 1; 种子（煤核）; 正模: BSC3.363（图版 1, 图 1－4; 图版 2; 图版 4, 图 1－3）; 副模: BSC3.102（图版 1, 图 5; 图版 3; 图版 4, 图 2）; 标本保存在中国科学院植物研究所; 山西太原西山; 晚石炭世太原组。

拟齿叶属 Genus *Paratingia* Zhang, 1987
（注: 后改名为拟丁氏蕨属）
1987　张泓, 200 页。
模式种: *Paratingia datongensis* Zhang, 1987
分类位置: 瓢叶纲（Noeggerathopsida）
分布与时代: 中国山西大同; 早二叠世。

大同拟齿叶 *Paratingia datongensis* Zhang, 1987
1987　张泓, 200 页, 图版 10, 图 1－4; 图版 11, 图 3, 4; 枝条状羽叶; 登记号: Mp-84061, Mp-84062, Mp-84063, Mp-84064; 山西大同七峰山; 早二叠世。

拟丁氏蕨穗属 Genus *Paratingiostachya* Sun,Deng,Cui et Shang,1999（中文和英文发表）
1999　孙克勤、邓胜徽、崔金钟、商平,1024,1025 页。
模式种：*Paratingiostachya cathaysiana* Sun,Deng,Cui et Shang,1999
分类位置：瓢叶纲（Noeggerathopsida）
分布与时代：中国内蒙古乌达地区；早二叠世。

华夏拟丁氏蕨穗 *Paratingiostachya cathaysiana* Sun,Deng,Cui et Shang,1999（中文和英文发表）
1999　孙克勤、邓胜徽、崔金钟、商平,1024,1025 页,图版 3,图 3,3a,4,4a；生殖器官；标本保存在中国地质大学（北京）；内蒙古乌达地区；早二叠世山西组。

雅蕨属 Genus *Pavoniopteris* Li et He,1986
1986　李佩娟、何元良,279 页。
模式种：*Pavoniopteris matonioides* Li et He,1986
分类位置：真蕨纲（Filicopsida）
分布与时代：中国青海都兰；晚三叠世。

马通蕨型雅蕨 *Pavoniopteris matonioides* Li et He,1986
1986　李佩娟、何元良,279 页,图版 2,图 1；图版 3,图 3－4；图版 4,图 1－1d；插图 1,2；营养蕨叶和生殖蕨叶；采集号：79PIVF22-3；登记号：PB10866,PB10869－PB10871；正模：PB10871（图版 4,图 1－1d）；标本保存在中国科学院南京地质古生物研究所；青海都兰八宝山；晚三叠世八宝山群下岩组。

箆囊属 Genus *Pectinangium* Gu et Zhi,1974
1974　中国科学院南京地质古生物研究所、植物研究所,见《中国古生代植物》,166 页。
模式种：*Pectinangium lanceolatum* Gu et Zhi,1974
分类位置：分类位置不明植物（Plantae incertae sedis）
分布与时代：中国江苏江宁；晚二叠世早期。

披针箆囊 *Pectinangium lanceolatum* Gu et Zhi,1974
（注：又名箆囊）
1974　中国科学院南京地质古生物研究所、植物研究所,见《中国古生代植物》,166 页,图版 128,图 9－12；插图 135；生殖羽叶；登记号：PB3807（图版 128,图 9）,PB3804（图版 128,图 10）；正模：PB3804；标本保存在中国科学院南京地质古生物研究所；江苏江宁；晚二叠世早期龙潭组。

雅观木属 Genus *Perisemoxylon* He et Zhang,1993
1993　何德长、张秀仪,262,264 页。
模式种:*Perisemoxylon bispirale* He et Zhang,1993
分类位置:苏铁纲苏铁目(Cycadales,Cycadopsida)
分布与时代:中国河南义马;中侏罗世。

双螺纹雅观木 *Perisemoxylon bispirale* He et Zhang,1993
1993　何德长、张秀仪,262,264 页,图版 1,图 1,2;图版 2,图 5;图版 4,图 3;丝炭化石;采集号:9001,9002;登记号:S006,S007;正模:S006(图版 1,图 1);副模:S007(图版 1,图 2);标本保存在煤炭科学研究总院西安分院;河南义马;中侏罗世。

此属创建时同时报道 2 种,除模式种外尚有:

雅观木(未定种) *Perisemoxylon* sp.
1993　*Perisemoxylon* sp.,何德长、张秀仪,263 页,图版 2,图 1－4;丝炭化石;河南义马;中侏罗世。

拟刺葵属 Genus *Phoenicopsis* Heer,1876
1876　Heer O,51 页。
模式种:*Phoenicopsis angustifolia* Heer,1876
分类位置:茨康目(Czekanowskiales)

狭叶拟刺葵 *Phoenicopsis angustifolia* Heer,1876
1876　Heer O,51 页,图版 1,图 1d;图版 2,图 3b;113 页,图版 31,图 7,8;叶;俄罗斯伊尔库茨克黑龙江上游;侏罗纪。

拟刺葵(斯蒂芬叶)亚属 Subgenus *Phoenicopsis* (*Stephenophyllum*) ex Li,1988
［注:此亚属名最早由李佩娟等(1988)应用,但未指明为新名和模式种］
1936　*Stephenophyllum* Florin,Florin R,82 页。
1988　李佩娟等,106 页。
模式种:*Phoenicopsis* (*Stephenophyllum*) *solmsi* (Seward)［注:*Stephenophyllum solmsi* (Seward) Florin 是 *Stephenophyllum* 属的模式种(Florin,1936)］
分类位置:茨康目(Czekanowskiales)
分布与时代:法兰士约瑟夫地和中国青海、新疆;侏罗纪。

索尔姆斯拟刺葵(斯蒂芬叶) *Phoenicopsis* (*Stephenophyllum*) *solmsi* (Seward)
1919　*Desmiophllum solmsi* Seward,Seward A C,71 页,图 662;叶的切面和气孔器;法兰士约瑟夫地;侏罗纪。
1936　*Stephenophyllum solmsi* (Seward) Florin,Florin R,82 页,图版 11,图 7－10;图版 12－16;插图 3,4;叶及角质层;法兰士约瑟夫地;侏罗纪。

此亚属创建时同时报道5种，除模式种外尚有：

美形拟刺葵（斯蒂芬叶）Phoenicopsis (Stephenophyllum) decorata Li, 1988

1988　李佩娟等, 106页, 图版68, 图5B; 图版79, 图4, 4a; 图版120, 图1—6; 叶及角质层; 采集号: 80LFu; 登记号: PB13630, PB13631; 正模: PB13631（图版79, 图4, 4a）; 标本保存在中国科学院南京地质古生物研究所; 青海绿草山绿草沟; 中侏罗世石门沟组 Nilssonia 层。

厄尼塞捷拟刺葵（斯蒂芬叶）Phoenicopsis (Stephenophyllum) enissejensis (Samylina) ex Li, 1988

[注: 此种名最早由李佩娟等（1988）应用, 但未指明为新名]

1972　Phoenicopsis (Phoenicopsis) enissejensis Samylina, 63页, 图版2, 图1, 2; 图版3, 图1—4; 图版4, 图1—5; 叶及角质层; 西西伯利亚; 中侏罗世。

1988　李佩娟等, 106页, 图版85, 图2, 2a; 图版86, 图1; 图版87, 图1; 图版121, 图1—6; 叶及角质层; 青海绿草山绿草沟; 中侏罗世石门沟组 Nilssonia 层。

特别拟刺葵（斯蒂芬叶）Phoenicopsis (Stephenophyllum) mira Li, 1988

1988　李佩娟等, 107页, 图版80, 图2—4a; 图版81, 图2; 图版122, 图5—6; 图版123, 图1—4; 图版136, 图5; 图版138, 图4; 叶及角质层; 采集号: 80DP$_1$F$_{89}$, 80DJ$_{2d}$Fu; 登记号: PB13635—PB13637; 正模: PB13635（图版81, 图5）; 标本保存在中国科学院南京地质古生物研究所; 青海柴达木盆地大煤沟; 中侏罗世饮马沟组 Coniopteris murrayana 层和大煤沟组 Tyrmia-Sphenobaiera 层。

塔什克斯拟刺葵（斯蒂芬叶）Phoenicopsis (Stephenophyllum) taschkessiensis (Krasser) ex Li, 1988

[注: 此种名最早由李佩娟等（1988）应用, 但未指明为新名]

1901　Phoenicopsis taschkessiensis Krasser, Krasser F, 150页, 图版4, 图2; 图版3, 图4t; 叶; 新疆哈密至吐鲁番之间的三道岭西南; 侏罗纪。

1988　李佩娟等, 3页。

贼木属 Genus Phoroxylon Sze, 1951

1951b　斯行健, 443, 451页。

模式种: Phoroxylon scalariforme Sze, 1951

分类位置: 本内苏铁目（Bennetittales）

分布与时代: 中国黑龙江鸡西; 晚白垩世。

梯纹状贼木 Phoroxylon scalariforme Sze, 1951

1951b　斯行健, 443, 451页, 图版5, 图2, 3; 图版6, 图1—4; 图版7, 图1—4; 插图3A—3E; 木化石; 黑龙江鸡西城子河; 晚白垩世。

叶茎属 Genus *Phylladendroid* He,Liang et Shen,1996（中文和英文发表）

1996 何锡麟、梁敦士、沈树忠,92,167 页。

模式种：*Phylladendroid jiangxiensis* He,Liang et Shen,1996

分类位置：分类位置不明植物（Plantae incertae sedis）

分布与时代：中国江西丰城；晚二叠世。

江西叶茎 *Phylladendroid jiangxiensis* He,Liang et Shen,1996（中文和英文发表）

1996 何锡麟、梁敦士、沈树忠,92,167 页,图版 74,图 3,4;茎干;登记号：X88311,X88312;合模：X88311（图版 74,图 3）,X88312（图版 74,图 4）;标本保存在中国矿业大学地质系古生物实验室;江西丰城建新煤矿;晚二叠世乐平组老山下亚段。

羽叶单网羊齿属 Genus *Pinnagigantonoclea* Yang,2006（中文和英文发表）

2006 杨关秀等,184,306 页。

模式种：*Pinnagigantonoclea zelkovoides* Yang,2006

分类位置：大羽羊齿目（Gigantopteridales）

分布与时代：中国河南禹州、贵州盘县、河北开平；二叠纪。

似榉羽叶单网羊齿 *Pinnagigantonoclea zelkovoides* Yang,2006（中文和英文发表）

2006 杨关秀等,184,307 页,图版 62,图 5,6;羽状复叶;合模：HEP3630（图版 62,图 5）,HEP3601（图版 62,图 6）;标本保存在中国地质大学（北京）;河南禹州方山滴水潭;晚二叠世云盖山组 6 煤段。

此属创建时同时报道 9 种,除模式种外尚有：

异常羽叶单网羊齿 *Pinnagigantonoclea heteroeura* Yang et Wang,2006（中文和英文发表）

2006 杨关秀、王洪山,见杨关秀等,185,307 页,图版 57,图 1,1a;羽状复叶;正模：HEP4236（图版 57,图 1,1a）;标本保存在中国地质大学（北京）;河南平顶山矿;中二叠世云小凤口组。

异脉羽叶单网羊齿 *Pinnagigantonoclea mira* (Gu et Zhi) Yang,2006（中文和英文发表）

1974 *Gigantonoclea mira* Gu et Zhi,中国科学院南京地质古生物研究所、植物研究所,见《中国古生代植物》,128 页,图版 97,图 1－5;蕨叶;河北开平;早二叠世—晚石炭世赵各庄群上部;辽宁本溪;早—中二叠世彩家群。

2006 杨关秀等,185,308 页,图版 55,图 1,2;羽状复叶;河南禹州大风口;中二叠世小凤口组。

贵州羽叶单网羊齿 *Pinnagigantonoclea guizhouensis* (Gu et Zhi) Yang,2006（中文和英文发表）

1974 *Gigantonoclea guizhouensis* Gu et Zhi,中国科学院南京地质古生物研究所、植物研究所,见《中国古生代植物》,127 页,图版 96,图 7－10;蕨叶;贵州盘县;晚二叠世宣威组。

2006 杨关秀等,186,308 页,图版 61,图 4,4a;羽状复叶;河南禹州方山滴水潭;晚二叠世云盖山组6煤段。

尖头羽叶单网羊齿 *Pinnagigantonoclea mucronata* Yang,2006（中文和英文发表）
2006 杨关秀等,186,308页,图版63,图7;图版64,图1,2;羽状复叶;合模:HEP0137(图版63,图7),HEP3608(图版64,图1),HEP3607(图版64,图2);标本保存在中国地质大学(北京);河南禹州云盖山、方山,新密马沟;晚二叠世云盖山组6煤段。

匙羽叶单网羊齿 *Pinnagigantonoclea spatulata*（Yang）Yang,2006（中文和英文发表）
1987 *Gigantonoclea spatulata* Yang,杨关秀等,187页,图版3,图1,2,7;插图8。
2006 杨关秀等,187,309页,图版58;图版59,图1;图版60,图3,4;插图7-9;羽状复叶;合模:HEP0185(图版58),HEP0183(图版59,图1),HEP0184(图版60,图3),HEP0186(图版60,图4;插图7-9.3);标本保存在中国地质大学(北京);河南禹州大风口、陈庄、李家门;晚二叠世云盖山组8煤段。

莲座羽叶单网羊齿 *Pinnagigantonoclea rosulata*（Gu et Zhi）Yang et Xie,2006（中文和英文发表）
1974 *Gigantonoclea rosulata* Gu et Zhi,中国科学院南京地质古生物研究所、植物研究所,见《中国古生代植物》,126页,图版96,图1-6;插图104;蕨叶;登记号:PB4969-PB4972;合模:PB4969-PB4972(图版96,图1-6);标本保存在中国科学院南京地质古生物研究所;河南禹县;晚二叠世上石盒子组。
2006 杨关秀、谢建华,见杨关秀等,188,309页,图版60,图1,2;图版75,图4;羽状复叶;河南禹州大风口;晚二叠世云盖山组8煤段;河南禹州方山滴水潭;晚二叠世云盖山组6煤段。

似槲羽叶单网羊齿 *Pinnagigantonoclea dryophylloides* Yang et Xie,2006（中文和英文发表）
2006 杨关秀、谢建华,见杨关秀等,188,310页,图版61,图5-7;羽状复叶;合模:HEP3377(图版61,图5),HEP3376(图版61,图6),HEP3378(图版61,图7);标本保存在中国地质大学(北京);河南临汝坡池;晚二叠世云盖山组8煤段。

多形羽叶单网羊齿 *Pinnagigantonoclea polymorpha* Xie,2006（中文和英文发表）
2006 谢建华,见杨关秀等,189,311页,图版62,图1-4;羽状复叶;合模:HEP3373(图版62,图1),HEP3374(图版62,图2),HEP3622(图版62,图3),HEP3375(图版62,图4);标本保存在中国地质大学(北京);河南临汝坡池;晚二叠世云盖山组8煤段。

羽叶大羽羊齿属 Genus *Pinnagigantopteris* Yang et Xie,2006（中文和英文发表）
2006 杨关秀、谢建华,见杨关秀等,197,317页。
模式种:*Pinnagigantopteris nicotianaefolia*（Gu et Zhi）Yang,2006
分类位置:大羽羊齿目（Gigantopteridales）
分布与时代:中国湖南耒阳,江苏句容、龙潭,浙江长兴,云南宣威,河南禹州;二叠纪。

烟叶羽叶大羽羊齿 *Pinnagigantopteris nicotianaefolia*（Gu et Zhi）Yang,2006（中文和英文发表）
1974 *Gigantopteris nicotianaefolia* Schenk,中国科学院南京地质古生物研究所、植物研究

所，见《中国古生代植物》,130 页,图版 100,图 2－4;图版 101,图 1;图版 102,图 7;插图 103－105,108;湖南耒阳,江苏句容、龙潭、浙江长兴;晚二叠世龙潭组;云南宣威;晚二叠世宣威组;河南平顶山(?);晚二叠世上盒子组。

2006 杨关秀等,197,317 页,图版 71,图 4－7;羽状复叶;河南禹州大风口、临汝坡池;晚二叠世云盖山组。

此属创建时同时报道 3 种,除模式种外尚有:

披针羽叶大羽羊齿 *Pinnagigantopteris lanceolatus* Yang et Xie,2006(中文和英文发表)

2006 杨关秀、谢建华,见杨关秀等,198,317 页,图版 64,图 3,3a;插图 7－14;羽状复叶;正模:HEP3613(图版 64,图 3,3a);标本保存在中国地质大学(北京);河南禹州方山滴水潭;晚二叠世云盖山组 6 煤段。

长圆羽叶大羽羊齿 *Pinnagigantopteris oblongus* Chen,2006(中文和英文发表)

2006 陈瑶,见杨关秀等,198,318 页,图版 71,图 1－3a;羽状复叶;合模:HEP3250(图版 71,图 1),HEP3259(图版 71,图 2),HEP3257(图版 71,图 3);标本保存在中国地质大学(北京);河南临汝坡池;中二叠世小风口组 3 煤段。

羽枝属 Genus *Pinnatiramosus* Geng,1986

1986 耿宝印,665,671 页。

模式种:*Pinnatiramosus qianensis* Geng,1986

分类位置:分类位置不明植物(Plantae incertae sedis)

分布与时代:中国贵州凤冈;中志留世。

黔羽枝 *Pinnatiramosus qianensis* Geng,1986

1986 耿宝印,665,671 页,图版 1－6;植物体羽状分枝;标本号:8196－8203;正模:8196(图版 1);标本保存在中国科学院植物研究所;贵州凤冈;中志留世秀山组。

拟斜羽叶属 Genus *Plagiozamiopsis* Sze,1943

1943 斯行健,511 页。

模式种:*Plagiozamiopsis podozamoides* Sze,1943

分类位置:苏铁类(Cycadophytes)

分布与时代:中国广东、江西;晚二叠世早期。

苏铁杉型拟斜羽叶 *Plagiozamiopsis podozamoides* Sze,1943

1943 斯行健,511 页,图版 1,图 1－10;叶;广东、江西;晚二叠世早期大羽羊齿煤系。

似远志属 Genus *Polygatites* Pan,1983(nom. nud.)

1983 潘广,1520 页。(中文)

1984　潘广,959页。(英文)

模式种:(没有种名)

分类位置:"原始被子植物类群"("primitive angiosperms")

分布与时代:华北燕辽地区;中侏罗世。

似远志(sp. indet.) *Polygatites* sp. indet.

(注:原文仅有属名,没有种名)

1983　*Polygatites* sp. indet.,潘广,1520页;华北燕辽地区东段(45°58′N,120°21′E);中侏罗世海房沟组。(中文)

1984　*Polygatites* sp. indet.,潘广,959页;华北燕辽地区东段(45°58′N,120°21′E);中侏罗世海房沟组。(英文)

似蓼属 Genus *Polygonites* Wu S Q,1999 (non Saporta,1865)(中文发表)

{注:此属名为 *Polygonites* Saporta,1865 的晚出同名,其模式种为 *Polygonites ulmaceus* Saporta,1865[Saporta,1865,92页,图版3,图14;翅籽;法国(St.-Jean-Garguier,France);第三纪]}

1999　吴舜卿,23页。

模式种:*Polygonites polyclonus* Wu S Q,1999

分类位置:单子叶植物纲蓼科(Polygonaceae,Monocotyledoneae)

分布与时代:中国辽西北票;晚侏罗世。

多小枝似蓼 *Polygonites polyclonus* Wu S Q,1999(中文发表)

1999　吴舜卿,23页,图版16,图4,4a;图版19,图1,1a,3A—4a;茎枝标本;营养枝;采集号:AEO-170,AEO-211,AEO-171,AEO-169;登记号:PB18319,PB18335—PB18337;正模:PB18337(图版19,图4);标本保存在中国科学院南京地质古生物研究所;辽西北票上园黄半吉沟;晚侏罗世义县组下部尖山沟层。(注:原文未指定模式种,本文暂把原文列在第一的种 *Polygonites polyclonus* Wu S Q 作为模式种编录)

此属创建时同时报道2种,除模式种外尚有:

扁平似蓼 *Polygonites planus* Wu S Q,1999(中文发表)

1999　吴舜卿,24页,图版19,图2;营养枝;采集号:AEO-122;登记号:PB18338;标本保存在中国科学院南京地质古生物研究所;辽西北票上园黄半吉沟;晚侏罗世义县组下部尖山沟层。

多瓣蕨属 Genus *Polypetalophyton* Geng,2003(英文发表)

1995　耿宝印,见李承森、崔金钟,5页。(裸名)

2003　耿宝印,见 Hilton J 等,795页。

模式种:*Polypetalophyton wufengensis* Geng,2003

分类位置:枝木蕨纲(Cladoxylopsida)

分布与时代:中国湖北五峰;晚泥盆世(Frasnian)。

五峰多瓣蕨 *Polypetalophyton wufengensis* Geng, 2003（英文发表）

1995　耿宝印，见李承森、崔金钟，29—35 页（仅有图）。（裸名）
2003　耿宝印，见 Hilton J 等，P，795 页，图 2—10；植物体多次分枝；标本号：CBP9796—CBP9798；正模：CBP9796（图 2）；标本保存在中国科学院植物研究所；湖北五峰；晚泥盆世（Frasnian）。

多囊枝属 Genus *Polythecophyton* Hao, Gensel et Wang, 2001（英文发表）

2001　郝守刚、Gensel P G、王德明，57 页。
模式种：*Polythecophyton demissum* Hao, Gensel et Wang, 2001
分类位置：始蕨植物门（Euphyllophytes）
分布与时代：中国云南文山；早泥盆世（Pragian）。

下弯多囊枝 *Polythecophyton demissum* Hao, Gensel et Wang, 2001（英文发表）

2001　郝守刚、Gensel P G、王德明，57 页，图版 1—3；插图 1—3；生殖枝；正模：BUPB-801（图版 1，图 4）；副模：BUPB-802，BUPB-806，BUPB-802′（图版 1，图 2；图版 2，图 1；图版 3）；标本保存在北京大学地质系；云南文山；早泥盆世（Pragian）坡松冲组。

始苏铁属 Genus *Primocycas* Zhu et Du, 1981

1981　朱家楠、杜贤铭，402 页。
模式种：*Primocycas chinensis* Zhu et Du, 1981
分类位置：苏铁类（Cycadophytes）
分布与时代：中国山西太原；早二叠世。

中国始苏铁 *Primocycas chinensis* Zhu et Du, 1981

1981　朱家楠、杜贤铭，402 页，图版 1，图 1—6；图版 2，图 1—4；插图 1—4；有胚珠和种子的大孢子叶；标本号：G001—G010；正模：G001—G010；山西太原东山；早二叠世下石盒子组。

此属创建时同时报道 2 种，除模式种外尚有：

帚状始苏铁 *Primocycas muscariformis* Zhu et Du, 1981

1927　*Norinia cucullata* Halle, Halle T G, 218 页，图版 56，图 8—12；叶部化石；山西中部（Ch'en-chia-yu）；晚二叠世早期上石盒子系。
1981　朱家楠、杜贤铭，401，403，404 页。

始拟银杏属 Genus *Primoginkgo* Ma et Du, 1989

1989　马洁、杜贤铭，1，2 页。
模式种：*Primoginkgo dissecta* Ma et Du, 1989
分类位置：银杏类？（Ginkgoaphytes?）

分布与时代：中国山西太原；早二叠世。

深裂始拟银杏 *Primoginkgo dissecta* Ma et Du,1989
1989　马洁、杜贤铭,1,2 页,图版 1；图版 2,图 1-3；叶；山西太原东山；早二叠世下石盒子组。

始查米苏铁属 Genus *Primozamia* Yang,2006（中文和英文发表）
2006　杨关秀等,142,285 页。
模式种：*Primozamia sinensis* Yang,2006
分类位置：苏铁植物门苏铁目(Cycadales,Cycadophyta)
分布与时代：中国河南禹州；中二叠世。

中国始查米苏铁 *Primozamia sinensis* Yang,2006（中文和英文发表）
2006　杨关秀等,142,285 页,图版 49,图 14；大孢子叶；正模：HEP0593（图版 49,图 14）；标本保存在中国地质大学（北京）；河南禹州大风口；中二叠世小风口组 4 煤段。

锯叶羊齿属 Genus *Prionophyllopteris* Mo,1980
1980　莫壮观,见赵修祜等,86 页。
模式种：*Prionophyllopteris spiniformis* Mo,1980
分类位置：分类位置不明植物(Plantae incertae sedis)
分布与时代：中国贵州；晚二叠世早期。

多刺锯叶羊齿 *Prionophyllopteris spiniformis* Mo,1980
1980　莫壮观,见赵修祜等,86 页,图版 19,图 9,10；叶；登记号：PB7081,PB7082；标本保存在中国科学院南京地质古生物研究所；贵州盘县；晚二叠世早期宣威组下段。

原苏铁属 Genus *Procycas* Zhang et Mo,1981
1981　张善桢、莫壮观,238 页。
模式种：*Procycas densinervioides* Zhang et Mo,1981
分类位置：苏铁类(Cycadophytes)
分布与时代：中国河南；早二叠世。

密脉原苏铁 *Procycas densinervioides* Zhang et Mo,1981
1981　张善桢、莫壮观,238 页,图版 1,图 1-6；图版 2,图 1-6；插图 1；蕨叶；登记号：PB8766－PB8772；合模：PB8766（图版 1,图 1）,PB8767（图版 1,图 2,3）,PB8768（图版 1,图 4）,PB8769（图版 1,图 6）,PB8770（图版 2,图 1,3）,PB8771（图版 2,图 2）,PB8772（图版 2,图 4）；标本保存在中国科学院南京地质古生物研究所；河南；早二叠世山西组上部。

原单网羊齿属 Genus *Progigantonoclea* Yang,2006(中文和英文发表)
2006　杨关秀等,178,302页。
模式种:*Progigantonoclea henanensis* (Yang) Yang,2006
分类位置:大羽羊齿目(Gigantopteridales)
分布与时代:中国河南禹州;中二叠世。

河南原单网羊齿 *Progigantonoclea henanensis* (Yang) Yang,2006(中文和英文发表)
1987b　*Emplecotopteris henanensis* Yang,杨关秀,183页,图版1,图1-4。
2006　杨关秀等,178,302页,图版52,图2-8;图版19,图4;插图7.5;蕨叶;标本号:HEP0114,HEP0116-HEP0119;合模:HEP0119(图版52,图2),HEP0118(图版52,图3),HEP0117(图版52,图4),HEP0114(图版52,图6),HEP0116(图版52,图8);标本保存在中国地质大学(北京);河南禹州大涧村;中二叠世神垕组。

原大羽羊齿属 Genus *Progigantopteris* Yang,1987
1987b　杨关秀,189,194页。
模式种:*Progigantopteris brevireticulatus* Yang,1987
分类位置:大羽羊齿类(Gigantopterides)
分布与时代:中国河南禹县;早二叠世。

短网原大羽羊齿 *Progigantopteris brevireticulatus* Yang,1987
1987b　杨关秀,190,194页,图版1,图10-12;插图10;蕨叶;登记号:HEP 226-HEP 228;合模:HEP 226-HEP 228(图版1,图10-12);标本保存在中国地质大学(北京);豫西禹县大涧村南;早二叠世小风口组3-4煤段。

原始银杏型木属 Genus *Proginkgoxylon* Zheng et Zhang,2008(英文发表)
2008　郑少林、张武,见郑少林等,43页。
模式种:*Proginkgoxylon benxiense* (Zheng et Zhang) Zheng et Zhang,2008
分类位置:银杏类(Ginkgophytes)
分布与时代:中国辽宁、内蒙古;二叠纪。

本溪原始银杏型木 *Proginkgoxylon benxiense* (Zheng et Zhang) Zheng et Zhang,2008(英文发表)
2000　*Protoginkgoxylon benxiense* Zheng et Zhang,郑少林、张武,121页,图版1,图1-6;图版2,图1-5;木化石;登记号:GJ6-21;正模:GJ6-21(图版1,图1-6;图版2,图1-5);标本保存在沈阳地质矿产研究所;辽宁本溪田师傅;早二叠世山西组。
2006　*Protoginkgoxylon benxiense* Zheng et Zhang,郑少林、张武,见张武等,43页,图版3-8,图A-F;图版3-9,图A-E;木化石;辽宁本溪田师傅;早二叠世山西组。
2008　郑少林、张武,见郑少林等,47页,图版3-8,图A-F;图版3-9,图A-E;木化石;登记号:GJ6-21;正模:GJ6-21(图版3-8,图 A-F;图版3-9,图 A-E);标本保存在沈阳地质矿产研究所;辽宁本溪田师傅;早二叠世山西组。

此属创建时同时报道 2 种,除模式种外尚有:
大青山原始银杏型木 Proginkgoxylon daqingshanense (Zheng et Zhang) Zheng et Zhang, 2008(英文发表)

2000　*Protoginkgoxylon daqingshanense* Zheng et Zhang,郑少林、张武,121 页,图版 2,图 6;图版 3,图 1-6;木化石;登记号:M56-114;正模:M56-114(图版 2,图 6;图版 3,图 1-6);标本保存在沈阳地质矿产研究所;内蒙古大青山;早二叠世"大青山组"。

2006　*Protoginkgoxylon daqingshanense* Zheng et Zhang,郑少林、张武,见张武等,47 页,图版 3-9,图 F;图版 3-10,图 A-F;木化石;内蒙古大青山;早二叠世"大青山组"。

2008　郑少林、张武,见郑少林等,47 页,图版 3-9,图 F;图版 3-10,图 A-F;木化石;登记号:M56-114;正模:M56-114(图版 3-9,图 F;图版 3-10,图 A-F);标本保存在沈阳地质矿产研究所;内蒙古大青山;早二叠世"大青山组"。

原始水松型木属 Genus Protoglyptostroboxylon He,1995

1995　何德长,8(中文),10(英文)页。

模式种:*Protoglyptostroboxylon giganteum* He,1995

分类位置:松柏纲(丝炭化石)[Coniferopsida (fusainized wood)]

分布与时代:中国内蒙古鄂温克旗;早白垩世。

巨大原始水松型木 Protoglyptostroboxylon giganteum He,1995

1995　何德长,8(中文),10(英文)页,图版 5,图 2-2c;图版 6,图 1-1e,2;图版 8,图 1-1d;丝炭化石;标本号:91363,91370;模式标本:91363;标本保存在煤炭科学研究总院西安分院;内蒙古鄂温克旗伊敏煤矿;早白垩世伊敏组 16 煤层。

此属创建时同时报道 2 种,除模式种外尚有:
伊敏原始水松型木 Protoglyptostroboxylon yimiense He,1995

1999　何德长,9(中文),11(英文)页,图版 1,图 3;图版 2,图 5;图版 7,图 1-1f;图版 8,图 2-2a,4-4a;图版 11,图 2;丝炭化石;标本号:91414,91403;模式标本:91403;标本保存在煤炭科学研究总院西安分院;内蒙古鄂温克旗伊敏煤矿;早白垩世伊敏组 16 煤层。

原始蕨属 Genus Protopteridophyton Li et Hsu,1987

1987　李承森、徐仁,120 页。

模式种:*Protopteridophyton devonicum* Li et Hsu,1987

分类位置:原始蕨类(Primitive ferns)

分布与时代:中国湖南、湖北;中泥盆世—晚泥盆世(Givetian-Frasnian)。

泥盆纪原始蕨 Protopteridophyton devonicum Li et Hsu,1987

1987　李承森、徐仁,120 页,图版 1-16;草本植物;正模:No. 8150a(图版 1,图 1),No. 8150b

(图版2,图1)(标本的正反面);标本保存在中国科学院植物研究所;湖南长沙,湖北汉阳;中泥盆世—晚泥盆世(Givetian—Frasnian)跳马池组—珞珈山组下部。

原始金松型木属 Genus *Protosciadopityoxylon* Zhang,Zheng et Ding,1999(英文发表)
1999　张武、郑少林、丁秋红,1314页。
模式种:*Protosciadopityoxylon liaoningensis* Zhang,Zheng et Ding,1999
分类位置:松柏纲杉科(木化石)[Taxodiaceae,Coniferopsida (fossil wood)]
分布与时代:中国辽宁义县;早白垩世。

辽宁原始金松型木 *Protosciadopityoxylon liaoningense* Zhang,Zheng et Ding,1999(英文发表)
1999　张武、郑少林、丁秋红,1314页,图版1—3;插图2;木化石;标本号:Sha.30;模式标本:Sha.30(图版1—3);标本保存在沈阳地质矿产研究所;辽宁义县毕家沟;早白垩世沙海组。

假耳蕨属 Genus *Pseudopolystichu* Deng et Chen,2001(中文和英文发表)
2001　邓胜徽、陈芬,153,229页。
模式种:*Pseudopolystichu cretaceum* Deng et Chen,2001
分类位置:真蕨纲(Filicopsida)
分布与时代:中国辽宁铁法盆地;早白垩世。

白垩假耳蕨 *Pseudopolystichu cretaceum* Deng et Chen,2001(中文和英文发表)
2001　邓胜徽、陈芬,153,229页,图版115,图1—4;图版116,图1—6;图版117,图1—9;图版118,图1—7;生殖羽片;标本号:TXQ-2520;标本保存在石油勘探开发科学研究院;辽宁铁法盆地;早白垩世小明安碑组。

假拟扇叶属 Genus *Pseudorhipidopsis* P'an,1937
(注:又名异叶属)
1937　潘钟祥,263页。
模式种:*Pseudorhipidopsis brevicaulis* (Kawasaki et Kon'no) P'an,1937
分类位置:银杏类?(Ginkgophytes?)
分布与时代:中国河南,朝鲜;晚二叠世早期。

宽叶假拟扇叶 *Pseudorhipidopsis brevicaulis* (Kawasaki et Kon'no) P'an,1937
(注:又名异叶)
1932　*Rhipidopsis brevicaulis* Kawasaki et Kon'no,Kawasaki S 和 Kon'no E,39页,图版51,图7,8;枝和叶;朝鲜平安南道;二叠纪平安系。
1937　潘钟祥,265页,图版1,图版1;图版2;图版3,图4,5;枝和叶;河南禹县;晚二叠世早期上石盒子组。

假带羊齿属 Genus *Pseudotaeniopteris* Sze, 1951

1951a 斯行健,83页。

模式种:*Pseudotaeniopteris piscatorius* Sze,1951

分类位置:疑问化石(Problematicum)

分布与时代:中国辽宁本溪;早白垩世。

鱼形假带羊齿 *Pseudotaeniopteris piscatorius* Sze, 1951

1951a 斯行健,83页,图版1,图1,2;疑问化石;辽宁本溪工源;早白垩世。

白豆杉型木属 Genus *Pseudotaxoxylon* Prakash, Du et Tripathi, 1995

1995 Prakash U、杜乃正、Tripathi P P,345页。

模式种:*Pseudotaxoxylon chinensis* Prakash,Du et Tripathi,1995

分类位置:红豆杉科(Taxaceae)

分布与时代:中国山东淄博;晚中新世。

中国白豆杉型木 *Pseudotaxoxylon chinensis* Prakash, Du et Tripathi, 1995

1995 Prakash U、杜乃正、Tripathi P P,345页,图7-12;木化石;标本保存在中国科学院植物研究所;山东淄博;晚中新世。

黄杉型木属 Genus *Pseudotsugxylon* Yang, 1994

1994 杨家驹,见陶君容等,111,112页。

模式种:*Pseudotsugxylon pingzhangensis* Yang,1994

分类位置:松杉目(Pinales)

分布与时代:中国内蒙古赤峰;中新世。

平庄黄杉型木 *Pseudotsugxylon pingzhangensis* Yang, 1994

1994 杨家驹,见陶君容等,111,113页,图版1,图1-8;木化石;标本号:48;内蒙古赤峰平庄煤矿;中新世。

假鳞杉属 Genus *Pseudoullmannia* He, Liang et Shen, 1996(中文和英文发表)

1996 何锡麟、梁敦士、沈树忠,86,168页。

模式种:*Pseudoullmannia frumentarioides* He,Liang et Shen,1996

分类位置:松柏纲(Coniferopsida)

分布与时代:中国江西;晚二叠世。

类麦假鳞杉 *Pseudoullmannia frumentarioides* He, Liang et Shen, 1996(中文和英文发表)

1996 何锡麟、梁敦士、沈树忠,87,168页,图版69,图1-5;图版70,图2;图版96;枝叶;登记号：X88287 — X88291；合模：X88287 — X88289（图版 69，图 1,4,5）；标本保存在中国矿业大学地质系古生物实验室；江西乐平鸣山煤矿和桥头丘煤矿、丰城建新煤矿、萍湖、高安八景煤矿；晚二叠世乐平组老山下亚段、王潘里段。

此属创建时同时报道 2 种，除模式种外尚有：

类纹假鳞杉 *Pseudoullmannia bronnioides* He, Liang et Shen, 1996（中文和英文发表）
1996　何锡麟、梁敦士、沈树忠，88，169 页，图版 69，图 6；枝叶；登记号：X88292；正模标本：X88292（图版 69，图 6）；标本保存在中国矿业大学地质系古生物实验室；江西丰城建新煤矿；晚二叠世乐平组老山下亚段。

拟蕨属 Genus *Pteridiopsis* Zheng et Zhang, 1983
1983　郑少林、张武，381 页。
模式种：*Pteridiopsis didaoensis* Zheng et Zhang, 1983
分类位置：真蕨纲蕨科（Pteridiaceae, Filicopsida）
分布与时代：中国黑龙江鸡西滴道；晚侏罗世。

滴道拟蕨 *Pteridiopsis didaoensis* Zheng et Zhang, 1983
1983　郑少林、张武，381 页，图版 1，图 1 — 3；插图 1a — 1c；营养羽片和生殖羽片；标本号：HDN021 — HDN023；正模：HDN021（图版 1，图 1 — 1d）；黑龙江鸡西滴道；晚侏罗世滴道组。

此属创建时同时报道 2 种，除模式种外尚有：

柔弱拟蕨 *Pteridiopsis tenera* Zheng et Zhang, 1983
1983　郑少林、张武，382 页，图版 2，图 1 — 3；插图 2c — 2f；营养羽片和生殖羽片；标本号：HDN036 — HDN038；正模：HDN036（图版 2，图 3 — 3c）；黑龙江鸡西滴道；晚侏罗世滴道组。

秦岭羊齿属 Genus *Qinlingopteris* Wu et Wang, 2004（中文和英文发表）
2004　吴秀元、王军，494，498 页。
模式种：*Qinlingopteris orientalis* Wu et Wang, 2004
分类位置：古羊齿类（Archaeopterides）
分布与时代：中国陕西镇安；早石炭世（Visean）。

东方秦岭羊齿 *Qinlingopteris orientalis* Wu et Wang, 2004（中文和英文发表）
2004　吴秀元、王军，494，498 页，图版 2，图 5,5a；蕨叶；登记号：PB20205；正模：PB20205（图版 2，图 5）；标本保存在中国科学院南京地质古生物研究所；陕西镇安茅坪东山梁；早石炭世（Visean）二峪河组。

此属创建时同时报道 2 种，除模式种外尚有：

秦岭羊齿（未定种）*Qinlingopteris* sp.
2004　*Qinlingopteris* sp.，吴秀元、王军，495 页，图版 2，图 6；羽片；登记号：PB20206；标本保

存在中国科学院南京地质古生物研究所;陕西镇安茅坪东山梁;早石炭世(Visean)二峪河组。

琼海叶属 Genus *Qionghaia* Zhou et Li,1979

1979　周志炎、厉宝贤,454页。

模式种:*Qionghaia carnosa* Zhou et Li,1979

分类位置:不明或本内苏铁类?(Incertae sedis or Bennettitales?)

分布与时代:中国海南琼海;早三叠世。

肉质琼海叶 *Qionghaia carnosa* Zhou et Li,1979

1979　周志炎、厉宝贤,454页,图版2,图21,21a;大孢子叶;登记号:PB7618;标本保存在中国科学院南京地质古生物研究所;海南琼海九曲江新华;早三叠世岭文群(九曲江组)。

辐叶属 Genus *Radiatifolium* Meng,1992

1992　孟繁松,705,707页。

模式种:*Radiatifolium magnusum* Meng,1992

分类位置:银杏类?(Ginkgophytes?)

分布与时代:中国湖北南漳;晚三叠世。

大辐叶 *Radiatifolium magnusum* Meng,1992

1992　孟繁松,705,707页,图版1,图1,2;图版2,图1,2;叶;登记号:P86020－P86024;正模:P86020(图版1,图1);标本保存在宜昌地质矿产研究所;湖北南漳东巩;晚三叠世九里岗组。

多枝蕨属 Genus *Ramophyton* Wang,2008(英文发表)

2008　王德明,1101页。

模式种:*Ramophyton givetianum* Wang,2008

分类位置:真蕨植物门枝木蕨目(Cladoxylates,Pteridophyta)

分布与时代:中国新疆;中泥盆世(Givetian)。

吉维特多枝蕨 *Ramophyton givetianum* Wang,2008(英文发表)

2008　王德明,1101页,图2－13;蕨类植物;正模:XZH25a(图3a),XZH25b(图3b);副模:XZH21b,XZH21a,XZH10b,XZH10a,XZH09b,XZH09a(图2a－2f),XZH06a,XZH06b,XZH5a,XZH5b(图5c,5e,5g,5i),XZH26a,XZH26b,XZH13a,XZH18a(图6b,6d－6f),XZH14a(图7a);标本保存在北京大学地质系;新疆西淮噶尔盆地和布克赛尔;中泥盆世(Givetian)呼吉尔斯特组。

毛茛叶属 Genus *Ranunculophyllum* ex Tao et Zhang,1990,emend Wu,1993
[注:此属名为陶君容、张川波(1990)首次使用,但未注明是新属名(吴向午,1993a,b)]
1990　陶君容、张川波,221,226页。
1993a　吴向午,31,232页。
1993b　吴向午,508,517页。
模式种:*Ranunculophyllum pinnatisctum* Tao et Zhang,1990
分类位置:双子叶植物毛茛科(Ranunculaceae,Dicotyledoneae)
分布与时代:中国吉林延吉;早白垩世。

羽状全裂毛茛叶 *Ranunculophyllum pinnatisctum* Tao et Zhang,1990
1990　陶君容、张川波,221,226页,图版2,图4;插图3;叶;标本号:$K_1 d_{41-9}$;标本保存在中国科学院植物研究所;吉林延吉;早白垩世大拉子组。
1993a　吴向午,31,232页。
1993b　吴向午,508,517页。

耙羊齿属 Genus *Rastropteris* Galtier,Wang,Li et Hilton,2001(英文发表)
2001　Galtier Jean、王士俊、李承森、Hilton Jason,436页。
模式种:*Rastropteris pingquanensis* Galtier,Wang,Li et Hilton,2001
分类位置:真蕨目(Filicales)
分布与时代:中国河北平泉;早二叠世(Sakmarian)。

平泉耙羊齿 *Rastropteris pingquanensis* Galtier,Wang,Li et Hilton,2001(英文发表)
2001　Galtier Jean、王士俊、李承森、Hilton Jason,436页,图1—20;茎;正模:CBP-PQ42(图1—20);标本保存在中国科学院植物研究所;河北平泉;早二叠世(Sakmarian)太原组上部。

热河似查米亚属 Genus *Rehezamites* Wu S,1999(中文发表)
1999　吴舜卿,15页。
模式种:*Rehezamites anisolobus* Wu S,1999
分类位置:苏铁纲本内苏铁目?(Bennettitales?,Cycadopsida)
分布与时代:中国辽宁北票;晚侏罗世。

不等裂热河似查米亚 *Rehezamites anisolobus* Wu S,1999(中文发表)
1999　吴舜卿,15页,图版8,图1,1a;羽叶;采集号:AEO-187;登记号:PB18265;标本保存在中国科学院南京地质古生物研究所;辽宁北票上园黄半吉沟;晚侏罗世义县组下部尖山沟层。

此属创建时同时报道2种,除模式种外尚有:
热河似查米亚(未定种) Rehezamites sp.
1999　*Rehezamites* sp.,吴舜卿,15页,图版7,图1,1a;羽叶;辽宁北票上园黄半吉沟;晚侏罗世义县组下部尖山沟层。

肾叶属 Genus *Renifolium* Li H et Lan,1982
1982　李汉民、蓝善先,见李汉民等,375页。
模式种:*Renifolium logipetiolatum* Li H et Lan,1982
分类位置:分类位置不明植物(Plantae incertae sedis)
分布与时代:中国江苏;晚二叠世。

长柄肾叶 *Renifolium logipetiolatum* Li H et Lan,1982
1982　李汉民、蓝善先,见李汉民等,375页,图版157,图3—10;生殖叶和营养叶;标本号:HP1594—HP1598;合模:HP1594(图版157,图3),HP1595(图版157,图4),HP1596(图版157,图6),HP1597(图版157,图7),HP1598(图版157,图8);江苏镇江;晚二叠世龙潭组。

网格蕨属 Genus *Reteophlebis* Lee et Tsao,1976
1976　李佩娟、曹正尧,见李佩娟等,102页。
模式种:*Reteophlebis simplex* Lee et Tsao,1976
分类位置:真蕨纲紫萁科(Osmundaceae,Filicopsida)
分布与时代:中国云南禄丰;晚三叠世。

单式网格蕨 *Reteophlebis simplex* Lee et Tsao,1976
1976　李佩娟、曹正尧,见李佩娟等,102页,图版10,图3—8;图版11;图版12,图4—5;插图3-2;裸羽片和实羽片;登记号:PB5203—PB5214,PB5218—PB5219;正模:PB5214(图版11,图8);标本保存在中国科学院南京地质古生物研究所;云南禄丰一平浪;晚三叠世一平浪组干海子段。

网延羊齿属 Genus *Reticalethopteris* Li,Shen et Wu,1993
1993　李星学、沈光隆、吴秀元,542,546页。
模式种:*Reticalethopteris yuani* (Sze) Li,Shen et Wu,1993
分类位置:分类位置不明植物(Plantae incertae sedis)
分布与时代:中国甘肃、宁夏、内蒙古;晚石炭世。

袁氏网延羊齿 *Reticalethopteris yuani* (Sze) Li,Shen et Wu,1993
1933　*Palaeoweichselia yuani* Sze,斯行健,59页,图版6,图1—12;图版7,图1—10;蕨叶;甘肃景泰;晚石炭世红土洼组。

1993 李星学、沈光隆、吴秀元,542,546 页,图版 1—4;蕨叶;甘肃景泰,宁夏中卫,内蒙古阿拉善左旗;晚石炭世红土洼组。

根状茎属 Genus *Rhizoma* Wu S Q,1999(中文发表)
1999 吴舜卿,24 页。
模式种:*Rhizoma elliptica* Wu S Q,1999
分类位置:双子叶植物纲睡莲科(Nymphaeaceae,Dicotyledoneae)
分布与时代:中国辽西北票;晚侏罗世。

椭圆形根状茎 *Rhizoma elliptica* Wu S Q,1999(中文发表)
1999 吴舜卿,24 页,图版 16,图 9,10;根状茎;采集号:AEO-100,AEO-197;登记号:PB18322,PB18323;标本保存在中国科学院南京地质古生物研究所;辽西北票上园黄半吉沟;晚侏罗世义县组下部尖山沟层。

拟根茎属 Genus *Rhizomopsis* Gothan et Sze,1933
(注:又名刺根茎属)
1933 Gothan W、斯行健,26 页。
模式种:*Rhizomopsis gemmifera* Gothan et Sze,1933
分类位置:分类位置不明植物(Plantae incertae sedis)
分布与时代:中国江苏龙潭;石炭纪。

具芽拟根茎 *Rhizomopsis gemmifera* Gothan et Sze,1933
(注:又名刺根茎)
1933 Gothan W、斯行健,26 页,图版 4,图 6;根茎化石(?);江苏龙潭;石炭纪。

菱羊齿属 Genus *Rhomboidopteris* Si,1989
1989 斯行健,51,192 页。
模式种:*Rhomboidopteris yongwolensis* (Kawasaki) Si,1989
分类位置:真蕨纲或种子蕨纲(Filices or Pteridospermopsida)
分布与时代:中国山西、内蒙古,朝鲜;二叠纪。

菱羊齿 *Rhomboidopteris yongwolensis* (Kawasaki) Si,1989
1931 *Neuroptridium? yongwolensis* Kawasaki,Kawasaki S,图版 50,图 129,129a;朝鲜;古生代平安系。
1934 *Neuroptridium? yongwolensis* Kawasaki,Kawasaki S,143 页。
1989 斯行健,51,192 页,图版 60,图 1—7;图版 61,图 1—3;蕨叶;山西河曲;二叠纪山西组;内蒙古准格尔旗;二叠纪石盒子群。

拟片叶苔属 Genus *Riccardiopsis* Wu et Li, 1992

1992　吴向午、厉宝贤, 268, 276 页。

模式种: *Riccardiopsis hsüi* Wu et Li, 1992

分类位置: 苔纲 (Hepaticae)

分布与时代: 中国河北蔚县; 中侏罗世。

徐氏拟片叶苔 *Riccardiopsis hsüi* Wu et Li, 1992

1992　吴向午、厉宝贤, 265, 275 页, 图版 4, 图 5, 6; 图版 5, 图 1—4A, 4a; 图版 6, 图 4—6a; 插图 5; 叶状体; 采集号: ADN41-03, ADN41-06, ADN41-07; 登记号: PB15472—PB15479; 正模: PB15475 (图版 5, 图 2); 标本保存在中国科学院南京地质古生物研究所; 河北蔚县涌泉庄附近; 中侏罗世乔儿涧组。

日蕨属 Genus *Rireticopteris* Hsu et Chu, 1974

1974　徐仁、朱家柟, 见徐仁等, 269 页。

模式种: *Rireticopteris microphylla* Hsu et Chu, 1974

分类位置: 真蕨纲紫萁科 (Osmundaceae, Filicopsida)

分布与时代: 中国云南永仁、四川渡口; 晚三叠世。

小叶日蕨 *Rireticopteris microphylla* Hsu et Chu, 1974

1974　徐仁、朱家柟, 见徐仁等, 269 页, 图版 1, 图 7—9; 图版 2, 图 1—4; 图版 3, 图 1; 插图 1; 蕨叶; 编号: No. 2785, No. 2839, No. 825, No. 830; 合模 1: No. 2785 (图版 1, 图 7); 合模 2: N. 2839 (图版 1, 图 8); 标本保存在中国科学院植物研究所; 云南永仁纳拉箐; 晚三叠世大荞地组; 四川渡口太平场; 晚三叠世大菁组底部。

轮叶蕨属 Genus *Rotafolia* Wang D M, Hao et Wang Q, 2005 (英文发表)

2005　王德明、郝守刚、王祺, 23 页。

模式种: *Rotafolia songziensis* (Feng) Wang D M, Hao et Wang Q, 2005

分类位置: 楔叶纲楔叶目 (Sphenophyllales, Sphenopsida)

分布与时代: 中国湖北松滋; 晚泥盆世 (Famennian)。

松滋轮叶蕨 *Rotafolia songziensis* (Feng) Wang D M, Hao et Wang Q, 2005 (英文发表)

1984　*Boumanite songziensis* Feng, 冯少南, 302 页, 图版 48, 图 4。

1991　*Sphenophyllostachys songziensis* Feng et Ma, 冯少南、马洁, 142 页, 图版 1, 图 1; 图版 2, 图 1; 插图 2。

2005　王德明、郝守刚、王祺, 23 页, 图 2—42; 孢子叶穗; 正模: A-076 (图 30) 也见冯少南、马洁, 1991, 142 页, 图版 1, 图 1; 图版 2, 图 1); 副模: Hu-27 (图 2), Hu-13 (图 4), Hu-20 (图 12), Hu-10 (图 14), Hu-06 (图 20), Hu-18 (图 23), Hu-B1 (图 26), Hu-19 (图 27), A-056 (图 32), Hu-22 (图 33), Hu-B2 (图 35), Hu-B1 (图 37), H-02 (图 43, 44); 标本

A-076，A-056 保存在宜昌地质矿产研究所，其余标本保存在北京大学地质系；湖北松滋；晚泥盆世（Famennian）写经寺组。

似圆柏属 Genus *Sabinites* Tan et Zhu，1982

1982　谭琳、朱家楠，153 页。
模式种：*Sabinites neimonglica* Tan et Zhu，1982
分类位置：松柏纲柏科（Cupressaceae，Coniferopsida）
分布与时代：中国内蒙古固阳；早白垩世。

内蒙古似圆柏 *Sabinites neimonglica* Tan et Zhu，1982

1982　谭琳、朱家楠，153 页，图版 39，图 2—6；小枝和球果；登记号：GR40，GR65，GR87，GR67，GR103；正模：GR87（图版 39，图 4，4a）；副模：GR65（图版 39，图 3，3a）；内蒙古固阳小三分子村东；早白垩世固阳组。

此属创建时同时报道 2 种，除模式种外尚有：

纤细似圆柏 *Sabinites gracilis* Tan et Zhu，1982

1982　谭琳、朱家楠，153 页，图版 40，图 1，2；小枝和球果；登记号：GR09，GR66；正模：GR09（图版 40，图 1）；副模：GR66（图版 40，图 2）；内蒙古固阳小三分子村东；早白垩世固阳组。

箭羽羊齿属 Genus *Sagittopteris* Zhang E et Xiao，1985（non Zhang S et Xiao，1987）

1985　张恩鹏、肖素珍，见肖素珍、张恩鹏，584 页。
模式种：*Sagittopteris belemnopteroides* Zhang E et Xiao，1985（non Zhang S et Xiao，1987）
分类位置：分类位置不明植物（Plantae incertae sedis）
分布与时代：中国山西太原、陵川、沁水；晚石炭世晚期一早二叠世。

戟形箭羽羊齿 *Sagittopteris belemnopteroides* Zhang E et Xiao，1985（non Zhang S et Xiao，1987）

1985　张恩鹏、肖素珍，见肖素珍、张恩鹏，584 页，图版 202，图 1—4；图版 203，图 1—2；羽叶；登记号：Sh361，Sh362，Sh363，Sh365，Sh366，Sh367；正模：Sh366（图版 202，图 1）；副模：Sh361（图版 202，图 3），Sh362（图版 203，图 1），Sh363（图版 203，图 3），Sh365（图版 202，图 4），Sh367（图版 202，图 2）；山西太原、陵川、沁水；晚石炭世晚期山西组。

箭羽羊齿属 Genus *Sagittopteris* Zhang S et Xiao，1987（non Zhang E et Xiao，1985）

（注：此属名为 *Sagittopteris* Zhang E et Xiao，1985 的晚出同名）
1987　张善桢、肖素珍，181，185 页。
模式种：*Sagittopteris belemnopteroides* Zhang S et Xiao，1987（non Zhang E et Xiao，1985）
分类位置：分类位置不明植物（Plantae incertae sedis）
分布与时代：中国山西太原、陵川；早二叠世。

戟形箭羽羊齿 *Sagittopteris belemnopteroides* Zhang S et Xiao,1987（non Zhang E et Xiao,1985）

（注：此种名为 *Sagittopteris belemnopteroides* Zhang E et Xiao,1985 的晚出同名）

1987　张善桢、肖素珍,181,185 页,图版 1,图 1－4；图版 2,图 1－6；羽叶；登记号：PB11272,PB11273,PB11274,PB11275,PB11276,PB11277；正模：PB11273（图版 1,图 2）；标本保存在中国科学院南京地质古生物研究所；登记号：SH672；标本保存在山西省地质局区域地质调查队；山西太原、陵川；早二叠世山西组。

拟裂鞘叶属 Genus *Schizoneuropsis* Yabe et Shimakura,1940

1940a　Yabe H 和 Shimakura M,177 页。

模式种：*Schizoneuropsis tokudae* Yabe et Shimakura,1940

分类位置：楔叶植物类（Sphenophytes）

分布与时代：中国安徽淮南；二叠纪。

德田拟裂鞘叶 *Schizoneuropsis tokudae* Yabe et Shimakura,1940

1940a　Yabe H 和 Shimakura M,177 页,图版 15,图 1－4；楔叶类茎叶；安徽淮南煤田；二叠纪。

小伞属 Genus *Sciadocillus* Geng,1992

1992a　耿宝印,197,206 页。

模式种：*Sciadocillus cuneifidus* Geng,1992

分类位置：苔纲？（Marchantiales?）

分布与时代：中国四川江油；早泥盆世。

楔裂小伞 *Sciadocillus cuneifidus* Geng,1992

1992a　耿宝印,197,206 页,图版 7,图 53－57；叶状体；标本号：8353,8354,8355；正模：8353〔图版 7,图 53,54（正反面）〕；标本保存在中国科学院植物研究所；四川江油；早泥盆世平驿铺组。

帚羽叶属 Genus *Scoparia* Wang,1993

1993　王庆之,223 页。

模式种：*Scoparia plumaria* Wang,1993

分类位置：苏铁类（Cycadophytes）

分布与时代：中国河北曲阳；二叠纪。

羽毛帚羽叶 *Scoparia plumaria* Wang,1993

1993　王庆之,223,225 页,图版 3,图 3,4；羽叶；采集号：7DPH250；登记号：3882；标本保存在上海自然博物馆；河北曲阳灵山；二叠纪下石盒子组。

翅籽属 Genus *Semenalatum* Dilcher,Mei et Du,1997（英文发表）
1997　Dilcher D L、梅美棠、杜美利,248 页。
模式种:*Semenalatum paucum* Dilcher,Mei et Du,1997
分类位置:分类位置不明植物(Plantae incertae sedis)
分布与时代:中国淮北;早二叠世。

珍贵翅籽 *Semenalatum paucum* Dilcher,Mei et Du,1997（英文发表）
1997　Dilcher D L、梅美棠、杜美利,248 页,图版 1;具翅种子;正模:UF♯14983(Specimen B)(图版 1,图 4－8);副模:UF♯14982(Specimen A)(图版 1,图 1－3);标本保存在佛罗里达自然博物馆古植物馆;淮北;早二叠世下石盒子组。

似狗尾草属 Genus *Setarites* Pan,1983（nom. nud.）
1983　潘广,1520 页。（中文）
1984　潘广,959 页。（英文）
模式种:（没有种名）
分类位置:"原始被子植物类群"("primitive angiosperms")
分布与时代:华北燕辽地区;中侏罗世。

似狗尾草(sp. indet.) *Setarites* sp. indet.
（注:原文仅有属名,没有种名）
1983　*Setarites* sp. indet.,潘广,1520 页;华北燕辽地区东段(45°58′N,120°21′E);中侏罗世海房沟组。（中文）
1984　*Setarites* sp. indet.,潘广,959 页;华北燕辽地区东段(45°58′N,120°21′E);中侏罗世海房沟组。（英文）

上园草属 Genus *Shangyuania* Zheng,Gao et Bo 2008（中文和英文发表）
2008　郑少林、高家俊、薄学,329,338 页,
模式种:*Shangyuania caii* Zheng,Gao et Bo,2008
分类位置:单子叶植物类?(Monocots?)
分布与时代:中国辽宁北票;早白垩世。

才氏上园草 *Shangyuania caii* Zheng,Gao et Bo,2008（中文和英文发表）
2008　郑少林、高家俊、薄学,329,339 页;插图 1.1A,3－8;插图 2.1A,4,6;具穗状花序的植物体;正模:LBY2001(插图 1.1A,3－8;插图 2.1A,4,6);标本由辽宁省锦州市化石收藏家才树仁先生妥善保管;辽宁北票上园黄半吉沟附近;早白垩世义县组尖山沟层。

山西枝属 Genus *Shanxicladus* Wang Z et Wang L,1990

1990b 王自强、王立新,308 页。

模式种:*Shanxicladus pastulosus* Wang Z et Wang L,1990

分类位置:真蕨纲? 或种子蕨纲? (Filicopsida? or Pteridospermae?)

分布与时代:中国山西武乡;中三叠世。

疹形山西枝 *Shanxicladus pastulosus* Wang Z et Wang L,1990

1990b 王自强、王立新,308 页,图版 5,图 1－2;枝干;标本号:No. 8407-4;正模:No. 8407-4（图版 5,图 1,2）;标本保存在中国科学院南京地质古生物研究所;山西武乡司庄;中三叠世二马营组底部。

山西木属 Genus *Shanxioxylon* Tian et Wang,1987

1987 田宝霖、王士俊,196,202 页。

模式种:*Shanxioxylon sinense* Tian et Wang,1987

分类位置:科达植物类（Cordaiphytes incertae sedis）

分布与时代:中国山西太原;晚石炭世。

中国山西木 *Shanxioxylon sinense* Tian et Wang,1987

1987 田宝霖、王士俊,196,202 页,图版 1,图 1－8;插图 1;茎的煤核;玻片号:T7-39A/1,T7-39A-R/1,T7-60B-L/5,TN-292;正模:T7-39A/1（图版 1,图 1）;副模:T7-60B-L/5（图版 1,图 3）;标本保存在中国矿业大学北京研究生部地质研究室;山西太原西山煤矿;晚石炭世太原组。

此属创建时同时报道 2 种,除模式种外尚有:

太原山西木 *Shanxioxylon taiyuanense* Tian et Wang,1987

1987 田宝霖、王士俊,198,202 页,图版 2,图 1－7;插图 2－4;茎的煤核;玻片号:T7-59A,T7-59A-L,T7-59-L,T7-59A,T7-59A/1,T7-59A/3;正模:T7-59A（图版 2,图 1）;标本保存在中国矿业大学北京研究生部地质研究室;山西太原西山煤矿;晚石炭世太原组。

沈氏蕨属 Genus *Shenea* Mathews,1947－1948

1947－1948 Mathews G B,240 页。

模式种:*Shenea hirschmeierii* Mathews,1947－1948

分类位置:不明（真蕨纲或种子蕨纲）[Plantae incertae sedis (Filicopsida or Pteridospermopsida)]

分布与时代:中国北京西山;二叠纪(?)、三叠纪(?)。

希氏沈氏蕨 *Shenea hirschmeierii* Mathews,1947－1948

1947－1948 Mathews G B,240 页,图 3;生殖叶印痕;北京西山;二叠纪(?)、三叠纪(?)双泉群。

沈括叶属 Genus *Shenkuoia* Sun et Guo,1992

1992 孙革、郭双兴,见孙革等,546 页。(中文)
1993 孙革、郭双兴,见孙革等,254 页。(英文)
模式种:*Shenkuoia caloneura* Sun et Guo,1992
分类位置:双子叶植物纲(Dicotyledoneae)
分布与时代:中国黑龙江鸡西;早白垩世。

美脉沈括叶 *Shenkuoia caloneura* Sun et Guo,1992

1992 孙革、郭双兴,见孙革等,547 页,图版 1,图 13,14;图版 2,图 1-6;叶及叶角质层;登记号:PB16775,PB16777;正模:PB16775(图版 1,图 13);标本保存在中国科学院南京地质古生物研究所;黑龙江鸡西城子河;早白垩世城子河组上部。(中文)
1993 孙革、郭双兴,见孙革等,254 页,图版 1,图 13,14;图版 2,图 1-6;叶及叶角质层;登记号:PB16775,PB16777;正模:PB16775(图版 1,图 13);标本保存在中国科学院南京地质古生物研究所;黑龙江鸡西城子河;早白垩世城子河组上部。(英文)

神州叶属 Genus *Shenzhouphyllum* Yang et Xie,2006(中文和英文发表)

2006 杨关秀、谢建华,见杨关秀等,128,275 页。
模式种:*Shenzhouphyllum undulatum*(Yang)Yang et Xie,2006
分类位置:种子蕨植物门盾籽目盾籽科(Peltaspermaceae,Peltaspermales,Pteridospermophyta)
分布与时代:中国河南禹州、登封、临汝;晚二叠世。

波缘神州叶 *Shenzhouphyllum undulatum*(Yang)Yang et Xie,2006(中文和英文发表)

1987 *Psygmophyllum undulatum* Yang,杨关秀,53 页,图版 14,图 6。
2006 杨关秀、谢建华,见杨关秀等,128,276 页,图版 32,图 2-6;单叶化石;合模:HEP3417(图版 32,图 2),HEP3408(图版 32,图 3),HEP3436(图版 32,图 4),HEP0702(图版 32,图 5),HEP3428(图版 32,图 6);标本保存在中国地质大学(北京);河南禹州大风口、登封磴槽、临汝坡池;晚二叠世云盖山组 7、8 段。

此属创建时同时报道 3 种,除模式种外尚有:

圆形神州叶 *Shenzhouphyllum rotundatum* Xie,2006(中文和英文发表)

2006 谢建华,见杨关秀等,129,277 页,图版 33,图 1-2;单叶化石;合模:HEP3468(图版 33,图 1),HEP3430(图版 33,图 2);标本保存在中国地质大学(北京);河南临汝坡池;晚二叠世云盖山组 8 段。

匙形神州叶 *Shenzhouphyllum spatulatum* Xie et Wu,2006(中文和英文发表)

2006 谢建华、吴跃晖,见杨关秀等,129,278 页,图版 33,图 3-8;图版 32,图 1;单叶化石;合模:HEP3149(图版 33,图 8),HEP3427(图版 33,图 5),HEP3388(图版 33,图 5);标本保存在中国地质大学(北京);河南登封磴槽、临汝坡池;晚二叠世云盖山组 8 段。

神州籽属 Genus *Shenzhouspermum* Yang, Xie et Wu, 2006（中文和英文发表）
2006　杨关秀、谢建华、吴跃晖，见杨关秀等，126，273 页。
模式种：*Shenzhouspermum trichotomum* Yang, Xie et Wu, 2006
分类位置：种子蕨植物门盾籽目盾籽科（Peltaspermaceae, Peltaspermales, Pteridospermophyta）
分布与时代：中国河南登封、临汝；晚二叠世。

三歧神州籽 *Shenzhouspermum trichotomum* Yang, Xie et Wu, 2006（中文和英文发表）
2006　杨关秀、谢建华、吴跃晖，见杨关秀等，126，273 页，图版 29，图 1－3；图版 31，图 5－7；插图 7-1；雌性生殖器官；合模：HEP3409，HEP3154，HEP3153（图版 29，图 1－3），HEP3425，HEP3156，HEP3135（图版 31，图 5－7）；标本保存在中国地质大学（北京）；河南登封磴槽、临汝坡池；晚二叠世云盖山组 8 段。

神州聚囊属 Genus *Shenzhoutheca* Yang et Wu, 2006（中文和英文发表）
2006　杨关秀、吴跃晖，见杨关秀等，127，275 页。
模式种：*Shenzhoutheca aspergilliformis* Yang et Wu, 2006
分类位置：种子蕨植物门盾籽目盾籽科（Peltaspermaceae, Peltaspermales, Pteridospermophyta）
分布与时代：中国河南登封、临汝；晚二叠世。

刷状神州聚囊 *Shenzhoutheca aspergilliformis* Yang et Wu, 2006（中文和英文发表）
2006　杨关秀、吴跃晖，见杨关秀等，127，275 页，图版 29，图 4－6；图版 31，图 4；孢子囊；合模：HEP3129，HEP3130，HEP3152（图版 29，图 4－6）；标本保存在中国地质大学（北京）；河南登封磴槽、临汝坡池；晚二叠世云盖山组 8 段。

双囊芦穗属 Genus *Shuangnangostachya* Gao et Thomas, 1991
1991　高志峰、Thomas B A，198 页。
模式种：*Shuangnangostachya gracilis* Gao et Thomas, 1991
分类位置：楔叶植物类（Sphenophytes）
分布与时代：中国山西太原；早二叠世。

细小双囊芦穗 *Shuangnangostachya gracilis* Gao et Thomas, 1991
1991　高志峰、Thomas B A，198 页，图版 1，图 1－7；插图 1；孢子叶穗；标本号：GP0108；正模：GP0108（图版 1，图 1－7）；标本保存在北京大学地质系；山西太原；早二叠世下石盒子组。

水城蕨属 Genus *Shuichengella* Li, 1993
1993a　李中明，53 页。

模式种：*Shuichengella primitiva*（Li）Li，1993
分类位置：紫萁目（Osmundales）
分布与时代：中国贵州水城；晚二叠世。

原始水城蕨 *Shuichengella primitiva*（Li）Li，1993

1983 *Palaeosmunda primitiva* Li，李中明，154 页，图版 9—13；茎（煤核化石）；合模：GP2.377-3-2/4-1，GP2.377-3-2/6-1，GP2.377-3-2/6-10；标本保存在中国科学院植物研究所；贵州水城汪家寨煤矿；晚二叠世汪家寨组。

1993a 李中明，53 页，图版 1—4；茎（煤核化石）；贵州水城汪家寨煤矿；晚二叠世汪家寨组。

斯氏鞘叶属 Genus *Siella* Yang，2006（中文和英文发表）

2006 杨关秀等，98，250 页。
模式种：*Siella leptocostata* Yang，2006
分类位置：楔叶目木贼科（Equisetaceae，Sphenophylates）
分布与时代：中国河南禹州；晚二叠世。

细肋斯氏鞘叶 *Siella leptocostata* Yang，2006（中文和英文发表）

2006 杨关秀等，98，250 页，图版 10，图 9；图版 11，图 4—6，6a；图版 73，图 1；楔叶类茎；合模：HEP0914（图版 10，图 9），HEP0915（图版 11，图 4），HEP0916（图版 11，图 5），HEP0917（图版 11，图 6）；标本保存在中国地质大学（北京）；河南禹州大风口、云盖山；晚二叠世云盖山组。

中华古果属 Genus *Sinocarpus* Leng et Friis，2003（英文发表）

2003 冷琴、Friis E M，79 页。
模式种：*Sinocarpus decussatus* Leng et Friis，2003
分类位置：被子植物类（Angiospermae incertae sedis）
分布时代：中国辽宁朝阳；早白垩世（Barremian or Aptian）。

下延中华古果 *Sinocarpus decussatus* Leng et Friis，2003（英文发表）

2003 冷琴、Friis E M，79 页，图 2—22；果实；正模：B0162［图 2 左（B0162A 正面），图 2 右（B0162B 反面），图 11—22 电镜照片］；标本保存在中国科学院古脊椎动物与古人类研究所；辽宁朝阳凌源大王杖子（41°15′N，119°15′E）；早白垩世（Barremian or Aptian）义县组大王杖子层。［注：此标本后改定为 *Hyrcantha decussata*（Leng et Friis）Dilcher，Sun，Ji et Li（David L Dilcher 等，2007）］

中国篦羽叶属 Genus *Sinoctenis* Sze，1931

1931 斯行健，14 页。
模式种：*Sinoctenis grabauiana* Sze，1931
分类位置：苏铁纲（Cycadopsida）

分布时代:中国江西萍乡;早侏罗世(Lias)。

葛利普中国箆羽叶 *Sinoctenis grabauiana* Sze,1931
1931　斯行健,14页,图版2,图1;图版4,图2;羽叶;江西萍乡;早侏罗世(Lias)。

中华缘蕨属 Genus *Sinodicotis* Pan,1983(nom. nud)
1983　潘广,1520页。(中文)
1984　潘广,958页。(英文)
模式种:(没有种名)
分类位置:"半被子植物类群"("hemiangiosperms")
分布时代:华北燕辽地区;中侏罗世。

中华缘蕨(sp. indet.) *Sinodicotis* sp. indet.
(注:原文仅有属名,没有种名)
1983　*Sinodicotis* sp. indet.,潘广,1520页;华北燕辽地区东段(45°58′N,120°21′E);中侏罗世海房沟组。(中文)
1984　*Sinodicotis* sp. indet.,潘广,958页;华北燕辽地区东段(45°58′N,120°21′E);中侏罗世海房沟组。(英文)

中国古螺纹木属 Genus *Sinopalaeospiroxylon* Zhang,Wang,Zheng,Yang,Li,Fu et Li,2007(英文发表)
2006　张武等,74页。(裸名)(中文)
2007　张武、王永栋、郑少林、杨小菊、李勇、傅小平、李楠,266页。
2008　张武,74页。(裸名)(英文)
模式种:*Sinopalaeospiroxylon baoligemiaoense* Zhang,Wang,Zheng,Yang,Li,Fu et Li,2007
分类位置:松柏类(Coniferophytes)
分布与时代:中国内蒙古、辽宁、河北;晚石炭世—早、中二叠世。

宝力格庙中国古螺纹木 *Sinopalaeospiroxylon baoligemiaoense* Zhang,Wang,Zheng,Yang,Li,Fu et Li,2007(英文发表)
2006　张武等,75页,图版3-31;图版3-32;插图3.2;木化石;内蒙古苏尼特左旗;晚石炭世宝力格庙组。(裸名)(中文)
2007　张武、王永栋、郑少林、杨小菊、李勇、傅小平、李楠,266页,图4A-D;图5A-G;图6A-G;木化石;玻片号:No. 3P2H4-2;正模:No. 3P2H4-2(图5A-G,6A-G);标本保存在沈阳地质矿产研究所;内蒙古苏尼特左旗宝力格庙;晚石炭世宝力格庙组。
2008　张武等,75页,图版3-31;图版3-32;插图3.2;木化石;内蒙古苏尼特左旗;晚石炭世宝力格庙组。(裸名)(英文)

此属最初报道时有3种,除模式种外尚有:

南票中国古螺纹木 Sinopalaeospiroxylon napiaoense Zhang et Zheng,2006 (2008)（中文和英文发表）

2006　张武、郑少林,见张武等,78页,图版3-33;图版3-34;木化石;标本号:GJ6-22;正模:GJ6-22(图版3-33;图版3-34);标本保存在沈阳地质矿产研究所;辽宁锦西南票煤矿;早二叠世太原组。(中文)

2008　张武、郑少林,见张武等,78页,图版3-33;图版3-34;木化石;标本号:GJ6-22;正模:GJ6-22(图版3-33;图版3-34);标本保存在沈阳地质矿产研究所;辽宁锦西南票煤矿;早二叠世太原组。(英文)

平泉中国古螺纹木 Sinopalaeospiroxylon pingquanense Zhang et Zheng,2006 (2008)（中文和英文发表）

2006　张武、郑少林,见张武等,78页,图版3-35;图版3-36;木化石;标本号:GJ6-3;正模:GJ6-3(图版3-35;图版3-36);标本保存在沈阳地质矿产研究所;河北平泉松树台山;中二叠世山西组。(中文)

2008　张武、郑少林,见张武等,78页,图版3-35;图版3-36;木化石;标本号:GJ6-3;正模:GJ6-3(图版3-35;图版3-36);标本保存在沈阳地质矿产研究所;河北平泉松树台山;中二叠世山西组。(英文)

中国叶属 Genus *Sinophyllum* Sze et Lee,1952

1952　斯行健、李星学,12,32页。

模式种:*Sinophyllum suni* Sze et Lee,1952

分类位置:银杏类?(Ginkgophytes?)

分布时代:中国四川巴县;早侏罗世。

孙氏中国叶 *Sinophyllum suni* Sze et Lee,1952

1952　斯行健、李星学,12,32页,图版5,图1;图版6,图1;插图2;叶;标本保存在中国科学院南京地质古生物研究所;四川巴县一品场;早侏罗世香溪群。

中国似查米亚属 Genus *Sinozamites* Sze,1956

1956a　斯行健,46,150页。

模式种:*Sinozamites leeiana* Sze,1956

分类位置:苏铁纲(Cycadopsida)

分布时代:中国陕西宜君;晚三叠世。

李氏中国似查米亚 *Sinozamites leeiana* Sze,1956

1956a　斯行健,47,151页,图版39,图1—3;图版50,图4;图版53,图5;羽叶;登记号:PB2447—PB2450;标本保存在中国科学院南京地质古生物研究所;陕西宜君杏树坪黄草湾;晚三叠世延长层上部。

管子麻黄属 Genus *Siphonospermum* Rydin et Friis,2010(英文发表)
2010　Rydin C 和 Friis E M,5 页。
模式种:*Siphonospermum simplex* Rydin et Friis,2010
分类位置:麻黄目(Gnetales)
分布与时代:中国东北;早白垩世。

简单管子麻黄 *Siphonospermum simplex* Rydin et Friis,2010(英文发表)
2010　Rydin C 和 Friis E M,5 页,图 1—3;茎直立,二歧分枝,顶生雌球果;正模:9880A(图 1a),9880B(图 1b);标本保存在中国科学院植物研究所;中国东北;早白垩世义县组。

太阳花属 Genus *Solaranthus* Zheng et Wang,2010(英文发表)
2010　郑少林、王鑫,896 页。
模式种:*Solaranthus daohugouensis* Zheng et Wang,2010
分类位置:被子植物(Angiosperms)
分布时代:中国内蒙古宁城;中侏罗世。

道虎沟太阳花 *Solaranthus daohugouensis* Zheng et Wang,2010(英文发表)
2010　郑少林、王鑫,896 页,图 2—4;"花序";登记号:PB21046,PB21107,B0179,B0201,No. 47—277,GBM3;正模:PB21046(图 2c,f,l—r);副模:B0179(图 2a,d,e,i),B0201(图 2b,j,k),PB21107(图 2g—h),No. 47—277,GBM3;标本 PB21046,PB21107 保存在中国科学院南京地质古生物研究所;B0179,B0201 保存在中国科学院古脊椎动物与古人类研究所;No. 47—277 保存在山东省天宇自然博物馆;GBM3 保存在深圳仙湖植物园古生物博物馆;内蒙古宁城道虎沟;中侏罗世九龙山组。

似卷囊蕨属 Genus *Speirocarpites* Yang,1978
1978　杨贤河,479 页。
模式种:*Speirocarpites virginiensis* (Fontaine) Yang,1978
分类位置:真蕨纲紫萁科(Osmundaceae,Filicopsida)
分布与时代:美国弗吉尼亚,中国四川渡口、云南祥云;晚三叠世。

弗吉尼亚似卷囊蕨 *Speirocarpites virginiensis* (Fontaine) Yang,1978
[注:此种后被叶美娜等(1986)改定为 *Cynepteris lasiophora* Ash]
1883　*Lonchopteris virginiensis* Fontaine,53 页,图版 28,图 1,2;图版 29,图 1—4;蕨叶;美国弗吉尼亚;晚三叠世。
1978　杨贤河,479 页;插图 101;美国弗吉尼亚;晚三叠世。

此属创建时同时报道 4 种,除模式种外尚有:
渡口似卷囊蕨 *Speirocarpites dukouensis* Yang,1978
[注:此种后被叶美娜等(1986)改定为 *Cynepteris lasiophora* Ash]

1978 杨贤河,480 页,图版 164,图 1—2;蕨叶及生殖羽片;标本号:Sp0044,Sp0045;正模:Sp0044(图版 164,图 1);标本保存在成都地质矿产研究所;四川渡口摩沙河;晚三叠世大荞地组;云南祥云;晚三叠世干海子组。

日蕨型似卷囊蕨 Speirocarpites rireticopteroides Yang, 1978
[注:此种后被叶美娜等(1986)改定为 Cynepteris lasiophora Ash]
1978 杨贤河,480 页,图版 164,图 3;蕨叶;标本号:Sp0046;正模:Sp0046(图版 164,图 3);标本保存在成都地质矿产研究所;四川渡口灰家所;晚三叠世大荞地组。

中国似卷囊蕨 Speirocarpites zhonguoensis Yang, 1978
[注:此种后被叶美娜等(1986)改定为 Cynepteris lasiophora Ash]
1978 杨贤河,481 页,图版 164,图 4—5;蕨叶及生殖羽片;标本号:Sp0047,Sp0048;正模:Sp0048(图版 164,图 5);标本保存在成都地质矿产研究所;四川渡口摩沙河;晚三叠世大荞地组。

楔叶拜拉花属 Genus *Sphenobaieroanthus* Yang, 1986
1986 杨贤河,54 页。
模式种:*Sphenobaieroanthus sinensis* Yang,1986
分类位置:银杏纲(Ginkgopsida)楔拜拉目(Sphenobaierales Yang,1986)楔拜拉科(Sphenobaieracea Yang,1986)
分布与时代:中国重庆大足;晚三叠世。

中国楔叶拜拉花 *Sphenobaieroanthus sinensis* Yang, 1986
1986 杨贤河,54 页,图版 1,图 1—2a;插图 2;带叶长枝、短枝和雄性花;采集号:H2-5;登记号:SP301(合模标本);标本保存在成都地质矿产研究所;重庆大足万古兴隆冉家湾;晚三叠世须家河组。

楔叶拜拉枝属 Genus *Sphenobaierocladus* Yang, 1986
1986 杨贤河,53 页。
模式种:*Sphenobaierocladus sinensis* Yang,1986
分类位置:银杏纲楔拜拉目楔拜科(Sphenobaieraceae,Sphenobaierales,Ginkgopsida)
分布时代:中国重庆大足;晚三叠世。

中国楔叶拜拉枝 *Sphenobaierocladus sinensis* Yang, 1986
1986 杨贤河,53 页,图版 1,图 1—2a;插图 2;带叶长枝、短枝和雄性花;采集号:H2-5;登记号:SP301(合模标本);标本保存在成都地质矿产研究所;重庆大足万古兴隆冉家湾;晚三叠世须家河组。

楔栉羊齿属 Genus *Sphenopecopteris* Zhang et Mo, 1985
1985 张善桢、莫壮观,173 页。

模式种：*Sphenopecopteris beaniata* Zhang et Mo,1985
分类位置：种子蕨纲(Pteridospermopsida)
分布与时代：中国河南；晚二叠世。

豆子楔栉羊齿 *Sphenopecopteris beaniata* Zhang et Mo,1985
1985　张善桢、莫壮观,173 页,图版 1,图 1－5；图版 2,图 1；着生种子的蕨叶；登记号：PB8106－PB8109；正模：PB8106(图版 1,图 1)；标本保存在中国科学院南京地质古生物研究所；河南；晚二叠世石盒子系。

斯芬克斯籽属 Genus *Sphinxia* Li,Hilton et Hemsley,1997（non Reid et Chandler,1933）(英文发表)
〔注：此属名为 *Sphinxia* Reid et Chandler,1933 的晚出同名,其模式种为 *Sphinxia ovalis* Reid et Chandler〔Reid E M and Chandler M E J,1933,397 页,图版 20,图 12－23；果实；英国；始新世(London Clay)〕〕
1997　李承森、Hilton J 和 Hemsley A R,139 页。
模式种：*Sphinxia wuhania* Li,Hilton et Hemsley,1997
分类位置：维管植物门(Tracheophyta incertae sedis)
分布与时代：中国湖北武汉；晚泥盆世(Frasnian)。

武汉斯芬克斯籽 *Sphinxia wuhania* Li,Hilton et Hemsley,1997(英文发表)
1997　李承森、Hilton J 和 Hemsley A R,139 页,图 1－24；种子状结构；标本号：CBMh 101－CBMh 147；正模：CBMh 105(图 6)；标本保存在中国科学院植物研究所；湖北武汉；晚泥盆世(Frasnian)。

仙籽属 Genus *Sphinxiocarpon* Wang,Xue et Prestianni,2007(英文发表)
1997　*Sphinxia* Li,Hilton et Hemsley,李承森、Hilton J 和 Hemsley A R,139 页。
2007　王祺、薛进庄、Prestianni C,393 页。
模式种：*Sphinxiocarpon wuhania*（Li,Hilton et Hemsley）Wang,Xue et Prestianni,2007
分类位置：维管植物门(Tracheophyta incertae sedis)
分布与时代：中国湖北武汉；晚泥盆世(Frasnian)。

武汉斯芬克斯仙籽 *Sphinxiocarpon wuhania*（Li,Hilton et Hemsley）Wang,Xue et Prestianni,2007(英文发表)
1997　*Sphinxia wuhania* Li,Hilton et Hemsley,李承森、Hilton J 和 Hemsley A R,139 页,图 1－24；种子状结构；标本号：CBMh 101－CBMh 147；正模：CBMh 105(图 6)；标本保存在中国科学院植物研究所；湖北武汉；晚泥盆世(Frasnian)。
2007　王祺、薛进庄、Prestianni C,393 页。

刺鳞木属 Genus *Spinolepidodendron* Chen,1999(中文和英文发表)
1999　陈其奭,17 页。

模式种：*Spinolepidodendron hangzhouense* Chen, 1999
分类位置：石松植物类（Lycopods incertae sedis）
分布与时代：中国浙江萧山；晚泥盆世。

杭州刺鳞木 *Spinolepidodendron hangzhouense* Chen, 1999（中文和英文发表）
1999　陈其奭, 17 页, 图版 2, 图 1—7；图版 3, 图 1, 1a；具叶座的茎干；标本号：M3587a, M3641, M3649c；标本保存在浙江自然博物院；浙江萧山虎山；晚泥盆世西湖组。

鳞籽属 Genus *Squamocarpus* Mo, 1980
1980　莫壮观, 见赵修祜等, 87 页。
模式种：*Squamocarpus papilioformis* Mo, 1980
分类位置：裸子植物门？（Gymnospermae?）
分布时代：中国云南富源；早三叠世。

蝶形鳞籽 *Squamocarpus papilioformis* Mo, 1980
1980　莫壮观, 见赵修祜等, 87 页, 图版 19, 图 13, 14（正反面）；种鳞；采集号：FQ-36；登记号：PB7085, PB7086；标本保存在中国科学院南京地质古生物所；云南富源庆云；早三叠世"卡以头层"。

翅鳞籽属 Genus *Squarmacarpus* Wang Z et Wang L, 1986
1986　王自强、王立新, 43 页。
模式种：*Squarmacarpus cuneiformis* Wang Z et Wang L, 1986
分类位置：分类位置不明植物（Plantae incertae sedis）
分布与时代：中国山西柳林；晚二叠世。

楔形翅鳞籽 *Squarmacarpus cuneiformis* Wang Z et Wang L, 1986
1986　王自强、王立新, 43 页, 图版 16, 图 1, 2（为图 1 标本的反面）；图版 1, 图 13(?)；插图 23；种鳞；标本号：8402-222, 8402-223, 8309-46；正模：8402-222（图版 16, 图 1）；等模：8402-223（图版 16, 图 2）；标本保存在中国科学院南京地质古生物研究所；山西柳林磨石沟；晚二叠世孙家沟组中段。

穗藓属 Genus *Stachybryolites* Wu X W, Wu X Y et Wang, 2000（英文发表）
2000　吴向午、吴秀元、王永栋, 168 页。
模式种：*Stachybryolites zhoui* Wu X W, Wu X Y et Wang, 2000
分类位置：真藓类（Bryiidae）
分布时代：中国新疆克拉玛依；早侏罗世。

周氏穗藓 *Stachybryolites zhoui* Wu X W, Wu X Y et Wang, 2000（英文发表）
2000　吴向午、吴秀元、王永栋, 168 页, 图版 1, 图 1—5；图版 2, 图 1—4；茎叶体；采集号：92-

T-22;登记号:PB17786－PB17796;合模1:PB17786(图版1,图1,1a,1b,1c);合模2:PB17791(图版2,图1);合模3:PB17796(图版2,图4);标本保存在中国科学院南京地质古生物研究所;新疆克拉玛依吐孜阿克内沟;早侏罗世八道湾组。

穗蕨属 Genus *Stachyophyton* Geng,1983

1983 耿宝印,574页。

模式种:*Stachyophyton yunnanense* Geng,1983

分类位置:裸蕨类(Psilophytes incertae sedis)

分布与时代:中国云南文山;早泥盆世。

云南穗蕨 *Stachyophyton yunnanense* Geng,1983

1983 耿宝印,574页,图版1,图1－9;图版2,图1－10;插图1;植物体和孢子叶穗;正模:8091(图版1,图1;插图1);标本保存在中国科学院植物研究所;云南文山古木;早泥盆世坡松冲组。

垂饰杉属 Genus *Stalagma* Zhou,1983

1983 周志炎,63页。

模式种:*Stalagma samara* Zhou,1983

分类位置:松柏纲罗汉松科(Podocarpaceae,Coniferopsida)

分布时代:中国湖南衡阳;晚三叠世。

翅籽垂饰杉 *Stalagma samara* Zhou,1983

1983 周志炎,63页,图版3,图7;图版4－11;插图3－6,7C,7I,7J;营养枝叶、生殖枝、雌球果、果鳞、种子、角质层、花粉;登记号:PB9586,PB9588,PB9592－PB9605;模式标本:PB9605(图版4,图4;插图3B);标本保存在中国科学院南京地质古生物研究所;湖南衡阳杉桥;晚三叠世杨柏冲组。

金藤叶属 Genus *Stephanofolium* Guo,2000(英文发表)

2000 郭双兴,233页。

模式种:*Stephanofolium ovatiphyllum* Guo,2000

分类位置:双子叶植物纲防己科(Menisspermaceae,Dicotyledoneae)

分布与时代:中国吉林珲春;晚白垩世。

卵形金藤叶 *Stephanofolium ovatiphyllum* Guo,2000(英文发表)

2000 郭双兴,233页,图版2,图8;图版6,图1－6;叶部化石;登记号:PB18630－PB18633;正模:PB18632(图版6,图1);标本保存在中国科学院南京地质古生物研究所;吉林珲春;晚白垩世珲春组。

刷囊属 Genus *Strigillotheca* Gu et Zhi,1974

1974 中国科学院南京地质古生物研究所、植物研究所,见《中国古生代植物》,167 页。

模式种:*Strigillotheca fasciculata* Gu et Zhi,1974

分类位置:分类位置不明植物(Plantae incertae sedis)

分布与时代:中国贵州盘县;晚二叠世早期。

束囊刷囊 *Strigillotheca fasciculata* Gu et Zhi,1974

(注:又名刷囊笸)

1974 中国科学院南京地质古生物研究所、植物研究所,见《中国古生代植物》,167 页,图版 129,图 5－7;生殖枝叶;登记号:PB4995－PB4996;合模:PB4995(图版 129,图 5),PB4996(图版 129,图 6);标本保存在中国科学院南京地质古生物研究所;贵州盘县;晚二叠世早期宣威组。

缝鞘杉属 Genus *Suturovagina* Chow et Tsao,1977

1977 周志炎、曹正尧,167 页。

模式种:*Suturovagina intermedia* Chow et Tsao,1977

分类位置:松柏纲掌鳞杉科(Cheirolepidiaceae,Coniferopsida)

分布与时代:中国江苏南京;晚白垩世。

过渡缝鞘杉 *Suturovagina intermedia* Chow et Tsao,1977

1977 周志炎、曹正尧,167 页,图版 2,图 1－14;插图 1;枝叶和叶角质层;登记号:PB6256－PB6260;正模:PB6256(图版 2,图 1);标本保存在中国科学院南京地质古生物研究所;江苏南京燕子矶;早白垩世葛村组。

束脉蕨属 Genus *Symopteris* Hsu,1979

1876 *Bernoullia* Heer,Heer O,88 页。

1979 徐仁等,17 页。

模式种:*Symopteris helvetica*(Heer)Hsu,1979

分类位置:真蕨纲合囊蕨科(Marattiaceae,Filicopsida)

分布与时代:瑞士和中国;晚三叠世。

瑞士束脉蕨 *Symopteris helvetica*(Heer)Hsu,1979

1876 *Bernoullia helvetica* Heer,Heer O,88 页,图版 38,图 1－6;蕨叶;瑞士;晚三叠世。

1979 徐仁等,17 页。

此属创建时同时报道 3 种,除模式种外尚有:

密脉束脉蕨 *Symopteris densinervis* Hsu et Tuan,1979

1979 徐仁、段淑英,见徐仁等,18 页,图版 6－7,图 4;图版 10,图 4－6;图版 58;图版 59,图

6;蕨叶;编号:814,829,831,839,846,885;标本保存在中国科学院植物研究所;四川宝鼎太平场;晚三叠世大箐组。

蔡耶束脉蕨 *Symopteris zeilleri* (Pan) Hsu,1979
1936　*Bernoullia zeilleri* P'an,潘钟祥,26页,图版9,图6,7;图版11,图3,3a,4,4a;图版14,图5,6,6a;裸羽片和实羽片;陕西延川清涧;晚三叠世延长层中部。
1979　徐仁等,17页。

天石蕨属 Genus *Szea* Yao et Taylor,1988
1988　姚兆奇、Taylor T N,123页。
模式种:*Szea sinensis* Yao et Taylor,1988
分类位置:真蕨纲里白科(Geicheniaceous,Filicopsida)
分布与时代:中国江苏南京、镇江;早二叠世。

中国天石蕨 *Szea sinensis* Yao et Taylor,1988
1988　姚兆奇、Taylor T N,123页,图版1—4;插图1—3;蕨叶、生殖羽片、囊群和孢子;登记号:PB9270,PB9271,PB9272,PB9273,PB9274,PB9275,PB9276,PB9277;正模:PB9270(图版1,图1;图版3,图1—6;图版4,图1—3,7—9);副模:PB9271—PB9277(图版1,图2—6;图版2,图1—5);标本保存在中国科学院南京地质古生物研究所;江苏南京、镇江间的伏牛山煤矿;早二叠世晚期龙潭组下部。

斯氏松属 Genus *Szecladia* Yao,Liu,Rothwell et Mapes,2000(英文发表)
2000　姚兆奇、刘陆军、Rothwell G W 和 Mapes G,525页。
模式种:*Szecladia multinervia* Yao,Liu,Rothwell et Mapes,2000
分类位置:松柏纲(Coniferopsida)
分布与时代:中国贵州安顺;晚二叠世。

多脉斯氏松 *Szecladia multinervia* Yao,Liu,Rothwell et Mapes,2000(英文发表)
2000　姚兆奇、刘陆军、Rothwell G W 和 Mapes G,525页,图3—5;松柏类枝;登记号:PB18129,PB18130;正模:PB18129(图3.1,3.2);副模:PB18130(图3.7,4.5,4.6);标本保存在中国科学院南京地质古生物研究所;贵州安顺;晚二叠世晚期大隆组。

斯氏木属 Genus *Szeioxylon* Wang,Jiang et Qin,1994
1994　王士俊、姜尧发、秦勇,194,195页。
模式种:*Szeioxylon xuzhouene* Wang,Jiang et Qin,1994
分类位置:松柏纲(木化石)(Coniferopsida)
分布与时代:中国江苏徐州;石炭纪。

徐州斯氏木 *Szeioxylon xuzhouene* Wang,Jiang et Qin,1994
1994　王士俊、姜尧发、秦勇,195页,图版1,图1—8;图版2,图1—8;木化石;标本号:XT-3;

江苏徐州;石炭纪太原组。

大箐羽叶属 Genus *Tachingia* Hu,1975
1975　胡雨帆,见徐仁等,75页。

模式种:*Tachingia pinniformis* Hu,1975

分类位置:裸子植物类或苏铁纲?(Gymnospermae incertae sedis or Cycadopsida?)

分布与时代:中国四川渡口太平场;晚三叠世。

大箐羽叶 *Tachingia pinniformis* Hu,1975
1975　胡雨帆,见徐仁等,75页,图版5,图1—4;羽叶;标本号:No.801;标本保存在中国科学院植物研究所;四川渡口太平场;晚三叠世大箐组底部。

拟带枝属 Genus *Taeniocladopsis* Sze,1956
1956a　斯行健,63,168页。

模式种:*Taeniocladopsis rhizomoides* Sze,1956

分类位置:楔叶纲木贼目(Equisetales,Sphenopsida)

分布与时代:中国陕西延长;晚三叠世。

假根茎型拟带枝 *Taeniocladopsis rhizomoides* Sze,1956
1956a　斯行健,63,168页,图版54,图1,1a;图版55,图1—4;根部化石(?);登记号:PB2494,PB2495—PB2499;标本保存在中国科学院南京地质古生物研究所;陕西延长周家湾;晚三叠世延长层。

太平场蕨属 Genus *Taipingchangella* Yang,1978
1978　杨贤河,489页。

模式种:*Taipingchangella zhongguoensis* Yang,1978

分类位置:真蕨纲太平场蕨科(Taipingchangellaceae,Filicopsida)[注:此科由杨贤河(1978)创立,包括 *Taipingchangella* 和 *Goeppertella* 两属]

分布与时代:中国四川渡口太平场;晚三叠世。

中国太平场蕨 *Taipingchangella zhongguoensis* Yang,1978
1978　杨贤河,489页,图版172,图4—6;图版170,图1b—2;图版171,图1;蕨叶;标本号:Sp0071—Sp0073,Sp0078(均为合模标本);标本保存在成都地质矿产研究所;四川渡口太平场;晚三叠世大荞地组。

太原蕨属 Genus *Taiyuanitheca* Gao et Thomas,1993
1993　高志峰、Thomas B A,82页。

模式种：*Taiyuanitheca tetralinea* Gao et Thomas,1993
分类位置：观音座莲目观音座莲科（Marattiaceae,Marattiales）
分布与时代：中国山西太原；早二叠世。

四线形太原蕨 *Taiyuanitheca tetralinea* Gao et Thomas,1993
1993　高志峰、Thomas B A,82页；插图1-2；正模：GP0112（插图2）；蕨叶；山西太原；早二叠世上石盒子组。

蛟河羽叶属 Genus *Tchiaohoella* Lee et Yeh ex Wang,1984（nom. nud.）
（注：此属名 *Tchiaohoella* 可能为 *chiaohoella* 的误拼）
1984　王自强,269页。
模式种：*Tchiaohoella mirabilis* Lee et Yeh,1964(MS) ex Wang,1984
分类位置：苏铁纲（Cycadopsida）
分布与时代：中国吉林蛟河、河北平泉；早白垩世。

奇异蛟河羽叶 *Tchiaohoella mirabilis* Lee et Yeh ex Wang,1984（nom. nud.）
1984　王自强,269页。

此属创建时同时报道2种,除模式种外尚有：
蛟河羽叶（未定种） *Tchiaohoella* sp.
1984　*Tchiaohoella* sp.,王自强,270页,图版149,图7；羽叶；河北平泉；早白垩世九佛堂组。

细轴始蕨属 Genus *Tenuisa* Wang,2007（英文发表）
2007　王德明,1342页。
模式种：*Tenuisa frasniana* Wang,2007
分类位置：始蕨植物门（Euphyllophytes）
分布与时代：中国湖南长沙；晚泥盆世（Frasnian）。

弗拉细轴始蕨 *Tenuisa frasniana* Wang,2007（英文发表）
2007　王德明,1342页,图3-5；正模：HNCS-01a（图3a），HNCS-01b（图3b），HNCS-02（图5a）；繁殖枝；标本保存在北京大学地质系；湖南长沙；晚泥盆世（Frasnian）云里岗组。

四叶属 Genus *Tetrafolia* Chu ex Feng et al.,1977
1977　冯少南等,673页。
模式种：*Tetrafolia changshaense*（Ngo）Chu ex Feng et al.,1977
分类位置：分类位置不明植物（Plantae incertae sedis）
分布与时代：中国湖南长沙；早石炭世。

长沙四叶 *Tetrafolia changshaense*（Ngo）Chu ex Feng et al.,1977
1963　?*Sphenophyllum changshaense* Ngo,敖振宽,610页,图1；楔叶化石；湖南长沙；早石

炭世。
1977　冯少南等,673 页,图版 235,图 5;叶;湖南长沙;早石炭世大塘组测水段。

哈瑞士叶属 Genus *Tharrisia* Zhou, Wu et Zhang, 2001(英文发表)
2001　周志炎、吴向午、章伯乐,99 页。
模式种:*Tharrisia dinosaurensis*(Harris)Zhou,Wu et Zhang,2001
分类位置:裸子植物类(Gymnospermae incertae sedis)
分布与时代:东格陵兰和中国;早侏罗世。

迪纳塞尔哈瑞士叶 *Tharrisia dinosaurensis*(Harris)Zhou,Wu et Zhang,2001(英文发表)
1932　*Stenopteris dinosaurensis* Harris,Harris T M,75 页,图版 8,图 4;插图 31;蕨叶和角质层;东格陵兰(Scoresby Sound,East Greenland);早侏罗世(*Thaumatopteris* Zone)。
1988　*Stenopteris dinosaurensis* Harris,李佩娟等,77 页,图版 53,图 1—2a;图版 102,图 3—5;图版 105,图 1—2;蕨叶和角质层;青海大柴旦大煤沟;早侏罗世甜水沟组 *Ephedrites* 层。
2001　周志炎、吴向午、章伯乐,99 页,图版 1,图 7—10;图版 3,图 2;图版 4,图 1—2;图版 5,图 1—5;图版 7,图 1—2;插图 3;叶和角质层;东格陵兰;早侏罗世(*Thaumatopteris* Zone);瑞典(?);早侏罗世;青海大柴旦大煤沟;早侏罗世甜水沟组 *Ephedrites* 层;陕西府谷殿儿湾;早侏罗世富县组。

此属创建时同时报道 3 种,除模式种外尚有:

侧生哈瑞士叶 *Tharrisia lata* Zhou et Zhang,2001(英文发表)
2001　周志炎、章伯乐,见周志炎等,103 页,图版 1,图 1—6;图版 3,图 1,3—8;图版 5,图 5—8;图版 6,图 1—8;插图 5;叶和角质层;登记号:PB18124—PB1828;正模:PH18124(图版 1,图 1);副模:PB18125—PB18128;标本保存在中国科学院南京地质古生物研究所;河南义马;中侏罗世义马组下部 4 层。

优美哈瑞士叶 *Tharrisia spectabilis*(Mi,Sun C,Sun Y,Cui,Ai et al.) Zhou,Wu et Zhang,2001(英文发表)
1996　*Stenopteris spectabilis* Mi,Sun C,Sun Y,Cui,Ai et al.,米家榕、孙春林、孙跃武、崔尚森、艾永亮等,101 页,图版 12,图 1,7—9;插图 5;蕨叶和角质层;辽宁北票台吉二井;早侏罗世北票组下段。
2001　周志炎、吴向午、章伯乐,101 页,图版 2,图 14;图版 4,图 3—7;图版 7,图 3—8;插图 4;叶和角质层;辽宁北票台吉二井;早侏罗世北票组下段。

奇异羽叶属 Genus *Thaumatophyllum* Yang,1978
1978　杨贤河,515 页。
模式种:*Thaumatophyllum ptilum*(Harris)Yang,1978
分类位置:苏铁纲本内苏铁目(Bennettiales,Cycadopsida)
分布与时代:东格陵兰和中国;早侏罗世。

羽毛奇异羽叶 *Thaumatophyllum ptilum* (Harris) Yang,1978

1932 *Pterophyllum ptilum* Harris,Harris T M,61页,图版5,图1—5,11;插图30,31;羽叶;东格陵兰;晚三叠世。
1954 *Pterophyllum ptilum* Harris,徐仁,58页,图版51,图2—4;羽叶;云南一平浪、江西安源、湖南石门口及四川等地;晚三叠世。
1978 杨贤河,515页,图版163,图14;羽叶;四川大邑太平;晚三叠世须家河组。

似金星蕨属 Genus *Thelypterites* Tao et Xiong,1986,emend Wu,1993

[注:此属名为陶君容、熊宪政最早(1986)使用,但未指明为新属。吴向午(1993)确认 *Thelypterites* 为新属名,指定 *Thelypterites* sp. A,Tao et Xiong,1986 为模式种]
1986　陶君容、熊宪政,122页。
1993a　吴向午,41,240页。
模式种:*Thelypterites* sp. A,Tao et Xiong,1986
分类位置:真蕨纲金星蕨科(Thelypteridaceae,Filicopsida)
分布与时代:中国黑龙江嘉荫;晚白垩世。

似金星蕨(未定种 A) *Thelypterites* sp. A

1986　*Thelypterites* sp. A,Tao et Xiong,陶君容、熊宪政,122页,图版5,图2b;生殖羽片;标本号:52701;标本保存在中国科学院植物研究所;黑龙江嘉荫;晚白垩世乌云组。
1993a　*Thelypterites* sp. A,吴向午,41,240页。

似金星蕨(未定种 B) *Thelypterites* sp. B

1986　*Thelypterites* sp. B,Tao et Xiong,陶君容、熊宪政,122页,图版6,图1;生殖羽片;标本号:52706;标本保存在中国科学院植物研究所;黑龙江嘉荫;晚白垩世乌云组。

田氏木属 Genus *Tianoxylon* Zhang et Zheng,2006(2008)(中文和英文发表)

2006　张武、郑少林,见张武等,114页。(中文)
2008　张武、郑少林,见张武等,114页。(英文)
模式种:*Tianoxylon duanmutouense* Zhang et Zheng,2006(中文),2008(英文)
分类位置:松柏类(Conifers incertae sedis)
分布与时代:中国辽宁朝阳;早三叠世。

段木头田氏木 *Tianoxylon duanmutouense* Zhang et Zheng,2006(2008)(中文和英文发表)

2006　张武、郑少林,见张武等,114页,图版4-11—4-13;木化石;正模:GJ6-46;副模:GJ6-47;标本保存在沈阳地质研究所;辽宁朝阳段木头沟;早三叠世红砬组。(中文)
2008　张武、郑少林,见张武等,114页,图版4-8—4-10;木化石;正模:GJ6-46;副模:GJ6-47;标本保存在沈阳地质研究所;辽宁朝阳段木头沟;早三叠世红砬组。(英文)

天山羊齿属 Genus *Tianshanopteris* Wu,1983

1983　吴绍祖,见窦亚伟等,599 页。
模式种:*Tianshanopteris wensuensis* Wu,1983
分类位置:楔羊齿类(Sphenopterides)
分布与时代:中国新疆;早二叠世。

温宿天山羊齿 *Tianshanopteris wensuensis* Wu,1983

1983　吴绍祖,见窦亚伟等,600 页,图版 217,图 1-5;蕨叶;采集号:75KH2-6-14;登记号:XPB-014-XPB-018;合模:XPB-014-XPB-018(图版 217,图 1-5);新疆温宿;早二叠世库尔干组。

天石枝属 Genus *Tianshia* Zhou et Zhang,1998(英文发表)

1998　周志炎、章伯乐,173 页。
模式种:*Tianshia patens* Zhou et Zhang,1998
分类位置:茨康目(Czekanowskiales)
分布与时代:中国河南义马;中侏罗世。

伸展天石枝 *Tianshia patens* Zhou et Zhang,1998(英文发表)

1998　周志炎、章伯乐,173 页,图版 2,图 1-6;图版 4,图 3,4,11;插图 3;枝、叶和角质层;登记号:PB17912,PB17913,PB17914;正模:PB17912(图版 2,图 1,4,5);标本保存在中国科学院南京地质古生物研究所;河南义马;中侏罗世义马组中部。

丁氏羊齿属 Genus *Tingia* Halle,1925

(注:又名齿叶属)
1925　Halle T G,5 页。
模式种:*Tingia carbonica* (Schenk) Halle,1925
分类位置:瓢叶目(Noeggerathiales)
分布与时代:中国山西;二叠纪。

石炭丁氏羊齿 *Tingia carbonica* (Schenk) Halle,1925

(注:又名华夏齿叶)
1883　*Pterophyllum carbonicum* Schenk,Schenk A,214 页,图版 44,图 4,5;山西太原;二叠纪。
1925　Halle T G,5 页,图版 1,图 1-4;山西太原;二叠纪。

托克逊蕨属 Genus *Toksunopteris* Wu S Q et Zhou,ap Wu X W,1993

1986　*Xinjiangopteris* Wu et Zhou (non Wu S Z,1983),吴舜卿、周汉忠,642,645 页。

1993b 吴舜卿、周汉忠,见吴向午,507,521 页。
模式种:*Toksunopteris ppsita*(Wu et Zhou) Wu S Q et Zhou,ap Wu X W,1993
分类位置:真蕨纲？或种子蕨纲？(Filicopsida? or Pteridospermopsida?)
分布与时代:中国新疆吐鲁番盆地;早侏罗世。

对生托克逊蕨 *Toksunopteris opposita*(Wu et Zhou) Wu S Q et Zhou,ap Wu X W,1993

1986 *Xinjiangopteris opposita* Wu et Zhou,吴舜卿、周汉忠,642,645 页,图版 5,图 1-8,10,10a;蕨叶;采集号:K215-K217,K219-K223,K228,K229;登记号:PB11780-PB11786,PB11793,PB11794;正模:PB11785(图版 5,图 10);标本保存在中国科学院南京地质古生物研究所;新疆吐鲁番盆地西北缘托克逊克尔碱地区;早侏罗世八道湾组。

1993b 吴舜卿、周汉忠,见吴向午,507,521 页;新疆吐鲁番盆地西北缘托克逊克尔碱地区;早侏罗世八道湾组。

铜川叶属 Genus *Tongchuanophyllum* Huang et Zhou,1980

1980 黄枝高、周惠琴,91 页。
模式种:*Tongchuanophyllum trigonus* Huang et Zhou,1980
分类位置:种子蕨纲(Pteridospermopsida)
分布与时代:中国陕西铜川、神木;中三叠世。

三角形铜川叶 *Tongchuanophyllum trigonus* Huang et Zhou,1980

1980 黄枝高、周惠琴,91 页,图版 17,图 2;图版 21,图 2,2a;蕨叶;登记号:OP3035,OP151;陕西铜川金锁关、神木枣坚;中三叠世铜川组上段下部。

此属创建时同时报道 3 种,除模式种外尚有:

优美铜川叶 *Tongchuanophyllum concinnum* Huang et Zhou,1980

1980 黄枝高、周惠琴,91 页,图版 16,图 4;图版 18,图 1-2;蕨叶;登记号:OP149,OP131;陕西铜川金锁关、神木枣坚;中三叠世铜川组上段下部。

陕西铜川叶 *Tongchuanophyllum shensiense* Huang et Zhou,1980

1980 黄枝高、周惠琴,91 页,图版 13,图 5;图版 14,图 3;图版 18,图 3;图版 21,图 1;图版 22,图 1;蕨叶;登记号:OP39,OP59,OP49,OP60;陕西铜川金锁关、神木枣坚;中三叠世铜川组下段。

桐庐籽属 Genus *Tonglucarpus* Chen et Zhu,1994

1994 陈其奭、朱德寿,6,8 页。
模式种:*Tonglucarpus spectabilis* Chen et Zhu,1994
分类位置:分类位置不明植物(Plantae incertae sedis)
分布与时代:中国浙江桐庐;晚二叠世。

奇丽桐庐籽 *Tonglucarpus spectabilis* Chen et Zhu,1994

1994 陈其奭、朱德寿,6,8 页,图版 1,图 1-4;插图 1;种子;标本号:75,76,78,79;标本保存在浙江自然博物院；浙江桐庐瑶琳；晚二叠世龙潭组。

钟囊属 Genus *Tongshania* Stockmans et Mathieu,1957
1957　Stockmans F 和 Mathieu F F,66 页。
模式种：*Tongshania dentate* Stockmans et Mathieu,1957
分类位置：分类位置不明植物（Plantae incertae sedis）
分布与时代：中国河北开平；晚石炭世。

齿状钟囊 *Tongshania dentate* Stockmans et Mathieu,1957
（注：又名钟囊）
1957　Stockmans F 和 Mathieu F F,66 页,图版 2,图 5－7a;图版 5,图 4,4a;孢子囊;河北开平;晚石炭世。

榧型枝属 Genus *Torreyocladus* Li et Ye,1980
1980　李星学、叶美娜,10 页。
模式种：*Torreyocladus spectabilis* Li et Ye,1980
分类位置：松柏纲（Coniferopsida）
分布与时代：中国吉林蛟河；早白垩世。

明显榧型枝 *Torreyocladus spectabilis* Li et Ye,1980
1980　李星学、叶美娜,10 页,图版 4,图 5；枝叶；登记号：PB8973；正模：PB8973（图版 4,图 5）；标本保存在中国科学院南京地质古生物研究所；吉林蛟河杉松；早白垩世磨石砬子组。[注：此标本后改定为 *Rhipidiocladus flabellata* Prynada（李星学等,1986）]

三网羊齿属 Genus *Tricoemplectopteris* Asama,1959
1959　Asama K,59 页。
模式种：*Tricoemplectopteris taiyuanensis* Asama,1959
分类位置：大羽羊齿类（Gigantopterides）
分布与时代：中国山西太原；晚古生代。

太原三网羊齿 *Tricoemplectopteris taiyuanensis* Asama,1959
1927　*Gigantopteris nicotianaefolia* Halle, Halle T G,164 页,图版 43－44,图 9。
1959　Asama K,59 页,图版 3,图 4；蕨叶；山西太原；晚古生代（石盒子系）。

三裂穗属 Genus *Tricrananthus* Wang Z Q et Wang L X,1990
1990a　王自强、王立新,137 页。
模式种：*Tricrananthus sagittatus* Wang Z Q et Wang L X,1990

分类位置:松柏纲(Coniferopsida)

分布与时代:中国山西榆社、和顺及蒲县;早三叠世。

箭头状三裂穗 *Tricrananthus sagittatus* Wang Z Q et Wang L X,1990

1990a 王自强、王立新,137页,图版21,图13—17;图版26,图6;雄性鳞片;标本号:Z16-418,Z16-422,Z16-17,Z16-426,Z16-422a,Iso19-29;模式标本:Z16-422(图版21,图15);标本保存在中国科学院南京地质古生物研究所;山西榆社屯村、和顺马坊;早三叠世和尚沟组底部。

此属创建时同时报道2种,除模式种外尚有:

瓣状三裂穗 *Tricrananthus lobatus* Wang Z Q et Wang L X,1990

1990a 王自强、王立新,137页,图版26,图5,10;雄性鳞片;标本号:Iso15-11,8304-3;合模:Iso15-11,8304-3(图版26,图5,10);标本保存在中国科学院南京地质古生物研究所;山西蒲县城关;早三叠世和尚沟组底部。

三脉蕨属 Genus *Trinerviopteris* Zhu,1995

1995 朱家楠、张秀生,316,317页。

模式种:*Trinerviopteris cardiophylla* (Zhu et Geng) Zhu,1995

分类位置:大羽羊齿类(Gigantopterides)

分布与时代:中国福建将乐;早二叠世晚期—晚二叠世早期。

心叶三脉蕨 *Trinerviopteris cardiophylla* (Zhu et Geng) Zhu,1995

1995 朱家楠、张秀生,316,317页,图版1;图版2,图1—7;插图1—4;蕨叶;正模:B. Y. Geng. B23(图版2,图6,7);副模:B. Y. Geng. B23(图版2,图1);标本保存在中国科学院植物研究所;福建将乐;早二叠世晚期—晚二叠世早期龙潭组。

三棱果属 Genus *Triqueteria* Stockmans et Mathieu,1957

1957 Stockmans F 和 Mathieu F F,68页。

模式种:*Triqueteria sinensis* Stockmans et Mathieu,1957

分类位置:分类位置不明植物(Plantae incertae sedis)

分布与时代:中国河北开平;晚石炭世。

中国三棱果 *Triqueteria sinensis* Stockmans et Mathieu,1957

1957 Stockmans F 和 Mathieu F F,68页,图版7,图6,6a;河北开平;晚石炭世。

蔡氏蕨属 Genus *Tsaia* Wang et Berry,2001(英文发表)

2001 王怿、Berry C M,82页。

模式种:*Tsaia denticulata* Wang et Berry,2001

分类位置:始蕨植物类(Euphyllophytes)

分布与时代：中国云南无定；中泥盆世(Givetian)。

细齿蔡氏蕨 *Tsaia denticulata* Wang et Berry, 2001 (英文发表)
2001　王怿、Berry C M, 82页, 图版1；图版2；插图2—5；草本植物；正模：PB18358(图版2, 图1；插图5a)；副模：PB18349—PB18357, PB18359—PB18370(图版1；图版2, 图2—13；插图2—4, 5b—5f)；标本保存在中国科学院南京地质古生物研究所；云南无定；中泥盆世(Givetian)西充组。

蛟河蕉羽叶属 Genus *Tsiaohoella* Lee et Yeh ex Zhang et al., 1980 (nom. nud.)
[注：此属名 *Tsiaohoella* 可能为 *Chiaohoella* 的误拼，分类位置为真蕨纲铁线蕨科 Adiantaceae(李星学等, 1986, 13页)]
1980　张武等, 279页。
模式种：*Tsiaohoella mirabilis* Lee et Yeh ex Zhang et al., 1980
分类位置：苏铁纲(Cycadopsida)
分布与时代：中国吉林蛟河杉松；早白垩世。

奇异蛟河蕉羽叶 *Tsiaohoella mirabilis* Lee et Yeh ex Zhang et al., 1980 (nom. nud.)
1980　张武等, 279页, 图版177, 图4—5；图版179, 图2, 4；羽叶；吉林蛟河杉松；早白垩世磨石砬子组。

此属创建时同时报道2种，除模式种外尚有：

新似查米亚型蛟河蕉羽叶 *Tsiaohoella neozamioides* Lee et Yeh ex Zhang et al., 1980 (nom. nud.)
1980　张武等, 79页, 图版179, 图1, 4；羽叶；吉林蛟河杉松；早白垩世磨石砬子组。

导管羊齿属 Genus *Vasovinea* Li et Taylor, 1999 (英文发表)
1999　李洪起、Taylor D W, 1564页。
模式种：*Vasovinea tianii* Li et Taylor, 1999
分类位置：大羽羊齿目(Gigantopteridales)
分布与时代：中国贵州盘县；晚二叠世。

田氏导管羊齿 *Vasovinea tianii* Li et Taylor, 1999 (英文发表)
1999　李洪起、Taylor D W, 1564页, 图1—32；茎干；标本号：PLY02, PLY03(图5, 7—20, 29—30)；正模：薄片号 Slides L9407-C-B2, L9407-C-B16, L9407-D-T2(图1—4, 6)；副模：薄片号 Slides PLY02-C10-1-1, PLY02-E1, PLY03-01, PLY03-06, PLY03-07, PLY03-11, PLY03-34, PLY04-B；标本保存在中国科学院植物研究所；贵州盘县；晚二叠世宣威组上部。

条叶属 Genus *Vittifoliolum* Zhou, 1984
1984　周志炎, 49页。

模式种：*Vittifoliolum segregatum* Zhou,1984

分类位置：银杏纲？或茨康目？(Ginkgopsida? or Czekanowskiales?)［注：原文将此属与 *Desmiophyllum*，*Cordaites*，*Yuccites*，*Bambusium*，*Phoenicopsis*，*Culgouweria*，*Windwardia*，*Pseudotorellia* 等属比较，认为可能属于银杏纲（周志炎，1984）；李佩娟等（1988）将此属归于银杏纲茨康目（?）］

分布与时代：中国湖南、青海；早侏罗世。

游离条叶 *Vittifoliolum segregatum* Zhou,1984

1984　周志炎,49页,图版29,图4—4d；图版30,图1—2b；图版31,图1—2a,4；插图12；叶及角质层；登记号：PB8938—PB8941,PB8943；正模：PB8937（图版30,图1）；标本保存在中国科学院南京地质古生物研究所；湖南祁阳、零陵、兰山、衡南、江永、永兴等地；早侏罗世观音滩组中、下部。

此属创建时同时报道3种,除模式种外尚有：

游离条叶脊条型 *Vittifoliolum segregatum* forma *costatum* Zhou,1984

1984　周志炎,50页,图版31,图3—3b；叶及角质层；登记号：PB8942；标本保存在中国科学院南京地质古生物研究所；湖南零陵黄阳司；早侏罗世观音滩组中（下?）部。

多脉条叶 *Vittifoliolum multinerve* Zhou,1984

1984　周志炎,50页,图版32,图1,2；叶及角质层；登记号：PB8944,PB8945；正模：PB8944（图版32,图1）；标本保存在中国科学院南京地质古生物研究所；湖南零陵黄阳司；早侏罗世观音滩组中（下?）部。

文山蕨属 Genus *Wenshania* Zhu et Kenrick,1999（英文发表）

1999　朱为庆、Kenrick P,112页。

模式种：*Wenshania zhichangensis* Zhu et Kenrick,1999

分类位置：工蕨类(Zosterophyllophytes)

分布与时代：中国云南文山；早泥盆世。

纸厂文山蕨 *Wenshania zhichangensis* Zhu et Kenrick,1999（英文发表）

1999　朱为庆、Kenrick P,112页,图版1,图1—6；插图1—3；植物体和孢子囊；标本号：MPB-Y 885-1,MPB-Y 885-3a,MPB-Y 885-3b,MPB-Y 885-4；正模：MPB-Y 885-1（图版1,图1；插图1）；标本保存在中国科学院植物研究所；云南文山古木；早泥盆世坡松冲组。

乌图布拉克蕨属 Genus *Wutubulaka* Wang,Fu,Xu et Hao,2007（英文发表）

2007　王怿、傅强、徐洪河、郝守刚,见王怿等,111页。

模式种：*Wutubulaka multidichotoma* Wang,Fu,Xu et Hao,2007

分类位置：分类位置不明植物(Plantae incertae sedis)

分布与时代：中国新疆准噶尔盆地；晚志留世(Late Pridoli)。

多叉乌图布拉克蕨 *Wutubulaka multidichotoma* Wang,Fu,Xu et Hao,2007（英文发表）

2007　王怿、傅强、徐洪河、郝守刚,113页,图1—4；植物体多次分叉；正模：PB20301（图1A,

1C);副模:PB20302,PB20304(图 4A-4B,4C);标本保存在中国科学院南京地质古生物研究所;新疆准噶尔盆地;晚志留世(Late Pridoli)乌图布拉克组中部。

无锡蕨属 Genus *Wuxia* Berry,Wang et Cai,2003(英文发表)
2003　Berry C M、王怿、蔡重阳,268 页。
模式种:*Wuxia bistrobilata* Berry,Wang et Cai,2003
分类位置:石松类(Pteridopsida incertae sedis)
分布与时代:中国江苏无锡;晚泥盆世(Famennian)。

双穗无锡蕨 *Wuxia bistrobilata* Berry,Wang et Cai,2003(英文发表)
2003　Berry C M、王怿、蔡重阳,268 页,图 2-6;石松类植物;正模:PB18870(图 3b);副模:PB18862b,PB18864,PB18866-PB18869,PB18871,PB18875-PB18877,PB18879(图 2,3a,3c-3f,5,6);标本保存在中国科学院南京地质古生物研究所;江苏无锡;晚泥盆世(Famennian)五通组。

乌云花属 Genus *Wuyunanthus* Wang,Li C,Li Z et Fu,2001(英文发表)
2001　王宇飞、李承森、李振宇、傅德志,325 页。
模式种:*Wuyunanthus hexapetalus* Wang,Li C,Li Z et Fu,2001
分类位置:被子植物双子叶纲蔷薇亚纲南蛇藤目卫矛科(Celastraceae,Celastrales,Rosidae)
分布与时代:中国黑龙江嘉阴;古新世。

六瓣乌云花 *Wuyunanthus hexapetalus* Wang,Li C,Li Z et Fu,2001(英文发表)
2001　王宇飞、李承森、李振宇、傅德志,325 页,图 1-3,5;花化石;标本号:wy-92-101a,wy-92-101b;正模:wy-92-101a,wy-92-101b(图 1,2);标本保存在中国科学院植物研究所;黑龙江嘉阴乌云煤田;古新世乌云组。

夏家街蕨属 Genus *Xiajiajienia* Sun et Zheng,2001(中文和英文发表)
2001　孙革、郑少林,见孙革等,77,187 页。
模式种:*Xiajiajienia mirabila* Sun et Zheng,2001
分类位置:真蕨纲(Filicopsida)
分布与时代:中国吉林辽源、辽宁北票;中—晚侏罗世。

奇异夏家街蕨 *Xiajiajienia mirabila* Sun et Zheng,2001(中文和英文发表)
2001　孙革、郑少林,见孙革等,77,187 页,图版 10,图 3-6;图版 39,图 1-10;图版 56,图 7;蕨叶;标本号:PB19025-PB19026,PB19028-PB19032,ZY3015;正模:PB19025(图版 10,图 3);标本保存在中国科学院南京地质古生物研究所;吉林辽源夏家街;中侏罗世夏家街组;辽宁北票上园黄半吉沟;晚侏罗世尖山沟组。

兴安叶属 Genus *Xinganphyllum* Huang,1977

1977　黄本宏,60 页。

模式种:*Xinganphyllum aequale* Huang,1977

分类位置:分类位置不明植物(Plantae incertae sedis)

分布与时代:中国黑龙江;晚二叠世。

对称兴安叶 *Xinganphyllum aequale* Huang,1977

1977　黄本宏,60 页,图版 6,图 1—2;图版 7,图 1—3;插图 20;叶部化石;登记号:PFH0238, PFH0240,PFH0234,PFH0236,PFH0241;标本保存在沈阳地质矿产研究所;黑龙江铁力三角山;晚二叠世三角山组。

此属创建时同时报道 3 种,除模式种外尚有:

不对称兴安叶 *Xinganphyllum inaequale* Huang,1977

1977　黄本宏,61 页,图版 27,图 2;插图 21;叶部化石;登记号:PFH0235;标本保存在沈阳地质矿产研究所;黑龙江铁力三角山;晚二叠世三角山组。

兴安叶(未定种) *Xinganphyllum* sp.

1977　*Xinganphyllum* sp.,黄本宏,62 页,图版 24,图 1;图版 38,图 6;插图 23;叶部化石;黑龙江铁力三角山;晚二叠世三角山组。

星学花属 Genus *Xingxueanthus* Wang X et Wang S,2010(英文发表)

2010　王鑫、王士俊,50 页。

模式种:*Xingxueanthus sinensis* Wang X et Wang S,2010

分类位置:被子植物(Angiosperms)

分布与时代:中国辽宁锦西;中侏罗世。

中国星学花 *Xingxueanthus sinensis* Wang X et Wang S,2010(英文发表)

2010　王鑫、王士俊,50 页,图 2—5;雌花"花序";正模:No. 8703a(图 2:a,d;图 3:a—d;图 4:e,h,i);副模:No. 8703b(图 2:b,c,e,g,i;图 3:e,f;图 4:a—d,f,g);标本保存在中国科学院植物研究所;辽宁锦西三角城村(120°21′E,40°58′N);中侏罗世海房沟组。

星学花序属 Genus *Xingxueina* Sun et Dilcher,1997(中文和英文发表)

1995a　孙革、Dilcher D L,见李星学,324 页。(中文)(裸名)

1995b　孙革、Dilcher D L,见李星学,429 页。(英文)(裸名)

1996　孙革、Dilcher D L,396 页。(裸名)

1997　孙革、Dilcher D L,137,141 页。

模式种:*Xingxueina heilongjiangensis* Sun et Dilcher,1997

分类位置:双子叶植物纲(Dicotyledoneae)

分布与时代：中国黑龙江鸡西；早白垩世。

黑龙江星学花序 *Xingxueina heilongjiangensis* Sun et Dilcher,1997（中文和英文发表）
1995a 孙革、Dilcher D L,见李星学,324 页；插图 9-2.8；花序及叶片；黑龙江鸡西城子河；早白垩世城子河组。（中文）（裸名）
1995b 孙革、Dilcher D L,见李星学,429 页；插图 9-2.8；花序及叶片；黑龙江鸡西城子河；早白垩世城子河组。（英文）（裸名）
1996 孙革、Dilcher D L,图版 2,图 1—6；插图 1E；花序及叶片；黑龙江鸡西城子河；早白垩世城子河组。（裸名）
1997 孙革、Dilcher D L,137,141 页,图版 1,图 1—7；图版 2,图 1—6；插图 2；花序及叶片；采集号：WR47—WR100；登记号：SC10025,SC10026；正模：SC10026（图版 5,图 1B,2；插图 4G）；标本保存在中国科学院南京地质古生物研究所；黑龙江鸡西城子河；早白垩世城子河组。

星学叶属 Genus *Xingxuephyllum* Sun et Dilcher,2002（英文发表）
2002 孙革、Dilcher D L,103 页。
模式种：*Xingxuephyllum jixiense* Sun et Dilcher,2002
分类位置：双子叶植物纲（Dicotyledoneae）
分布与时代：中国黑龙江鸡西；早白垩世。

鸡西星学叶 *Xingxuephyllum jixiense* Sun et Dilcher,2002（英文发表）
2002 孙革、Dilcher D L,103 页,图版 5,图 1B,2；插图 4G；叶部化石；标本号：SC10025,SC10026；模式标本：SC10026（图版 5,图 1B,2；插图 4G）；黑龙江鸡西城子河；早白垩世城子河组。

新疆木属 Genus *Xinjiangophyton* Sun,1983
1983 孙喆华,见窦亚伟等,581 页。
模式种：*Xinjiangophyton spinosum* Sun,1983
分类位置：石松类（Lycopsida incertae sedis）
分布与时代：中国新疆准噶尔盆地；中泥盆世。

刺状新疆木 *Xinjiangophyton spinosum* Sun,1983
1983 孙喆华,见窦亚伟等,581 页,图版 207,图 1—3；茎；采集号：63-7G-1-3341-b；登记号：XPA167—XPA169；正模：XPA167（图版 207,图 1）；新疆奇台北塔山；中泥盆世。

新疆蕨属 Genus *Xinjiangopteris* Wu S Q et Zhou,1986（non Wu S Z,1983）
［注：此属名为 *Xinjiangopteris* Wu S Z,1983 的晚出同名（吴向午,1993a,1993b）］
1986 吴舜卿、周汉忠,642,645 页。
模式种：*Xinjiangopteris opposita* Wu S Q et Zhou,1986

分类位置：真蕨类或种子蕨类(Filicopsida or Pteridospermopsida)

分布与时代：中国新疆吐鲁番盆地；早侏罗世。

对生新疆蕨 *Xinjiangopteris opposita* Wu S Q et Zhou, 1986

［注：此种后改定为 *Toksunopteris opposita* Wu S Q et Zhou(吴向午，1993a)］

1986　吴舜卿、周汉忠，642，645 页，图版 5，图 1—8，10，10a；蕨叶；采集号：K215—K217，K219—K223，K228，K229；登记号：PB11780—PB11786，PB11793，PB11794；正模：PB11785(图版 5，图 10)；标本保存在中国科学院南京地质古生物研究所；新疆吐鲁番盆地西北缘托克逊克尔碱地区；早侏罗世八道湾组。

新疆蕨属 Genus *Xinjiangopteris* Wu S Z, 1983（non Wu S Q et Zhou, 1986）

［注：此属名另有晚出同名 *Xinjiangopteris* Wu S Q et Zhou, 1986(吴向午，1993a，1993b)］

1983　吴绍祖，见窦亚伟等，607 页。

模式种：*Xinjiangopteris toksunensis* Wu S Z, 1983

分类位置：种子蕨纲(Pteridospermopsida)

分布与时代：中国新疆和静；晚二叠世。

托克逊新疆蕨 *Xinjiangopteris toksunensis* Wu S Z, 1983

1983　吴绍祖，见窦亚伟等，607 页，图版 223，图 1—6；蕨叶；采集号：73KH1-6a；登记号：XPB-032—XPB-037；合模：XPB-032—XPB-037(图版 223，图 1—6)；新疆和静艾乌尔沟；晚二叠世。

新龙叶属 Genus *Xinlongia* Yang, 1978

1978　杨贤河，516 页。

模式种：*Xinlongia pterophylloides* Yang, 1978

分类位置：苏铁纲本内苏铁目(Bennettiales, Cycadopsida)

分布与时代：中国四川新龙；晚三叠世。

侧羽叶型新龙叶 *Xinlongia pterophylloides* Yang, 1978

1978　杨贤河，516 页，图版 182，图 1；插图 118；羽叶；标本号：Sp0116；正模：Sp0116(图版 182，图 1)；标本保存在地质部成都地质矿产研究所；四川新龙雄龙；晚三叠世喇嘛垭组。

此属创建时同时报道 2 种，除模式种外尚有：

和恩格尔新龙叶 *Xinlongia hoheneggeri* (Schenk) Yang, 1978

1869　*Podozamites hoheneggeri* Schenk, Schenk A, 9 页, 图版 11, 图 3—6。

1978　杨贤河，516 页，图版 178，图 7；羽叶；四川广元须家河；晚三叠世须家河组。

新龙羽叶属 Genus *Xinlongophyllum* Yang, 1978

1978　杨贤河，505 页。

模式种：*Xinlongophyllum ctenopteroides* Yang,1978
分类位置：种子蕨纲（Pteridospermopsida）
分布与时代：中国四川新龙；晚三叠世。

篦羽羊齿型新龙羽叶 *Xinlongophyllum ctenopteroides* Yang,1978

1978　杨贤河,505 页,图版 182,图 2;羽叶;标本号：Sp0117;正模：Sp0117（图版 182,图 2）;标本保存在成都地质矿产研究所;四川新龙雄龙;晚三叠世喇嘛垭组。

此属创建时同时报道 2 种,除模式种外尚有：

多条纹新龙羽叶 *Xinlongophyllum multilineatum* Yang,1978

1978　杨贤河,506 页,图版 182,图 3－4;羽叶;标本号：Sp0118,Sp0119;合模 1:Sp0118（图版 182,图 3）;合模 2:Sp00119（图版 182,图 4）;标本保存在成都地质矿产研究所;四川新龙雄龙;晚三叠世喇嘛垭组。

西屯蕨属 Genus *Xitunia* Xue,2009（英文发表）

2009　薛进庄,505 页。

模式种：*Xitunia spinitheca* Xue,2009
分类位置：工蕨类（Zosterophyllophytes）
分布与时代：中国云南曲靖；早泥盆世。

刺囊西屯蕨 *Xitunia spinitheca* Xue,2009（英文发表）

2009　薛进庄,505 页,图版 1,图 1－5;插图 1a,1b;植物体和孢子囊;标本号：PKU-XH200a,PKU-XH200b;正模：PKU-XH200a（图版 1,图 1）;标本保存在北京大学地质系;云南曲靖；早泥盆世西屯组。

杨氏木属 Genus *Yangzunyia* Yang,2006（中文和英文发表）

2006　杨关秀等,79,236 页。

模式种：*Yangzunyia henanensis* Yang,2006
分类位置：石松植物门鳞木目（Protolepidodendrales,Lycophyta）
分布与时代：中国河南禹州；中二叠世。

河南杨氏木 *Yangzunyia henanensis* Yang,2006（中文和英文发表）

2006　杨关秀等,79,236 页,图版 2,图 1－4;具叶座的茎干;合模 1：HEP0364（图版 2,图 3）;合模 2：HEP0365（图版 2,图 4）;标本保存在中国地质大学（北京）;河南禹州大风口;中二叠世小风口组。

延吉叶属 Genus *Yanjiphyllum* Zhang,1980

1980　张志诚,338 页。

模式种：*Yanjiphyllum ellipticum* Zhang,1980

分类位置：双子叶植物纲（Dicotyledoneae）

分布与时代：中国吉林延吉；早白垩世。

椭圆延吉叶 *Yanjiphyllum ellipticum* Zhang，1980

1980 张志诚，338页，图版192，图7，7a；叶；登记号：D631；标本保存在沈阳地质矿产研究所；吉林延吉大拉子；早白垩世大拉子组。

燕辽杉属 Genus *Yanliaoa* Pan，1977

1977 潘广，70页。

模式种：*Yanliaoa sinensis* Pan，1977

分类位置：松柏纲杉科（Taxodiaceae，Coniferopsida）

分布与时代：中国辽西锦西；中－晚侏罗世。

中国燕辽杉 *Yanliaoa sinensis* Pan，1977

1977 潘广，70页，图版5；营养枝、生殖枝（包括花、果枝）；登记号：L0064，L0027，L0034，L0040A；标本保存在辽宁煤田地质勘探公司；辽西锦西；中－晚侏罗世。

义马果属 Genus *Yimaia* Zhou et Zhang，1988

1988a 周志炎、章伯乐，217页。（中文）

1988b 周志炎、章伯乐，1202页。（英文）

模式种：*Yimaia recurva* Zhou et Zhang，1988

分类位置：银杏目（Ginkgoales）

分布与时代：中国河南义马；中侏罗世。

外弯义马果 *Yimaia recurva* Zhou et Zhang，1988

1988a 周志炎、章伯乐，217页，图3；生殖枝；登记号：PB14193；正模：PB14193（图3）；标本保存在中国科学院南京地质古生物研究所；河南义马；中侏罗世。（中文）

1988b 周志炎、章伯乐，1202页，图3；生殖枝；登记号：PB14193；正模：PB14193（图3）；标本保存在中国科学院南京地质古生物研究所；河南义马；中侏罗世。（英文）

义县叶属 Genus *Yixianophyllum* Zheng，Li N，Li Y，Zhang et Bian，2005（英文发表）

2005 郑少林、李楠、李勇、张武、边雄飞，585页。

模式种：*Yixianophyllum jinjiagouensie* Zheng，Li N，Li Y，Zhang et Bian，2005

分类位置：苏铁目（Cycadales）

分布与时代：中国辽宁义县；晚侏罗世。

金家沟义县叶 *Yixianophyllum jinjiagouensie* Zheng，Li N，Li Y，Zhang et Bian，2005（英文发表）

2004 *Taeniopteris* sp.（gen. et sp. nov.），王五力等，232页，图版30，图2－5；单叶；辽宁义县

2005 郑少林、李楠、李勇、张武、边雄飞，585 页，图版 1—2；图 2,3A,3B,4A,5J；叶和角质层；采集号：JJG-7—JJG-11；正模：JJG-7（图版 1，图 1）；副模：JJG-8—JJG-10（图版 1，图 3,5,6）；标本保存在沈阳地质矿产研究所；辽宁义县金家沟；晚侏罗世义县组下部。

卵叶属 Genus *Yuania* Sze, 1953
1953 斯行健，13,18 页。
模式种：*Yuania striata* Sze, 1953
分类位置：瓢叶目（Noeggerathiales）
分布与时代：中国陕西；晚二叠世。

条纹卵叶 *Yuania striata* Sze, 1953
（注：又名卵叶）
1953 斯行健，13,18 页，图版 1，图 6—7a；插图 1；羽状枝叶；陕西麟游磨子沟；晚二叠世石千峰系。

玉光蕨属 Genus *Yuguangia* Hao, Xue, Wang et Liu, 2007（英文发表）
2007 郝守刚、薛进庄、王祺、刘振锋，1163 页。
模式种：*Yuguangia ordinata* Hao, Xue, Wang et Liu, 2007
分类位置：石松目（Lycopodiales）
分布与时代：中国云南沾益；中泥盆世（Late Givetian）。

规则玉光蕨 *Yuguangia ordinata* Hao, Xue, Wang et Liu, 2007（英文发表）
2007 郝守刚、薛进庄、王祺、刘振锋，1163 页，图 2,5—7；草本植物；正模：BUP. H-y07（图 5d）；副模：BUP. H-y01, BUH-y1-t1-6, BUH-y1-11（图 5c,5e,5g,6）, BUP. H-y02, BUP. H-y03, BUP. H-y04, BUP. H-y05, BUP. H-y06（图 2a—2c），SEM-y. t-01,02, SEM-y. 1-03,04（图 2k—2n），SEM-ypo-04,06（图 7d,7g），LM-yspo-01,02（图 7e,7h—7j）；标本保存在北京大学地质系；云南沾益；中泥盆世（Late Givetian）海口组。

永仁叶属 Genus *Yungjenophyllum* Hsu et Chen, 1974
1974 徐仁、陈晔，见徐仁等，275 页。
模式种：*Yungjenophyllum grandifolium* Hsu et Chen, 1974
分类位置：分类位置不明植物（Plantae incertae sedis）
分布与时代：中国云南永仁；晚三叠世。

大叶永仁叶 *Yungjenophyllum grandifolium* Hsu et Chen, 1974
1974 徐仁、陈晔，见徐仁等，275，图版 8，图 1—3；单叶；编号：No. 2883；标本保存在中国科学院植物研究所；云南永仁宝鼎；晚三叠世大荞地组中部。

云蕨属 Genus *Yunia* Hao et Beck,1991

1991b 郝守刚、Beck C B,191 页。
模式种:*Yunia dichotoma* Hao et Beck,1991
分类位置:裸蕨类(Psilophytes incertae sedis)
分布与时代:中国云南文山;早泥盆世。

二叉云蕨 *Yunia dichotoma* Hao et Beck,1991

1991b 郝守刚、Beck C B,192 页,图版 1,图 1－6;图版 2,图 7－13;图版 3,图 14－23;图版 4,图 24－33;插图 1－3;茎轴(原生木质部横切面)及孢子囊;正模:BUHB-1101(图版 1,图 1);副模:BUHB-1102(图版 1,图 2),BUHB-1103(图版 1,图 4,5),HS-11-1(图版 2,图 7),HB4-2,4(图版 2,图 8,9),HB1-2(图版 2,图 10),HB5-4,5,6(图版 3,图 14－16),HB6-2,4(图版 3,图 17,18),HS11-2(图版 3,图 19);标本保存在北京大学地质系;云南文山;早泥盆世坡松冲组。

蔡耶羊齿属 Genus *Zeillerpteris* Koidzumi,1936

1936 Koidzumi G,135 页。
模式种:*Zeillerpteris yunnanensis* Koidzumi,1936
分类位置:大羽羊齿类种子蕨纲(Pteridospermopsida,Gigantopterides)
分布与时代:中国云南;二叠纪－石炭纪。

云南蔡耶羊齿 *Zeillerpteris yunnanensis* Koidzumi,1936

1907 *Gigantopteris nicotinaefolia* Zeiller,Zeiller R,480 页,图版 14,图 15,15a。
1936 Koidzumi G,135 页;云南(Sine-si-kou);二叠纪－石炭纪。

郑氏叶属 Genus *Zhengia* Sun et Dilcher,2002(英文发表)

1996 孙革、Dilcher D L,图版 1,图 15;图版 2,图 7－9。(裸名)
2002 孙革、Dilcher D L,103 页。
模式种:*Zhengia chinensis* Sun et Dilcher,2002
分类位置:双子叶植物纲(Dicotyledonae)
分布与时代:中国黑龙江鸡西;早白垩世。

中国郑氏叶 *Zhengia chinensis* Sun et Dilcher,2002(英文发表)

1996 孙革、Dilcher D L,图版 1,图 15;图版 2,图 7－9;叶及角质层;黑龙江鸡西城子河;早白垩世城子河组。(裸名)
2002 孙革、Dilcher D L,103 页,图版 4,图 1－7;叶及角质层;标本号:JS10004,SC10023,SC01996;正模:SC10023(图版 4,图 1,3－6);标本保存在中国科学院南京地质古生物研究所;黑龙江鸡西城子河;早白垩世城子河组。

正理蕨属 Genus *Zhenglia* Hao,Wang D,Wang Q et Xue,2006（英文发表）
2006　郝守刚、王德明、王祺、薛进庄,13 页。
模式种:*Zhenglia radiata* Hao,Wang D,Wang Q et Xue,2006
分类位置:石松类（Lycopsida）
分布与时代:中国云南文山;早泥盆世。

辐射正理蕨 *Zhenglia radiata* Hao,Wang D,Wang Q et Xue,2006（英文发表）
2006　郝守刚、王德明、王祺、薛进庄,13 页,图版 1,图 1—12;插图 2;草本植物;正模:PKU-HW-Ly.06（图版 1,图 7;插图 2d）;副模:PKU-HW-Ly.01,02,04,07,08,09（图版 1,图 1,4,6,9,11,12）;云南文山;早泥盆世坡松冲组。

中州籽属 Genus *Zhongzhoucarus* Yang,2006（中文和英文发表）
2006　杨关秀等,171,327 页。
模式种:*Zhongzhoucarus deltatus* Yang,2006
分类位置:裸子植物种子（Gymnospermarum）
分布与时代:中国河南禹州;晚二叠世。

三角中州籽 *Zhongzhoucarus deltatus* Yang,2006（中文和英文发表）
2006　杨关秀等,171,327 页,图版 44,图 9;种子;正模:HEP0595（图版 44,图 9）;标本保存在中国地质大学（北京）;河南禹州大风口;晚二叠世云盖山组。

中州蕨属 Genus *Zhongzhoupteris* Yang,2006（中文和英文发表）
2006　杨关秀等,105,258 页。
模式种:*Zhongzhoupteris cathaysicus* Yang,2006
分类位置:真蕨目紫萁科?（Osmundaceae?,Filicales）
分布与时代:中国河南禹州;中二叠世。

尾羽中州蕨 *Zhongzhoupteris cathaysicus* Yang,2006（中文和英文发表）
2006　杨关秀等,105,258 页,图版 17,图 6;图版 18;图版 19,图 1,2;蕨叶;合模:HEP0260（图版 18,图 1）,HEP0262（图版 19,图 1）,HEP0263（图版 19,图 2）,HEP0261（图版 17,图 6）;标本保存在中国地质大学（北京）;河南禹州大涧村;中二叠世神垕组。

周氏籽属 Genus *Zhouia* Zheng,Gao et Bo,2008（中文和英文发表）
2008　郑少林、高家俊、薄学,见郑少林等,331,340 页。
模式种:*Zhouia beipiaoensis* Zheng,Gao et Bo,2008
分类位置:单子叶植物类?（Monocots?）

分布与时代：中国辽宁北票；早白垩世。

北票周氏籽 *Zhouia beipiaoensis* Zheng, Gao et Bo, 2008（中文和英文发表）
2008　郑少林、高家俊、薄学，见郑少林等，332,340页；插图1.1B；插图2.1B,3,5；种子化石；正模：LBY2001（插图1.1B；插图2.1B,3,5）；标本由辽宁省锦州市化石收藏家才树仁先生妥善保管；辽宁北票上园黄半吉沟村附近，早白垩世义县组尖山沟层。

朱氏囊蕨属 Genus *Zhutheca* Liu, Li et Hilton, 2000（英文发表）
2000　刘照华、李承森、Hilton J, 150页。
模式种：*Zhutheca densata* (Gu et Zhi) Liu, Li et Hilton, 2000
分类位置：观音座莲目合囊蕨科（Marattaceae, Marattiales）
分布与时代：中国云南、江苏、贵州和西藏；晚二叠世。

密囊朱氏囊蕨 *Zhutheca densata* (Gu et Zhi) Liu, Li et Hilton, 2000（英文发表）
1974　*Fascipteris* (*Ptychocarpus*) *densata* Gu et Zhi,中国科学院南京地质古生物研究所、植物研究所，见《中国古生代植物》，100页，图版69，图8-14；插图85-86；生殖羽叶；登记号：PB686,PB688,PB690；合模：PB686（图版69，图8），PB688（图版69，图10），PB690（图版69，图12）；标本保存在中国科学院南京地质古生物研究所；江苏江宁；晚二叠世龙潭组；贵州盘县；晚二叠世宣威组；西藏昌都；晚二叠世。
2000　刘照华、李承森、Hilton J, 150页，图版1，图1-5；图版2，图1-4；图版3，图1-6；插图1；生殖羽叶；正模：PB686,PB688,PB690；标本保存在中国科学院南京地质古生物研究所；标本号：9782,9783,9784；副模：9782,9784（图版1，图1,2,4）；标本保存在中国科学院植物研究所；云南宣威；晚二叠世宣威组。

附 录

附录 1　属名索引

[按中文名称的汉语拼音升序排列,属名后为页码(中文记录页码/英文记录页码)]

A

凹尖枝属 *Koilosphenus* ··· 51/230

B

白豆杉型木属 *Pseudotaxoxylon* ·· 90/273
白果叶属 *Baiguophyllum* ·· 9/184
苞片蕨属 *Bracteophyton* ·· 13/188
抱囊蕨属 *Amplectosporangium* ··· 6/181
鲍斯木属 *Boseoxylon* ·· 12/188
北票果属 *Beipiaoa* ·· 10/185
贝叶属 *Conchophyllum* ·· 22/198
本内缘蕨属 *Bennetdicotis* ·· 11/186
本溪羊齿属 *Benxipteris* ··· 11/186
篦囊属 *Pectinangium* ·· 78/260
扁囊蕨属 *Demersatheca* ·· 26/202
变态鳞木属 *Metalepidodendron* ··· 64/244

C

蔡氏蕨属 *Tsaia* ··· 120/306
蔡耶羊齿属 *Zeillerpteris* ··· 130/316
缠绕蕨属 *Helicophyton* ·· 42/220
长穗属 *Longostachys* ·· 58/237
长武蕨属 *Changwuia* ··· 18/194
长阳木属 *Changyanophyton* ·· 19/194
朝阳序属 *Chaoyangia* ·· 19/195
城子河叶属 *Chengzihella* ·· 19/195
翅鳞籽属 *Squarmacarpus* ·· 109/293
翅叶属 *Neurophyllites* ··· 70/251

附录　133

翅籽属 *Semenalatum* ·· 99/282
垂饰杉属 *Stalagma* ·· 110/294

D

大囊蕨属 *Gigantotheca* ·· 38/216
大箐羽叶属 *Tachingia* ·· 113/297
大舌羊齿属 *Macroglossopteris* ·· 60/240
大同叶属 *Datongophyllum* ·· 25/201
大叶羊齿属 *Megalopteris* Schenk,1883 (non Andrews E B,1875) ·· 62/242
大羽羊齿属 *Gigantopteris* ·· 38/216
带囊蕨属 *Gigantonomia* ·· 38/215
带状鳞穗属 *Loroderma* ·· 59/239
单网羊齿属 *Gigantonoclea* ·· 37/215
单叶大羽羊齿属 *Monogigantopteris* ·· 68/248
单叶单网羊齿属 *Monogigantonoclea* ·· 67/247
导管羊齿属 *Vasovinea* ·· 121/306
道虎沟叶状体属 *Daohugouthallus* ·· 24/201
灯笼蕨属 *Denglongia* ·· 26/203
登封籽属 *Dengfengia* ·· 26/203
碟囊属 *Lopadiangium* ·· 58/238
蝶叶属 *Papilionifolium* ·· 75/256
丁氏羊齿属 *Tingia* ·· 117/302
东北穗属 *Manchurostachys* ·· 60/240
渡口痕木属 *Dukouphyton* ·· 29/206
渡口叶属 *Dukouphyllum* ·· 29/206
对枝柏属 *Ditaxocladus* ·· 28/205
多瓣蕨属 *Polypetalophyton* ·· 84/267
多囊枝属 *Polythecophyton* ·· 85/267
多枝蕨属 *Ramophyton* ·· 92/275

E

耳叶属 *Otofolium* ·· 73/254

F

榧型枝属 *Torreyocladus* ·· 119/304
缝鞘杉属 *Suturovagina* ·· 111/295
辐射叶属 *Actinophyllus* Xiao,1985 (non *Actinophyllum* Phillips,1848) ·· 3/177
辐叶属 *Radiatifolium* ·· 92/275
福建羊齿属 *Fujianopteris* ·· 36/214
副开通尼亚属 *Paracaytonia* ·· 75/256
副镰羽叶属 *Paradrepanozamites* ·· 76/257

副球果属 *Paraconites* ⋯⋯⋯⋯⋯⋯⋯⋯⋯⋯⋯⋯⋯⋯⋯⋯⋯⋯⋯⋯⋯⋯⋯⋯⋯⋯⋯⋯⋯⋯⋯ 75/257

G

甘肃芦木属 *Gansuphyllite* ⋯⋯⋯⋯⋯⋯⋯⋯⋯⋯⋯⋯⋯⋯⋯⋯⋯⋯⋯⋯⋯⋯⋯⋯⋯⋯⋯ 37/215
赣囊蕨属 *Jiangxitheca* ⋯⋯⋯⋯⋯⋯⋯⋯⋯⋯⋯⋯⋯⋯⋯⋯⋯⋯⋯⋯⋯⋯⋯⋯⋯⋯⋯⋯⋯ 47/226
根状茎属 *Rhizoma* ⋯⋯⋯⋯⋯⋯⋯⋯⋯⋯⋯⋯⋯⋯⋯⋯⋯⋯⋯⋯⋯⋯⋯⋯⋯⋯⋯⋯⋯⋯⋯ 95/278
钩蕨属 *Hamatophyton* ⋯⋯⋯⋯⋯⋯⋯⋯⋯⋯⋯⋯⋯⋯⋯⋯⋯⋯⋯⋯⋯⋯⋯⋯⋯⋯⋯⋯⋯ 41/220
古果属 *Archaefructus* ⋯⋯⋯⋯⋯⋯⋯⋯⋯⋯⋯⋯⋯⋯⋯⋯⋯⋯⋯⋯⋯⋯⋯⋯⋯⋯⋯⋯⋯⋯ 7/182
古买麻藤属 *Palaeognetaleaana* ⋯⋯⋯⋯⋯⋯⋯⋯⋯⋯⋯⋯⋯⋯⋯⋯⋯⋯⋯⋯⋯⋯⋯⋯⋯ 73/255
古木蕨属 *Gumuia* ⋯⋯⋯⋯⋯⋯⋯⋯⋯⋯⋯⋯⋯⋯⋯⋯⋯⋯⋯⋯⋯⋯⋯⋯⋯⋯⋯⋯⋯⋯⋯ 40/218
古银杏型木属 *Palaeoginkgoxylon* ⋯⋯⋯⋯⋯⋯⋯⋯⋯⋯⋯⋯⋯⋯⋯⋯⋯⋯⋯⋯⋯⋯⋯ 73/254
古舟藤属 *Palaeoskapha* ⋯⋯⋯⋯⋯⋯⋯⋯⋯⋯⋯⋯⋯⋯⋯⋯⋯⋯⋯⋯⋯⋯⋯⋯⋯⋯⋯⋯ 74/255
管子麻黄属 *Siphonospermum* ⋯⋯⋯⋯⋯⋯⋯⋯⋯⋯⋯⋯⋯⋯⋯⋯⋯⋯⋯⋯⋯⋯⋯⋯⋯ 106/289
广南蕨属 *Guangnania* ⋯⋯⋯⋯⋯⋯⋯⋯⋯⋯⋯⋯⋯⋯⋯⋯⋯⋯⋯⋯⋯⋯⋯⋯⋯⋯⋯⋯⋯ 39/217
广西叶属 *Guangxiophyllum* ⋯⋯⋯⋯⋯⋯⋯⋯⋯⋯⋯⋯⋯⋯⋯⋯⋯⋯⋯⋯⋯⋯⋯⋯⋯⋯ 39/217
贵州木属 *Guizhouoxylon* ⋯⋯⋯⋯⋯⋯⋯⋯⋯⋯⋯⋯⋯⋯⋯⋯⋯⋯⋯⋯⋯⋯⋯⋯⋯⋯⋯ 40/218

H

哈勒角籽属 *Hallea* Mathews,1947－1948（non Yang et Wu,2006） ⋯⋯⋯⋯⋯⋯ 40/219
哈瑞士叶属 *Tharrisia* ⋯⋯⋯⋯⋯⋯⋯⋯⋯⋯⋯⋯⋯⋯⋯⋯⋯⋯⋯⋯⋯⋯⋯⋯⋯⋯⋯⋯ 115/299
哈氏蕨属 *Halleophyton* ⋯⋯⋯⋯⋯⋯⋯⋯⋯⋯⋯⋯⋯⋯⋯⋯⋯⋯⋯⋯⋯⋯⋯⋯⋯⋯⋯ 41/219
和丰孢穗属 *Hefengistrobus* ⋯⋯⋯⋯⋯⋯⋯⋯⋯⋯⋯⋯⋯⋯⋯⋯⋯⋯⋯⋯⋯⋯⋯⋯⋯ 42/220
河南羊齿属 *Henanopteris* ⋯⋯⋯⋯⋯⋯⋯⋯⋯⋯⋯⋯⋯⋯⋯⋯⋯⋯⋯⋯⋯⋯⋯⋯⋯⋯ 43/221
河南叶属 *Henanophyllum* ⋯⋯⋯⋯⋯⋯⋯⋯⋯⋯⋯⋯⋯⋯⋯⋯⋯⋯⋯⋯⋯⋯⋯⋯⋯⋯ 42/221
赫勒单网羊齿属 *Hallea* Yang et Wu,2006（non Mathews,1947－1948） ⋯⋯ 41/219
湖北蕨属 *Hubeiia* ⋯⋯⋯⋯⋯⋯⋯⋯⋯⋯⋯⋯⋯⋯⋯⋯⋯⋯⋯⋯⋯⋯⋯⋯⋯⋯⋯⋯⋯⋯ 45/223
湖北叶属 *Hubeiophyllum* ⋯⋯⋯⋯⋯⋯⋯⋯⋯⋯⋯⋯⋯⋯⋯⋯⋯⋯⋯⋯⋯⋯⋯⋯⋯⋯ 45/224
湖南木贼属 *Hunanoequisetum* ⋯⋯⋯⋯⋯⋯⋯⋯⋯⋯⋯⋯⋯⋯⋯⋯⋯⋯⋯⋯⋯⋯⋯ 46/224
花穗杉果属 *Amentostrobus* ⋯⋯⋯⋯⋯⋯⋯⋯⋯⋯⋯⋯⋯⋯⋯⋯⋯⋯⋯⋯⋯⋯⋯⋯⋯⋯ 6/180
华丽羊齿属 *Abrotopteris* ⋯⋯⋯⋯⋯⋯⋯⋯⋯⋯⋯⋯⋯⋯⋯⋯⋯⋯⋯⋯⋯⋯⋯⋯⋯⋯⋯ 1/175
华脉蕨属 *Abropteris* ⋯⋯⋯⋯⋯⋯⋯⋯⋯⋯⋯⋯⋯⋯⋯⋯⋯⋯⋯⋯⋯⋯⋯⋯⋯⋯⋯⋯⋯ 1/175
华美木属 *Decoroxylon* ⋯⋯⋯⋯⋯⋯⋯⋯⋯⋯⋯⋯⋯⋯⋯⋯⋯⋯⋯⋯⋯⋯⋯⋯⋯⋯⋯ 25/202
华网蕨属 *Areolatophyllum* ⋯⋯⋯⋯⋯⋯⋯⋯⋯⋯⋯⋯⋯⋯⋯⋯⋯⋯⋯⋯⋯⋯⋯⋯⋯⋯ 8/183
华夏木属 *Cathaysiodendron* ⋯⋯⋯⋯⋯⋯⋯⋯⋯⋯⋯⋯⋯⋯⋯⋯⋯⋯⋯⋯⋯⋯⋯⋯⋯ 15/191
华夏苏铁属 *Cathaysiocycas* ⋯⋯⋯⋯⋯⋯⋯⋯⋯⋯⋯⋯⋯⋯⋯⋯⋯⋯⋯⋯⋯⋯⋯⋯⋯ 15/191
华夏穗属 *Cathayanthus* ⋯⋯⋯⋯⋯⋯⋯⋯⋯⋯⋯⋯⋯⋯⋯⋯⋯⋯⋯⋯⋯⋯⋯⋯⋯⋯⋯ 15/190
华夏羊齿属 *Cathaysiopteris* ⋯⋯⋯⋯⋯⋯⋯⋯⋯⋯⋯⋯⋯⋯⋯⋯⋯⋯⋯⋯⋯⋯⋯⋯⋯ 17/192
华夏叶属 *Cathaysiophyllum* ⋯⋯⋯⋯⋯⋯⋯⋯⋯⋯⋯⋯⋯⋯⋯⋯⋯⋯⋯⋯⋯⋯⋯⋯⋯ 16/192
黄杉型木属 *Pseudotsugxylon* ⋯⋯⋯⋯⋯⋯⋯⋯⋯⋯⋯⋯⋯⋯⋯⋯⋯⋯⋯⋯⋯⋯⋯⋯ 90/273

J

鸡西叶属 *Jixia*	48/226
吉林羽叶属 *Chilinia*	20/196
汲清羊齿属 *Huangia*	45/223
假带羊齿属 *Pseudotaeniopteris*	90/272
假耳蕨属 *Pseudopolystichu*	89/272
假鳞杉属 *Pseudoullmannia*	90/273
假拟扇叶属 *Pseudorhipidopsis*	89/272
尖籽属 *Muricosperma*	68/249
间羽蕨属 *Mixopteris*	66/247
间羽叶属 *Mixophylum*	66/246
箭羽羊齿属 *Sagittopteris* Zhang E et Xiao,1985 (non Zhang S et Xiao,1987)	97/280
箭羽羊齿属 *Sagittopteris* Zhang S et Xiao,1987 (non Zhang E et Xiao,1985)	97/280
江西叶属 *Jiangxifolium*	47/225
蛟河蕉羽叶属 *Tsiaohoella*	121/306
蛟河羽叶属 *Tchiaohoella*	114/298
睫囊蕨属 *Fimbriotheca*	35/213
今野羊齿属 *Konnoa*	52/230
金藤叶属 *Stephanofolium*	110/295
荆棘果属 *Batenburgia*	10/185
荆门叶属 *Jingmenophyllum*	48/226
锯叶羊齿属 *Prionophyllopteris*	86/268

K

开平木属 *Kaipingia*	50/229
孔山羊齿属 *Kongshania*	51/230
宽甸叶属 *Kuandiania*	52/231

L

拉萨木属 *Lhassoxylon*	53/232
刺蕨属 *Acanthopteris*	2/176
刺鳞木属 *Spinolepidodendron*	108/293
刺羊齿属 *Aculeovinea*	3/178
刺枝属 *Acanthocladus*	2/176
乐平蕨属 *Lopinopteris*	59/238
乐平苏铁属 *Lepingia*	53/231
类紫杉籽属 *Parataxospermum*	77/259
李氏蕨属 *Lixotheca*	57/236
李氏木属 *Lioxylon*	56/236
李氏苏铁属 *Liella*	55/234

李氏穗属 *Leeites*	52/231
丽花属 *Callianthus*	13/189
连山草属 *Lianshanus*	53/232
两瓣叶属 *Bilobphyllum*	12/187
两列羊齿属 *Distichopteris*	28/204
辽宁木属 *Liaoningoxylon*	54/233
辽宁缘蕨属 *Liaoningdicotis*	54/232
辽宁枝属 *Liaoningocladus*	54/233
辽西草属 *Liaoxia*	55/234
鳞籽属 *Squamocarpus*	109/293
灵乡叶属 *Lingxiangphyllum*	56/235
菱羊齿属 *Rhomboidopteris*	95/278
柳林果属 *Liulinia*	57/236
六叶属 *Hexaphyllum*	43/222
龙井叶属 *Longjingia*	58/237
龙蕨属 *Dracopteris*	29/205
鹿角蕨属 *Cervicornus*	17/193
卵叶属 *Yuania*	129/315
轮叶蕨属 *Rotafolia*	96/279
裸囊穗属 *Nudasporestrobus*	71/252
吕蕨属 *Luereticopteris*	60/239

M

毛茛叶属 *Ranunculophyllum*	93/275
梅氏叶属 *Meia*	63/243
密叶属 *Myriophyllum*	68/249
膜质叶属 *Membranifolia*	63/243

N

那琳壳斗属 *Norinia*	71/252
南票叶属 *Nanpiaophyllum*	69/249
南漳叶属 *Nanzhangophyllum*	69/250
拟爱博拉契蕨属 *Eboraciopsis*	30/206
拟安杜鲁普蕨属 *Amdrupiopsis*	5/180
拟瓣轮叶属 *Lobatannulariopsis*	57/237
拟齿叶属 *Paratingia*	77/259
拟刺葵属 *Phoenicopsis*	79/261
拟刺葵(斯蒂芬叶)亚属 *Phoenicopsis* (*Stephenophyllum*)	79/261
拟带枝属 *Taeniocladopsis*	113/297
拟丁氏蕨穗属 *Paratingiostachya*	78/259
拟根茎属 *Rhizomopsis*	95/278
拟蕨属 *Pteridiopsis*	91/274

拟裂鞘叶属 *Schizoneuropsis* ·········· 98/281
拟片叶苔属 *Riccardiopsis* ·········· 96/278
拟斯托加枝属 *Parastorgaardis* ·········· 77/258
拟苏铁籽属 *Cycadeoidispermum* ·········· 23/199
拟楔羊齿属 *Parasphenopteris* ·········· 76/258
拟楔叶属 *Parasphenophyllum* ·········· 76/258
拟斜羽叶属 *Plagiozamiopsis* ·········· 83/265
宁夏叶属 *Ningxiaphyllum* ·········· 71/252

P

耙羊齿属 *Rastropteris* ·········· 93/276
潘广叶属 *Pankuangia* ·········· 74/256
潘氏果属 *Pania* ·········· 74/255
盘囊蕨属 *Discalis* ·········· 27/204
普通蕨属 *Coenosophyton* ·········· 22/198

Q

奇脉叶属 *Mironeura* ·········· 66/246
奇羊齿属 *Aetheopteris* ·········· 4/179
奇叶属 *Acthephyllum* ·········· 3/177
奇异木属 *Allophyton* ·········· 5/180
奇异羊齿属 *Mirabopteris* ·········· 65/245
奇异羊齿属 *Paradoxopteris* Mi et Liu,1977 (non Hirmer,1927) ·········· 75/257
奇异叶属 *Adoketophyllum* ·········· 4/178
奇异羽叶属 *Thaumatophyllum* ·········· 115/300
契丹穗属 *Khitania* ·········· 50/229
钱耐果属 *Chaneya* ·········· 18/194
黔囊属 *Guizhoua* ·········· 39/217
秦岭羊齿属 *Qinlingopteris* ·········· 91/274
琼海叶属 *Qionghaia* ·········· 92/274
全泽米属 *Holozamites* ·········· 44/222

R

热河似查米亚属 *Rehezamites* ·········· 93/276
日蕨属 *Rireticopteris* ·········· 96/279

S

三棱果属 *Triqueteria* ·········· 120/305
三裂穗属 *Tricrananthus* ·········· 119/305
三脉蕨属 *Trinerviopteris* ·········· 120/305

三网羊齿属 Tricoemplectopteris	119/304
山西木属 Shanxioxylon	100/283
山西枝属 Shanxicladus	100/283
上园草属 Shangyuania	99/282
神州聚囊属 Shenzhoutheca	102/285
神州叶属 Shenzhouphyllum	101/284
神州籽属 Shenzhouspermum	102/285
沈括叶属 Shenkuoia	101/284
沈氏蕨属 Shenea	100/284
肾叶属 Renifolium	94/277
始查米苏铁属 Primozamia	86/268
始鳞木属 Eolepidodendron	31/209
始木兰属 Archimagnolia	8/182
始拟银杏属 Primoginkgo	85/268
始苹果属 Archimalus	8/183
始水松属 Eoglyptostrobus	30/207
始苏铁属 Primocycas	85/267
始团扇蕨属 Eogonocormus Deng,1995 (non Deng,1997)	30/207
始团扇蕨属 Eogonocormus Deng,1997 (non Deng,1995)	31/208
始叶蕨属 Eophyllophyton	32/210
始叶羊齿属 Eophyllogonium	32/209
始羽蕨属 Eogymnocarpium	31/208
似八角属 Illicites	46/225
似百合属 Lilites	55/235
似叉苔属 Metzgerites	65/245
似齿囊蕨属 Odontosorites	72/253
似杜仲属 Eucommioites	33/210
似狗尾草属 Setarites	99/282
似画眉草属 Eragrosites	33/210
似金星蕨属 Thelypterites	116/201
似茎状地衣属 Foliosites	36/213
似卷囊蕨属 Speirocarpites	106/290
似克鲁克蕨属 Klukiopsis	51/229
似兰属 Orchidites	72/253
似蓼属 Polygonites Wu S Q,1999 (non Saporta,1865)	84/266
似轮叶属 Annularites	7/181
似木麻黄属 Casuarinites	14/189
似南五味子属 Kadsurrites	50/228
似槭树属 Acerites	2/176
似提灯藓属 Mnioites	66/247
似铁线莲叶属 Clematites	21/197
似乌头属 Aconititis	2/177
似阴地蕨属 Botrychites	13/188
似雨蕨属 Gymnogrammitites	40/218

似圆柏属 *Sabinites* ⋯⋯ 97/280
似远志属 *Polygatites* ⋯⋯ 83/266
梳囊属 *Lophotheca* ⋯⋯ 58/238
束脉蕨属 *Symopteris* ⋯⋯ 111/296
束羊齿属 *Fascipteris* ⋯⋯ 34/211
刷囊属 *Strigillotheca* ⋯⋯ 111/295
双列囊蕨属 *Bifariusotheca* ⋯⋯ 12/187
双囊芦穗属 *Shuangnangostachya* ⋯⋯ 102/286
双生叶属 *Geminofoliolum* ⋯⋯ 37/215
双网羊齿属 *Bicoemplectopteris* ⋯⋯ 12/187
水城蕨属 *Shuichengella* ⋯⋯ 102/286
水杉属 *Metasequoia* ⋯⋯ 64/244
斯芬克斯籽属 *Sphinxia* Li, Hilton et Hemsley, 1997 (non Reid et Chandler, 1933) ⋯⋯ 108/292
斯氏木属 *Szeioxylon* ⋯⋯ 112/297
斯氏鞘叶属 *Siella* ⋯⋯ 103/286
斯氏松属 *Szecladia* ⋯⋯ 112/297
四叶属 *Tetrafolia* ⋯⋯ 114/299
苏铁鳞叶属 *Cycadolepophyllum* ⋯⋯ 24/200
苏铁缘蕨属 *Cycadicotis* ⋯⋯ 23/199
苏铁籽属 *Dioonocarpus* ⋯⋯ 27/204
穗蕨属 *Stachyophyton* ⋯⋯ 110/294
穗藓属 *Stachybryolites* ⋯⋯ 109/204

T

塔状木属 *Minarodendron* ⋯⋯ 65/245
太平场蕨属 *Taipingchangella* ⋯⋯ 113/298
太阳花属 *Solaranthus* ⋯⋯ 106/290
太原蕨属 *Taiyuanitheca* ⋯⋯ 113/298
天山羊齿属 *Tianshanopteris* ⋯⋯ 117/302
天石蕨属 *Szea* ⋯⋯ 112/296
天石枝属 *Tianshia* ⋯⋯ 117/302
田氏木属 *Tianoxylon* ⋯⋯ 116/301
条形蕨属 *Lorophyton* ⋯⋯ 59/239
条叶属 *Vittifoliolum* ⋯⋯ 121/307
铁花属 *Cycadostrobilus* ⋯⋯ 24/200
桐庐籽属 *Tonglucarpus* ⋯⋯ 118/303
铜川叶属 *Tongchuanophyllum* ⋯⋯ 118/303
托克逊蕨属 *Toksunopteris* ⋯⋯ 117/302

W

网格蕨属 *Reteophlebis* ⋯⋯ 94/277
网延羊齿属 *Reticalethopteris* ⋯⋯ 94/277

网叶属 *Linophyllum*	56/235
文山蕨属 *Wenshania*	122/308
乌图布拉克蕨属 *Wutubulaka*	122/308
乌云花属 *Wuyunanthus*	123/309
无锡蕨属 *Wuxia*	123/308

X

西屯蕨属 *Xitunia*	127/313
细毛蕨属 *Ciliatopteris*	20/197
细轴始蕨属 *Tenuisa*	114/299
夏家街蕨属 *Xiajiajienia*	123/309
仙籽属 *Sphinxiocarpon*	108/292
先骕蕨属 *Huia*	46/224
纤木属 *Chamaedendron*	18/193
纤细蕨属 *Filiformorama*	35/213
香溪叶属 *Hsiangchiphyllum*	44/222
小蛟河蕨属 *Chiaohoella*	20/196
小伞属 *Sciadocillus*	98/281
楔叶拜拉花属 *Sphenobaieroanthus*	107/291
楔叶拜拉枝属 *Sphenobaierocladus*	107/291
楔栉羊齿属 *Sphenopecopteris*	107/292
新孢穗属 *Neostachya* Wang,1977	70/251
新常富籽属 *Nystroemia*	72/253
新疆蕨属 *Xinjiangopteris* Wu S Q et Zhou,1986 (non Wu S Z,1983)	125/311
新疆蕨属 *Xinjiangopteris* Wu S Z,1983 (non Wu S Q et Zhou,1986)	126/312
新疆木属 *Xinjiangophyton*	125/311
新科达属 *Neocordaites*	70/250
新龙叶属 *Xinlongia*	126/312
新龙羽叶属 *Xinlongophyllum*	126/312
新轮叶属 *Neoannularia*	69/250
新准大羽羊齿属 *Neogigantopteridium*	70/251
星壳斗属 *Astrocupulites*	9/184
星学花属 *Xingxueanthus*	124/310
星学花序属 *Xingxueina*	124/310
星学叶属 *Xingxuephyllum*	125/311
兴安叶属 *Xinganphyllum*	124/309
袖套杉属 *Manica*	61/240
袖套杉(长岭杉)亚属 *Manica* (*Chanlingia*)	61/240
袖套杉(袖套杉)亚属 *Manica* (*Manica*)	61/240
徐氏蕨属 *Hsuea*	44/223

Y

牙羊齿属 *Dentopteris* ·· 27/203
雅观木属 *Perisemoxylon* ··· 79/260
雅蕨属 *Pavoniopteris* ··· 78/259
亚洲羊齿属 *Asiopteris* ··· 9/184
亚洲叶属 *Asiatifolium* ··· 8/183
延吉叶属 *Yanjiphyllum* ·· 127/314
燕辽杉属 *Yanliaoa* ·· 128/314
羊齿缘蕨属 *Filicidicotis* ··· 35/212
杨氏木属 *Yangzunyia* ··· 127/313
耶氏蕨属 *Jaenschea* ·· 47/225
叶茎属 *Phylladendroid* ·· 81/262
疑麻黄属 *Amphiephedra* ··· 6/181
义马果属 *Yimaia* ·· 128/314
义县叶属 *Yixianophyllum* ·· 128/315
异麻黄属 *Alloephedra* ··· 5/179
异枝蕨属 *Metacladophyton* ··· 64/244
隐囊蕨属 *Celathega* ··· 17/193
隐羊齿属 *Cryptonoclea* ··· 23/199
缨囊属 *Distichotheca* ·· 28/205
永仁叶属 *Yungjenophyllum* ······································· 129/316
羽叶大羽羊齿属 *Pinnagigantopteris* ······························ 82/264
羽叶单网羊齿属 *Pinnagigantonoclea* ······························ 81/263
羽枝属 *Pinnatiramosus* ··· 83/265
玉光蕨属 *Yuguangia* ·· 129/315
豫囊蕨属 *Henanotheca* ··· 43/221
原大羽羊齿属 *Progigantopteris* ··································· 87/269
原单网羊齿属 *Progigantonoclea* ·································· 87/269
原始金松型木属 *Protosciadopityoxylon* ························· 89/271
原始蕨属 *Protopteridophyton* ····································· 88/271
原始水松型木属 *Protoglyptostroboxylon* ························ 88/271
原始银杏型木属 *Proginkgoxylon* ································· 87/270
原苏铁属 *Procycas* ··· 86/269
云蕨属 *Yunia* ··· 130/316

Z

贼木属 *Phoroxylon* ·· 80/262
窄叶属 *Angustiphyllum* ··· 6/181
粘合囊蕨属 *Cohaerensitheca* ······································· 22/198
掌裂蕨属 *Catenalis* ··· 14/189
正理蕨属 *Zhenglia* ··· 131/317

正三角籽属 *Deltoispermum* ·········· 26/202
郑氏叶属 *Zhengia* ·········· 130/317
枝带羊齿属 *Cladotaeniopteris* ·········· 21/197
织羊齿属 *Emplectopteris* ·········· 30/207
中国篦羽叶属 *Sinoctenis* ·········· 103/287
中国古螺纹木属 *Sinopalaeospiroxylon* ·········· 104/288
中国似查米亚属 *Sinozamites* ·········· 105/289
中国羊齿属 *Cathaiopteridium* ·········· 14/190
中国叶属 *Sinophyllum* ·········· 105/289
中华古果属 *Sinocarpus* ·········· 103/287
中华羊齿属 *Cathaysiopteridium* ·········· 16/192
中华缘蕨属 *Sinodicotis* ·········· 104/287
中间苏铁属 *Mediocycas* ·········· 62/242
中州蕨属 *Zhongzhoupteris* ·········· 131/318
中州籽属 *Zhongzhoucarus* ·········· 131/317
钟囊属 *Tongshania* ·········· 119/304
周氏籽属 *Zhouia* ·········· 131/318
帚羽叶属 *Scoparia* ·········· 98/281
朱氏囊蕨属 *Zhutheca* ·········· 132/318
侏罗木兰属 *Juramagnolia* ·········· 49/228
侏罗球果属 *Jurastrobus* ·········· 49/228
侏罗缘蕨属 *Juradicotis* ·········· 49/227
准爱河羊齿属 *Aipteridium* ·········· 4/179
准噶尔蕨属 *Junggaria* ·········· 48/227
准束羊齿属 *Fascipteridium* ·········· 33/211
准枝脉蕨属 *Cladophlebidium* ·········· 21/197

附录2 种名索引

[按中文名称的汉语拼音升序排列,属名或种名后为页码(中文记录页码/英文记录页码)]

A

凹尖枝属 *Koilosphenus* ········ 51/230
  楔裂凹尖枝 *Koilosphenus cuneifolius* ········ 51/230
  ?凹尖枝(未定种) ?*Koilosphenus* sp. ········ 51/230

B

白豆杉型木属 *Pseudotaxoxylon* ········ 90/273
  中国白豆杉型木 *Pseudotaxoxylon chinensis* ········ 90/273
白果叶属 *Baiguophyllum* ········ 9/184
  利剑白果叶 *Baiguophyllum lijianum* ········ 10/184
苞片蕨属 *Bracteophyton* ········ 13/188
  变异苞片蕨 *Bracteophyton variatum* ········ 13/188
抱囊蕨属 *Amplectosporangium* ········ 6/181
  江油抱囊蕨 *Amplectosporangium jiagyouense* ········ 6/181
鲍斯木属 *Boseoxylon* ········ 12/188
  安德鲁斯鲍斯木 *Boseoxylon andrewii* ········ 13/188
北票果属 *Beipiaoa* ········ 10/185
  强刺北票果 *Beipiaoa spinosa* ········ 10/185
  小北票果 *Beipiaoa parva* ········ 10/185
  圆形北票果 *Beipiaoa rotunda* ········ 10/185
贝叶属 *Conchophyllum* ········ 22/198
  李氏贝叶 *Conchophyllum richthofenii* ········ 22/199
本内缘蕨属 *Bennetdicotis* ········ 11/186
  本内缘蕨(sp. indet.) *Bennetdicotis* sp. indet. ········ 11/186
本溪羊齿属 *Benxipteris* ········ 11/186
  尖叶本溪羊齿 *Benxipteris acuta* ········ 11/186
  密脉本溪羊齿 *Benxipteris densinervis* ········ 11/186
  裂缺本溪羊齿 *Benxipteris partita* ········ 11/186
  多态本溪羊齿 *Benxipteris polymorpha* ········ 11/187
篦囊属 *Pectinangium* ········ 78/260
  披针篦囊 *Pectinangium lanceolatum* ········ 78/260
扁囊蕨属 *Demersatheca* ········ 26/202
  紧贴扁囊蕨 *Demersatheca contigua* ········ 26/202
变态鳞木属 *Metalepidodendron* ········ 64/244
  中国变态鳞木 *Metalepidodendron sinensis* ········ 64/244
  下板城变态鳞木 *Metalepidodendron xiabanchengensis* ········ 64/244

C

蔡氏蕨属 *Tsaia* ··· 120/306
    细齿蔡氏蕨 *Tsaia denticulata* ··· 121/306
蔡耶羊齿属 *Zeillerpteris* ··· 130/316
    云南蔡耶羊齿 *Zeillerpteris yunnanensis* ··· 130/316
缠绕蕨属 *Helicophyton* ·· 42/220
    二叉缠绕蕨 *Helicophyton dichotomum* ·· 42/220
长穗属 *Longostachys* ·· 58/237
    宽叶长穗 *Longostachys latisporophyllus* ··· 58/238
长武蕨属 *Changwuia* ··· 18/194
    施魏策尔长武蕨 *Changwuia schweitzeri* ·· 18/194
长阳木属 *Changyanophyton* ··· 19/194
    湖北长阳木 *Changyanophyton hupeiense* ·· 19/195
朝阳序属 *Chaoyangia* ·· 19/195
    梁氏朝阳序 *Chaoyangia liangii* ··· 19/195
城子河叶属 *Chengzihella* ··· 19/195
    倒卵城子河叶 *Chengzihella obovata* ··· 19/195
翅鳞籽属 *Squarmacarpus* ·· 109/293
    楔形翅鳞籽 *Squarmacarpus cuneiformis* ··· 109/293
翅叶属 *Neurophyllites* ·· 70/251
    栉状翅叶 *Neurophyllites pecopteroides* ··· 71/251
翅籽属 *Semenalatum* ··· 99/282
    珍贵翅籽 *Semenalatum paucum* ··· 99/282
垂饰杉属 *Stalagma* ··· 110/294
    翅籽垂饰杉 *Stalagma samara* ··· 110/294

D

大囊蕨属 *Gigantotheca* ·· 38/216
    奇异大囊蕨 *Gigantotheca paradoxa* ·· 38/216
大箐羽叶属 *Tachingia* ·· 113/297
    大箐羽叶 *Tachingia pinniformis* ·· 113/297
大舌羊齿属 *Macroglossopteris* ··· 60/240
    李氏大舌羊齿 *Macroglossopteris leeiana* ··· 60/240
大同叶属 *Datongophyllum* ·· 25/201
    长柄大同叶 *Datongophyllum longipetiolatum* ·· 25/201
    大同叶 (未定种) *Datongophyllum* sp. ··· 25/201
大叶羊齿属 *Megalopteris* Schenk, 1883 (non Andrews E B, 1875) ·· 62/242
    烟叶大叶羊齿 *Megalopteris nicotianaefolia* ·· 63/243
大羽羊齿属 *Gigantopteris* ··· 38/216
    烟叶大羽羊齿 *Gigantopteris nicotianaefolia* ·· 38/216
带囊蕨属 *Gigantonomia* ·· 38/215

福建带囊蕨（单网羊齿）*Gigantonomia* (*Gigatonoclea*) *fukienensis* ········· 38/216
带状鳞穗属 *Loroderma* ········· 59/239
    河南带状鳞穗 *Loroderma henania* ········· 59/239
单网羊齿属 *Gigantonoclea* ········· 37/215
    波缘单网羊齿 *Gigantonoclea lagrelii* ········· 37/215
单叶大羽羊齿属 *Monogigantopteris* ········· 68/248
    格网单叶大羽羊齿 *Monogigantopteris clathroreticulatus* ········· 68/248
    密网单叶大羽羊齿 *Monogigantopteris densireticulatus* ········· 68/248
单叶单网羊齿属 *Monogigantonoclea* ········· 67/247
    芋叶单叶单网羊齿 *Monogigantonoclea colocasifolia* ········· 67/247
    圆形单叶单网羊齿 *Monogigantonoclea rotundifolia* ········· 67/248
    阔卵单叶单网羊齿 *Monogigantonoclea latiovata* ········· 67/248
    巨齿单叶单网羊齿 *Monogigantonoclea grandidenia* ········· 67/248
    似槭单叶单网羊齿 *Monogigantonoclea aceroides* ········· 67/248
导管羊齿属 *Vasovinea* ········· 121/306
    田氏导管羊齿 *Vasovinea tianii* ········· 121/307
道虎沟叶状体属 *Daohugouthallus* ········· 24/201
    细毛道虎沟叶状体 *Daohugouthallus ciliiferus* ········· 25/201
灯笼蕨属 *Denglongia* ········· 26/203
    湖北灯笼蕨 *Denglongia hubeiensis* ········· 27/203
登封籽属 *Dengfengia* ········· 26/203
    双翅登封籽 *Dengfengia bifurcata* ········· 26/203
碟囊属 *Lopadiangium* ········· 58/238
    齿缘碟囊 *Lopadiangium acmodontum* ········· 58/238
蝶叶属 *Papilionifolium* ········· 75/256
    徐氏蝶叶 *Papilionifolium hsui* ········· 75/256
丁氏羊齿属 *Tingia* ········· 117/302
    石炭丁氏羊齿 *Tingia carbonica* ········· 117/302
东北穗属 *Manchurostachys* ········· 60/240
    裂鞘叶东北穗 *Manchurostachys manchuriensis* ········· 60/240
渡口痕木属 *Dukouphyton* ········· 29/206
    较小渡口痕木 *Dukouphyton minor* ········· 29/206
渡口叶属 *Dukouphyllum* ········· 29/206
    诺格拉齐蕨型渡口叶 *Dukouphyllum noeggerathioides* ········· 29/206
对枝柏属 *Ditaxocladus* ········· 28/205
    扁叶对枝柏 *Ditaxocladus planiphyllus* ········· 29/205
多瓣蕨属 *Polypetalophyton* ········· 84/267
    五峰多瓣蕨 *Polypetalophyton wufengensis* ········· 85/267
多囊枝属 *Polythecophyton* ········· 85/267
    下弯多囊枝 *Polythecophyton demissum* ········· 85/267
多枝蕨属 *Ramophyton* ········· 92/275
    吉维特多枝蕨 *Ramophyton givetianum* ········· 92/275

E

耳叶属 *Otofolium* 73/254
    多形耳叶 *Otofolium polymorphum* 73/254
    卵耳叶 *Otofolium ovatum* 73/254

F

榧型枝属 *Torreyocladus* 119/304
    明显榧型枝 *Torreyocladus spectabilis* 119/304
缝鞘杉属 *Suturovagina* 111/295
    过渡缝鞘杉 *Suturovagina intermedia* 111/295
辐射叶属 *Actinophyllus* Xiao, 1985 (non *Actinophyllum* Phillips, 1848) 3/177
    科达状辐射叶 *Actinophyllus cordaioides* 3/178
辐叶属 *Radiatifolium* 92/275
    大辐叶 *Radiatifolium magnusum* 92/275
福建羊齿属 *Fujianopteris* 36/214
    闽福建羊齿 *Fujianopteris fukianensis* 36/214
    狭角福建羊齿 *Fujianopteris angustiangla* 36/214
    枝脉福建羊齿 *Fujianopteris cladonervis* 36/214
    中间福建羊齿 *Fujianopteris intermedia* 37/214
副开通尼亚属 *Paracaytonia* 75/256
    洪涛副开通尼亚 *Paracaytonia hongtaoi* 75/256
副镰羽叶属 *Paradrepanozamites* 76/257
    大道场副镰羽叶 *Paradrepanozamites dadaochangensis* 76/257
副球果属 *Paraconites* 75/257
    伸长副球果 *Paraconites longifolius* 75/257

G

甘肃芦木属 *Gansuphyllite* 37/215
    多脉甘肃芦木 *Gansuphyllite multivervis* 37/215
赣囊蕨属 *Jiangxitheca* 47/226
    新安赣囊蕨 *Jiangxitheca xinanensis* 47/226
根状茎属 *Rhizoma* 95/278
    椭圆形根状茎 *Rhizoma elliptica* 95/278
钩蕨属 *Hamatophyton* 41/220
    轮生钩蕨 *Hamatophyton verticillatum* 42/220
古果属 *Archaefructus* 7/182
    辽宁古果 *Archaefructus liaoningensis* 7/182
古买麻藤属 *Palaeognetaleaana* 73/255
    吉祥古买麻藤 *Palaeognetaleaana auspicia* 74/255
古木蕨属 *Gumuia* 40/218

曲轴古木蕨 *Gumuia zyzzata* ·· 40/218
古银杏型木属 *Palaeoginkgoxylon* ·· 73/254
    周氏古银杏型木 *Palaeoginkgoxylon zhoui* ·· 73/254
古舟藤属 *Palaeoskapha* ··· 74/255
    四川古舟藤 *Palaeoskapha sichuanensis* ·· 74/255
管子麻黄属 *Siphonospermum* ··· 106/289
    简单管子麻黄 *Siphonospermum simplex* ·· 106/290
广南蕨属 *Guangnania* ·· 39/217
    楔形广南蕨 *Guangnania cuneata* ··· 39/217
广西叶属 *Guangxiophyllum* ·· 39/217
    上思广西叶 *Guangxiophyllum shangsiense* ·· 39/217
贵州木属 *Guizhouoxylon* ··· 40/218
    大河边贵州木 *Guizhouoxylon dahebianense* ··· 40/218

H

哈勒角籽属 *Hallea* Mathews, 1947—1948（non Yang et Wu, 2006）······················· 40/219
    北京哈勒角籽 *Hallea pekinensis* ··· 41/219
哈瑞士叶属 *Tharrisia* ·· 115/299
    迪纳塞尔哈瑞士叶 *Tharrisia dinosaurensis* ·· 115/299
    侧生哈瑞士叶 *Tharrisia lata* ·· 115/300
    优美哈瑞士叶 *Tharrisia spectabilis* ·· 115/300
哈氏蕨属 *Halleophyton* ·· 41/219
    纸厂哈氏蕨 *Halleophyton zhichangense* ·· 41/219
和丰孢穗属 *Hefengistrobus* ··· 42/220
    二歧和丰孢穗 *Hefengistrobus bifurcus* ·· 42/220
河南羊齿属 *Henanopteris* ·· 43/221
    披针河南羊齿 *Henanopteris lanceolatus* ·· 43/221
河南叶属 *Henanophyllum* ·· 42/221
    掌河南叶 *Henanophyllum palamifolium* ·· 43/221
赫勒单网羊齿属 *Hallea* Yang et Wu, 2006（non Mathews, 1947—1948）················ 41/219
    登封赫勒单网羊齿 *Hallea dengfengensis* ··· 41/219
湖北蕨属 *Hubeiia* ··· 45/223
    叉叶湖北蕨 *Hubeiia dicrofollia* ··· 45/223
湖北叶属 *Hubeiophyllum* ··· 45/224
    楔形湖北叶 *Hubeiophyllum cuneifolium* ·· 45/224
    狭细湖北叶 *Hubeiophyllum angustum* ·· 45/224
湖南木贼属 *Hunanoequisetum* ··· 46/224
    浏阳湖南木贼 *Hunanoequisetum liuyangense* ··· 46/225
花穗杉果属 *Amentostrobus* ··· 6/180
    花穗杉果(sp. indet.) *Amentostrobus* sp. indet. ··· 6/180
华丽羊齿属 *Abrotopteris* ·· 1/175
    贵州华丽羊齿 *Abrotopteris guizhouensis* ··· 1/175
华脉蕨属 *Abropteris* ··· 1/175

弗吉尼亚华脉蕨 Abropteris virginiensis	1/175
永仁华脉蕨 Abropteris yongrenensis	1/175
华美木属 Decoroxylon	25/202
朝阳华美木 Decoroxylon chaoyangense	25/202
华网蕨属 Areolatophyllum	8/183
青海华网蕨 Areolatophyllum qinghaiense	8/183
华夏木属 Cathaysiodendron	15/191
不定华夏木 Cathaysiodendron incertum	16/191
朱森华夏木 Cathaysiodendron chuseni	16/191
南票华夏木 Cathaysiodendron nanpiaoense	16/192
华夏苏铁属 Cathaysiocycas	15/191
直脉华夏苏铁 Cathaysiocycas rectanervis	15/191
华夏穗属 Cathayanthus	15/190
少鳞华夏穗 Cathayanthus ramentrarus	15/190
中国华夏穗 Cathayanthus sinensis	15/190
华夏羊齿属 Cathaysiopteris	17/192
怀特华夏羊齿 Cathaysiopteris whitei	17/193
华夏叶属 Cathaysiophyllum	16/192
裂瓣华夏叶 Cathaysiophyllum lobifolium	16/192
黄杉型木属 Pseudotsugxylon	90/273
平庄黄杉型木 Pseudotsugxylon pingzhangensis	90/273

J

鸡西叶属 Jixia	48/226
羽裂鸡西叶 Jixia pinnatipartita	48/227
吉林羽叶属 Chilinia	20/196
篦羽叶型吉林羽叶 Chilinia ctenioides	20/196
汲清羊齿属 Huangia	45/223
椭圆汲清羊齿 Huangia elliptica	45/223
假带羊齿属 Pseudotaeniopteris	90/272
鱼形假带羊齿 Pseudotaeniopteris piscatorius	90/272
假耳蕨属 Pseudopolystichu	89/272
白垩假耳蕨 Pseudopolystichu cretaceum	89/272
假鳞杉属 Pseudoullmannia	90/273
类麦假鳞杉 Pseudoullmannia frumentarioides	90/273
类纹假鳞杉 Pseudoullmannia bronnioides	91/273
假拟扇叶属 Pseudorhipidopsis	89/272
宽叶假拟扇叶 Pseudorhipidopsis brevicaulis	89/272
尖籽属 Muricosperma	68/249
贵州尖籽 Muricosperma guizhouensis	68/249
间羽蕨属 Mixopteris	66/247
插入间羽蕨 Mixopteris intercalaris	66/247
间羽叶属 Mixophylum	66/246

简单间羽叶 *Mixophylum simplex* ……………………………………………… 66/246
箭羽羊齿属 *Sagittopteris* Zhang E et Xiao,1985（non Zhang S et Xiao,1987）……… 97/280
　　　戟形箭羽羊齿 *Sagittopteris belemnopteroides* Zhang E et Xiao,1985
　　　（non Zhang S et Xiao,1987） ……………………………………………… 97/280
箭羽羊齿属 *Sagittopteris* Zhang S et Xiao,1987（non Zhang E et Xiao,1985）……… 97/280
　　　戟形箭羽羊齿 *Sagittopteris belemnopteroides* Zhang S et Xiao,1987
　　　（non Zhang E et Xiao,1985） ……………………………………………… 98/281
江西叶属 *Jiangxifolium* …………………………………………………………… 47/225
　　　短尖头江西叶 *Jiangxifolium mucronatum* ………………………………… 47/225
　　　细齿江西叶 *Jiangxifolium denticulatum* …………………………………… 47/226
蛟河蕉羽叶属 *Tsiaohoella* ………………………………………………………… 121/306
　　　奇异蛟河蕉羽叶 *Tsiaohoella mirabilis* ……………………………………… 121/306
　　　新似查米亚型蛟河蕉羽叶 *Tsiaohoella neozamioides* ……………………… 121/306
蛟河羽叶属 *Tchiaohoella* ………………………………………………………… 114/298
　　　奇异蛟河羽叶 *Tchiaohoella mirabilis* ……………………………………… 114/299
　　　蛟河羽叶（未定种）*Tchiaohoella* sp. ……………………………………… 114/299
睫囊蕨属 *Fimbriotheca* …………………………………………………………… 35/213
　　　毛状睫囊蕨 *Fimbriotheca tomentosa* ……………………………………… 35/213
今野羊齿属 *Konnoa* ……………………………………………………………… 52/230
　　　高丽今野羊齿 *Konnoa koraiensis* …………………………………………… 52/231
　　　本溪今野羊齿 *Konnoa penchihuensis* ……………………………………… 52/231
金藤叶属 *Stephanofolium* ………………………………………………………… 110/295
　　　卵形金藤叶 *Stephanofolium ovatiphyllum* ………………………………… 110/295
荆棘果属 *Batenburgia* …………………………………………………………… 10/185
　　　萨克马尔荆棘果 *Batenburgia sakmarica* …………………………………… 10/185
荆门叶属 *Jingmenophyllum* ……………………………………………………… 48/226
　　　西河荆门叶 *Jingmenophyllum xiheense* …………………………………… 48/226
锯叶羊齿属 *Prionophyllopteris* …………………………………………………… 86/268
　　　多刺锯叶羊齿 *Prionophyllopteris spiniformis* ……………………………… 86/268

K

开平木属 *Kaipingia* ……………………………………………………………… 50/229
　　　中国开平木 *Kaipingia sinica* ………………………………………………… 50/229
孔山羊齿属 *Kongshania* ………………………………………………………… 51/230
　　　类连生孔山羊齿 *Kongshania synangioides* ………………………………… 51/230
宽甸叶属 *Kuandiania* …………………………………………………………… 52/231
　　　粗茎宽甸叶 *Kuandiania crassicaulis* ………………………………………… 52/231

L

拉萨木属 *Lhassoxylon* …………………………………………………………… 53/232
　　　阿普特拉萨木 *Lhassoxylon aptianum* ……………………………………… 53/232
刺蕨属 *Acanthopteris* ……………………………………………………………… 2/176

高腾刺蕨 *Acanthopteris gothani*	2/176
刺鳞木属 *Spinolepidodendron*	108/293
杭州刺鳞木 *Spinolepidodendron hangzhouense*	109/293
刺羊齿属 *Aculeovinea*	3/178
云贵刺羊齿 *Aculeovinea yunguiensis*	4/178
刺枝属 *Acanthocladus*	2/176
木质刺枝 *Acanthocladus xyloides*	2/176
乐平蕨属 *Lopinopteris*	59/238
插入乐平蕨 *Lopinopteris intercalata*	59/239
乐平苏铁属 *Lepingia*	53/231
缺顶乐平苏铁 *Lepingia emarginata*	53/232
类紫杉籽属 *Parataxospermum*	77/259
太原类紫杉籽 *Parataxospermum taiyuanesis*	77/259
李氏蕨属 *Lixotheca*	57/236
二叠李氏蕨（枝脉蕨）*Lixotheca* (*Cladophlebis*) *permica*	57/237
李氏木属 *Lioxylon*	56/236
辽宁李氏木 *Lioxylon liaoningense*	56/236
李氏苏铁属 *Liella*	55/234
奇异李氏苏铁 *Liella mirabilis*	55/234
李氏穗属 *Leeites*	52/231
椭圆李氏穗 *Leeites oblongifolis*	52/231
丽花属 *Callianthus*	13/189
迪拉丽花 *Callianthus dilae*	14/189
连山草属 *Lianshanus*	53/232
连山草（sp. indet.）*Lianshanus* sp. indet.	53/232
两瓣叶属 *Bilobphyllum*	12/187
丰城两瓣叶 *Bilobphyllum fengchengensis*	12/187
两列羊齿属 *Distichopteris*	28/204
异常两列羊齿 *Distichopteris heteropinna*	28/205
辽宁木属 *Liaoningoxylon*	54/233
朝阳辽宁木 *Liaoningoxylon chaoyangehse*	54/233
辽宁缘蕨属 *Liaoningdicotis*	54/232
辽宁缘蕨（sp. indet.）*Liaoningdicotis* sp. indet.	54/233
辽宁枝属 *Liaoningocladus*	54/233
薄氏辽宁枝 *Liaoningocladus boii*	54/233
辽西草属 *Liaoxia*	55/234
陈氏辽西草 *Liaoxia chenii*	55/234
鳞籽属 *Squamocarpus*	109/293
蝶形鳞籽 *Squamocarpus papilioformis*	109/293
灵乡叶属 *Lingxiangphyllum*	56/235
首要灵乡叶 *Lingxiangphyllum princeps*	56/235
菱羊齿属 *Rhomboidopteris*	95/278
菱羊齿 *Rhomboidopteris yongwolensis*	95/278
柳林果属 *Liulinia*	57/236

条裂柳林果 *Liulinia lacinulata*	57/236
六叶属 *Hexaphyllum*	43/222
中国六叶 *Hexaphyllum sinense*	43/222
龙井叶属 *Longjingia*	58/237
细叶龙井叶 *Longjingia gracilifolia*	58/237
龙蕨属 *Dracopteris*	29/205
辽宁龙蕨 *Dracopteris liaoningensis*	29/206
鹿角蕨属 *Cervicornus*	17/193
文山鹿角蕨 *Cervicornus wenshanensis*	17/193
卵叶属 *Yuania*	129/315
条纹卵叶 *Yuania striata*	129/315
轮叶蕨属 *Rotafolia*	96/279
松滋轮叶蕨 *Rotafolia songziensis*	96/279
裸囊穗属 *Nudasporestrobus*	71/252
宁夏裸囊穗 *Nudasporestrobus ningxicus*	71/252
吕蕨属 *Luereticopteris*	60/239
大叶吕蕨 *Luereticopteris megaphylla*	60/240

M

毛茛叶属 *Ranunculophyllum*	93/275
羽状全裂毛茛叶 *Ranunculophyllum pinnatisctum*	93/276
梅氏叶属 *Meia*	63/243
鸣山梅氏叶 *Meia mingshanensis*	63/243
大叶梅氏叶 *Meia magnifolia*	63/243
密叶属 *Myriophyllum*	68/249
山西密叶 *Myriophyllum shanxiense*	68/249
膜质叶属 *Membranifolia*	63/243
奇异膜质叶 *Membranifolia admirabilis*	63/243

N

那琳壳斗属 *Norinia*	71/252
僧帽状那琳壳斗 *Norinia cucullata*	71/252
南票叶属 *Nanpiaophyllum*	69/249
心形南票叶 *Nanpiaophyllum cordatum*	69/249
南漳叶属 *Nanzhangophyllum*	69/250
东巩南漳叶 *Nanzhangophyllum donggongense*	69/250
拟爱博拉契蕨属 *Eboraciopsis*	30/206
三裂叶拟爱博拉契蕨 *Eboraciopsis trilobifolia*	30/207
拟安杜鲁普蕨属 *Amdrupiopsis*	5/180
楔羊齿型拟安杜鲁普蕨 *Amdrupiopsis sphenopteroides*	5/180
拟瓣轮叶属 *Lobatannulariopsis*	57/237
云南拟瓣轮叶 *Lobatannulariopsis yunnanensis*	57/237

拟齿叶属 *Paratingia* 77/259
 大同拟齿叶 *Paratingia datongensis* 77/259
拟刺葵属 *Phoenicopsis* 79/261
 狭叶拟刺葵 *Phoenicopsis angustifolia* 79/261
 拟刺葵(斯蒂芬叶)亚属 *Phoenicopsis* (*Stephenophyllum*) 79/261
  索尔姆斯拟刺葵(斯蒂芬叶) *Phoenicopsis* (*Stephenophyllum*) *solmsi* 79/261
  美形拟刺葵(斯蒂芬叶) *Phoenicopsis* (*Stephenophyllum*) *decorata* 80/261
  厄尼塞捷拟刺葵(斯蒂芬叶) *Phoenicopsis* (*Stephenophyllum*) *enissejensis* 80/261
  特别拟刺葵(斯蒂芬叶) *Phoenicopsis* (*Stephenophyllum*) *mira* 80/262
  塔什克斯拟刺葵(斯蒂芬叶) *Phoenicopsis* (*Stephenophyllum*) *taschkessiensis* 80/262
拟带枝属 *Taeniocladopsis* 113/297
 假根茎型拟带枝 *Taeniocladopsis rhizomoides* 113/298
拟丁氏蕨穗属 *Paratingiostachya* 78/259
 华夏拟丁氏蕨穗 *Paratingiostachya cathaysiana* 78/259
拟根茎属 *Rhizomopsis* 95/278
 具芽拟根茎 *Rhizomopsis gemmifera* 95/278
拟蕨属 *Pteridiopsis* 91/274
 滴道拟蕨 *Pteridiopsis didaoensis* 91/274
 柔弱拟蕨 *Pteridiopsis tenera* 91/274
拟裂鞘叶属 *Schizoneuropsis* 98/281
 德田拟裂鞘叶 *Schizoneuropsis tokudae* 98/281
拟片叶苔属 *Riccardiopsis* 96/278
 徐氏拟片叶苔 *Riccardiopsis hsüi* 96/279
拟斯托加枝属 *Parastorgaardis* 77/258
 门头沟拟斯托加枝 *Parastorgaardis mentoukouensis* 77/258
拟苏铁籽属 *Cycadeoidispermum* 23/199
 具柄拟苏铁籽 *Cycadeoidispermum petiolatum* 23/199
拟楔羊齿属 *Parasphenopteris* 76/258
 东方拟楔羊齿 *Parasphenopteris orientalis* 77/258
拟楔叶属 *Parasphenophyllum* 76/258
 山西拟楔叶 *Parasphenophyllum shansiense* 76/258
拟斜羽叶属 *Plagiozamiopsis* 83/265
 苏铁杉型拟斜羽叶 *Plagiozamiopsis podozamoides* 83/266
宁夏叶属 *Ningxiaphyllum* 71/252
 三裂宁夏叶 *Ningxiaphyllum trilobatum* 71/252

P

耙羊齿属 *Rastropteris* 93/276
 平泉耙羊齿 *Rastropteris pingquanensis* 93/276
潘广叶属 *Pankuangia* 74/256
 海房沟潘广叶 *Pankuangia haifanggouensis* 74/256
潘氏果属 *Pania* 74/255
 似苏铁潘氏果 *Pania cycadina* 74/255

盘囊蕨属 *Discalis* ·· 27/204
　　长柄盘囊蕨 *Discalis longistipa* ·· 28/204
普通蕨属 *Coenosophyton* ·· 22/198
　　三叉普通蕨 *Coenosophyton tristichus* ·· 22/198

Q

奇脉叶属 *Mironeura* ·· 66/246
　　大坑奇脉叶 *Mironeura dakengensis* ··· 66/246
奇羊齿属 *Aetheopteris* ··· 4/179
　　坚直奇羊齿 *Aetheopteris rigida* ·· 4/179
奇叶属 *Acthephyllum* ··· 3/177
　　开县奇叶 *Acthephyllum kaixianense* ··· 3/177
奇异木属 *Allophyton* ·· 5/180
　　丁青奇异木 *Allophyton dengqenensis* ··· 5/180
奇异羊齿属 *Mirabopteris* ··· 65/245
　　浑江奇异羊齿 *Mirabopteris hunjiangensis* ··· 65/246
奇异羊齿属 *Paradoxopteris* Mi et Liu, 1977 (non Hirmer, 1927) ······························ 75/257
　　浑江奇异羊齿 *Paradoxopteris hunjiangensis* ··· 76/257
奇异叶属 *Adoketophyllum* ··· 4/178
　　亚轮生奇异叶 *Adoketophyllum subverticillatum* ·· 4/178
奇异羽叶属 *Thaumatophyllum* ·· 115/300
　　羽毛奇异羽叶 *Thaumatophyllum ptilum* ·· 116/300
契丹穗属 *Khitania* ··· 50/229
　　柱状契丹穗 *Khitania columnispicata* ··· 50/229
钱耐果属 *Chaneya* ··· 18/194
　　细小钱耐果 *Chaneya tenuis* ··· 18/194
　　科干钱耐果 *Chaneya kokangensis* ··· 18/194
黔囊属 *Guizhoua* ·· 39/217
　　堆黔囊 *Guizhoua gregalis* ·· 39/217
秦岭羊齿属 *Qinlingopteris* ·· 91/274
　　东方秦岭羊齿 *Qinlingopteris orientalis* ··· 91/274
　　秦岭羊齿（未定种）*Qinlingopteris* sp. ·· 91/274
琼海叶属 *Qionghaia* ·· 92/274
　　肉质琼海叶 *Qionghaia carnosa* ·· 92/275
全泽米属 *Holozamites* ··· 44/222
　　洪涛全查米亚 *Holozamites hongtaoi* ··· 44/222

R

热河似查米亚属 *Rehezamites* ·· 93/276
　　不等裂热河似查米亚 *Rehezamites anisolobus* ··· 93/276
　　热河似查米亚（未定种）*Rehezamites* sp. ·· 94/276
日蕨属 *Rireticopteris* ··· 96/279

小叶日蕨 *Rireticopteris microphylla* ·· 96/279

S

三棱果属 *Triqueteria* ·· 120/305
    中国三棱果 *Triqueteria sinensis* ·· 120/306
三裂穗属 *Tricrananthus* ··· 119/305
    箭头状三裂穗 *Tricrananthus sagittatus* ··· 120/305
    瓣状三裂穗 *Tricrananthus lobatus* ··· 120/305
三脉蕨属 *Trinerviopteris* ··· 120/305
    心叶三脉蕨 *Trinerviopteris cardiophylla* ·· 120/305
三网羊齿属 *Tricoemplectopteris* ··· 119/304
    太原三网羊齿 *Tricoemplectopteris taiyuanensis* ·· 119/304
山西木属 *Shanxioxylon* ·· 100/283
    中国山西木 *Shanxioxylon sinense* ··· 100/283
    太原山西木 *Shanxioxylon taiyuanense* ·· 100/283
山西枝属 *Shanxicladus* ·· 100/283
    疹形山西枝 *Shanxicladus pastulosus* ··· 100/283
上园草属 *Shangyuania* ··· 99/282
    才氏上园草 *Shangyuania caii* ··· 99/283
神州聚囊属 *Shenzhoutheca* ··· 102/285
    刷状神州聚囊 *Shenzhoutheca aspergilliformis* ··· 102/285
神州叶属 *Shenzhouphyllum* ·· 101/284
    波缘神州叶 *Shenzhouphyllum undulatum* ·· 101/284
    圆形神州叶 *Shenzhouphyllum rotundatum* ·· 101/285
    匙形神州叶 *Shenzhouphyllum spatulatum* ··· 101/285
神州籽属 *Shenzhouspermum* ··· 102/285
    三歧神州籽 *Shenzhouspermum trichotomum* ··· 102/285
沈括叶属 *Shenkuoia* ·· 101/284
    美脉沈括叶 *Shenkuoia caloneura* ·· 101/284
沈氏蕨属 *Shenea* ··· 100/284
    希氏沈氏蕨 *Shenea hirschmeierii* ·· 100/284
肾叶属 *Renifolium* ··· 94/277
    长柄肾叶 *Renifolium logipetiolatum* ··· 94/277
始查米苏铁属 *Primozamia* ·· 86/268
    中国始查米苏铁 *Primozamia sinensis* ·· 86/268
始鳞木属 *Eolepidodendron* ·· 31/209
    句容始鳞木 *Eolepidodendron jurongense* ·· 32/209
    无锡始鳞木 *Eolepidodendron wusihense* ·· 32/209
    无锡始鳞木（比较种） *Eolepidodendron* cf. *wusihense* ······································ 32/209
    始鳞木（未定种） *Eolepidodendron* sp. ··· 32/209
始木兰属 *Archimagnolia* ·· 8/182
    喙柱始木兰 *Archimagnolia rostrato-stylosa* ·· 8/182
始拟银杏属 *Primoginkgo* ·· 85/268

深裂始拟银杏 *Primoginkgo dissecta*	86/268
始苹果属 *Archimalus*	8/183
大萼始苹果 *Archimalus calycina*	8/183
始水松属 *Eoglyptostrobus*	30/207
清风藤型始水松 *Eoglyptostrobus sabioides*	30/207
始苏铁属 *Primocycas*	85/267
中国始苏铁 *Primocycas chinensis*	85/267
蓠状始苏铁 *Primocycas muscariformis*	85/268
始团扇蕨属 *Eogonocormus* Deng, 1995 (non Deng, 1997)	30/207
白垩始团扇蕨 *Eogonocormus cretaceum* Deng, 1995 (non Deng, 1997)	30/207
线形始团扇蕨 *Eogonocormus linearifolium*	31/207
始团扇蕨属 *Eogonocormus* Deng, 1997 (non Deng, 1995)	31/208
白垩始团扇蕨 *Eogonocormus cretaceum* Deng, 1997 (non Deng, 1995)	31/208
始叶蕨属 *Eophyllophyton*	32/210
优美始叶蕨 *Eophyllophyton bellum*	33/210
始叶羊齿属 *Eophyllogonium*	32/209
华夏始叶羊齿 *Eophyllogonium cathayense*	32/209
始羽蕨属 *Eogymnocarpium*	31/208
中国始羽蕨 *Eogymnocarpium sinense*	31/208
似八角属 *Illicites*	46/225
似八角(sp. indet.) *Illicites* sp. indet.	46/225
似百合属 *Lilites*	55/235
热河似百合 *Lilites reheensis*	55/235
似叉苔属 *Metzgerites*	65/245
蔚县似叉苔 *Metzgerites yuxinanensis*	65/245
似齿囊蕨属 *Odontosorites*	72/253
海尔似齿囊蕨 *Odontosorites heerianus*	72/253
似杜仲属 *Eucommioites*	33/210
东方似杜仲 *Eucommioites orientalis*	33/211
似狗尾草属 *Setarites*	99/282
似狗尾草(sp. indet.) *Setarites* sp. indet.	99/282
似画眉草属 *Eragrosites*	33/210
常氏似画眉草 *Eragrosites changii*	33/210
似金星蕨属 *Thelypterites*	116/301
似金星蕨(未定种 A) *Thelypterites* sp. A	116/301
似金星蕨(未定种 B) *Thelypterites* sp. B	116/301
似茎状地衣属 *Foliosites*	36/213
美丽似茎状地衣 *Foliosites formosus*	36/213
似卷囊蕨属 *Speirocarpites*	106/290
弗吉尼亚似卷囊蕨 *Speirocarpites virginiensis*	106/290
渡口似卷囊蕨 *Speirocarpites dukouensis*	106/290
日蕨型似卷囊蕨 *Speirocarpites rireticopteroides*	107/291
中国似卷囊蕨 *Speirocarpites zhonguoensis*	107/291
似克鲁克蕨属 *Klukiopsis*	51/229

侏罗似克鲁克蕨 *Klukiopsis jurassica*	51/229
似兰属 *Orchidites*	72/253
线叶似兰 *Orchidites linearifolius*	72/253
披针叶似兰 *Orchidites lancifolius*	73/254
似蓼属 *Polygonites* Wu S Q,1999 (non Saporta,1865)	84/266
多小枝似蓼 *Polygonites polyclonus*	84/266
扁平似蓼 *Polygonites planus*	84/266
似轮叶属 *Annularites*	7/181
剑瓣似轮叶 *Annularites ensilolius*	7/182
舌形似轮叶 *Annularites lingulatus*	7/182
平安似轮叶 *Annularites heianensis*	7/182
中国似轮叶 *Annularites sinensis*	7/182
似木麻黄属 *Casuarinites*	14/189
似木麻黄(sp. indet.) *Casuarinites* sp. indet.	14/189
似南五味子属 *Kadsurrites*	50/228
似南五味子(sp. indet.) *Kadsurrites* sp. indet.	50/228
似槭树属 *Acerites*	2/176
似槭树(sp. indet.) *Acerites* sp. indet.	2/176
似提灯藓属 *Mnioites*	66/247
短叶杉型似提灯藓 *Mnioites brachyphylloides*	67/247
似铁线莲叶属 *Clematites*	21/197
披针似铁线莲叶 *Clematites lanceolatus*	22/198
似乌头属 *Aconititis*	2/177
似乌头(sp. indet.) *Aconititis* sp. indet.	3/177
似阴地蕨属 *Botrychites*	13/188
热河似阴地蕨 *Botrychites reheensis*	13/188
似雨蕨属 *Gymnogrammitites*	40/218
鲁福德似雨蕨 *Gymnogrammitites ruffordioides*	40/218
似圆柏属 *Sabinites*	97/280
内蒙古似圆柏 *Sabinites neimonglica*	97/280
纤细似圆柏 *Sabinites gracilis*	97/280
似远志属 *Polygatites*	83/266
似远志(sp. indet.) *Polygatites* sp. indet.	84/266
梳囊属 *Lophotheca*	58/238
盘县梳囊 *Lophotheca panxianensis*	59/238
束脉蕨属 *Symopteris*	111/296
瑞士束脉蕨 *Symopteris helvetica*	111/296
密脉束脉蕨 *Symopteris densinervis*	111/296
蔡耶束脉蕨 *Symopteris zeilleri*	112/296
束羊齿属 *Fascipteris*	34/211
赫勒束羊齿 *Fascipteris hallei*	34/211
垂束羊齿 *Fascipteris recta*	34/211
中国束羊齿 *Fascipteris sinensis*	34/212
密囊束羊齿(皱囊蕨) *Fascipteris (Ptychocarpus) densata*	34/212

狭束羊齿 *Fascipteris stena*	34/212
刷囊属 *Strigillotheca*	111/295
束囊刷囊 *Strigillotheca fasciculata*	111/295
双列囊蕨属 *Bifariusotheca*	12/187
晴隆双列囊蕨 *Bifariusotheca qinglongensis*	12/187
双囊芦穗属 *Shuangnangostachya*	102/286
细小双囊芦穗 *Shuangnangostachya gracilis*	102/286
双生叶属 *Geminofoliolum*	37/215
纤细双生叶 *Geminofoliolum gracilis*	37/215
双网羊齿属 *Bicoemplectopteris*	12/187
赫勒双网羊齿 *Bicoemplectopteris hallei*	12/187
水城蕨属 *Shuichengella*	102/286
原始水城蕨 *Shuichengella primitiva*	103/286
水杉属 *Metasequoia*	64/244
水松型水杉 *Metasequoia glyptostroboides*	64/244
二列水杉 *Metasequoia disticha*	64/245
斯芬克斯籽属 *Sphinxia* Li, Hilton et Hemsley, 1997 (non Reid et Chandler, 1933)	108/292
武汉斯芬克斯籽 *Sphinxia wuhania*	108/292
斯氏木属 *Szeioxylon*	112/297
徐州斯氏木 *Szeioxylon xuzhouene*	112/297
斯氏鞘叶属 *Siella*	103/286
细肋斯氏鞘叶 *Siella leptocostata*	103/287
斯氏松属 *Szecladia*	112/297
多脉斯氏松 *Szecladia multinervia*	112/297
四叶属 *Tetrafolia*	114/299
长沙四叶 *Tetrafolia changshaense*	114/299
苏铁鳞叶属 *Cycadolepophyllum*	24/200
较小苏铁鳞叶 *Cycadolepophyllum minor*	24/200
等形苏铁鳞叶 *Cycadolepophyllum aequale*	24/200
苏铁缘蕨属 *Cycadicotis*	23/199
蕉羽叶脉苏铁缘蕨 *Cycadicotis nissonervis*	23/200
苏铁缘蕨(sp. indet.) *Cycadicotis* sp. indet.	24/200
苏铁籽属 *Dioonocarpus*	27/204
卵形苏铁籽 *Dioonocarpus ovatus*	27/204
穗蕨属 *Stachyophyton*	110/294
云南穗蕨 *Stachyophyton yunnanense*	110/294
穗藓属 *Stachybryolites*	109/294
周氏穗藓 *Stachybryolites zhoui*	109/294

T

塔状木属 *Minarodendron*	65/245
华夏塔状木 *Minarodendron cathaysiense*	65/245
太平场蕨属 *Taipingchangella*	113/298

中国太平场蕨 *Taipingchangella zhongguoensis*	113/298
太阳花属 *Solaranthus*	106/290
道虎沟太阳花 *Solaranthus daohugouensis*	106/290
太原蕨属 *Taiyuanitheca*	113/298
四线形太原蕨 *Taiyuanitheca tetralinea*	114/298
天山羊齿属 *Tianshanopteris*	117/302
温宿天山羊齿 *Tianshanopteris wensuensis*	117/302
天石蕨属 *Szea*	112/296
中国天石蕨 *Szea sinensis*	112/296
天石枝属 *Tianshia*	117/302
伸展天石枝 *Tianshia patens*	117/302
田氏木属 *Tianoxylon*	116/301
段木头田氏木 *Tianoxylon duanmutouense*	116/301
条形蕨属 *Lorophyton*	59/239
高氏条形蕨 *Lorophyton goense*	60/239
条叶属 *Vittifoliolum*	121/307
游离条叶 *Vittifoliolum segregatum*	122/307
游离条叶脊条型 *Vittifoliolum segregatum* forma *costatum*	122/307
多脉条叶 *Vittifoliolum multinerve*	122/307
铁花属 *Cycadostrobilus*	24/200
古生铁花 *Cycadostrobilus paleozoicus*	24/201
桐庐籽属 *Tonglucarpus*	118/303
奇丽桐庐籽 *Tonglucarpus spectabilis*	118/304
铜川叶属 *Tongchuanophyllum*	118/303
三角形铜川叶 *Tongchuanophyllum trigonus*	118/303
优美铜川叶 *Tongchuanophyllum concinnum*	118/303
陕西铜川叶 *Tongchuanophyllum shensiense*	118/303
托克逊蕨属 *Toksunopteris*	117/302
对生托克逊蕨 *Toksunopteris opposita*	118/303

W

网格蕨属 *Reteophlebis*	94/277
单式网格蕨 *Reteophlebis simplex*	94/277
网延羊齿属 *Reticalethopteris*	94/277
袁氏网延羊齿 *Reticalethopteris yuani*	94/277
网叶属 *Linophyllum*	56/235
宣威网叶 *Linophyllum xuanweiense*	56/235
文山蕨属 *Wenshania*	122/308
纸厂文山蕨 *Wenshania zhichangensis*	122/308
乌图布拉克蕨属 *Wutubulaka*	122/308
多叉乌图布拉克蕨 *Wutubulaka multidichotoma*	122/308
乌云花属 *Wuyunanthus*	123/309
六瓣乌云花 *Wuyunanthus hexapetalus*	123/309

无锡蕨属 Wuxia 123/308
    双穗无锡蕨 Wuxia bistrobilata 123/308

X

西屯蕨属 Xitunia 127/313
    刺囊西屯蕨 Xitunia spinitheca 127/313
细毛蕨属 Ciliatopteris 20/197
    栉齿细毛蕨 Ciliatopteris pecotinata 21/197
细轴始蕨属 Tenuisa 114/299
    弗拉细轴始蕨 Tenuisa frasniana 114/299
夏家街蕨属 Xiajiajienia 123/309
    奇异夏家街蕨 Xiajiajienia mirabila 123/309
仙籽属 Sphinxiocarpon 108/292
    武汉斯芬克斯仙籽 Sphinxiocarpon wuhania 108/292
先骕蕨属 Huia 46/224
    回弯先骕蕨 Huia recurvata 46/224
纤木属 Chamaedendron 18/193
    异囊纤木 Chamaedendron multisporangiatum 18/193
纤细蕨属 Filiformorama 35/213
    简单纤细蕨 Filiformorama simplexa 35/213
香溪叶属 Hsiangchiphyllum 44/222
    三脉香溪叶 Hsiangchiphyllum trinerve 44/222
小蛟河蕨属 Chiaohoella 20/196
    奇异小蛟河蕨 Chiaohoella mirabilis 20/196
    新查米叶型小蛟河蕨 Chiahooella neozamioide 20/196
小伞属 Sciadocillus 98/281
    楔裂小伞 Sciadocillus cuneifidus 98/281
楔叶拜拉花属 Sphenobaieroanthus 107/291
    中国楔叶拜拉花 Sphenobaieroanthus sinensis 107/291
楔叶拜拉枝属 Sphenobaierocladus 107/291
    中国楔叶拜拉枝 Sphenobaierocladus sinensis 107/291
楔栉羊齿属 Sphenopecopteris 107/292
    豆子楔栉羊齿 Sphenopecopteris beaniata 108/292
新孢穗属 Neostachya 70/251
    陕西新孢穗 Neostachya shanxiensis 70/251
新常富籽属 Nystroemia 72/253
    篦形新常富籽 Nystroemia pectiniformis 72/253
新疆蕨属 Xinjiangopteris Wu S Q et Zhou,1986 (non Wu S Z,1983) 125/311
    对生新疆蕨 Xinjiangopteris opposita 126/311
新疆蕨属 Xinjiangopteris Wu S Z,1983 (non Wu S Q et Zhou,1986) 126/312
    托克逊新疆蕨 Xinjiangopteris toksunensis 126/312
新疆木属 Xinjiangophyton 125/311
    刺状新疆木 Xinjiangophyton spinosum 125/311

新科达属 Neocordaites	70/250
披针新科达 Neocordaites lanceolatus	70/250
新龙叶属 Xinlongia	126/312
侧羽叶型新龙叶 Xinlongia pterophylloides	126/312
和恩格尔新龙叶 Xinlongia hoheneggeri	126/312
新龙羽叶属 Xinlongophyllum	126/312
篦羽羊齿型新龙羽叶 Xinlongophyllum ctenopteroides	127/313
多条纹新龙羽叶 Xinlongophyllum multilineatum	127/313
新轮叶属 Neoannularia	69/250
陕西新轮叶 Neoannularia shanxiensis	69/250
川滇新轮叶 Neoannularia chuandianensis	69/250
新准大羽羊齿属 Neogigantopteridium	70/251
具刺新准大羽羊齿 Neogigantopteridium spiniferum	70/251
星壳斗属 Astrocupulites	9/184
渐尖星壳斗 Astrocupulites acuminatus	9/184
星学花属 Xingxueanthus	124/310
中国星学花 Xingxueanthus sinensis	124/310
星学花序属 Xingxueina	124/310
黑龙江星学花序 Xingxueina heilongjiangensis	125/310
星学叶属 Xingxuephyllum	125/311
鸡西星学叶 Xingxuephyllum jixiense	125/311
兴安叶属 Xinganphyllum	124/309
对称兴安叶 Xinganphyllum aequale	124/309
不对称兴安叶 Xinganphyllum inaequale	124/310
兴安叶(未定种) Xinganphyllum sp.	124/310
袖套杉属 Manica	61/240
希枝袖套杉 Manica parceramosa	61/240
袖套杉(长岭杉)亚属 Manica (Chanlingia)	61/241
穹孔袖套杉(长岭杉) Manica (Chanlingia) tholistoma	61/241
袖套杉(袖套杉)亚属 Manica (Manica)	61/241
希枝袖套杉(袖套杉) Manica (Manica) parceramosa	61/241
大拉子袖套杉(袖套杉) Manica (Manica) dalatzensis	61/241
窝穴袖套杉(袖套杉) Manica (Manica) foveolata	62/241
乳突袖套杉(袖套杉) Manica (Manica) papillosa	62/242
徐氏蕨属 Hsuea	44/223
粗壮徐氏蕨 Hsuea robusta	44/223

Y

牙羊齿属 Dentopteris	27/203
窄叶牙羊齿 Dentopteris stenophylla	27/203
宽叶牙羊齿 Dentopteris platyphylla	27/204
雅观木属 Perisemoxylon	79/260
双螺纹雅观木 Perisemoxylon bispirale	79/260

雅观木（未定种）*Perisemoxylon* sp. ... 79/260
雅蕨属 *Pavoniopteris* ... 78/259
　　马通蕨型雅蕨 *Pavoniopteris matonioides* ... 78/260
亚洲羊齿属 *Asiopteris* ... 9/184
　　怀仁亚洲羊齿 *Asiopteris huairenensis* ... 9/184
亚洲叶属 *Asiatifolium* ... 8/183
　　雅致亚洲叶 *Asiatifolium elegans* ... 9/183
延吉叶属 *Yanjiphyllum* ... 127/314
　　椭圆延吉叶 *Yanjiphyllum ellipticum* ... 128/314
燕辽杉属 *Yanliaoa* ... 128/314
　　中国燕辽杉 *Yanliaoa sinensis* ... 128/314
羊齿缘蕨属 *Filicidicotis* ... 35/212
　　羊齿缘蕨（sp. indet.）*Filicidicotis* sp. indet. ... 35/212
杨氏木属 *Yangzunyia* ... 127/313
　　河南杨氏木 *Yangzunyia henanensis* ... 127/313
耶氏蕨属 *Jaenschea* ... 47/225
　　中国耶氏蕨 *Jaenschea sinensis* ... 47/225
叶茎属 *Phylladendroid* ... 81/262
　　江西叶茎 *Phylladendroid jiangxiensis* ... 81/262
疑麻黄属 *Amphiephedra* ... 6/181
　　鼠李型疑麻黄 *Amphiephedra rhamnoides* ... 6/181
义马果属 *Yimaia* ... 128/314
　　外弯义马果 *Yimaia recurva* ... 128/314
义县叶属 *Yixianophyllum* ... 128/315
　　金家沟义县叶 *Yixianophyllum jinjiagouensie* ... 128/315
异麻黄属 *Alloephedra* ... 5/179
　　星学异麻黄 *Alloephedra xingxuei* ... 5/179
异枝蕨属 *Metacladophyton* ... 64/244
　　四木质柱异枝蕨 *Metacladophyton tetraxylum* ... 64/244
隐囊蕨属 *Celathega* ... 17/193
　　贝氏隐囊蕨 *Celathega beckii* ... 17/193
隐羊齿属 *Cryptonoclea* ... 23/199
　　原始隐羊齿 *Cryptonoclea primitiva* ... 23/199
缨囊属 *Distichotheca* ... 28/205
　　具边缨囊 *Distichotheca crossothecoides* ... 28/205
永仁叶属 *Yungjenophyllum* ... 129/316
　　大叶永仁叶 *Yungjenophyllum grandifolium* ... 129/316
羽叶大羽羊齿属 *Pinnagigantopteris* ... 82/264
　　烟叶羽叶大羽羊齿 *Pinnagigantopteris nicotianaefolia* ... 82/265
　　披针羽叶大羽羊齿 *Pinnagigantopteris lanceolatus* ... 83/265
　　长圆羽叶大羽羊齿 *Pinnagigantopteris oblongus* ... 83/265
羽叶单网羊齿属 *Pinnagigantonoclea* ... 81/263
　　似榉羽叶单网羊齿 *Pinnagigantonoclea zelkovoides* ... 81/263
　　异常羽叶单网羊齿 *Pinnagigantonoclea heteroeura* ... 81/263

异脉羽叶单网羊齿 *Pinnagigantonoclea mira* 81/263
　　贵州羽叶单网羊齿 *Pinnagigantonoclea guizhouensis* 81/263
　　尖头羽叶单网羊齿 *Pinnagigantonoclea mucronata* 82/263
　　匙羽叶单网羊齿 *Pinnagigantonoclea spatulata* 82/264
　　莲座羽叶单网羊齿 *Pinnagigantonoclea rosulata* 82/264
　　似槲羽叶单网羊齿 *Pinnagigantonoclea dryophylloides* 82/264
　　多形羽叶单网羊齿 *Pinnagigantonoclea polymorpha* 82/264
羽枝属 *Pinnatiramosus* 83/265
　　黔羽枝 *Pinnatiramosus qianensis* 83/265
玉光蕨属 *Yuguangia* 129/315
　　规则玉光蕨 *Yuguangia ordinata* 129/315
豫囊蕨属 *Henanotheca* 43/221
　　卵豫囊蕨（楔羊齿） *Henanotheca (Sphenopteris) ovata* 43/221
原大羽羊齿属 *Progigantopteris* 87/269
　　短网原大羽羊齿 *Progigantopteris brevireticulatus* 87/269
原单网羊齿属 *Progigantonoclea* 87/269
　　河南原单网羊齿 *Progigantonoclea henanensis* 87/269
原始金松型木属 *Protosciadopityoxylon* 89/271
　　辽宁原始金松型木 *Protosciadopityoxylon liaoningense* 89/272
原始蕨属 *Protopteridophyton* 88/271
　　泥盆纪原始蕨 *Protopteridophyton devonicum* 88/271
原始水松型木属 *Protoglyptostroboxylon* 88/271
　　巨大原始水松型木 *Protoglyptostroboxylon giganteum* 88/271
　　伊敏原始水松型木 *Protoglyptostroboxylon yimiense* 88/271
原始银杏型木属 *Proginkgoxylon* 87/270
　　本溪原始银杏型木 *Proginkgoxylon benxiense* 87/270
　　大青山原始银杏型木 *Proginkgoxylon daqingshanense* 88/270
原苏铁属 *Procycas* 86/269
　　密脉原苏铁 *Procycas densinervioides* 86/269
云蕨属 *Yunia* 130/316
　　二叉云蕨 *Yunia dichotoma* 130/316

Z

贼木属 *Phoroxylon* 80/262
　　梯纹状贼木 *Phoroxylon scalariforme* 80/262
窄叶属 *Angustiphyllum* 6/181
　　腰埠窄叶 *Angustiphyllum yaobuense* 7/181
粘合囊蕨属 *Cohaerensitheca* 22/198
　　沙尼粘合囊蕨 *Cohaerensitheca sahnii* 22/198
掌裂蕨属 *Catenalis* 14/189
　　指状掌裂蕨 *Catenalis digitata* 14/189
正理蕨属 *Zhenglia* 131/317
　　辐射正理蕨 *Zhenglia radiata* 131/317

正三角籽属 *Deltoispermum* 26/202
　　河南正三角籽 *Deltoispermum henanense* 26/202
郑氏叶属 *Zhengia* 130/317
　　中国郑氏叶 *Zhengia chinensis* 130/317
枝带羊齿属 *Cladotaeniopteris* 21/197
　　陕西枝带羊齿 *Cladotaeniopteris shaanxiensis* 21/197
织羊齿属 *Emplectopteris* 30/207
　　三角织羊齿 *Emplectopteris trangularis* 30/207
中国篦羽叶属 *Sinoctenis* 103/287
　　葛利普中国篦羽叶 *Sinoctenis grabauiana* 104/287
中国古螺纹木属 *Sinopalaeospiroxylon* 104/288
　　宝力格庙中国古螺纹木 *Sinopalaeospiroxylon baoligemiaoense* 104/288
　　南票中国古螺纹木 *Sinopalaeospiroxylon napiaoense* 105/288
　　平泉中国古螺纹木 *Sinopalaeospiroxylon pingquanense* 105/289
中国似查米亚属 *Sinozamites* 105/289
　　李氏中国似查米亚 *Sinozamites leeiana* 105/289
中国羊齿属 *Cathaiopteridium* 14/190
　　细小中国羊齿 *Cathaiopteridium minutum* 14/190
中国叶属 *Sinophyllum* 105/289
　　孙氏中国叶 *Sinophyllum suni* 105/289
中华古果属 *Sinocarpus* 103/287
　　下延中华古果 *Sinocarpus decussatus* 103/287
中华羊齿属 *Cathaysiopteridium* 16/192
　　束脉中华羊齿 *Cathaysiopteridium fasciculatum* .. 17/192
中华缘蕨属 *Sinodicotis* 104/287
　　中华缘蕨（sp. indet.）*Sinodicotis* sp. indet. 104/288
中间苏铁属 *Mediocycas* 62/242
　　喀左中间苏铁 *Mediocycas kazuoensis* 62/242
中州蕨属 *Zhongzhoupteris* 131/318
　　尾羽中州蕨 *Zhongzhoupteris cathaysicus* 131/318
中州籽属 *Zhongzhoucarus* 131/317
　　三角中州籽 *Zhongzhoucarus deltatus* 131/317
钟囊属 *Tongshania* 119/304
　　齿状钟囊 *Tongshania dentate* 119/304
周氏籽属 *Zhouia* 131/318
　　北票周氏籽 *Zhouia beipiaoensis* 132/318
帚羽叶属 *Scoparia* 98/281
　　羽毛帚羽叶 *Scoparia plumaria* 98/282
朱氏囊蕨属 *Zhutheca* 132/318
　　密囊朱氏囊蕨 *Zhutheca densata* 132/318
侏罗木兰属 *Juramagnolia* 49/228
　　侏罗木兰（sp. indet.）*Juramagnolia* sp. indet. 49/228
侏罗球果属 *Jurastrobus* 49/228
　　陈氏侏罗球果 *Jurastrobus chenii* 49/228

侏罗缘蕨属 *Juradicotis* ·· 49/227
  侏罗缘蕨(sp. indet.) *Juradicotis* sp. indet. ·· 49/227
准爱河羊齿属 *Aipteridium* ·· 4/179
  羽状准爱河羊齿 *Aipteridium pinnatum* ·· 5/179
准噶尔蕨属 *Junggaria* ·· 48/227
  刺状准噶尔蕨 *Junggaria spinosa* ·· 48/227
准束羊齿属 *Fascipteridium* ·· 33/211
  椭圆准束羊齿 *Fascipteridium ellipticum* ·· 34/211
准枝脉蕨属 *Cladophlebidium* ·· 21/197
  翁氏准枝脉蕨 *Cladophlebidium wongi* ·· 21/197

Supported by Special Research Program of
Basic Science and Technology of the Ministry
of Science and Technology (2013FY113000)

Record of Megafossil Plants from China

Ⅸ Record of Megafossil Plant New Generic Names Established for Chinese Specimens (1865 — 2010)

Compiled by
WU Xiangwu

University of Science and Technology of China Press

Brief Introduction

This book is the one volume of *Record of Megafossil Plants from China*. There are two parts of both Chinese and English versions, mainly documents complete data on the megafossil plant new generic names established for Chinese specimens between the years 1865 to 2010. This index records totally 380 generic names (among them 195 are established based on Palaeozoic specimens; 177 are established based on Mesozoic specimens and 8 are established based on Cenozoic specimens). Each record of the generic taxon include: author(s) who established the genus, establishing year, synonym, type species, taxonomic status. For those generic names or specific names established based on Chinese specimens, the type specimens and their depository institutions have also been recorded. Each part attaches one appendixe (Index of generic and specific names). At the end of the book, there are references.

This book is a complete collection and an easy reference document that compiled based on extensive survey of both Chinese and abroad literatures and a systematic data collections of palaeobotany. It is suitable for reading for those who are working on research, education and data base related to palaeobotany, life sciences and earth sciences.

GENERAL FOREWORD

As a branch of sciences studying organisms of the geological history, palaeontology relies utterly on the fossil record, so does the palaeobotany as a branch of palaeontology. The compilation and editing of fossil plant data started early in the 19 century. F. Unger published *Synopsis Plantarum Fossilium* and *Genera et Species Plantarium Fossilium* in 1845 and 1850 respectively, not long after the introduction of C. von Linné's binomial nomenclature to the study of fossil plants by K. M. von Sternberg in 1820. Since then, indices or catalogues of fossil plants have been successively compiled by many professional institutions and specialists. Amongst them, the most influential are catalogues of fossil plants in the Geological Department of British Museum written by A. C. Seward and others, *Fossilium Catalogus II : Palantae* compiled by W. J. Jongmans and his successor S. J. Dijkstra, *The Fossil Record (Volume 1)* and *The Fossil Revord (Volume 2)* chief-edited by W. B. Harland and others and afterwards by M. J. Benton, and *Index of Generic Names of Fossil Plants* compiled by H. N. Andrews Jr. and his successors A. D. Watt, A. M. Blazer and others. Based partly on Andrews' index, the digital database "Index Nominum Genericorum (ING)" was set up by the joint efforts of the International Association of Plant Taxonomy and the Smithsonian Institution. There are also numerous catalogues or indices of fossil plants of specific regions, periods or institutions, such as catalogues of Cretaceous and Tertiary plants of North America compiled by F. H. Knowlton, L. F. Ward and R. S. La Motte, Upper Triassic plants of the western United States by S. Ash, Carboniferous, Permian and Jurassic plants by M. Boersma and L. M. Broekmeyer, Indian fossil plants by R. N. Lakhanpal, and fossil record of plants by S. V. Meyen and index of sporophytes and gymnosperm referred to USSR by V. A. Vachrameev. All these have no doubt benefited to the academic exchanges between palaeobotanists from different countries, and contributed considerably to the development of palaeobotany.

Although China is amongst the countries with widely distributed terrestrial deposits and rich fossil resources, scientific researches on fossil plants began much later in our country than in many other countries. For a quite long time, in our country, there were only few researchers, who are engaged in palaeobotanical studies. Since the 1950s, especially the beginning of Reform and Opening to the outside world in the late 1980s, palaeobotany became blooming in our country as other disciplines of science and technology. During the development and construction of the country, both palaeobotanists and publications have been markedly increased. The editing and compilation of the fossil plant record has also been put on the agenda to meet the needs of increasing academic activities, along with participation in the "Plant Fossil Record (PFR)" project sponsored by the International Organization of Palaeobotany. Professor Wu is one of the few pioneers who have paid special attention to data accumulation and compilation of the fossil plant record in China. Back in 1993, He published *Record of Generic Names of Mesozoic Megafossil Plants from China (1865 — 1990)* and *Index of New Generic Names Founded on Mesozoic and Cenozoic Specimens from China (1865 — 1990)*. In 2006, he published the generic names after 1990. *Catalogue of the Cenozoic Megafossil Plants of China* was also Published by Liu and others (1996).

It is a time consuming task to compile a comprehensive catalogue containing the fossil records of all plant groups in the geological history. After years of hard work, all efforts finally bore fruits, and are able to publish separately according to classification and geological distribution, as well as the progress of data accumulating and editing. All data will eventually be incorporated into the databases of all China fossil records: "Palaeontological and Stratigraphical Database of China" and "Geobiodiversity Database (GBDB)".

The pubilication of *Record of Megafossil Plants from China* is one of the milestones in the development of palaeobotany, undoubtedly it will provide a good foundation and platform for the further development of this discipline. As an aged researcher in palaeobotany, I look eagerly forward to seeing the publication of the serial fossil catalogues of China.

INTRODUCTION

In China, there is a long history of plant fossil discovery, as it is well documented in ancient literatures. Among them the voluminous work *Mengxi Bitan (Dream Pool Essays)* by Shen Kuo (1031 — 1095) in the Beisong (Northern Song) Dynasty is probably the earliest. In its 21st volume, fossil stems [later identified as stems of *Equisctites* or pith-casts of *Neocalamites* by Deng (1976)] from Yongningguan, Yanzhou, Shaanxi (now Yanshuiguan of Yanchuan County, Yan'an City, Shaanxi Province) were named "bamboo shoots" and described in details, which based on an interesting interpretation on palaeogeography and palaeoclimate was offered.

Like the living plants, the binary nomenclature is the essential way for recognizing, naming and studying fossil plants. The binary nomenclature (nomenclatura binominalis) was originally created for naming living plants by Swedish explorer and botanist Carl von Linné in his *Species Plantarum* firstly published in 1753. The nomenclature was firstly adopted for fossil plants by the Czech mineralogist and botanist K. M. von Sternberg in his *Versuch einer Geognostisch, Botanischen Darstellung der Flora der Vorwelt* issued since 1820. The *International Code of Botanical Nomenclature* thus set up the beginning year of modern botanical and palaeobotanical nomenclature as 1753 and 1820 respectively. Our series volumes of Chinese megafossil plants also follows this rule, compile generic and specific names of living plants set up in and after 1753 and of fossil plants set up in and after 1820. As binary nomenclature was firstly used for naming fossil plants found in China by J. S. Newberry [1865 (1867)] at the Smithsonian Institute, USA, his paper *Description of Fossil Plants from the Chinese Coal-bearing Rocks* naturally becomes the starting point of the compiling of Chinese megafossil plant records of the current series.

China has a vast territory covers well developed terrestrial strata, which yield abundant fossil plants. During the past one and over a half centuries, particularly after the two milestones of the founding of PRC in 1949 and the beginning of Reform and Opening to the outside world in late 1970s, to meet the growing demands of the development and construction of the country, various scientific disciplines related to geological prospecting and meaning have been remarkably developed, among which palaeobotanical studies have been also well-developed with lots of fossil materials being accumulated. Preliminary statistics has shown that during 1865 (1867) — 2000, more than 2000 references related to Chinese megafossil plants had been published [Zhou and Wu (chief compilers), 2002]; 525 genera of Mesozoic megafossil plants discovered in China had been reported during 1865 (1867) — 1990 (Wu, 1993a), while 281 genera of Cenozoic megafossil plants found in China had been documented by 1993 (Liu et al., 1996); by the year of 2000, totally about 154 generic names have been established based on Chinese fossil plant material for the Mesozoic and Cenozoic deposits (Wu, 1993b, 2006). The above-mentioned megafossil plant records were published scatteredly in various periodicals or scientific magazines in different languages, such as Chinese, English, German, French, Japanese, Russian, etc., causing much inconvenience for the use and exchange of colleagues of palaeobotany and related fields both at home and abroad.

To resolve this problem, besides bibliographies of palaeobotany [Zhou and Wu (chief compilers), 2002], the compilation of all fossil plant records is an efficient way, which has already obtained enough attention in China since the 1980s (Wu, 1993a, 1993b, 2006). Based on the previous compilation as well as extensive searching for the bibliographies and literatures, now we are planning to publish series volumes of *Record of Megafossil Plants from China* which is tentatively scheduled to comprise volumes of bryophytes, lycophytes, sphenophytes, filicophytes, cycadophytes, ginkgophytes, coniferophytes, angiosperms and others. These volumes are mainly focused on the Mesozoic megafossil plant data that were published from 1865 to 2005.

In each volume, only records of the generic and specific ranks are compiled, with higher ranks in the taxonomical hierarchy, e. g., families, orders, only mentioned in the item of "taxonomy" under each record. For a complete compilation and a well understanding for geological records of the megafossil plants, those genera and species with their type species and type specimens not originally described from China are also included in the volume.

Records of genera are organized alphabetically, followed by the items of author(s) of genus, publishing year of genus, type species (not necessary for genera originally set up for living plants), and taxonomy and others.

Under each genus, the type species (not necessary for genera originally set up for living plants) is firstly listed, and other species are then organized alphabetically. Every taxon with symbols of "aff.""Cf.""cf.""ex gr." or "?" and others in its name is also listed as an individual record but arranged after the species without any symbol. Undetermined species (sp.) are listed at the end of each genus entry. If there are more than one undetermined species (spp.), they will be arranged chronologically. In every record of species (including undetermined species) items of author of species, establishing year of species, and so on, will be included.

Under each record of species, all related reports (on species or specimens) officially published are covered with the exception of those shown solely as names with neither description nor illustration. For every report of the species or specimen, the following items are included: publishing year, author(s) or the person(s) who identify the specimen (species), page(s) of the literature, plate(s), figure(s), preserved organ(s), locality(ies), horizon(s) or stratum(a) and age(s). Different reports of the same specimen (species) is (are) arranged chronologically, and then alphabetically by authors' names, which may further classified into a, b, etc., if the same author(s) published more than one report within one year on the same species.

Records of generic and specific names founded on Chinese specimen(s) is (are) marked by the symbol "△". Information of these records are documented as detailed as possible based on their original publication.

To completely document *Record of Megafossil Plants from China*, we compile all records faithfully according to their original publication without doing any delection or modification, nor offering annotations. However, all related modification and comments published later are included under each record, particularly on those with obvious problems, e.g., invalidly published naked names (nom. nud.).

According to *International Code of Botanical Nomenclature (Vienna Code)* article 36.3, in order to be validly published, a name of a new taxon of fossil plants published on or after January 1st, 1996 must be accompanied by a Latin or English description or diagnosis or by a reference to a previously and effectively published Latin or English description or diagnosis (McNeill and others, 2006; Zhou, 2007; Zhou Zhiyan, Mei Shengwu, 1996; *Brief News of Palaeobotany in China*, No. 38). The current series follows article 36.3 and the original language(s) of description and/or diagnosis is (are) shown in the records for those published on or after January 1st, 1996.

For the convenience of both Chinese speaking and non-Chinese speaking colleagues, every record in this series is compiled as two parts that are of essentially the same contents, in Chinese and English respectively. All cited references are listed only in western language (mainly English) strictly following the format of the English part of Zhou and Wu (chief compilers) (2002).

The publication of series volumes of *Record of Megafossil Plants from China* is the necessity for the discipline accumulation and development. It provides further references for understanding the plant fossil biodiversity evolution and radiation of major plant groups through the geological ages. We hope that the publication of these volumes will be helpful for promoting the professional exchange at home and abroad of palaeobotany.

This book is the one volume of *Record of Megafossil Plants from China*. There are two parts of both Chinese and English versions, mainly documents complete data on the megafossil plant new generic names established for Chinese specimens between the years 1865 to 2010. Each record of the generic taxon include: author(s) who established the genus, establishing year, synonym, type species, taxonomic status. For those generic names or specific names established based on Chinese specimens, the type specimens and their depository institutions have also been recorded. This volume records totally 380 new generic names (among them 195 are established based on Palaeozoic specimens; 177 are established based on Mesozoic specimens and 8 are established based on Cenozoic specimens). The dispersed pollen grains are not included in this book. We are grateful to receive further comments and suggestions form readers and colleagues.

This work is jointly supported by the Basic Work of Science and Technology (2013FY113000) and the State Key Program of Basic Research (2012CB822003) of the Ministry of Science and Technology, the National Natural Sciences Foundation of China (No. 41272010), the State Key Laboratory of Palaeobiology and Stratigraphy (No. 103115), the Important Directional Project (ZKZCX2-YW-154) and the Information Construction Project (INF105-SDB-1-42) of Knowledge Innovation Program of the Chinese Academy of Sciences.

We thank many colleagues and experts from the Department of Palaeobotany and Palynology of Nanjing Institute of Geology and Palaeontology (NIGPS), CAS for helpful suggestions and support. Special thanks are due to Acad. Zhou Zhiyan for his kind help and support for this work, and writing the "General Foreword" of this book. We also acknowledge our sincere thanks to Prof. Zhan Renbin, Prof. Wang Jun (the director of NIGPAS), Acad. Rong Jiayu, and Prof. Yuan Xunlai (the head of the State Key Laboratory of Palaeobiology and Stratigraphy) for their support for successful compilation and publication of this book. Ms. Feng Man, Ms. Chu Cunyin and Sir. Yuan Daojun from the Liboratory of NIGPAS are appreciated for assistances of books and literatures collections.

Editor

SYSTEMATIC RECORDS

Genus *Abropteris* Lee et Tsao, 1976
1976 Lee P C and Tsao Chengyao, in Lee P C and others, p. 100.

Type species: *Abropteris virginiensis* (Fontaine) Lee et Tsao, 1976

Taxonomic status: Osmundaceae, Filicopsida

Distribution and Age: Virginia of USA and Dukou in Sichuan of China; Late Triassic.

Abropteris virginiensis (Fontaine) Lee et Tsao, 1976
1883 *Lonchopteris virginiensis* Fontaine, Fontaine W M, p. 53, pl. 28, figs. 1, 2; pl. 29, figs. 1—4; frond; Virginia, USA; Late Triassic.

1976 Lee P C and Tsao Chengyao, in Lee P C and others, p. 100; frond; Virginia, USA; Late Triassic.

The other species:

Abropteris yongrenensis Lee et Tsao, 1976
1976 Lee P C and Tsao Chengyao, in Lee P C and others, p. 102, pl. 12, figs. 1—3; pl. 13, figs. 6, 10, 11; text-fig. 3-1; sterile pinna; Reg. No. : PB5215 — PB5217, PB5220 — PB5222; Holotype: PB5215 (pl. 12, fig. 1); Repository: Nanjing Institute of Geology and Palaeontology, Chinese Academy of Sciences; Moshahe of Dukou, Sichuan; Late Triassic Daqiaodi Member of Nalaqing Formation.

Genus *Abrotopteris* Mo, 1980
1980 Mo Zhuangguan, in Zhao Xiuhu and others, p. 84.

Type species: *Abrotopteris guizhouensis* (Gu et Zhi) Mo, 1980

Taxonomic status: Plantae incertae sedis

Distribution and Age: Panxian of Guizhou, China; early Late Permian.

Abrotopteris guizhouensis (Gu et Zhi) Mo, 1980
1974 *Glossopteris guizhouensis* Gu et Zhi, Nanjing Institute of Geology and Palaeontology, Institute of Botany, Chinese Academy of Sciences, in *Palaeozoic Plants from China*, p. 137, pl. 110, figs. 3, 4; leaves; Reg. No. : PB4983, PB4984; Syntypes[①]: PB4983, PB4984 (pl. 110, figs. 3, 4); Repository: Nanjing Institute of Geology and Palaeontology, Chinese

---

[①] Notes: According to *International Code of Botanical Nomenclature* (*Vienna Code*) article 37.2, from the year 1958, the holotype type specimen should be unique.

Academy of Sciences; Panxian, Guizhou; early Late Permian lower member of Xuanwei Formation.
1980 Mo Zhuangguan, in Zhao Xiuhu and others, p. 84, pl. 19, figs. 1 — 7,8a (?); leaves; Reg. No. : PB4983, PB4984, PB7075 — PB7080; Repository: Nanjing Institute of Geology and Palaeontology, Chinese Academy of Sciences; Panxian, Guizhou; early Late Permian lower member of Xuanwei Formation.

Genus *Acanthocladus* Yang, 2006 (in Chinese and English)
2006 Yang Guanxiu and others, pp. 165, 321.
Type species: *Acanthocladus xyloides* Yang, 2006
Taxonomic status: Plantae incertae sedis
Distribution and Age: Yuzhou of western Henan, China; Late Permian.

Acanthocladus xyloides Yang, 2006 (in Chinese and English)
2006 Yang Guanxiu and others, pp. 165, 321, pl. 38, fig. 7; stem or shoot; Holotype: HEP0948 (pl. 38, fig. 7); Repository: China University of Geosciences, Beijing; Dafengkou and Yungaishan of Yuzhou, western Henan; Late Permian member 6 of Yungaishan Formation.

Genus *Acanthopteris* Sze, 1931
1931 Sze H C, p. 53.
Type species: *Acanthopteris gothani* Sze, 1931
Taxonomic status: Dicksoniaceae, Filicopsida
Distribution and Age: Fuxin of Liaoning, China; Early Jurassic (Lias).

Acanthopteris gothani Sze, 1931
1931 Sze H C, p. 53, pl. 7, figs. 2 — 4; frond; Sunjiagou of Fuxin, Liaoning; Early Jurassic (Lias).

Genus *Acerites* Pan, 1983 (nom. nud.)
1983 Pan Guang, p. 1520. (in Chinese)
1984 Pan Guang, p. 959. (in English)
Type species: (without specific name)
Taxonomic status: "primitive angiosperms"
Distribution and Age: western Liaoning, China; Middle Jurassic.

Acerites sp. indet.
[Notes: Generic name was given only, but without specific name (or type species) in the original paper]

1983 *Acerites* sp. indet. ,Pan Guang,p. 1520; western Liaoning (about 45°58′N,120°21′E); Middle Jurassic Haifanggou Formation. (in Chinese)

1984 *Acerites* sp. indet. ,Pan Guang,p. 959; western Liaoning (about 45°58′N,120°21′E); Middle Jurassic Haifanggou Formation. (in English)

Genus *Aconititis* Pan,1983 (nom. nud.)

1983 Pan Guang,p. 1520. (in Chinese)

1984 Pan Guang,p. 959. (in English)

Type species:(without specific name)

Taxonomic status:"primitive angiosperms"

Distribution and Age:western Liaoning,China;Middle Jurassic.

Aconititis sp. indet.

[Notes:Generic name was given only, but without specific name (or type species) in the original paper]

1983 *Aconititis* sp. indet. ,Pan Guang,p. 1520; western Liaoning (about 45°58′N,120°21′E); Middle Jurassic Haifanggou Formation. (in Chinese)

1984 *Aconititis* sp. indet. ,Pan Guang,p. 959; western Liaoning (about 45°58′N,120°21′E); Middle Jurassic Haifanggou Formation. (in English)

Genus *Acthephyllum* Duan et Chen,1982

1982 Duan Shuying and Chen Ye,p. 510.

Type species:*Acthephyllum kaixianense* Duan et Chen,1982

Taxonomic status:Gymnospermae incertae sedis

Distribution and Age:Kaixian of Sichuan,China;Late Triassic.

Acthephyllum kaixianense Duan et Chen,1982

1982 Duan Shuying and Chen Ye,p. 510,pl. 11,figs. 1 − 5; fern-like leaves; Reg. No. : No. 7173 − No. 7176, No. 7219; Holotype: No. 7219 (pl. 11, fig. 3); Repository: Institute of Botany, Chinese Academy of Sciences; Tongshuba of Kaixian, Sichuan; Late Triassic Hsuchiaho Formation.

Genus *Actinophyllus* Xiao,1985 (non *Actinophyllum* Phillips,1848)

[Notes:The type species is *Actinophyllum plicatum* Phillips,1848 (in Phillips J and Salter J W,1848,p. 386,pl. 30,fig. 4;alga?;Compared *Actabulum*;near Stoke Edith,Woolhope district, Scotland;Devonian)]

1985 Xiao Suzhen,in Xiao Suzhen and Zhang Enpeng,p. 585.

Type species:*Actinophyllus cordaioides* Xiao,1985

Taxonomic status: Plantae incertae sedis

Distribution and Age: Taiyuan of Shanxi, China; Early Permian.

Actinophyllus cordaioides Xiao, 1985

1985　Xiao Suzhen, in Xiao Suzhen and Zhang Enpeng, p. 585, pl. 205, figs. 1 — 3; text-fig. 39; leaves; Reg. No. : SH721, SH722, SH725; Holotype: SH722 (pl. 205, fig. 1); Paratypes: SH721 (pl. 205, fig. 2), SH725 (pl. 205, fig. 3); Taiyuan, Shanxi; Early Permian Lower Shihezi (Shihhotse) Formation.

Genus *Aculeovinea* Li et Taylor, 1998 (in English)

1998　Li Hongqi and Taylor D W, p. 1024.

Type species: *Aculeovinea yunguiensis* Li et Taylor, 1998

Taxonomic status: Gigantopteridaceae, Gigantopteridales

Distribution and Age: Panxian of Guizhou, China; Late Permian.

Aculeovinea yunguiensis Li et Taylor, 1998 (in English)

1998　Li Hongqi and Taylor D W, p. 1024, figs. 1 — 3; fig. 4. 25; stems bear prickles; Holotype: PLY02 (rock sample), Slides PLY02♯C10 (L)-1, PLY02♯C10 (L)-2, PLY02♯C10-4S1, PLY02♯C10-4S2 (figs. 1. 1 — 1. 3); Paratype: PLY02 (rock sample), Slides PLY02♯A, PLY02♯A1-1-1-3, PLY02♯A1-1-2-2, PLY02♯A1-2-1-2, PLY02♯C7-6T, PLY02♯C10 (L)-2, PLY02♯C15 (R-A)-2, PLY02♯A2-3-3-2 (figs. 1. 4, 2. 10 — 3. 23); Repository: National Museum of Plant History of China, Institute of Botany, Chinese Academy of Sciences; Yueliangtian Coal Mine of Panxian, Guizhou; Late Permian Xuanwei Formation.

Genus *Adoketophyllum* Li et Edwards, 1992

1992　Li Chengsen and Edwards D, p. 259.

Type species: *Adoketophyllum subverticillatum* (Li et Cai) Li et Edwards, 1992

Taxonomic status: Psilophytes incertae sedis

Distribution and Age: Wenshan of Yunnan, China; Early Devonian.

Adoketophyllum subverticillatum (Li et Cai) Li et Edwards, 1992

1977　*Zosterophyllum subverticillatum* Li et Cai, Li Xingxue and Cai Chongyang, p. 24, pl. 3, figs. 1 — 3, 3a; text-fig. 8; strobili; Col. No. : ACE187 (pl. 3, figs. 1, 2), ACE188 (pl. 3, figs. 3, 3a); Reg. No. : PB6464 — PB6466; Holotype: PB6466 (pl. 3, figs. 3, 3a); Repository: Nanjing Institute of Geology and Palaeontology, Chinese Academy of Sciences; Wenshan, Yunnan; Early Devonian Posongchong Formation.

1992　Li Chengsen and Edwards D, p. 259, pls. 1 — 4; text-fig. 2; strobili; Wenshan, Yunnan; Early Devonian Posongchong Formation.

Genus *Aetheopteris* Chen G X et Meng, 1984
1984 Chen Gongxin and Meng Fansong, in Chen Gongxin, p. 587.
Type species: *Aetheopteris rigida* Chen G X et Meng, 1984
Taxonomic status: Gymnospermae incertae sedis
Distribution and Age: Jingmen of Hubei, China; Late Triassic.

Aetheopteris rigida Chen G X et Meng, 1984
1984 Chen Gongxin and Meng Fansong, in Chen Gongxin, p. 587, pl. 261, figs. 3, 4; pl. 262, fig. 3; text-fig. 133; fern-like leaf; Reg. No. : EP685; Holotype: EP685 (pl. 262, fig. 3); Repository: Geological Bureau of Hubei Province; Paratype: pl. 261, figs. 3, 4; Repository: Yichang Institute of Geology and Mineral Resources; Fenshuiling of Jingmen, Hubei; Late Triassic Jiuligang Formation.

Genus *Aipteridium* Li et Yao, 1983
1983 Li Xingxue and Yao Zhaoqi, p. 322.
Type species: *Aipteridium pinnatum* (Sixtel) Li et Yao, 1983
Taxonomic status: Pteridospermopsida
Distribution and Age: South Fergana; Late Triassic. China; Palaeozoic.

Aipteridium pinnatum (Sixtel) Li et Yao, 1983
1961 *Aipteris pinnatum* Sixtel, Sixtel T A, p. 153, pl. 3; South Fergana; Late Triassic.
1983 Li Xingxue and Yao Zhaoqi, p. 322.

Genus *Alloephedra* Tao et Yang, 2003 (in Chinese and English)
2003 Tao Junrong and Yang Yong, pp. 209, 212.
Type species: *Alloephedra xingxuei* Tao et Yang, 2003
Taxonomic status: Ephedraceae, Gnetales
Distribution and Age: Yanbian Basin of Jilin, China; Early Cretaceous.

Alloephedra xingxuei Tao et Yang, 2003 (in Chinese and English)
2003 Tao Junrong and Yang Yong, pp. 209, 212, pls. 1, 2; stems, branches and female cones terminate to the branchlets; No. : No. 54018a, No. 54018b; Holotypes: No. 54018a, No. 54018b (pl. 1, fig. 1); Repository: National Museum of Plant History of China, Institute of Botany, Chinese Academy of Sciences; Yanbian Basin, Jilin; Early Cretaceous Dalazi Formation.

Genus *Allophyton* Wu, 1982

1982   Wu Xiangwu, p. 53.

Type species: *Allophyton dengqenensis* Wu, 1982

Taxonomic status: Filicopsida?

Distribution and Age: Dengqen of Xizang (Tibet), China; Late Triassic?.

Allophyton dengqenensis Wu, 1982

1982   Wu Xiangwu, p. 53, pl. 6, fig. 1; pl. 7, figs. 1, 2; fern-like stem; Col. No.: RN0038, RN0040, RN0045; Reg. No.: PB7263 — PB7265; Holotype: PB7263 (pl. 6, fig. 1); Repository: Nanjing Institute of Geology and Palaeontology, Chinese Academy of Sciences; Dengqen, Xizang (Tibet); Mesozoic coal-bearing strata (Late Triassic?).

Genus *Amdrupiopsis* Sze et Lee, 1952

1952   Sze H C and Lee H H, pp. 6, 24.

Type species: *Amdrupiopsis sphenopteroides* Sze et Lee, 1952

Taxonomic status: Gymnospermae incertae sedis

Distribution and Age: Weiyuan of Sichuan, China; Early Jurassic.

Amdrupiopsis sphenopteroides Sze et Lee, 1952

1952   Sze H C and Lee H H, pp. 6, 24, pl. 3, figs. 7 — 7b; text-fig. 1; fern-like leaves; Repository: Nanjing Institute of Geology and Palaeontology, Chinese Academy of Sciences; Aishanzi of Weiyuan, Sichuan; Early Jurassic. [Notes: This specimen was referred as *Amdrupia stenodonta* Harris (Hsu J, 1954) or as *Amdrupia sphenopteroides* (Sze et Lee) Lee (Sze H C, Lee H H and others, 1963)]

Genus *Amentostrobus* Pan, 1983 (nom. nud.)

1983   Pan Guang, p. 1520. (in Chinese)

1984   Pan Guang, p. 958. (in English)

Type species: (only generic name)

Taxonomic status: Coniferopsida

Distribution and Age: western Liaoning, China; Middle Jurassic.

Amentostrobus sp. indet.

[Notes: Generic name was given only, but without specific name (or type species) in the original paper]

1983   *Amentostrobus* sp. indet., Pan Guang, p. 1520; western Liaoning (about 45°58′N, 120°21′E); Middle Jurassic Haifanggou Formation. (in Chinese)

1984   *Amentostrobus* sp. indet., Pan Guang, p. 958; western Liaoning (about 45°58′N, 120°21′E);

Middle Jurassic Haifanggou Formation. (in English)

Genus *Amphiephedra* Miki, 1964
1964   Miki S, pp. 19, 21.
Type species: *Amphiephedra rhamnoides* Miki, 1964
Taxonomic status: Ephedraceae, Gnetopsida
Distribution and Age: Lingyuan of western Liaoning, China; Late Jurassic.

Amphiephedra rhamnoides Miki, 1964
1964   Miki S, pp. 19, 21, pl. 1, fig. F; shoot with normal leaves; Lingyuan, western Liaoning; Late Jurassic *Lycoptera* Bed.

Genus *Amplectosporangium* Geng, 1992
1992b   Geng Baoyin, p. 451.
Type species: *Amplectosporangium jiagyouense* Geng, 1992
Taxonomic status: Psilophytes incertae sedis
Distribution and Age: Jiangyou of Sichuan, China; Early Devonian.

Amplectosporangium jiagyouense Geng, 1992
1992b   Geng Baoyin, p. 451, pl. 1, figs. 1 — 7; pl. 2, figs. 1 — 8; text-fig. 1; fertile branch-systems and sporangia; No. : 8255 — 8258, 8356 — 8360; Holotype: 8356 (pl. 1, fig. 1); Paratypes: 8357 — 8360, 8255 — 8258; Repository: Institute of Botany, Chinese Academy of Sciences; Jiangyou, Sichuan; Early Devonian Pingyipu Formation.

Genus *Angustiphyllum* Huang, 1983
1983   Huang Qisheng, p. 33.
Type species: *Angustiphyllum yaobuense* Huang, 1983
Taxonomic status: Pteridospermopsida
Distribution and Age: Huaining of Anhui, China; Early Jurassic.

Angustiphyllum yaobuense Huang, 1983
1983   Huang Qisheng, p. 33, pl. 4, figs. 1 — 7; leaves; Reg. No. : Ahe8132, Ahe8134 — Ahe8138, Ahe8140; Holotypes: Ahe8132, Ahe8134 (pl. 4, figs. 1, 2); Repository: Palaeontological Section, Wuhan Institute of Geology; Lalijian of Huaining, Anhui; Early Jurassic lower part of Xiangshan Group.

Genus *Annularites* Halle, 1927
1927   Halle T G, p. 19.

Type species: *Annularites ensilolius* Halle, 1927

Taxonomic status: Equisetales

Distribution and Age: central Shanxi, China; late Early Permian.

Annularites ensilolius Halle, 1927
1927   Halle T G, p. 19, pl. 1, figs. 1 — 5; pl. 2, figs. 1, 2; pl. 3, figs. 1 — 4; pl. 4, figs. 1 — 3; foliage; Ch'en-chia-yu, central Shanxi; late Early Permian Lower Shihhotse Series.

The other species:
Annularites lingulatus Halle, 1927
1927   Halle T G, p. 26, pl. 1, fig. 6; pl. 2, figs. 3, 4; pl. 6, fig. 4; foliage; Ch'en-chia-yu, central Shanxi; late Early Permian Upper Shihhotse Series.

Annularites heianensis (Kodaira) Halle, 1927
1924   *Schizoneura heianensis* Kodaira, Kodaira R, p. 163, pl. 23.
1927   Halle T G, p. 27, pl. 5, figs. 13, 14; foliage; Ch'en-chia-yu, central Shanxi; late Early Permian Upper Shihhotse Series.

Annularites sinensis Halle, 1927
1927   Halle T G, p. 28, pl. 5, figs. 6 — 11, 12 (?); pl. 6 (?), figs. 8, 9; foliage; Ch'en-chia-yu, central Shanxi; late Early Permian Upper Shihhotse Series.

Genus *Archaefructus* Sun, Dilcher, Zheng et Zhou, 1998 (in English)
1998   Sun Ge, Dilcher D L, Zheng Shaolin and Zhou Zhekun, p. 1692.

Type species: *Archaefructus liaoningensis* Sun, Dilcher, Zheng et Zhou, 1998

Taxonomic status: Dicotyledoneae

Distribution and Age: Beipiao of western Liaoning, China; Late Jurassic.

Archaefructus liaoningensis Sun, Dilcher, Zheng et Zhou, 1998 (in English)
1998   Sun Ge, Dilcher D L, Zheng Shaolin and Zhou Zhekun, p. 1692, figs. 2A — 2C; angiosperm fruiting axes and cuticles of seed-coats; Reg. No.: SZ0916; Holotype: SZ0916 (fig. 2A); Huangbanjigou near Shangyuan of Beipiao, western Liaoning; Late Jurassic Yixian Formation.

Genus *Archimagnolia* Tao et Zhang, 1992
1992   Tao Junrong and Zhang Chuanbo, pp. 423, 424.

Type species: *Archimagnolia rostrato-stylosa* Tao et Zhang, 1992

Taxonomic status: Dicotyledoneae

Distribution and Age: Yanji of Jilin, China; Early Cretaceous.

Archimagnolia rostrato-stylosa Tao et Zhang, 1992
1992   Tao Junrong and Zhang Chuanbo, pp. 423, 424, pl. 1, figs. 1 — 6; an impression of froral

axis; No.: 503882; Holotype: 503882 (pl. 1, figs. 1－6); Repository: National Museum of Plant History of China, Institute of Botany, Chinese Academy of Sciences; Yanji, Jilin; Early Cretaceous Dalazi Formation.

Genus *Archimalus* Tao, 1992
1992   Tao Junrong, pp. 240, 241.
Type species: *Archimalus calycina* Tao, 1992
Taxonomic status: Maloidea, Rosaceae, Dicotyledoneae
Distribution and Age: Shanwang of Shandong, China; Middle Miocene.

Archimalus calycina Tao, 1992
1992   Tao Junrong, pp. 240, 241, pl. 1, figs. 6－8; fossil flower of angiosperm; No.: 053880; Repository: National Museum of Plant History of China, Institute of Botany, Chinese Academy of Sciences; Shanwang, Shandong; Middle Miocene.

Genus *Areolatophyllum* Li et He, 1979
1979   Li Peijuan and He Yuanliang, in He Yuanliang and others, p. 137.
Type species: *Areolatophyllum qinghaiense* Li et He, 1979
Taxonomic status: Dipteridaceae, Filicopsida
Distribution and Age: Dulan of Qinghai, China; Late Triassic.

Areolatophyllum qinghaiense Li et He, 1979
1979   Li Peijuan and He Yuanliang, in He Yuanliang and others, p. 137, pl. 62, figs. 1, 1a, 2, 2a; frond; Col. No.: 58-7a-12; Reg. No.: PB6327, PB6328; Holotype: PB6328 (pl. 62, figs. 1, 1a); Paratype: PB6327 (pl. 62, figs. 2, 2a); Repository: Nanjing Institute of Geology and Palaeontology, Chinese Academy of Sciences; Babaoshan of Dulan, Qinghai; Late Triassic Babaoshan Group.

Genus *Asiatifolium* Sun, Guo et Zheng, 1992
1992   Sun Ge, Guo Shuangxing and Zheng Shaolin, in Sun Ge and others, p. 546. (in Chinese)
1993   Sun Ge, Guo Shuangxing and Zheng Shaolin, in Sun Ge and others, p. 253. (in English)
Type species: *Asiatifolium elegans* Sun, Guo et Zheng, 1992
Taxonomic status: Dicotyledoneae
Distribution and Age: Jixi of Heilongjiang, China; Early Cretaceous.

Asiatifolium elegans Sun, Guo et Zheng, 1992
1992   Sun Ge, Guo Shuangxing and Zheng Shaolin, in Sun Ge and others, p. 546, pl. 1, figs. 1－3; leaves; Reg. No.: PB16766, PB16767; Holotype: PB16766 (pl. 1, fig. 1); Repository: Nanjing Institute of Geology and Palaeontology, Chinese Academy of Sciences; Chengzihe

1993　Sun Ge,Guo Shuangxing and Zheng Shaolin,in Sun Ge and others,p. 253,pl. 1,figs. 1 — 3; leaves; Reg. No. : PB16766, PB16767; Holotype: PB16766 (pl. 1, fig. 1); Repository: Nanjing Institute of Geology and Palaeontology,Chinese Academy of Sciences;Chengzihe of Jixi,Heilongjiang;Early Cretaceous upper part of Chengzihe Formation. (in English)

Genus *Asiopteris* Zhang,1987
1987　Zhang Hong,p. 203.
Type species:*Asiopteris huairenensis* Zhang,1987
Taxonomic status:Filices or Pteridospermopsida
Distribution and Age:Huairen of Shanxi,China;Early Permian.

Asiopteris huairenensis Zhang,1987
1987　Zhang Hong,p. 203,pl. 17,figs. 1 — 5; fronds; No. : Hs-No. 5; Reg. No. : Mp-85152 — Mp-85154; Huairen, Shanxi; Early Permian.

Genus *Astrocupulites* Halle,1927
1927　Halle T G,p. 219.
Type species:*Astrocupulites acuminatus* Halle,1927
Taxonomic status:Plantae incertae sedis
Distribution and Age:Central Shanxi,China;late Early Permian.

Astrocupulites acuminatus Halle,1927
1927　Halle T G,p. 219,pl. 48,figs. 10,11; cupules; Ch'en-chia-yu, central Shanxi; late Early Permian Lower Shihhotse Series.

Genus *Baiguophyllum* Duan,1987
1987　Duan Shuying,p. 52.
Type species:*Baiguophyllum lijianum* Duan,1987
Taxonomic status:Czekanowskiales
Distribution and Age:West Hill of Beijing,China;Middle Jurassic.

Baiguophyllum lijianum Duan,1987
1987　Duan Shuying,p. 52,pl. 16,figs. 4,4a;pl. 17,fig. 1; text-fig. 14; leaves, long shoots and dwarf shoots;No. :S-PA-86-680 (1),S-PA-86-680 (2);Holotype:S-PA-86-680 (2) (pl. 17, fig. 1); Repository: Department of Palaeobotany, Swedish Museum of Natural History;Zhaitang of West Hill,Beijing;Middle Jurassic Mentougou Coal Series.

Genus *Batenburgia* Hilton et Geng,1998 (in English)
1998　Hilton J and Geng Baoyin,p. 265.

Type species:*Batenburgia sakmarica* Hilton et Geng,1998

Taxonomic status:Coniferales

Distribution and Age:Henan,China;Early Permian (Sakmarian).

Batenburgia sakmarica Hilton et Geng,1998 (in English)
1998　Hilton J and Geng Baoyin, p. 265, pls. 1 — 6; text-figs. 3, 4; cone; No. : CBP9199 — CBP9200; Holotypes:CBP9199a (pl. 1, fig. 1),CBP9199b (counterpart) (pl. 1, fig. 2); Repository:Institute of Botany,Chinese Academy of Sciences;Guanyintan, Henan; Early Permian (Sakmarian) Shanxi Formation.

Genus *Beipiaoa* Dilcher,Sun et Zheng,2001 (in Chinese and English)
2001　Dilcher D L,Sun Ge and Zheng Shaolin,in Sun Ge and others,pp. 25,151.

Type species:*Beipiaoa spinosa* Dilcher,Sun et Zheng,2001

Taxonomic status:Angiospermae?

Distribution and Age:Beipiao of western Liaoning,China;Late Jurassic.

Beipiaoa spinosa Dilcher,Sun et Zheng,2001 (in Chinese and English)
2001　Dilcher D L,Sun Ge and Zheng Shaolin,in Sun Ge and others,pp. 26,152,pl. 5,figs. 1 — 4,5 (?); pl. 33, figs. 11 — 19; text-fig. 4. 7G; fruits; Reg. No. : PB18959 — PB18962, PB18966 — PB18967, ZY3004 — ZY3006; Holotype: PB18959 (pl. 5, fig. 1); Huangbanjigou in Shangyuan of Beipiao, western Liaoning; Late Jurassic Jianshangou Formation.

The other species:

Beipiaoa parva Dilcher,Sun et Zheng,2001 (in Chinese and English)
1999　*Trapa*? sp. , Wu Shunqing, p. 22, pl. 16, figs. 1 — 2a, 6 (?), 6a (?), 8 (?); fruits; Huangbanjigou in Shangyuan of Beipiao, western Liaoning; Late Jurassic Jianshangou Bed in lower part of Yixian Formation.

2001　Dilcher D L,Sun Ge and Zheng Shaolin,in Sun Ge and others,pp. 25, 151,pl. 5,fig. 7; pl. 33, figs. 1 — 8, 21; text-fig. 4. 7A; fruits; Reg. No. : PB18953, ZY3001 — ZY3003; Holotype: PB18953 (pl. 5, fig. 7); Huangbanjigou in Shangyuan of Beipiao, western Liaoning; Late Jurassic Jianshangou Formation.

Beipiaoa rotunda Dilcher,Sun et Zheng,2001 (in Chinese and English)
2001　Dilcher D L,Sun Ge and Zheng Shaolin,in Sun Ge and others,pp. 25,151,pl. 5,figs. 8,6 (?); pl. 33, figs. 10, 9 (?); text-fig. 4. 7B; fruits; Reg. No. : PB18958, ZY3001 — ZY3003; Holotype: PB18958 (pl. 5, fig. 8); Huangbanjigou in Shangyuan of Beipiao, western Liaoning; Late Jurassic Jianshangou Formation.

Genus *Bennetdicotis* Pan, 1983 (nom. nud.)

1983   Pan Guang, p. 1520. (in Chinese)
1984   Pan Guang, p. 958. (in English)
Type species: (without specific name)
Taxonomic status: "hemiangiosperms"
Distribution and Age: western Liaoning, China; Middle Jurassic.

Bennetdicotis sp. indet.

[Notes: Generic name was given only, but without specific name (or type species) in the original paper]

1983   *Bennetdicotis* sp. indet., Pan Guang, p. 1520; western Liaoning (about 45°58′N, 120°21′E); Middle Jurassic Haifanggou Formation. (in Chinese)
1984   *Bennetdicotis* sp. indet., Pan Guang, p. 958; western Liaoning (about 45°58′N, 120°21′E); Middle Jurassic Haifanggou Formation. (in English)

Genus *Benxipteris* Zhang et Zheng, 1980

1980   Zhang Wu and Zheng Shaolin, in Zhang Wu and others, p. 263.
Type species: *Benxipteris acuta* Zhang et Zheng, 1980 [Notes: The type species was not designated in the original paper; *Benxipteris acuta* Zhang et Zheng was designated by Wu Xiangwu (1993) as the type species]
Taxonomic status: Pteridospermopsida
Distribution and Age: Benxi of Liaoning, China; Middle Triassic.

Benxipteris acuta Zhang et Zheng, 1980

1980   Zhang Wu and Zheng Shaolin, in Zhang Wu and others, p. 263, pl. 108, figs. 1—13; text-fig. 193; sterile and fertile leaves; Reg. No.: D323 — D335; Repository: Shenyang Institute of Geology and Mineral Resources; Linjiawaizi of Benxi, Liaoning; Middle Triassic Linjia Formation.

The other species:

Benxipteris densinervis Zhang et Zheng, 1980

1980   Zhang Wu and Zheng Shaolin, in Zhang Wu and others, p. 264, pl. 107, figs. 3 — 6; text-fig. 194; sterile and fertile leaves; Reg. No.: D319 — D322; Repository: Shenyang Institute of Geology and Mineral Resources; Linjiawaizi of Benxi, Liaoning; Middle Triassic Linjia Formation.

Benxipteris partita Zhang et Zheng, 1980

1980   Zhang Wu and Zheng Shaolin, in Zhang Wu and others, p. 265, pl. 107, figs. 7 — 9; pl. 109, figs. 6 — 7; fern-like leaf; Reg. No.: D344 — D346, D336 — D337; Repository: Shenyang Institute of Geology and Mineral Resources; Linjiawaizi of Benxi, Liaoning;

Middle Triassic Linjia Formation.

Benxipteris polymorpha Zhang et Zheng, 1980
1980　Zhang Wu and Zheng Shaolin, in Zhang Wu and others, p. 265, pl. 109, figs. 1 — 5; fern-like leaf; Reg. No. : D338 — D342; Linjiawaizi of Benxi, Liaoning; Middle Triassic Linjia Formation.

Genus *Bicoemplectopteris* Asama, 1959
1959　Asama K, p. 57.
Type species: *Bicoemplectopteris hallei* Asama, 1959
Taxonomic status: Gigantopterides
Distribution and Age: Benxi (Penchi) of Liaoning, Taiyuan of Shanxi and Fujian of China and Korea; Late Palaeozoic.

Bicoemplectopteris hallei Asama, 1959
1927　*Gigantopteris nicotianaefolia* Halle, Halle T G, p. 162, pls. 43 — 44, figs. 1 — 13; pl. 45, figs. 1 — 5; pl. 46, fig. 1; pl. 47, fig. 10; pl. 48, figs. 1 (?) — 6 (?), 7.
1959　Asama K, p. 57, pl. 5, figs. 1, 6; pl. 6, figs. 1 — 4; pl. 7, figs. 1 — 3; fronds and seeds; Benxi (Penchi) of Liaoning, Taiyuan of Shanxi and Fujian of China and Korea; Late Palaeozoic.

Genus *Bifariusotheca* Zhao, 1980
1980　Zhao Xiuhu and others, p. 80.
Type species: *Bifariusotheca qinglongensis* Zhao, 1980
Taxonomic status: Pecopterides
Distribution and Age: Guizhou, China; early Late Permian.

Bifariusotheca qinglongensis Zhao, 1980
1980　Zhao Xiuhu and others, p. 81, pl. 12, figs. 5, 5a, 6; fertile fronds; Col. No. : H11-7; Reg. No. : PB7032, PB7033; Repository: Nanjing Institute of Geology and Palaeontology, Chinese Academy of Sciences; Qinglong, Guizhou; early Late Permian.

Genus *Bilobphyllum* He, Liang et Shen, 1996 (in Chinese and English)
1996　He Xilin, Liang Dunshi and Shen Shuzhong, pp. 78, 165.
Type species: *Bilobphyllum fengchengensis* He, Liang et Shen, 1996
Taxonomic status: Filices or Pteridospermopsida
Distribution and Age: Jiangxi, China; Late Permian.

Bilobphyllum fengchengensis He, Liang et Shen, 1996 (in Chinese and English)
1996　He Xilin, Liang Dunshi and Shen Shuzhong, pp. 78, 165, pl. 79, figs. 1, 2; pl. 80, figs. 1 — 5;

leaves; Reg. No.: X88324 — X88330; Syntypes: X88324 — X88330 (pl. 79, figs. 1, 2; pl. 80, figs. 1 — 5); Repository: Department of Geology, China University of Mining and Technology; Fengcheng, Jiangxi; Late Permian Lower Laoshan Submember of Loping Formation.

Genus *Boseoxylon* Zheng et Zhang, 2005 (in Chinese and English)
2005   Zheng Shaolin and Zhang Wu, in Zheng Shaolin and others, pp. 209, 212.
Type species: *Boseoxylon andrewii* (Bose et Sah) Zheng et Zhang, 2005
Taxonomic status: Cycadophytes
Distribution and Age: Rajmahal Hills of Behar, India; Jurassic.

Boseoxylon andrewii (Bose et Sah) Zheng et Zhang, 2005 (in Chinese and English)
1954   *Sahnioxylon andrewii* Bose et Sah, Bose M N and Sah S C D, p. 4, pl. 2, figs. 11 — 18; wood of cycadophytes; Rajmahal Hills of Behar, India; Jurassic.
2005   Zheng Shaolin and Zhang Wu, in Zheng Shaolin and others, pp. 209, 212; Rajmahal Hills of Behar, India; Jurassic.

Genus *Botrychites* Wu S, 1999 (in Chinese)
1999   Wu Shunqing, p. 13.
Type species: *Botrychites reheensis* Wu S, 1999
Taxonomic status: Botrychiaceae?, Filicopsida
Distribution and Age: Beipiao of western Liaoning, China; Late Jurassic.

Botrychites reheensis Wu S, 1999 (in Chinese)
1999a   Wu Shunqing, p. 13, pl. 4, figs. 8 — 10A, 10a; pl. 6, figs. 1 — 3a; sterile frond and fertile frond; Col. No.: AEO-233, AEO-65, AEO-66, AEO-233a, AEO-233, AEO-117, AEO-119; Reg. No.: PB18248 — PB18253; Holotype: PB18257 (pl. 6, fig. 2); Repository: Nanjing Institute of Geology and Palaeontology, Chinese Academy of Sciences; Huangbanjigou in Shangyuan of Beipiao, western Liaoning; Late Jurassic Jianshangou Bed in lower part of Yixian Formation.

Genus *Bracteophyton* Wang et Hao, 2004 (in English)
2004   Wang Deming and Hao Shougang, p. 337.
Type species: *Bracteophyton variatum* Wang et Hao, 2004
Taxonomic status: Psilophytes incertae sedis
Distribution and Age: Qujing of Yunnan, China; Early Devonian.

Bracteophyton variatum Wang et Hao, 2004 (in English)
2004   Wang Deming and Hao Shougang, p. 337, figs. 1 — 5; fertile axes; Holotypes: WH9901A

(fig. 1a), WH9901B2 (fig. 3a); Paratype: WH9902 (fig. 2); Repository: Department of Geology, Peking University; Xujiachong of Qujing, Yunnan; Early Devonian Xujiachong Formation.

Genus *Callianthus* Wang et Zheng, 2009 (in English)
2009　Wang Xin and Zheng Shaolin, p. 800.
Type species: *Callianthus dilae* Wang et Zheng, 2009
Taxonomic status: Angiosperms
Distribution and Age: Beipiao of western Liaoning, China; Early Cretaceous.

Callianthus dilae Wang et Zheng, 2009 (in English)
2009　Wang Xin and Zheng Shaolin, p. 800, figs. 1 — 5; bisexual flower; Reg. No.: PB21047a, PB21047b, PB18320, PB21091a, PB21091b, PB21092; Holotypes: PB21047a (figs. 1-A, C, D), PB21047b (figs. 1-B-E, G-L); Repository: Nanjing Institute of Geology and Palaeontology, Chinese Academy of Sciences; Huangbanjigou of Beipiao, Liaoning (41°12′N, 119°22′E); Early Cretaceous (Barremian) Yixian Formation.

Genus *Casuarinites* Pan, 1983 (nom. nud.)
1983　Pan Guang, p. 1520. (in Chinese)
1984　Pan Guang, p. 959. (in English)
Type species: (without specific name)
Taxonomic status: "primitive angiosperms"
Distribution and Age: western Liaoning, China; Middle Jurassic.

Casuarinites sp. indet.
[Notes: Generic name was given only, but without specific name (or type species) in the original paper]
1983　*Casuarinites* sp. indet., Pan Guang, p. 1520; western Liaoning (about 45°58′N, 120°21′E); Middle Jurassic Haifanggou Formation. (in Chinese)
1984　*Casuarinites* sp. indet., Pan Guang, p. 959; western Liaoning (about 45°58′N, 120°21′E); Middle Jurassic Haifanggou Formation. (in English)

Genus *Catenalis* Hao et Beck, 1991
1991a　Hao Shougang and Beck C B, p. 874.
Type species: *Catenalis dichotoma* Hao et Beck, 1991
Taxonomic status: Psilophytes incertae sedis
Distribution and Age: Wenshan of Yunnan, China; Early Devonian.

Catenalis digitata Hao et Beck, 1991
1991a　Hao Shougang and Beck C B, p. 874, figs. 1 — 24; vegetative and fertile branchlets;

Holotype: PUH-Ch 801 (fig. 1); Paratypes: PUH-Ch 803 (fig. 6), PUH-Ch 807 (fig. 7), PUH-Ch 806 (fig. 8); Repository: Department of Geology, Peking University; Wenshan, Yunnan; Early Devonian Posongchong Formation.

Genus *Cathaiopteridium* Obrhel, 1966

1966   Obrhel Jiri, p. 442.

Type species: *Cathaiopteridium minutum* (Halle) Obrhel, 1966

Taxonomic status: Primofilices

Distribution and Age: Yunnan, China; Early Devonian.

Cathaiopteridium minutum (Halle) Obrhel, 1966

1936   *Protopteridium minutum* Halle, Halle T G, p. 16, pls. 4, 5; text-fig. 2; sporangium-bearing branches; Yunnan; Early Devonian.

1966   Obrhel Jiri, p. 442.

Genus *Cathayanthus* Wang, Tian et Galtier, 2003 (in English)

2003   Wang Shijun, Tian Baolin and Galtier J, p. 98.

Type species: *Cathayanthus ramentrarus* (Wang et Tian) Wang, Tian et Galtier, 2003

Taxonomic status: Cordaitalean

Distribution and Age: Taiyuan of Shanxi, China; Early Permian.

Cathayanthus ramentrarus (Wang et Tian) Wang, Tian et Galtier, 2003 (in English)

1991   *Cardaianthus ramentrarus* Wang et Tian, Wang Shijun and Tian Baolin, p. 743, pls. 1 — 3; text-figs. 1 — 4; male cordaitean reproductive organs; Xishan Coal Field of Taiyuan, Shanxi; Early Permian Taiyuan Formation.

2003   Wang Shijun, Tian Baolin and Galtier J, p. 98, figs. 7 — 9; fertile shoot systems and male shoot systems; Holotype: Coal balls TN8 Slide W113, T7-70B W133; Paratypes: Coal ball T7-90C, Slide W137; Repository: Beijing Graduate School, China University of Mining and Technology; Xishan Coal Field of Taiyuan, Shanxi; early Early Permian coal bed 7 of Taiyuan Formation.

The other species:

Cathayanthus sinensis (Wang et Tian) Wang, Tian et Galtier, 2003 (in English)

1993   *Cardaianthus sinensis* Wang et Tian, Wang Shijun and Tian Baolin, p. 760, pl. 1, figs. 1 — 3; pl. 2, figs. 1 — 4; female cordaitean reproductive organs; Xishan Coal Field of Taiyuan, Shanxi; early Early Permian coal bed 7 of Taiyuan Formation.

2003   Wang Shijun, Tian Baolin and Galtier J, p. 101, figs. 10d — 10h, 11a — 11h; ovulat shoot systems; Holotype: Coal ball N61B, Slide W402 — W406 (figs. 10g, 11a); Repository: Beijing Graduate School, China University of Mining and Technology; Xishan Coal Field of Taiyuan, Shanxi; early Early Permian coal bed 7 of Taiyuan Formation.

Genus *Cathaysiocycas* Yang, 1990

1990   Yang Guanxiu, pp. 38, 41.

Type species: *Cathaysiocycas rectanervis* Yang, 1990

Taxonomic status: Cycadophytes

Distribution and Age: Yuxian of Henan, China; Early Permian.

Cathaysiocycas rectanervis Yang, 1990

1990   Yang Guanxiu, pp. 39, 41, pl. 1, figs. 1 — 5; text-figs. 1, 2; frond of cycadophytes; No. : HEP890 — HEP894; Syntypes: HEP890 (pl. 1, fig. 1), HEP891 (pl. 1, fig. 2), HEP892 (pl. 1, fig. 3), HEP893 (pl. 1, fig. 4), HEP894 (pl. 1, fig. 5); Repository: China University of Geosciences, Beijing; Yuxian, Henan; Early Permian Shenhou Formation ("Shanxi Formation").

Genus *Cathaysiodendron* Lee, 1963

1963   Lee H H, pp. 22, 126.

Type species: *Cathaysiodendron incertum* (Sze et Lee) Lee, 1963

Taxonomic status: Lycopodiales

Distribution and Age: Ningxia, Shanxi, Hebei, Liaoning and Inner Mongolia, China; Carboniferous — Permian.

Cathaysiodendron incertum (Sze et Lee) Lee, 1963

1945   *Lepidodendron*? *incertum* Sze et Lee, Sze H C and Lee H H, p. 243, pl. 1, figs. 1 — 3; the stems with leaf cushion; Hulusitai and Alanshan, Ningxia; Carboniferous lower part of Taiyuan Formation (?).

1963   Lee H H, pp. 23, 127, pl. 19, fig. 6 (?); pl. 20, fig. 8 (?); pl. 21, figs. 1 — 6 (?); the stems with leaf cushions; Reg. No. : PB3089 (pl. 19, fig. 6), PB3096 (pl. 20, fig. 8), PB3097, PB855 — PB857 (pl. 21, figs. 1 — 6); Repository: Nanjing Institute of Geology and Palaeontology, Chinese Academy of Sciences; Peitayu of Fengfeng, Hebei; Permian lower part of Shanxi Formation (?); Xishan of Taiyuan, Shanxi; Carboniferous lower and middle parts of Taiyuan Formation (?); Hulusitai and Alanshan, Ningxia; Carboniferous lower part of Taiyuan Formation (?).

The other species:

Cathaysiodendron chuseni Lee, 1963

1963   Lee H H, pp. 24, 128, pl. 21, figs. 7, 8; the stem with leaf cushions; Reg. No. : PB3098 (pl. 21, figs. 7, 8); Repository: Nanjing Institute of Geology and Palaeontology, Chinese Academy of Sciences; Weitzukou of Nanpiao, Liaoning; Permian upper part of Shanxi Formation (Upper Hunglohsien Formation).

Cathaysiodendron nanpiaoense Lee, 1963

1963   Lee H H, pp. 24, 129, pl. 17, fig. 5; pl. 21, figs. 9, 10; the stems with leaf cushions; Reg. No. : PB3082 (pl. 17, fig. 5), PB3099 (pl. 21, figs. 9, 10); Repository: Nanjing Institute of Geology and Palaeontology, Chinese Academy of Sciences; Sachiatzu of Nanpiao, Liaoning; Carboniferous lower part of Taiyuan Formation (Lower Hunglohsien Formation); Shilipu of Ordos, Inner Mongolia; Carboniferous middle part (?) of Taiyuan Formation.

Genus *Cathaysiophyllum* Lan, Li H et Wang, 1982

1982   Lan Shanxian, Li Hanmin and Wang Guoping, in Li Hanmin and others, p. 376.

Type species: *Cathaysiophyllum lobifolium* (Yang et Chen) Lan, Li H et Wang, 1982

Taxonomic status: Plantae incertae sedis

Distribution and Age: Jiangsu, Fujian and Guangdong, China; Late Permian.

Cathaysiophyllum lobifolium (Yang et Chen) Lan, Li H et Wang, 1982

1979   *Adiantites*? *lobifolius* Yang et Chen, Yang Guanxiu and Chen Fen, p. 110, pl. 19, figs. 5, 5a; frond; Guangdong; Late Permian.

1982   Lan Shanxian, Li Hanmin and Wang Guoping, in Li Hanmin and others, p. 376, pl. 147, figs. 4 — 12; text-fig. 92; fronds; Zhenjiang, Jiangsu; Late Permian Longtan Formation; Longyan, Fujian; Late Permian Cuipingshan Formation.

Genus *Cathaysiopteridium* Li (MS) ex Mei et Li, 1989

1989   Mei Meitang and Li Shengsheng, in Huang Lianmeng and others, p. 46.

Type species: *Cathaysiopteridium fasciculatum* Li (MS) ex Mei et Li, 1989

Taxonomic status: Pteridospermopsida or Gigantopterides

Distribution and Age: Anxi and Yongding of Fujian, China; Early Permian.

Cathaysiopteridium fasciculatum Li (MS) ex Mei et Li, 1989

1989   Mei Meitang and Li Shengsheng, in Huang Lianmeng and others, p. 47, pl. 26, figs. 1 — 4; pl. 27, fig. 15; pteridosperm foliage; Anxi and Yongding, Fujian; Early Permian member 3 of Tongziyan Formation.

Genus *Cathaysiopteris* Koidzumi, 1934

1934   Koidzumi G, p. 113.

Type species: *Cathaysiopteris whitei* (Halle) Koidzumi, 1934

Taxonomic status: Pteridospermopsida or Gigantopterides

Distribution and Age: central Shanxi, China; late Early Permian.

Cathaysiopteris whitei (Halle) **Koidzumi, 1934**

1927 *Gigantopteris whitei* Halle, Halle T G, p. 162, pl. 47, fig. 10; pteridosperm foliage; central Shanxi; late Early Permian Upper Shihhotse Series.

1934 Koidzumi G, p. 113.

Genus *Celathega* Hao et Gensel, 1995

1995 Hao Shougang and Gensel P G, p. 897.

Type species: *Celathega beckii* Hao et Gensel, 1995

Taxonomic status: Psilophytes incertae sedis

Distribution and Age: Wenshan of Yunnan, China; Early Devonian.

Celathega beckii Hao et Gensel, 1995

1995 Hao Shougang and Gensel P G, p. 897, figs. 1 — 35; axes; Holotype: BUHG-110 (fig. 10); Paratypes: BUHG-101, 102, 105, 116, 116', 119, 126 (figs. 2 — 6, 13); Repository: Department of Geology, Peking University; Wenshan, Yunnan; Early Devonian Posongchong Formation.

Genus *Cervicornus* Li et Hueber, 2000 (in English)

2000 Li Chengsen and Hueber F M, p. 116.

Type species: *Cervicornus wenshanensis* Li et Hueber, 2000

Taxonomic status: Protolepidodendrales, Lycophyta

Distribution and Age: Wenshan of Yunnan, China; Early Devonian (Siegenian).

Cervicornus wenshanensis Li et Hueber, 2000 (in English)

2000 Li Chengsen and Hueber F M, p. 116, pl. 1, figs. 1 — 8; text-fig. 1; herbaceous plant; Holotype: CBYn8802001 (pl. 1, fig. 1); Repository: Institute of Botany, Chinese Academy of Sciences; near Zichang Village of Wenshan, Yunnan; Early Devonian (Siegenian) Posongchong Formation.

Genus *Chamaedendron* Schweitzer et Li, 1996 (in English)

1996 Schweitzer H J and Li Chengsen, pp. 45, 50.

Type species: *Chamaedendron multisporangiatum* Schweitzer et Li, 1996

Taxonomic status: Protolepidodendrales, Lycophyta

Distribution and Age: Wuhan of Hubei, China; Late Devonian.

Chamaedendron multisporangiatum Schweitzer et Li, 1996 (in English)

1996 Schweitzer H J and Li Chengsen, pp. 45, 50, pls. 1 — 4; text-figs. 1 — 13; small tree-shaped plant; No.: CBHb 8912001a, CBHb 8912002a, CBHb 8912003a, CBHb 8912003b, CBHb 8912004a, CBHb 8912006b, CBHb 8912007, CBHb 8912009b, CBHb

8912010a,CBHb 8912011,CBHb 8912013,CBHb 8912014,CBHb 8912015,CBHb 8912016,CBHb 8912017,CBHb 8912018,CBHb 8912019,CBHb 8912020,CBHb 8912021,CBHb 8912022,CBHb 8912024; Repository: Institute of Botany, Chinese Academy of Sciences; Wuhan, Hubei; Late Devonian.

Genus *Chaneya* Wang et Manchester, 2000 (in English)
2000　Wang Yufei and Manchester R, p. 169.

Type species: *Chaneya tenuis* (Lesquereux) Wang et Manchester, 2000

Taxonomic status: Dicotyledoneae

Distribution and Age: North American, Korea, Shandong and Heilongjiang of China; Oligocene (or Late Eocene) and Miocene.

Chaneya tenuis (Lesquereux) Wang et Manchester, 2000 (in English)
1883　*Porana tenuis* Lesquereux, Lesquereux L, p. 173.
2000　Wang Yufei and Manchester R, p. 169, figs. 2, 3; winged fruit; North American; Oligocene or Late Eocene; Yilan, Heilongjiang; Eocene.

The other species:
Chaneya kokangensis (Endo) Wang et Manchester, 2000 (in English)
1939　*Porana kokangensis* Endo, Endo S, p. 346, pl. 23, fig. 6.
2000　Wang Yufei and Manchester R, p. 173, figs. 4, 5; winged fruit; Tyosen of Korea and Shanwang in Linju of Shandong; Miocene.

Genus *Changwuia* Hilton et Li, 2000 (in English)
2000　Hilton J and Li Chengsen, p. 10.

Type species: *Changwuia schweitzeri* Hilton et Li, 2000

Taxonomic status: Plantae incertae sedis

Distribution and Age: Changwu of Guangxi, China; middle Early Devonian (Siegenian).

Changwuia schweitzeri Hilton et Li, 2000 (in English)
2000　Hilton J and Li Chengsen, p. 10, pl. 1, figs. 1—6; text-fig. 1; the axis helically bearing lateral branching; Holotypes: CBG9805001a (part), CBG9805001b (counterpart) (pl. 1, figs. 1—6; text-fig. 1); Repository: Institute of Botany, Chinese Academy of Sciences; Shiqiao of Changwu, Guangxi; middle Early Devonian (Siegenian) Shiqiao Group.

Genus *Changyanophyton* Sze, 1952
1952a　Sze H C, p. 22. (in Chinese)
1952b　Sze H C, p. 185. (in English)

Type species: *Changyanophyton hupeiense* Sze, 1952

Taxonomic status: Protolepidodendrales?

Distribution and Age: Changyang of Hubei, China; Late Devonian.

Changyanophyton hupeiense Sze, 1952

1952a   Sze H C, p. 22, pl. 5, figs. 1 — 3; pl. 6, fig. 1; stems; Changyang, Hubei; Late Devonian. (in Chinese)

1952b   Sze H C, p. 185, pl. 4, figs. 1 — 3a; stems; Changyang, Hubei; Late Devonian. (in English)

Genus *Chaoyangia* Duan, 1997 (1998) (in Chinese and English)

1997   Duan Shuying, p. 519. (in Chinese)

1998   Duan Shuying, p. 15. (in English)

Type species: *Chaoyangia liangii* Duan, 1997 (1998)

Taxonomic status: Angiosperm [Notes: The type species of the genus was later referred to Chlamydospermopsida (Gnetales) (Guo Shuangxing, Wu Xiangwu, 2000; Wu Shunqing, 1999)]

Distribution and Age: Chaoyang of Liaoning, China; Late Jurassic.

Chaoyangia liangii Duan, 1997 (1998) (in Chinese and English)

1997   Duan Shuying, p. 519, figs. 1 — 4; fossil plant female reproductive organs; Holotype: 9341 comprising fossil part (fig. 1) and counterpart (fig. 2); Chaoyang, Liaoning; Late Jurassic Yixian Formation. (in Chinese)

1998   Duan Shuying, p. 15, figs. 1 — 4; fossil plant female reproductive organs; Holotype: 9341 comprising fossil part (fig. 1) and counterpart (fig. 2); Chaoyang, Liaoning; Late Jurassic Yixian Formation. (in English)

Genus *Chengzihella* Guo et Sun, 1992

1992   Guo Shuangxing and Sun Ge, in Sun Ge and others, p. 546. (in Chinese)

1993   Guo Shuangxing and Sun Ge, in Sun Ge and others, p. 254. (in English)

Type species: *Chengzihella obovata* Guo et Sun, 1992

Taxonomic status: Dicotyledoneae

Distribution and Age: Jixi of Heilongjiang, China; Early Cretaceous.

Chengzihella obovata Guo et Sun, 1992

1992   Guo Shuangxing and Sun Ge, in Sun Ge and others, p. 546, pl. 1, figs. 4 — 9; leaves; Reg. No.: PB16768 — PB16772; Holotype: PB16768 (pl. 1, fig. 4); Repository: Nanjing Institute of Geology and Palaeontology, Chinese Academy of Sciences; Chengzihe of Jixi, Heilongjiang; Early Cretaceous upper part of Chengzihe Formation. (in Chinese)

1993   Guo Shuangxing and Sun Ge, in Sun Ge and others, p. 546, pl. 1, figs. 4 — 9; leaves; Reg. No.: PB16768 — PB16772; Holotype: PB16768 (pl. 1, fig. 4); Repository: Nanjing

Institute of Geology and Palaeontology, Chinese Academy of Sciences; Chengzihe of Jixi, Heilongjiang; Early Cretaceous upper part of Chengzihe Formation. (in English)

Genus *Chiaohoella* Li et Ye, 1980

1978   *Chiaohoella* Lee et Yeh, in Yang Xuelin and others, pl. 3, figs. 2 — 4. (nom. nud.)
1980   Li Xingxue and Ye Meina, p. 7.

Type species: *Chiaohoella mirabilis* Li et Ye, 1980

Taxonomic status: Adiantaceae, Filicopsida

Distribution and Age: Jiaohe of Jilin, China; Early Cretaceous.

Chiaohoella mirabilis Li et Ye, 1980

1978   *Chiaohoella mirabilis* Lee et Yeh, in Yang Xuelin and others, pl. 3, figs. 2 — 4; frond; Shansong of Jiaohe Basin, Jilin; Early Cretaceous Moshilazi Formation. (nom. nud.)
1980   Li Xingxue and Ye Meina, p. 7, pl. 2, fig. 7; pl. 4, figs. 1 — 3; frond; Reg. No.: PB8970, PB4606, PB4608; Holotype: PB4606 (pl. 4, fig. 1); Repository: Nanjing Institute of Geology and Palaeontology, Chinese Academy of Sciences; Shansong of Jiaohe, Jilin; middle — late Early Cretaceous Shansong Formation.

The other species:

Chiaohoella neozamioides Li et Ye, 1980

1980   Li Xingxue and Ye Meina, p. 8, pl. 3, fig. 1; frond; Reg. No.: PB8971; Holotype: PB8971 (pl. 3, fig. 1); Repository: Nanjing Institute of Geology and Palaeontology, Chinese Academy of Sciences; Shansong of Jiaohe, Jilin; middle — late Early Cretaceous Shansong Formation.

Genus *Chilinia* Li et Ye, 1980

1980   Li Xingxue and Ye Meina, p. 7.

Type species: *Chilinia ctenioides* Li et Ye, 1980

Taxonomic status: Cycadales, Cycadopsida

Distribution and Age: Jiaohe of Jilin, China; Early Cretaceous.

Chilinia ctenioides Li et Ye, 1980

1980   Zhang Wu and others, p. 273, pl. 171, fig. 3; pl. 172, figs. 4, 4a; cycadophyte leaf; Shansong of Jiaohe, Jilin; Early Cretaceous Moshilazi Formation. (nom. nud.)
1980   Li Xingxue and Ye Meina, p. 7, pl. 2, figs. 1 — 6; cycadophyte leaf and cuticle; Reg. No.: PB8966 — PB8969; Holotype: PB8966 (pl. 2, fig. 1); Repository: Nanjing Institute of Geology and Palaeontology, Chinese Academy of Sciences; Shansong of Jiaohe, Jilin; middle — late Early Cretaceous Shansong Formation.

Genus *Ciliatopteris* Wu X W, 1979

1979   Wu Xiangwu, in He Yuanliang and others, p. 139.

Type species: *Ciliatopteris pecotinata* Wu X W, 1979

Taxonomic status: Dicksoniaceae?, Filicopsida

Distribution and Age: Gangcha of Qinghai, China; Early — Middle Jurassic.

Ciliatopteris pecotinata Wu X W, 1979

1979   Wu Xiangwu, in He Yuanliang and others, p. 139, pl. 63, figs. 3 — 6; text-fig. 9; sterile pinna and fertile pinna; Col. No. : 002, 003; Reg. No. : PB6339 — PB6342; Holotype: PB6340 (pl. 63, fig. 4); Paratypes: PB6339 (pl. 63, fig. 4), PB6342 (pl. 63, fig. 6); Repository: Nanjing Institute of Geology and Palaeontology, Chinese Academy of Sciences; Haide'er of Gangcha, Qinghai; Early — Middle Jurassic Jiangcang Formation of Muli Group.

Genus *Cladophlebidium* Sze, 1931

1931   Sze H C, p. 4.

Type species: *Cladophlebidium wongi* Sze, 1931

Taxonomic status: Filicopsida

Distribution and Age: Pingxiang of Jiangxi, China; Early Jurassic (Lias).

Cladophlebidium wongi Sze, 1931

1931   Sze H C, p. 4, pl. 2, fig. 4; frond; Pingxiang, Jiangxi; Early Jurassic (Lias).

Genus *Cladotaeniopteris* Zhang et Mo, 1981

1981   Zhang Shanzhen and Mo Zhuangguan, p. 240.

Type species: *Cladotaeniopteris shaanxiensis* Zhang et Mo, 1981

Taxonomic status: Cycadophytes

Distribution and Age: Hancheng of Shaanxi, China; Early Permian.

Cladotaeniopteris shaanxiensis Zhang et Mo, 1981

1981   Zhang Shanzhen and Mo Zhuangguan, p. 240, pl. 3, figs. 1 — 4; text-fig. 2; leaves taeniopterid; Holotype: PB8773 (pl. 3, fig. 1); Paratype: S51-19 (pl. 3, fig. 4); Repository: Nanjing Institute of Geology and Palaeontology, Chinese Academy of Sciences; Hancheng, Shaanxi; Early Permian Lower Shihhotse Formation.

Genus *Clematites* ex Tao et Zhang, 1990, emend Wu, 1993

[Notes: The generic name was originally not mentioned clearly as a new generic name (Wu

Xiangwu,1993a)]
1990   in Tao Junrong and Zhang Chuanbo,pp. 221,226.
1993a   Wu Xiangwu,pp. 12,217.
1993b   Wu Xiangwu,pp. 508,511.
Type species:*Clematites lanceolatus* Tao et Zhang,1990
Taxonomic status:Ranunculaceae?,Dicotyledoneae
Distribution and Age:Yanji of Jilin,China;Early Cretaceous.

Clematites lanceolatus Tao et Zhang,1990
1990   Tao Junrong and Zhang Chuanbo,pp. 221,226,pl. 1,fig. 9;text-fig. 4;leaf;No. :K_1d_{41-3}; Repository:National Museum of Plant History of China,Institute of Botany,Chinese Academy of Sciences;Yanji,Jilin;Early Cretaceous Dalazi Formation.
1993a   Wu Xiangwu,pp. 12,217.
1993b   Wu Xiangwu,pp. 508,511.

Genus *Coenosophyton* Wang et Xu,2003 (in English)
2003   Wang Yi and Xu Honghe,p. 78.
Type species:*Coenosophyton tristichus* Wang et Xu,2003
Taxonomic status:Pteridophyta incertae sedis
Distribution and Age:Chaohu of Anhui,China;Early Carboniferous.

Coenosophyton tristichus Wang et Xu,2003 (in English)
2003   Wang Yi and Xu Honghe,p. 78,figs. 2 — 7;foliage shoots;Holotype:PB19285 (fig. 2d); Paratypes:PB19291,PB19292b,PB19294 (figs. 4i,4k,6a);Repository:Nanjing Institute of Geology and Palaeontology,Chinese Academy of Sciences;Beishan of Chaohu,Anhui; Early Carboniferous Wutong Formation.

Genus *Cohaerensitheca* Liu et Yao,2006 (in English)
2006   Liu Lujun and Yao Zhaoqi,p. 69.
Type species:*Cohaerensitheca sahnii* (Hsu) Liu et Yao,2006
Taxonomic status:Marattiaceae,Marattiales
Distribution and Age:Zhenjiang of Jiangsu,China;Middle Permian.

Cohaerensitheca sahnii (Hsu) Liu et Yao,2006 (in English)
1952   *Pecopteris sahnii* Hsu,Hsu J,p. 250,pl. 3,figs. 30,33.
2006   Liu Lujun and Yao Zhaoqi,p. 70,figs. 1 — 4;foliage shoots;Zhenjiang,Jiangsu;Middle Permian Longtan Formation.

Genus *Conchophyllum* Schenk,1883
1883   Schenk A,p. 223.

Type species: *Conchophyllum richthofenii* Schenk, 1883
Taxonomic status: Noeggerathiales or Cordaitales?
Distribution and Age: Kaiping of Hebei, China; Carboniferous.

Conchophyllum richthofenii Schenk, 1883
1883   Schenk A, p. 223, pl. 42, figs. 21 — 26; foliage shoots; Kaiping, Hebei; Carboniferous.

Genus *Cryptonoclea* Li, 1992
1992   Li Zhongming, p. 162.
Type species: *Cryptonoclea primitiva* Li, 1992
Taxonomic status: Gigantopteridales
Distribution and Age: Shuicheng of Guizhou, China; Late Permian.

Cryptonoclea primitiva Li, 1992
1992   Li Zhongming, p. 162, figs. 1 — 57; whole fossil plant includes seeds, pollen-bearing organs and compound leaves; Holotype and Syntypes: Slides and peels from coal balls GP2. 309-4-3, GP2. 312-6-2, GP2. 713-2; Repository: Institute of Botany, Chinese Academy of Sciences; Wangjiazhai Mine of Shuicheng, Guizhou; Late Permian Wangjiazhai Formation.

Genus *Cycadeoidispermum* Hu et Zhu, 1982
1982   Hu Yufan and Zhu Jianan, in Zhu Jianan and others, p. 80.
Type species: *Cycadeoidispermum petiolatum* Hu et Zhu, 1982
Taxonomic status: Cycadophytes
Distribution and Age: Taiyuan of Shanxi, China; Late Permian.

Cycadeoidispermum petiolatum Hu et Zhu, 1982
1982   Hu Yufan and Zhu Jianan, in Zhu Jianan and others, p. 80, pl. 2, fig. 8; text-fig. 3; seed; No. : B69a; Holotype: B69a (pl. 2, fig. 8); Repository: Institute of Botany, Chinese Academy of Sciences; Taiyuan, Shanxi; early Late Permian Upper Shihezi (Shihhotse) Formation.

Genus *Cycadicotis* Pan, 1983 (nom. nud.)
[Notes: The genus was later referred by Kimura and others (1994) to *Pankuangia*, and the type species *Cycadicotis nissonervis* to *Pankuangia haifanggouensis* Kimura, Ohana, Zhao et Geng]
1983   Pan Guang, p. 1520. (in Chinese)
1983   Pan Guang, in Li Jieru, p. 22.
1984   Pan Guang, p. 958. (in English)

Type species: *Cycadicotis nissonervis* Pan (MS) ex Li, 1983

Taxonomic status: Sinodicotiaceae, "hemiangiosperms" or Cycadopsida?

Distribution and Age: western Liaoning, China; Middle Jurassic.

Cycadicotis nissonervis **Pan (MS) ex Li, 1983** (nom. nud.)

1983 Pan Guang, in Li Jieru, p. 22, pl. 2, fig. 3; leaf and reproductive organ-like appendage; No.: Jp1h2-30; Repository: Regional Geological Surveying Team, Liaoning Geological Bureau; western Liaoning (about 45°58′N, 120°21′E); Middle Jurassic Haifanggou Formation.

The other species:

Cycadicotis **sp. indet.**

[Notes: Generic name was given only, but without specific name (or type species) in the original paper]

1983 *Cycadicotis* sp. indet., Pan Guang, p. 1520; western Liaoning (about 45°58′N, 120°21′E); Middle Jurassic Haifanggou Formation. (in Chinese)

1984 *Cycadicotis* sp. indet., Pan Guang, p. 958; western Liaoning (about 45°58′N, 120°21′E); Middle Jurassic Haifanggou Formation. (in English)

Genus *Cycadolepophyllum* Yang, 1978

1978 Yang Xianhe, p. 510.

Type species: *Cycadolepophyllum minor* Yang, 1978

Taxonomic status: Bennettiales, Cycadopsida

Distribution and Age: Changning of Sichuan, China; Late Triassic.

Cycadolepophyllum minor **Yang, 1978**

1978 Yang Xianhe, p. 510, pl. 163, fig. 11; pl. 175, fig. 4; cycadophyte leaf; No.: Sp0041; Holotype: Sp0041 (pl. 163, fig. 11); Repository: Chengdu Institute of Geology and Mineral Resources; Shuanghe of Changning, Sichuan; Late Triassic Hsuchiaho Formation.

The other species:

Cycadolepophyllum aequale **(Brongniart) Yang, 1978**

1942 *Pterophyllum aequale* (Brongniart) Nathorst, Sze H C, p. 189, pl. 1, figs. 1 — 4; cycadophyte leaf; Lechang, Guangdong; Late Triassic — Early Jurassic.

1978 Yang Xianhe, p. 510; cycadophyte leaf; Lechang, Guangdong; Late Triassic.

Genus *Cycadostrobilus* Zhu, 1994

1994 Zhu Jianan and others, p. 341.

Type species: *Cycadostrobilus paleozoicus* Zhu, 1994

Taxonomic status: Cycadophytes

Distribution and Age: Taiyuan of Shanxi, China; Early Permian.

Cycadostrobilus paleozoicus Zhu, 1994

1994　Zhu Jianan and others, p. 342, pl. 1, figs. 1 — 6; macrosporophyll; No.: PZ4006; Holotype: PZ4006 (pl. 1, fig. 1); Repository: Institute of Botany, Chinese Academy of Sciences; Dongshan of Taiyuan, Shanxi; Early Permian Lower Shihezi (Shihhotse) Formation.

Genus *Daohugouthallus* Wang, Krings et Taylor, 2010 (in English)

2010　Wang Xin, Krings M and Taylor T N, p. 592.

Type species: *Daohugouthallus ciliiferus* Wang, Krings et Taylor, 2010

Taxonomic position: Lichenes

Distribution and Age: Ningcheng of Inner Mongolia, Northeast China; Middle Jurassic.

Daohugouthallus ciliiferus Wang, Krings et Taylor, 2010 (in English)

2010　Wang Xin, Krings M and Taylor T N, p. 592, pl. 1, figs. 1 — 5; pl. 2, figs. 1 — 6; pl. 3, figs. 1 — 4; thallus; Reg. No.: PB21398 — PB21400; Holotype: PB21398 (pl. 1, fig. 1); Paratypes: PB21399 (pl. 1, fig. 5), PB21400 (pl. 3, fig. 1); Repository: Nanjing Institute of Geology and Palaeontology, Chinese Academy of Sciences; Daohugoucun of Ningcheng, Inner Mongolia (119°14.318′E, 41°18.979′N); Middle Jurassic Jiulongshan Formation.

Genus *Datongophyllum* Wang, 1984

1984　Wang Ziqiang, p. 281.

Type species: *Datongophyllum longipetiolatum* Wang, 1984

Taxonomic status: Ginkgoales incertae sedis

Distribution and Age: Huairen of Shanxi, China; Early Jurassic.

Datongophyllum longipetiolatum Wang, 1984

1984　Wang Ziqiang, p. 218, pl. 130, figs. 5 — 13; foliage twig and fertile twig; 7 species; Reg. No.: P0174, P01759 (Syntype), P0176, P0177 (Syntype), P0182, P0179, P0180 (Syntype); Repository: Nanjing Institute of Geology and Palaeontology, Chinese Academy of Sciences; Huairen, Shanxi; Early Jurassic Yongdingzhuang Formation.

The other species:

Datongophyllum sp.

1984　*Datongophyllum* sp., Wang Ziqiang, p. 282, pl. 130, fig. 14; twigs; Huairen, Shanxi; Early Jurassic Yongdingzhuang Formation.

Genus *Decoroxylon* Zhang et Zheng, 2006 (2008) (in Chinese and English)

2006　Zhang Wu and Zheng Shaolin, in Zhang Wu and others, p. 94. (in Chinese)
2008　Zhang Wu and Zheng Shaolin, in Zhang Wu and others, p. 97. (in English)
Type species: *Decoroxylon chaoyangense* Zhang et Zheng, 2006 (in Chinese), 2008 (in English)
Taxonomic status: Plantae incertae sedis
Distribution and Age: Liaoning, China; Early Permian.

Decoroxylon chaoyangense Zhang et Zheng, 2006 (2008) (in Chinese and English)

2006　Zhang Wu and Zheng Shaolin, in Zhang Wu and others, p. 97, pl. 3-46; pl. 3-47; pl. 3-48; fossil wood; No.: Xt-8; Holotype: Xt-8 (pl. 3-46; pl. 3-47; pl. 3-48); Repository: Shenyang Institute of Geology and Mineral Resources; Chaoyang, Liaoning; Early Permian Taiyuan Formation. (in Chinese)
2008　Zhang Wu and Zheng Shaolin, in Zhang Wu and others, p. 97, pl. 3-46; pl. 3-47; pl. 3-48; fossil wood; No.: Xt-8; Holotype: Xt-8 (pl. 3-46; pl. 3-47; pl. 3-48); Repository: Shenyang Institute of Geology and Mineral Resources; Chaoyang, Liaoning; Early Permian Taiyuan Formation. (in English)

Genus *Deltoispermum* Yang, 2006 (in Chinese and English)

2006　Yang Guanxiu and others, pp. 175, 330.
Type species: *Deltoispermum henanense* Yang, 2006
Taxonomic status: Gymnospermarum
Distribution and Age: Linru of western Henan, China; Late Permian.

Deltoispermum henanense Yang, 2006 (in Chinese and English)

2006　Yang Guanxiu and others, pp. 175, 330, pl. 57, fig. 3; seed; Holotype: HEP3441 (pl. 57, fig. 3); Repository: China University of Geosciences, Beijing; Pochi of Linru, western Henan; Late Permian member 8 of Yungaishan Formation.

Genus *Demersatheca* Li et Edwards, 1996 (in English)

1996　Li Chengsen and Edwards D, p. 79.
Type species: *Demersatheca contigua* (Li et Cai) Li et Edwards, 1996
Taxonomic status: Zosterophyllophytes
Distribution and Age: Wenshan of Yunnan, China; Early Devonian.

Demersatheca contigua (Li et Cai) Li et Edwards, 1996 (in English)

1977　*Zosterophyllum contiguum* Li et Cai, Li Xingxue and Cai Chongyang, p. 24, pl. 3, figs. 4, 5, 7, 9; text-fig. 7; strobili; Col. No.: ACE194 (pl. 3, fig. 4), ACE186 (pl. 3, fig. 5), ACE192 (pl. 3, figs. 7, 9); Reg. No.: PB6467 — PB6470; Repository: Nanjing Institute of

Geology and Palaeontology, Chinese Academy of Sciences; Wenshan, Yunnan; Early Devonian Posongchong Formation.

1996　Li Chengsen and Edwards D, p. 79, pls. 1 — 4; text-figs. 1, 2; strobili; Wenshan, Yunnan; Early Devonian Posongchong Formation.

Genus *Dengfengia* Yang, 2006 (in Chinese and English)

2006　Yang Guanxiu and others, pp. 172, 328.

Type species: *Dengfengia bifurcata* Yang, 2006

Taxonomic status: Gymnospermarum

Distribution and Age: Dengfeng of western Henan, China; Middle Permian.

Dengfengia bifurcata Yang, 2006 (in Chinese and English)

2006　Yang Guanxiu and others, pp. 172, 328, pl. 51, figs. 8, 9; pl. 74, figs. 4, 5; bract-scale complex; Holotype: HEP3068 (pl. 51, fig. 8); Paratype: HEP3069 (pl. 51, fig. 9); Repository: China University of Geosciences, Beijing; Dengcao of Dengfeng, western Henan; Middle Permian coal member 4 of Xiaofengkou Formation.

Genus *Denglongia* Xue et Hao, 2008 (in English)

2008　Xue Jinzhuang and Hao Shougang, p. 1315.

Type species: *Denglongia hubeiensis* Xue et Hao, 2008

Taxonomic status: Cladoxylates, Pteridophyta

Distribution and Age: Hubei, China; Late Devonian (Frasnian).

Denglongia hubeiensis Xue et Hao, 2008 (in English)

2008　Xue Jinzhuang and Hao Shougang, p. 1315, figs. 1 — 14; plant; Holotype: PKU-XH120 (figs. 8a, 8b); Paratypes: PKU-XH110, 113a, 122b, 140, 144 (figs. 2a, 2b; 4a, 4c, 4f; 11a); Repository: Department of Geology, Peking University; Changyang, Hubei; Late Devonian (Frasnian) Huangjiadeng Formation.

Genus *Dentopteris* Huang, 1992

1992　Huang Qisheng, p. 179.

Type species: *Dentopteris stenophylla* Huang, 1992

Taxonomic status: Gymnospermae incertae sedis

Distribution and Age: Tieshan in Daxian of Sichuan, China; Late Triassic.

Dentopteris stenophylla Huang, 1992

1992　Huang Qisheng, p. 179, pl. 18, figs. 1, 1a; fern-like leaves; Reg. No.: SD87001; Repository: Department of Geology, China University of Geosciences, Wuhan; Tieshan of Daxian, Sichuan; Late Triassic member 7 of Hsuchiaho Formation.

The other species:
Dentopteris platyphylla Huang, 1992
1992   Huang Qisheng, p. 179, pl. 19, figs. 3, 5, 7; pl. 20, fig. 13; fern-like leaves; Col. No. : SD5; Reg. No. : SD87003 — SD87005; Holotype: SD87003 (pl. 19, fig. 7); Repository: Department of Geology, China University of Geosciences, Wuhan; Tieshan of Daxian, Sichuan; Late Triassic member 3 of Hsuchiaho Formation.

Genus *Dioonocarpus* Hu et Zhu, 1982
1982   Hu Yufan and Zhu Jianan, in Zhu Jianan and others, p. 79.
Type species: *Dioonocarpus ovatus* Hu et Zhu, 1982
Taxonomic status: Cycadophytes
Distribution and Age: Taiyuan of Shanxi, China; Late Permian.

Dioonocarpus ovatus Hu et Zhu, 1982
1982   Hu Yufan and Zhu Jianan, in Zhu Jianan and others, p. 79, pl. 2, fig. 7; text-fig. 1; seed; No. : B69; Holotype: B69 (pl. 2, fig. 7); Repository: Institute of Botany, Chinese Academy of Sciences; Taiyuan, Shanxi; early Late Permian Upper Shihezi (Shihhotse) Formation.

Genus *Discalis* Hao, 1989
1989a   Hao Shougang, p. 158.
Type species: *Discalis longistipa* Hao, 1989
Taxonomic status: Zosterophyllophytes
Distribution and Age: Yunnan, China; Early Devonian (Siegenian).

Discalis longistipa Hao, 1989
1989a   Hao Shougang. p. 158, pl. 1, figs. 1 — 4; pl. 2, figs. 1 — 6; pl. 3, figs. 1 — 11; pl. 4, figs. 1 — 16; text-figs. 1 — 5; sterile axes and fertile axes; No. : PUH301 — PUH311; Holotype: PUH301 (pl. 1, fig. 1); Paratypes: PUH302 — PUH306; Repository: Department of Geology, Peking University; Wenshan, Yunnan; Early Devonian (Siegenian) Posongchong Formation.

Genus *Distichopteris* Yabe et Shimakura, 1940
1940b   Yabe H and Shimakura M, p. 179.
Type species: *Distichopteris heteropinna* Yabe et Shimakura, 1940
Taxonomic status: Pteridopsida incertae sedis
Distribution and Age: Jiangning of Jiangsu, China; Middle Permian.

Distichopteris heteropinna Yabe et Shimakura, 1940

1940b Yabe H and Shimakura M, p. 179, pl. 16, figs. 1 — 7; frond (or pinna?); Longtan (Lungtan) Coal Mine, Jiangsu (Kiangsu); Middle Permian Longtan (Lungtan) Formation. [Notes: This specimen lately was referred as *Pecopteris heteropinna* (Yabe et Shimakura) Gu et Zhi (Nanjing Institute of Geology and Palaeontology, Institute of Botany, Chinese Academy of Sciences, 1974, p. 95)]

Genus *Distichotheca* Gu et Zhi, 1974

1974 Nanjing Institute of Geology and Palaeontology, Institute of Botany, Chinese Academy of Sciences (Gu et Zhi), in *Palaeozoic Plants from China*, p. 167.

Type species: *Distichotheca crossothecoides* Gu et Zhi, 1974

Taxonomic status: Plantae incertae sedis

Distribution and Age: Jiangning of Jiangsu, China; early Late Permian.

Distichotheca crossothecoides Gu et Zhi, 1974

1974 Nanjing Institute of Geology and Palaeontology, Institute of Botany, Chinese Academy of Sciences, in *Palaeozoic Plants from China*, p. 167, pl. 129, figs. 1 — 4; fertile pinnae; Reg. No.: PB3803 (pl. 129, fig. 1), PB3800 (pl. 129, fig. 3); Holotype: PB3800; Repository: Nanjing Institute of Geology and Palaeontology, Chinese Academy of Sciences; Jiangning, Jiangsu; early Late Permian Longtan Formation.

Genus *Ditaxocladus* Guo et Sun, 1984

1984 Guo Shuangxing and Sun Zhehua, in Guo Shuangxing and others, p. 126.

Type species: *Ditaxocladus planiphyllus* Guo et Sun, 1984

Taxonomic status: Cupressaceae, Coniferopsida

Distribution and Age: Altai of Xinjiang, China; Paleocene.

Ditaxocladus planiphyllus Guo et Sun, 1984

1984 Guo Shuangxing and Sun Zhehua, in Guo Shuangxing and others, p. 128, pl. 1, figs. 5, 5a, 6, 6a, 8, 8a; pl. 6, fig. 8; leafy shoots; Col. No.: 790HK-3-4, 790HK-3-3, 790HK-3-42, 790HK-3-40; Reg. No.: PB9863, PB9864, PB9866, PB9867; Holotype: PB9863 (pl. 1, fig. 5); Paratypes: PB9864 (pl. 1, fig. 6), PB9866 (pl. 1, fig. 8), PB9867 (pl. 6, fig. 8); Repository: Nanjing Institute of Geology and Palaeontology, Chinese Academy of Sciences; Altai, Xinjiang; Paleocene.

Genus *Dracopteris* Deng, 1994

1994 Deng Shenghui, p. 18.

Type species: *Dracopteris liaoningensis* Deng, 1994

Taxonomic status: Filicopsida

Distribution and Age: Fuxin and Tiefa of Liaoning, China; Early Cretaceous.

Dracopteris liaoningensis Deng, 1994

1994 Deng Shenghui, p. 18, pl. 1, figs. 1—8; pl. 2, figs. 1—15; pl. 3, figs. 1—9; pl. 4, figs. 1—9; text-fig. 2; fronds, fertile pinnae, sori and sporangia; Col. No.: Fxt5-086 — Fxt5-090, TDMe622; Holotype: Fxt5-087 (pl. 1, fig. 6); Repository: Research Institute of Petroleum Exploration and Development; Fuxin Basin and Tiefa Basin, Liaoning; Early Cretaceous Fuxin Formation and Xiaoming'anbei Formation.

Genus *Dukouphyllum* Yang, 1978

1978 Yang Xianhe, p. 525.

Type species: *Dukouphyllum noeggerathioides* Yang, 1978

Taxonomic status: Cycadopsida [Notes: This genus lately was referred in Sphenobaieraceae, Ginkgoales (Yang Xianhe, 1982)]

Distribution and Age: Dukou of Sichuan, China; Late Triassic.

Dukouphyllum noeggerathioides Yang, 1978

1978 Yang Xianhe, p. 525, pl. 186, figs. 1—3; pl. 175, fig. 3; leaf; Reg. No.: Sp0134 — Sp0137; Syntypes: Sp0134 — Sp0137; Repository: Chengdu Institute of Geology and Mineral Resources; Moshahe of Dukou, Sichuan; Late Triassic Daqiaodi Formation.

Genus *Dukouphyton* Yang, 1978

1978 Yang Xianhe, p. 518.

Type species: *Dukouphyton minor* Yang, 1978

Taxonomic status: Bennettiales, Cycadopsida

Distribution and Age: Dukou of Sichuan, China; Late Triassic.

Dukouphyton minor Yang, 1978

1978 Yang Xianhe, p. 518, pl. 160, fig. 2; cycadophyte leaf; Reg. No.: Sp0021; Holotype: Sp0021 (pl. 160, fig. 2); Repository: Chengdu Institute of Geology and Mineral Resources; Moshahe of Dukou, Sichuan; Late Triassic Daqiaodi Formation.

Genus *Eboraciopsis* Yang, 1978

1978 Yang Xianhe, p. 495.

Type species: *Eboraciopsis trilobifolia* Yang, 1978

Taxonomic status: Filicopsida

Distribution and Age: Dukou of Sichuan, China; Late Triassic.

Eboraciopsis trilobifolia Yang,1978

1978 Yang Xianhe, p. 495, pl. 163, fig. 6; pl. 175, fig. 5; frond; Reg. No. : Sp0036; Holotype: Sp0036 (pl. 163, fig. 6); Repository: Chengdu Institute of Geology and Mineral Resources; Taipingchang of Dukou, Sichuan; Late Triassic Daqiaodi Formation.

Genus *Emplectopteris* Halle, 1927

1927 Halle T G, p. 119.

Type species: *Emplectopteris trangularis* Halle, 1927

Taxonomic status: Pteridospermopsida

Distribution and Age: central Shanxi, China; late Early Permian.

Emplectopteris trangularis Halle, 1927

1927 Halle T G, p. 122, pl. 31; pteridosperm foliage; central Shanxi; late Early Permian Lower Shihhotse Series.

Genus *Eoglyptostrobus* Miki, 1964

1964 Miki S, pp. 14, 21.

Type species: *Eoglyptostrobus sabioides* Miki, 1964

Taxonomic status: Coniferales, Coniferopsida

Distribution and Age: Lingyuan of western Liaoning, China; Late Jurassic.

Eoglyptostrobus sabioides Miki, 1964

1964 Miki S, pp. 14, 21, pl. 1, fig. E; shoot with leaves; Lingyuan, western Liaoning; Late Jurassic *Lycoptera* Bed.

Genus *Eogonocormus* Deng, 1995 (non Deng, 1997)

1995b Deng Shenghui, pp. 14, 108.

Type species: *Eogonocormus cretaceum* Deng, 1995

Taxonomic status: Hymenophyllaceae, Filicopsida

Distribution and Age: Huolinhe Basin of Inner Mongolia, China; Early Cretaceous.

Eogonocormus cretaceum Deng, 1995 (non Deng, 1997)

1995b Deng Shenghui, pp. 14, 108, pl. 3, figs. 1 — 2; pl. 4, figs. 1 — 2, 6 — 8; pl. 5, figs. 1 — 6; text-fig. 4; sterile frond and fertile frond; No. : H17-431; Repository: Research Institute of Petroleum Exploration and Development; Huolinhe Basin, Inner Mongolia; Early Cretaceous Huolinhe Formation.

The other species:

Eogonocormus linearifolium (Deng) Deng, 1995

1993 *Hymenophyllites linearifolius* Deng, Deng Shenghui, p. 256, pl. 1, figs. 5 — 7; text-figs.

d — f; frond and fertile pinnae; Huolinhe Basin, Inner Mongolia; Early Cretaceous Huolinhe Formation.

1995b   Deng Shenghui, pp. 17, 108, pl. 3, figs. 3 — 4; sterile frond and fertile frond; No. : H14-509, H14-510; Repository: Research Institute of Petroleum Exploration and Development; Huolinhe Basin, Inner Mongolia; Early Cretaceous Huolinhe Formation.

Genus *Eogonocormus* Deng, 1997 (non Deng, 1995) (in English)

(Notes: This generic name *Eogonocormus* Deng, 1997 is a later isonym of *Eogonocormus* Deng, 1995)

1997   Deng Shenghui, p. 60.

Type species: *Eogonocormus cretaceum* Deng, 1997

Taxonomic status: Hymenophyllaceae, Filicopsida

Distribution and Age: Huolinhe Basin of Inner Mongolia, China; Early Cretaceous.

Eogonocormus cretaceum Deng, 1997 (non Deng, 1995) (in English)

(Notes: This specific name *Eogonocormus cretaceum* Deng, 1997 is a later isonym of *Eogonocormus cretaceum* Deng, 1995)

1997   Deng Shenghui, p. 60, figs. 2 — 5; sterile frond and fertile frond; No. : H17-431; Holotype: H17-431 (fig. 3a); Repository: Research Institute of Petroleum Exploration and Development; Huolinhe Basin, Inner Mongolia; Early Cretaceous Huolinhe Formation.

Genus *Eogymnocarpium* Li, Ye et Zhou, 1986

1986   Li Xingxue, Ye Meina and Zhou Zhiyan, p. 14.

Type species: *Eogymnocarpium sinense* Li, Ye et Zhou, 1986

Taxonomic status: Athyriaceae, Filicopsida

Distribution and Age: Jiaohe of Jilin, China; Early Cretaceous.

Eogymnocarpium sinense (Li et Ye) Li, Ye et Zhou, 1986

1978   *Dryopterites sinense* Lee et Yeh, in Yang Xuelin and others, pl. 2, figs. 3, 4; pl. 3, fig. 7; frond; Shansong of Jiaohe, Jilin; Early Cretaceous Moshilazi Formation. (nom. nud.)

1980   *Dryopterites sinense* Li et Ye, Li Xingxue and Ye Meina, p. 6, pl. 1, figs. 1 — 5; fertile pinnae; Shansong of Jiaohe, Jilin; middle — late Early Cretaceous Shansong Formation.

1986   Li Xingxue, Ye Meina and Zhou Zhiyan, p. 14, pl. 12; pl. 13; pl. 14, figs. 1 — 6; pl. 15, figs. 5 — 7a; pl. 16, fig. 3; pl. 40, fig. 4; pl. 45, figs. 1 — 3; text-figs. 4A, 4B; fertile frond; Shansong of Jiaohe, Jilin (127°15′E, 43°30′N); Early Cretaceous Jiaohe Group.

Genus *Eolepidodendron* Wu et Zhao, 1981

1981   Wu Xiuyuan and Zhao Xiuhu, p. 54.

Type species: *Eolepidodendron jurongense* Wu et Zhao, 1981

Taxonomic status: Protolepidodendrales, Lycophyta

Distribution and Age: Jurong and Wuxi of Jiangsu, China; Early Carboniferous.

Eolepidodendron jurongense Wu et Zhao, 1981

1981   Wu Xiuyuan and Zhao Xiuhu, p. 54, pl. 1, figs. 5, 5a; stem; Col. No.: MWS-66; Reg. No.: PB7494; Repository: Nanjing Institute of Geology and Palaeontology, Chinese Academy of Sciences; Jurong, Jiangsu; Early Carboniferous Kaolishan Formation.

The other species:

Eolepidodendron wusihense (Sze) Wu et Zhao, 1981

1936   *Lepidodendron* aff. *leeianum* Gothan et Sze (? n. sp.), Sze H C, p. 141, pl. 2, figs. 1 – 6; pl. 3, figs. 1, 2; pl. 5, figs. 1, 2; stems; Wuxi, Jiangsu; Early Carboniferous Wutung Quartzites.

1943   *Lepidodendron wusihense* Sze, Sze H C, p. 63.

1956c   *Sublepidodendron wusihense* (Sze) Sze, Sze H C, p. 49.

1981   Wu Xiuyuan and Zhao Xiuhu, p. 54

Eolepidodendron cf. *wusihense* (Sze) Wu et Zhao, 1981

1981   Wu Xiuyuan and Zhao Xiuhu, p. 54, pl. 1, fig. 7; stem; Jurong, Jiangsu; Early Carboniferous Kaolishan Formation.

Eolepidodendron sp.

1981   *Eolepidodendron* sp., Wu Xiuyuan and Zhao Xiuhu, p. 54, pl. 1, fig. 6; stem; Jurong, Jiangsu; Early Carboniferous Kaolishan Formation.

Genus *Eophyllogonium* Mei, Dilcher et Wan, 1992

1992   Mei Meitang, Dilcher D L and Wan Zhihui, p. 99.

Type species: *Eophyllogonium cathayense* Mei, Dilcher et Wan, 1992

Taxonomic status: Taeniopterides

Distribution and Age: Leping of Jiangxi, China; Permian.

Eophyllogonium cathayense Mei, Dilcher et Wan, 1992

1992   Mei Meitang, Dilcher D L and Wan Zhihui, p. 99, pl. 1, figs. 1 – 5; pl. 2, figs. 1 – 4; pl. 3, fig. 2; pl. 4, figs. 1 – 5; pl. 5, figs. 1 – 4; text-fig. 1; seed-bearing leaf; Holotype: X 9-005 (pl. 1, fig. 3); Paratypes: X 9-147 (pl. 2, fig. 3), X 9-282 (pl. 2, fig. 1); Repository: Beijing Graduate School, China University of Mining and Technology; Leping, Jiangxi; Permian Guanshan Member of Leping Formation.

Genus *Eophyllophyton* Hao, 1988

1988 Hao Shougang, p. 442.

Type species: *Eophyllophyton bellum* Hao, 1988

Taxonomic status: Psilophytes incertae sedis

Distribution and Age: Wenshan of Yunnan, China; Early Devonian.

Eophyllophyton bellum Hao, 1988

1988 Hao Shougang, p. 442, pls. 1 — 3; text-figs. 1, 2; plant; Holotype: BUPb101 (pl. fig. 1); Paratypes: BUPb113, 152, 131 (pl. 1, figs. 2, 6, 7), BUPb110, 127 (pl. 2, figs. 1, 2, 11), BUPb137, 116, 121, 112 (pl. 3, figs. 1, 2, 4, 6); Repository: Department of Geology, Peking University; Wenshan, Yunnan; Early Devonian Posongchong Formation.

Genus *Eragrosites* Cao et Wu S Q, 1997 (1998) (in Chinese and English)

[Notes: The type species of the genus lately was referred into Gnetales or Chlamydospermopsida and named as *Ephedrites chenii* (Cao et Wu S Q) Guo et Wu X W (Guo Shuangxing, Wu Xiangwu, 2000), or into Gnetales, as *Liaoxia chenii* (Cao et Wu S Q) Wu S Q (Wu Shunqing, 1999)]

1997 Cao Zhengyao and Wu Shunqing, in Cao Zhengyao and others, p. 1765. (in Chinese)

1998 Cao Zhengyao and Wu Shunqing, in Cao Zhengyao and others, p. 231. (in English)

Type species: *Eragrosites changii* Cao et Wu S Q, 1997 (1998)

Taxonomic status: Gramineae, Monocotyledoneae

Distribution and Age: Beipiao of western Liaoning, China; Late Jurassic.

Eragrosites changii Cao et Wu S Q, 1997 (1998) (in Chinese and English)

1997 Cao Zhengyao and Wu Shunqing, in Cao Zhengyao and others, p. 1765, pl. 2, figs. 1 — 3; text-fig. 1; herbaceous plant; Reg. No. : PB17802 — PB17801; Holotype: PB17803 (pl. 2, fig. 2); Repository: Nanjing Institute of Geology and Palaeontology, Chinese Academy of Sciences; Shangyuan of Beipiao, western Liaoning; Late Jurassic Jianshangou Bed of Yixian Formation. (in Chinese)

1998 Cao Zhengyao and Wu Shunqing, in Cao Zhengyao and others, p. 231, pl. 2, figs. 1 — 3; text-fig. 1; herbaceous plant; Reg. No. : PB17802 — PB17801; Holotype: PB17803 (pl. 2, fig. 2); Repository: Nanjing Institute of Geology and Palaeontology, Chinese Academy of Sciences; Shangyuan of Beipiao, western Liaoning; Late Jurassic Jianshangou Bed of Yixian Formation. (in English)

Genus *Eucommioites* ex Tao et Zhang, 1992

(Notes: The generic name was originally not mentioned clearly as a new generic name, only specific name *Eucommioites orientalis* Tao et Zhang, 1992)

1992   in Tao Junrong and Zhang Chuanbo, pp. 423, 425.

Type species: *Eucommioites orientalis* Tao et Zhang, 1992

Taxonomic status: Dicotyledoneae

Distribution and Age: Yanji of Jilin, China; Early Cretaceous.

Eucommioites orientalis Tao et Zhang, 1992

1992   Tao Junrong and Zhang Chuanbo, pp. 423, 425, pl. 1, figs. 7 — 9; samara; No. : 503882; Holotype: 503883 (pl. 1, figs. 7 — 9); Repository: National Museum of Plant History of China, Institute of Botany, Chinese Academy of Sciences; Yanji, Jilin; Early Cretaceous Dalazi Formation.

Genus *Fascipteridium* Zhang et Mo, 1985

1985   Zhang Shanzhen and Mo Zhuangguan, p. 175.

Type species: *Fascipteridium ellipticum* Zhang et Mo, 1985

Taxonomic status: Pteridospermopsida

Distribution and Age: Henan, China; Permian.

Fascipteridium ellipticum Zhang et Mo, 1985

1985   Zhang Shanzhen and Mo Zhuangguan, p. 175, pl. 3, figs. 1 — 5; pinna with seed; Reg. No. : PB8114 — PB8117; Syntypes: PB8114 (pl. 3, fig. 1), PB8117 (pl. 3, fig. 5); Repository: Nanjing Institute of Geology and Palaeontology, Chinese Academy of Sciences; Henan; Late Permian Upper Shihhotse Series.

Genus *Fascipteris* Gu et Zhi, 1974

1974   Nanjing Institute of Geology and Palaeontology, Institute of Botany, Chinese Academy of Sciences, in *Palaeozoic Plants from China*, p. 99.

Type species: *Fascipteris hallei* (Kawasaki) Gu et Zhi, 1974

Taxonomic status: Filices or Pteridospermopsida

Distribution and Age: Hebei, Shanxi and Jiangsu, China; Late Permian.

Fascipteris hallei (Kawasaki) Gu et Zhi, 1974

1939   *Validopteris hallei* (Kawasaki) Stockmans et Mathieu, Stockmans F and Mathieu F F, p. 75, pl. 34, fig. 1; Kaiping, Hebei; early Late Permian.

1974   Nanjing Institute of Geology and Palaeontology, Institute of Botany, Chinese Academy of Sciences, in *Palaeozoic Plants from China*, p. 99, pl. 68, figs. 8 — 12; text-fig. 83. 1; pinnae; Taiyuan, Shanxi; early Late Permian Upper Shihhotse Formation; Kaiping, Hebei; early Late Permian Kuyeh Formation.

The other species:

Fascipteris recta Gu et Zhi, 1974

1927   *Pecopteris* sp. a, Halle T H, p. 100, pl. 23, figs. 4 — 13; central Shanxi; Late Permian

Upper Shihhotse Formation.

1974 Nanjing Institute of Geology and Palaeontology, Institute of Botany, Chinese Academy of Sciences, in *Palaeozoic Plants from China*, p. 100, pl. 69, figs. 1 — 4; pinnae; Taiyuan, Shanxi; early Late Permian Upper Shihhotse Formation.

Fascipteris sinensis (Stockmans et Mathieu) Gu et Zhi, 1974

1957 *Validopteris sinensis* Stockmans et Mathieu, Stockmans F and Mathieu F F, p. 22, pl. 13, figs. 2, 3; pinnae; Kaiping, Hebei; Late Permian.

1974 Nanjing Institute of Geology and Palaeontology, Institute of Botany, Chinese Academy of Sciences, in *Palaeozoic Plants from China*, p. 100, pl. 69, figs. 5 — 7; text-fig. 84. 1; pinnae; Kaiping, Hebei; early Late Permian Kuyeh Formation.

Fascipteris (Ptychocarpus) densata Gu et Zhi, 1974

1974 Nanjing Institute of Geology and Palaeontology, Institute of Botany, Chinese Academy of Sciences, in *Palaeozoic Plants from China*, p. 100, pl. 69, figs. 8 — 14; text-figs. 85 — 86; pinnae; Reg. No.: PB3686, PB3688, PB3690; Syntypes: PB3686 (pl. 69, fig. 8), PB3688 (pl. 69, fig. 10), PB3690 (pl. 69, fig. 12); Repository: Nanjing Institute of Geology and Palaeontology, Chinese Academy of Sciences; Jiangning, Jiangsu; early Late Permian Longtan Formation.

Fascipteris stena Gu et Zhi, 1974

1974 Nanjing Institute of Geology and Palaeontology, Institute of Botany, Chinese Academy of Sciences, in *Palaeozoic Plants from China*, p. 101, pl. 69, figs. 15 — 17; text-fig. 84. 2; pinnae; Reg. No.: PB3697, PB3699; Holotype: PB3699 (pl. 69, fig. 15); Repository: Nanjing Institute of Geology and Palaeontology, Chinese Academy of Sciences; Jiangning, Jiangsu; early Late Permian Longtan Formation.

Genus *Filicidicotis* Pan, 1983 (nom. nud.)

1983 Pan Guang, p. 1520. (in Chinese)
1984 Pan Guang, p. 958. (in English)

Type species: (without specific name)
Taxonomic status: "hemiangiosperms"
Distribution and Age: western Liaoning, China; Middle Jurassic.

Filicidicotis sp. indet.

[Notes: Generic name was given only, but without specific name (or type species) in the original paper]

1983 *Filicidicotis* sp. indet., Pan Guang, p. 1520; western Liaoning (about 45°58′N, 120°21′E); Middle Jurassic Haifanggou Formation. (in Chinese)

1984 *Filicidicotis* sp. indet., Pan Guang, p. 958; western Liaoning (about 45°58′N, 120°21′E); Middle Jurassic Haifanggou Formation. (in English)

Genus *Filiformorama* Wang, Hao et Cai, 2006 (in English)
2006   Wang Yi, Hao Shougang and Cai Chongyang, in Wang Yi and others, p. 25.

Type species: *Filiformorama simplexa* Wang, Hao et Cai, 2006

Taxonomic status: ? Rhynophytoid incertae sedis

Distribution and Age: Junggar Basin of Xinjiang, China; Late Silurian (Late Pridoli).

Filiformorama simplexa Wang, Hao et Cai, 2006 (in English)
2006   Wang Yi, Hao Shougang and Cai Chongyang, in Wang Yi and others, p. 25, figs. 1 — 4; smooth dichotomously branching plant; Holotype: PB20338 (figs. 1C, 1D); Paratypes: PB20336, PB20337, PB20339 — PB20346 (figs. 1A, F — K; figs. 3A — J); Repository: Nanjing Institute of Geology and Palaeontology, Chinese Academy of Sciences; Hoboksar Mongolia Autonomous County of west Junggar Basin, Xinjiang; Late Silurian (Late Pridoli) middle part of Wutubulake Formation.

Genus *Fimbriotheca* Zhu et Chen, 1981
1981   Zhu Jianan and Chen Gongxin, p. 488.

Type species: *Fimbriotheca tomentosa* Zhu et Chen, 1981

Taxonomic status: Marattiales, Filicopsida

Distribution and Age: Hubei, China; Late Permian.

Fimbriotheca tomentosa Zhu et Chen, 1981
1981   Zhu Jianan and Chen Gongxin, p. 488, pl. 1, figs. 1 — 7; text-fig. 1; fertile pinna; No.: 2004/55-13, 2004/55-14; Yangxin, Hubei; early Late Permian.

Genus *Foliosites* Ren, 1989
[Notes: This genus was initially described as a Lichenes, but was later suggested as a Bryophyta? (Wu Xiangwu and Li Baoxin, 1992)]

1989   Ren Shouqin, in Ren Shouqin and Chen Fen, pp. 634, 639.

1992   Wu Xiangwu and Li Baoxin, p. 272.

Type species: *Foliosites formosus* Ren, 1989

Taxonomic status: Lichenes? or Bryophyta?

Distribution and Age: Hulun Buir League of Inner Mongolia, China; Early Cretaceous.

Foliosites formosus Ren, 1989
1989   Ren Shouqin, in Ren Shouqin and Chen Fen, pp. 634, 639, pl. 1, figs. 1 — 4; text-fig. 1; thallus; Reg. No.: HW043, HW044, HWS012; Holotype: HW043 (pl. 1, fig. 1); Repository: China University of Geology (Beijing); Wujiu Coal Mine of Hulun Buir League, Inner Mongolia; Early Cretaceous Damoguaihe Formation.

1992　Wu Xiangwu and Li Baoxin, p. 272.

Genus *Fujianopteris* Liu et Yao, 2004 (in Chinese and English)
2004　Liu Lujun and Yao Zhaoqi, pp. 474, 480.
Type species: *Fujianopteris fukianensis* (Yabe et Ôishi) Liu et Yao, 2004
Taxonomic status: Pteridospermopsida or Gigantopterides
Distribution and Age: Fujian and Guangdong, China; Permian.

Fujianopteris fukianensis (Yabe et Ôishi) Liu et Yao, 2004 (in Chinese and English)
1938　*Gigantopteris fukianensis* Yabe et Ôishi, Yabe H and Ôishi S, p. 231, pl. 32, figs. 8, 9, 12C; text-fig. 9 (non 10); Longyan, Fujian (Fukien); Permian.
2004　Liu Lujun and Yao Zhaoqi, pp. 478, 484, pl. 1, figs. 2, 7 — 9; pl. 2, fig. 1; text-fig. 6; frond pinnate; Subang of Longyan and Changta of Nanjing, Fujian; Middle Permian (Guadalupian) upper member of Tongziyan Formation.

The other species:

Fujianopteris angustiangla (Yang et Chen) Liu et Yao, 2004 (in Chinese and English)
1979　*Gigantonoclea angustiangla* Yang et Chen, Yang Guanxiu and Chen Fen, p. 127, pl. 37, fig. 2; text-fig. 40; frond pinnate; Gedingzhai of Renhua, Guangdong; Late Permian Longtan Formation.
2004　Liu Lujun and Yao Zhaoqi, pp. 475, 482; text-fig. 2; frond pinnate; Gedingzhai of Renhua, Guangdong; "Late Permian Longtan Formation" [equivalent to Middle Permian (Guadalupian) upper member of Tongziyan Formation].

Fujianopteris cladonervis (S Li) Liu et Yao, 2004 (in Chinese and English)
1989　*Gigantonoclea cladonervis* S Li, Li Shengsheng, in Huang Lianmeng, p. 49, pl. 29, figs. 1 — 3; pl. 30, figs. 1 — 3; frond pinnate; Jiandou of Anxi, Fujian; Early Permian.
2004　Liu Lujun and Yao Zhaoqi, pp. 476, 482; text-fig. 3; frond pinnate; Changta of Nanjing and Jiandou of Anxi, Fujian; Middle Permian (Guadalupian) upper member of Tongziyan Formation.

Fujianopteris intermedia Liu et Yao, 2004 (in Chinese and English)
2004　Liu Lujun and Yao Zhaoqi, pp. 477, 483, pl. 1, figs. 1, 3 — 6; pl. 2, figs. 2 — 5; text-figs. 4, 5; frond pinnate; Reg. No.: PB9154, PB9155, PB9156, PB9157, PB9158, PB9160, PB9161; Holotype: PB9158 (pl. 1, fig. 5); Paratypes: PB9154 (pl. 1, fig. 1), PB9155 (pl. 2, fig. 2), PB9156 (pl. 1, fig. 3), PB9157 (counter part of the specimen PB9156), PB9160 (pl. pl. 1, fig. 1), PB9161 (pl. 2, fig. 3); Repository: Nanjing Institute of Geology and Palaeontology, Chinese Academy of Sciences; Kengbing of Longyan and Niulanshan of Yongding, Fujian; Middle Permian (Guadalupian) upper member of Tongziyan Formation.

Genus *Gansuphyllite* Xu et Shen, 1982

1982　Xu Fuxiang and Shen Guanglong, in Liu Zijin, p. 118.

Type species: *Gansuphyllite multivervis* Xu et Shen, 1982

Taxonomic status: Equisetales, Sphenopsida

Distribution and Age: Wudu of Gansu, China; Middle Jurassic.

Gansuphyllite multivervis Xu et Shen, 1982

1982　Xu Fuxiang and Shen Guanglong, in Liu Zijin, p. 118, pl. 58, fig. 5; calamitean stem and leaf whorl; No. : LP00013-3; Dalinggou of Wudu, Gansu; Middle Jurassic upper part of Longjiagou Formation.

Genus *Geminofoliolum* Zeng, Shen et Fan, 1995

1995　Zeng Yong, Shen Shuzhong and Fan Bingheng, pp. 49, 76.

Type species: *Geminofoliolum gracilis* Zeng, Shen et Fan, 1995

Taxonomic status: Calamariaceae, Sphenopsida

Distribution and Age: Yima of Henan, China; Middle Jurassic.

Geminofoliolum gracilis Zeng, Shen et Fan, 1995

1995　Zeng Yong, Shen Shuzhong and Fan Bingheng, pp. 49, 76, pl. 7, figs. 1 − 2; text-fig. 9; calamitean stems; Col. No. : No. 117146, No. 117144; Reg. No. : YM94031, YM94032; Holotype: YM94032 (pl. 7, fig. 2); Paratype: YM94031 (pl. 7, fig. 1); Repository: Department of Geology, China University of Mining and Technology; Yima, western Henan; Middle Jurassic Yima Formation.

Genus *Gigantonoclea* Koidzumi, 1936

1936　Koidzumi G, p. 138.

Type species: *Gigantonoclea lagrelii* (Halle) Koidzumi, 1938

Taxonomic status: Pteridospermopsida, Gigantopterides

Distribution and Age: central Shanxi, China; late Early Permian.

Gigantonoclea lagrelii (Halle) Koidzumi, 1936

1927　*Gigantopteris lagrelii* Halle, Halle T G, p. 170, pl. 46; teridosperm foliage; central Shanxi; late Early Permian Upper Shihhotse Series.

1936　Koidzumi G, p. 138.

Genus *Gigantonomia* Li et Yao, 1983

1983　Li Xingxue and Yao Zhaoqi, p. 14.

Type species:*Gigantonomia (Gigatonoclea) fukienensis* (Yabe et Ôishi) Li et Yao,1983

Taxonomic status:Pteridospermopsida,Gigantopterides

Distribution and Age:South China;Permian.

Gigantonomia (Gigatonoclea) fukienensis (Yabe et Ôishi) Li et Yao,1983

1938　*Gigatonoclea fukienensis* Yabe et Ôishi,Yabe H and Ôishi S,p. 231,pl. 1,figs. 8,9, 12C;text-figs. 9,12;Longyan,Fujian (Fukien);Permian.

1983　Li Xingxue and Yao Zhaoqi,p. 14,pl. 1,figs. 1 — 3;pl. 2,figs. 1 — 4;pl. 3,figs. 1 — 5; text-figs. 1 — 4;fructifications of gigantopterids;South China;Permian.

Genus *Gigantopteris* Schenk ex Whit D,1912

1883　*Megalopteris* Schenk,Schenk A,p. 238.

1912　in Whit D,p. 494.

Type species:*Gigantopteris nicotianaefolia* Schenk ex Whit D,1912

Taxonomic status:Pteridospermopsida,Gigantopterides

Distribution and Age:Hunan, Jiangsu, Yunnan and Henan, China; Late Carboniferous (Late Permian).

Gigantopteris nicotianaefolia Schenk ex Whit D,1912

1883　*Megalopteris nicotianaefolia* Schenk,Schenk A,p. 238,pl. 32,figs. 6 — 8;pl. 33,figs. 1 — 3;pl. 35,fig. 6;foliage,affinity uncertain;Leiyang,Hunan;Late Carboniferous.

1912　in Whit D,p. 494.

1974　Nanjing Institute of Geology and Palaeontology,Institute of Botany,Chinese Academy of Sciences,in *Palaeozoic Plants from China*,p. 130,pl. 100,figs. 2 — 4;pl. 101,fig. 1;pl. 102,fig. 7;text-figs. 103 — 105,108;foliage;Hunan,Jiangsu,Yunnan and Henan;Late Permian.

Genus *Gigantotheca* Li et Yao,1983

1983　Li Xingxue and Yao Zhaoqi,p. 19.

Type species:*Gigantotheca paradoxa* Li et Yao,1983

Taxonomic status:Pteridospermopsida,Gigantopterides

Distribution and Age:South China;Permian.

Gigantotheca paradoxa Li et Yao,1983

1983　Li Xingxue and Yao Zhaoqi,p. 20,pl. 4,figs. 1 — 3;pl. 5,figs. 1 — 4;pl. 6,figs. 1 — 5; text-figs. 5,6;fructifications of gigantopterids;Reg. No. :PB9059 (pl. 4,fig. 1),PB9077 (pl. 4,fig. 2),PB9067 (pl. 4,fig. 3),PB9073 (pl. 5,fig. 1),PB9060 (pl. 5,fig. 2), PB9078 (pl. 5,fig. 3),PB9074 (pl. 5,fig. 4),PB9062 (pl. 6,fig. 1),PB9063 (pl. 6,fig. 2),PB9076 (pl. 6,fig. 3),PB9071 (pl. 6,fig. 4),PB9066 (pl. 6,fig. 5);South China; Permian.

Genus *Guangnania* Wang et Hao, 2002 (in English)
2002　Wang Deming and Hao Shougang, p. 14.
Type species: *Guangnania cuneata* Wang et Hao, 2002
Taxonomic status: Zosterophyllophytes
Distribution and Age: Yunnan, China; Early Devonian.

Guangnania cuneata Wang et Hao, 2002 (in English)
2002　Wang Deming and Hao Shougang, p. 15, pls. 1 — 3; text-figs. 1 — 3; strobili; No. : WH-D01 — WH-D13, WH-G, WH-X01 — WH-X03; Holotype: WH-D01 (pl. 1, fig. 1); Repository: Department of Geology, Peking University; Guangnan, Yunnan; Early Devonian Posongchong Formation; Qujing, Yunnan; Early Devonian Xujiachong Formation.

Genus *Guangxiophyllum* Feng, 1977
1977　Feng Shaonan and others, p. 247.
Type species: *Guangxiophyllum shangsiense* Feng, 1977
Taxonomic status: Gymnospermae incertae sedis
Distribution and Age: Shangsi of Guangxi, China; Late Triassic.

Guangxiophyllum shangsiense Feng, 1977
1977　Feng Shaonan and others, p. 247, pl. 95, fig. 1; cycadophyte leaf; No. : P25281; Holotype: P25281 (pl. 95, fig. 1); Repository: Hubei Institute of Geological Sciences; Wangmen of Shangsi, Guangxi; Late Triassic.

Genus *Guizhoua* Zhao, 1980
1980　Zhao Xiuhu and others, p. 89.
Type species: *Guizhoua gregalis* Zhao, 1980
Taxonomic status: Plantae incertae sedis
Distribution and Age: Guizhou, China; early Late Permian.

Guizhoua gregalis Zhao, 1980
1980　Zhao Xiuhu and others, p. 89, pl. 23, figs. 1, 1a, 1b, 2 — 4; fertile frond and sporangia; Col. No. : PZ2-7; Reg. No. : PB7107 — PB7110; Repository: Nanjing Institute of Geology and Palaeontology, Chinese Academy of Sciences; Panxian, Guizhou; early Late Permian lower member of Xuanwei Formation.

Genus *Guizhouoxylon* Tian et Li, 1992

1992   Tian Baolin and Li Hongqi, pp. 336, 343.

Type species: *Guizhouoxylon dahebianense* Tian et Li, 1992

Taxonomic status: Plantae incertae sedis

Distribution and Age: Shuicheng of Guizhou, China; Late Permian.

Guizhouoxylon dahebianense Tian et Li, 1992

1992   Tian Baolin and Li Hongqi, pp. 337, 343, pl. 1, figs. 1 — 9; pl. 2, figs. 1 — 10; pl. 3, figs. 1 — 10; pl. 4, figs. 1 — 10; petrified stem; Repository: Beijing Graduate School, China University of Mining and Technology; Shuicheng, Guizhou; Late Permian Longtan Formation.

Genus *Gumuia* Hao, 1989

1989b   Hao Shougang, pp. 954, 955.

Type species: *Gumuia zyzzata* Hao, 1989

Taxonomic status: Zosterophyllophytes

Distribution and Age: Yunnan, China; Early Devonian.

Gumuia zyzzata Hao, 1989

1989b   Hao Shougang, p. 955, pl. 1, figs. 1 — 3; pl. 2, figs. 1 — 11; text-figs. 1 — 3; No.: Bupb601, 602, 603, 604, 604', 605, 606, 607, 608, 608', 609; Holotype: Bupb602 (pl. 2, fig. 1); Repository: Department of Geology, Peking University; Wenshan, Yunnan; Early Devonian Posongchong Formation.

Genus *Gymnogrammitites* Sun et Zheng, 2001 (in Chinese and English)

2001   Sun Ge and Zheng Shaolin, in Sun Ge and others, pp. 75, 185.

Type species: *Gymnogrammitites ruffordioides* Sun et Zheng, 2001

Taxonomic status: Filicopsida

Distribution and Age: Beipiao of western Liaoning, China; Late Jurassic.

Gymnogrammitites ruffordioides Sun et Zheng, 2001 (in Chinese and English)

2001   Sun Ge and Zheng Shaolin, in Sun Ge and others, pp. 75, 185, pl. 7, fig. 6; pl. 9, figs. 1 — 2; pl. 40, figs. 5 — 8; frond; No.: PB19020, PB19020A (counterpart); Holotype: PB19020 (pl. 7, fig. 6); Repository: Nanjing Institute of Geology and Palaeontology, Chinese Academy of Sciences; Huangbanjigou in Shangyuan of Beipiao, western Liaoning; Late Jurassic Jianshangou Formation.

Genus *Hallea* Mathews, 1947 — 1948 (non Yang et Wu, 2006)
1947 — 1948　Mathews G B, p. 241.

Type species: *Hallea pekinensis* Mathews, 1947 — 1948

Taxonomic status: Incertae sedis

Distribution and Age: West Hill of Beijing, China; Permian (?) or Triassic (?).

Hallea pekinensis Mathews, 1947 — 1948
1947 — 1948　Mathews G B, p. 241, fig. 4; seed; West Hill, Beijing; Permian (?) or Triassic (?) Shuantsuang Series.

Genus *Hallea* Yang et Wu, 2006 (non Mathews, 1947 — 1948) (in Chinese and English)
[Notes: This generic name *Hallea* Yang et Wu, 2006 is a late homomum (homomum junius) of *Hallea* Mathews, 1947 — 1948]

2006　Yang Guanxiu and Wu Yuehui, in Yang Guanxiu and others, pp. 194, 314.

Type species: *Hallea dengfengensis* Yang et Wu, 2006

Taxonomic status: Gigantopteridales

Distribution and Age: Dengfeng of western Henan, China; Late Permian.

Hallea dengfengensis Yang et Wu, 2006 (in Chinese and English)
2006　Yang Guanxiu and Wu Yuehui, in Yang Guanxiu and others, pp. 194, 314, pl. 56, figs. 3, 3a; leaf pinnte compound (?); Holotype: HEP3150 (pl. 56, figs. 3, 3a); Repository: China University of Geosciences, Beijing; Dengcao of Dengfeng, western Henan; Late Permian coal member 8 of Yungaishan Formation.

Genus *Halleophyton* Li et Edwards, 1997 (in English)
1997　Li Chengsen and Edwards D, p. 1447.

Type species: *Halleophyton zhichangense* Li et Edwards, 1997

Taxonomic status: Drepanophycales, Lycophyta

Distribution and Age: Wenshan of Yunnan, China; Early Devonian.

Halleophyton zhichangense Li et Edwards, 1997 (in English)
1997　Li Chengsen and Edwards D, p. 1448, figs. 1 — 29; shoots and sporangia; No.: CBYn9003001a, CBYn9003001b, CBYn9003002a, CBYn9003002b, CBYn9003003, CBYn9003003, CBYn9003004b, CBYn9003005a, CBYn9003008, CBYn9003012a, CBYn9003014a, CBYn9003015a, CBYn9003015b, CBYn9003018, CBYn9003021, CBYn9003023b, CBYn9003030; Holotypes: CBYn9003001a, CBYn9003001b (figs. 17, 18); Repository: Institute of Botany, Chinese Academy of Sciences; Wenshan, Yunnan; Early Devonian Posongchong Formation.

Genus *Hamatophyton* Gu et Zhi, 1974

1974   Nanjing Institute of Geology and Palaeontology, Institute of Botany, Chinese Academy of Sciences, in *Palaeozoic Plants from China*, p. 38.

Type species: *Hamatophyton verticillatum* Gu et Zhi, 1974

Taxonomic status: Hyeniales, Sphenopsida

Distribution and Age: Jiangsu, Zhejiang and Jiangxi, China; Late Devonian.

Hamatophyton verticillatum Gu et Zhi, 1974

1974   Nanjing Institute of Geology and Palaeontology, Institute of Botany, Chinese Academy of Sciences, in *Palaeozoic Plants from China*, p. 38, pl. 19, figs. 3 — 5; pl. 20, figs. 1 — 6; shoots with showing the crooked hook-like leaves; Reg. No.: PB4893 — PB4897; Syntypes: PB4893, PB4894 (pl. 19, figs. 3, 4), PB4895 — PB4897 (pl. 20, figs. 1, 3, 4); Repository: Nanjing Institute of Geology and Palaeontology, Chinese Academy of Sciences; Jiangyin, Longtan, Jurong and Zhenze of Jiangsu, Changxing of Zhejiang; Late Devonian Wutong Formation (Group); Lianhua and Yongxin, Jiangxi; Middle — Late Devonian upper part of Xiashan Group; Hangzhou, Zhejiang; Late Devonian Qianligang Group.

Genus *Hefengistrobus* Xu et Wang, 2002 (in Chinese and English)

2002   Xu Honghe and Wang Yi, pp. 251, 256.

Type species: *Hefengistrobus bifurcus* Xu et Wang, 2002

Taxonomic status: Lycopsida

Distribution and Age: Junggar Basin of Xinjiang, China; Late Devonian.

Hefengistrobus bifurcus Xu et Wang, 2002 (in Chinese and English)

2002   Xu Honghe and Wang Yi, pp. 252, 256, pl. 1, figs. 1 — 9; pl. 2, fig. 11; text-fig. 2; heterosporous lycopsid plant; Reg. No.: PB19243 — PB19250; Holotype: PB19245 (pl. 1, fig. 6); Paratypes: PB19244A, PB19244B (pl. 1, figs. 2, 3), PB19247A, PB19247B (pl. 2, figs. 1, 2); Repository: Nanjing Institute of Geology and Palaeontology, Chinese Academy of Sciences; Junggar Basin, Xinjiang; Late Devonian Hongguleleng Formation.

Genus *Helicophyton* Wang et Xu, 2002 (in English)

2002   Wang Yi and Xu Honghe, p. 475.

Type species: *Helicophyton dichotomum* Wang et Xu, 2002

Taxonomic status: Pteridopsida incertae sedis

Distribution and Age: Nanjing of Jiangsu, China; Early Carboniferous.

Helicophyton dichotomum Wang et Xu, 2002 (in English)

2002   Wang Yi and Xu Honghe, p. 475, figs. 1 — 6, plant with main axis and two orders of branches; Holotype: PB19210 (fig. 3a); Paratypes: PB19206 (fig. 1a), PB19208 (fig. 1c); Repository: Nanjing Institute of Geology and Palaeontology, Chinese Academy of Sciences; Kongshan of Nanjing, Jiangsu; Early Carboniferous upper part of upper member of Wutong Formation.

Genus *Henanophyllum* Xi et Feng, 1977
1977　Xi Yunhong and Feng Shaonan, in Feng Shaonan and others, p. 673.
Type species: *Henanophyllum palamifolium* Xi et Feng, 1977
Taxonomic status: Plantae incertae sedis
Distribution and Age: Henan, China; Late Permian.

Henanophyllum palamifolium Xi et Feng, 1977
1977　Xi Yunhong and Feng Shaonan, in Feng Shaonan and others, p. 674, pl. 248, figs. 6, 7; leaves; No. : P0021, P0022; Syntypes: P0021 (pl. 248, fig. 6), P0022 (pl. 248, fig. 7); Yuxian, Henan; Late Permian Upper Shihhotse Formation.

Genus *Henanopteris* Yang, 1987
1987a　Yang Guanxiu, p. 52.
Type species: *Henanopteris lanceolatus* Yang, 1987
Taxonomic status: Pteridospermopsida
Distribution and Age: Henan, China; Late Permian.

Henanopteris lanceolatus Yang, 1987
1987a　Yang Guanxiu, p. 52, pl. 15, figs. 2 — 5; pinnae; Reg. No. : HEP728, HEP729, HEP730, HEP731; Syntype 1: HEP728 (pl. 15, fig. 2); Syntype 2: HEP729 (pl. 15, fig. 3); Syntype 3: HEP730 (pl. 15, fig. 4); Syntype 4: HEP731 (pl. 15, fig. 5); Repository: China University of Geosciences, Beijing; Dafengkou of Yuxian, Henan; Late Permian Upper Shihhotse Formation.

Genus *Henanotheca* Yang, 2006 (in Chinese and English)
2006　Yang Guanxiu and others, pp. 107, 260.
Type species: *Henanotheca (Sphenopteris) ovata* Yang, 2006
Taxonomic status: Gleicheniaceae, Filicales
Distribution and Age: Yuzhou of western Henan, China; Middle Permian.

Henanotheca (Sphenopteris) ovata Yang, 2006 (in Chinese and English)
2006　Yang Guanxiu and others, pp. 107, 260, pl. 16, figs. 10, 10a, 10b; pl. 24, fig. 5; pinnae; Holotype: HEP0908 (pl. 16, fig. 10); Repository: China University of Geosciences, Beijing; Dafengkou of Yuzhou, western Henan; Middle Permian Xiaofengkou Formation.

Genus *Hexaphyllum* Ngo,1956

1956　Ngo C K,p. 25.

Type species: *Hexaphyllum sinense* Ngo,1956

Taxonomic status: Plantae incertae sedis or Equisetales?

Distribution and Age: Xiaoping in Guangzhou of Guangdong, China; Late Triassic.

Hexaphyllum sinense Ngo,1956

1956　Ngo C K, p. 25, pl. 1, fig. 2; pl. 6, figs. 1, 2; text-fig. 3; leaf whorls; No. : A4; Reg. No. : 0015; Repository: Palaeontology Section, Department of Geology, Central-South Institute of Mining and Metallurgy; Xiaoping of Guangzhou, Guangdong; Late Triassic Siaoping Coal Series. [Notes: This specimen lately was referred as *Annulariopsis*? *sinensis* (Ngo) Lee (Sze H C, Lee H H and others, 1963)]

Genus *Holozamites* Wang X, Li, Wang Y et Zheng, 2009 (nom. nud.)

2009a　Wang Xin, Li Nan, Wang Yongdong and Zheng Shaolin, p. 1937. (in Chinese) (only name)

2009b　Wang Xin, Li Nan, Wang Yongdong and Zheng Shaolin, p. 3116. (in English) (only name)

Type species: *Holozamites hongtaoi* Wang X, Li, Wang Y et Zheng, 2009

Taxonomic status: Angiospermae?

Distribution and Age: Beipiao of Liaoning, China; Late Jurassic.

Holozamites hongtaoi Wang X, Li, Wang Y et Zheng, 2009 (nom. nud.)

2009a　Wang Xin, Li Nan, Wang Yongdong and Zheng Shaolin, p. 1937; Huangbanjigou of Beipiao, Liaoning; Late Jurassic. (in Chinese) (only name)

2009b　Wang Xin, Li Nan, Wang Yongdong and Zheng Shaolin, p. 3116; Huangbanjigou of Beipiao, Liaoning; Late Jurassic. (in English) (only name)

Genus *Hsiangchiphyllum* Sze,1949

1949　Sze H C, p. 28.

Type species: *Hsiangchiphyllum trinerve* Sze,1949

Taxonomic status: Cycadopsida

Distribution and Age: Xiangxi in Zigui of Hubei, China; Early Jurassic.

Hsiangchiphyllum trinerve Sze,1949

1949　Sze H C, p. 28, pl. 7, fig. 6; pl. 8, fig. 1; cycadophyte leaf; Xiangxi in Zigui of Hubei; Early Jurassic Hsiangchi Coal Series.

Genus *Hsuea* Li, 1982

1982   Li Chengsen, pp. 331, 341.

Type species: *Hsuea robusta* (Li et Cai) Li, 1982

Taxonomic status: Cooksoniaceae, Rhynales, Psilophyta

Distribution and Age: Qujing of Yunnan, China; Early Devonian.

Hsuea robusta (Li et Cai) Li, 1982

1978   *Cooksonia zhanyiensis* Li et Cai, Li Xingxue and Cai Chongyang, p. 10, pl. 2, fig. 6; psilophytic plant; Qujing, Yunnan; Early Devonian.

1978   *Taeniocrada robusta* Li et Cai, Li Xingxue and Cai Chongyang, p. 10, pl. 2, figs. 7 — 14; psilophytic plant; Qujing, Yunnan; Early Devonian.

1982   Li Chengsen, pp. 331, 341, pls. 3 — 10; psilophytic plant; No. : 7771 — 7775, 7787, 7788, 7791, 7793, 7796, 7799, 7800, 7802, 7853, 7863, 7870, 7874, 7880, 7888, 7895 and others; Holotype: 7771 (pl. 3, fig. 1); Repository: Institute of Botany, Chinese Academy of Sciences; Qujing, Yunnan; Early Devonian Xujiachong Formation.

Genus *Huangia* Si, 1989

1989   Si Xingjian, pp. 55, 196.

Type species: *Huangia elliptica* Si, 1989

Taxonomic status: Filices or Pteridospermopsida

Distribution and Age: Jungar Banner of Inner Mongolia, China; Early Permian.

Huangia elliptica Si, 1989

1989   Si Xingjian, pp. 55, 196, pl. 63, figs. 1 — 4; pl. 64, fig. 1; frond; Col. No. : YF545, YF546, YF549; Reg. No. : PB4219, PB4218, PB4217, PB4221 (pl. 63, figs. 1 — 4), PB4220 (pl. 64, fig. 1); Holotype: PB4221 (pl. 63, fig. 4); Repository: Nanjing Institute of Geology and Palaeontology, Chinese Academy of Sciences; Heidaigou of Jungar Banner, Inner Mongolia; Early Permian Shanxi (Shansi) Formation? (or Taiyuan Formation?).

Genus *Hubeiia* Xue, Hao, Wang et Liu, 2005 (in English)

2005   Xue Jinzhuang, Hao Shougang, Wang Deming and Liu Zhenfeng, p. 520.

Type species: *Hubeiia dicrofollia* Xue, Hao, Wang et Liu, 2005

Taxonomic status: Protolepidodendraceae, Protolepidodendrales, Lycophyta

Distribution and Age: Changyang of Hubei, China; Late Devonian (Famennian).

Hubeiia dicrofollia Xue, Hao, Wang et Liu, 2005 (in English)

2005   Xue Jinzhuang, Hao Shougang, Wang Deming and Liu Zhenfeng, p. 520, figs. 2 — 6; herbaceous lycopsid; Holotype: PKU-X-11a (fig. 2a); Paratypes: PKU-X-5b, 7, 19,

PKU-X-16-3, PKU-X-17-3, 17-4a (figs. 2h — k, fig. 4a, figs. 5a, 5d); Repository: Department of Geology, Peking University; Changyang, Hubei; Late Devonian (Famennian) Xiejingsi Formation.

Genus *Hubeiophyllum* Feng, 1977

1977　Feng Shaonan and others, p. 247.

Type species: *Hubeiophyllum cuneifolium* Feng, 1977

Taxonomic status: Gymnospermae incertae sedis

Distribution and Age: Tieluwan in Yuan'an of Hubei, China; Late Triassic.

Hubeiophyllum cuneifolium Feng, 1977

1977　Feng Shaonan and others, p. 247, pl. 100, figs. 1 — 4; leaf; Reg. No. : P25298 — P25301; Syntypes: P25298 — P25301 (pl. 100, figs. 1 — 4); Repository: Hubei Institute of Geological Sciences; Tieluwan of Yuan'an, Hubei; Late Triassic Lower Coal Formation of Hsiangchi Group.

The other species:

Hubeiophyllum angustum Feng, 1977

1977　Feng Shaonan and others, p. 247, pl. 100, figs. 5 — 7; leaf; Reg. No. : P25302 — P25304; Syntypes: P25298 — P25301 (pl. 100, figs. 5 — 7); Repository: Hubei Institute of Geological Sciences; Tieluwan of Yuan'an, Hubei; Late Triassic Lower Coal Formation of Hsiangchi Group.

Genus *Huia* Geng, 1985

1985　Geng Baoyin, pp. 419, 425.

Type species: *Huia recurvata* Geng, 1985

Taxonomic status: Taeniocradaceae, Rhyniophytina, Psilophyta

Distribution and Age: Wenshan of Yunnan, China; Early Devonian.

Huia recurvata Geng, 1985

1985　Geng Baoyin, pp. 420, 425, pls. 1, 2; psilophytic plant; No. : 8122 — 8133; Holotype: 8122 (pl. 1, fig. 10); Paratypes: 8123 — 8133; Repository: Institute of Botany, Chinese Academy of Sciences; Wenshan, Yunnan; Early Devonian Posongchong Formation.

Genus *Hunanoequisetum* Zhang, 1986

1986　Zhang Caifan, p. 191.

Type species: *Hunanoequisetum liuyangense* Zhang, 1986

Taxonomic status: Equisetales, Sphenopsida

Distribution and Age: Yuelong in Liuyang of Hunan, China; Early Jurassic.

Hunanoequisetum liuyangense Zhang, 1986
1986   Zhang Caifan, p. 191, pl. 4, figs. 4—4a, 5; text-fig. 1; calamitean stem; Reg. No.: PH472, PH473; Holotype: PH472 (pl. 4, fig. 4); Repository: Geology Museum of Hunan Province; Yuelong of Liuyang, Hunan; Early Jurassic Yuelong Formation.

Genus *Illicites* Pan, 1983 (nom. nud.)
1983   Pan Guang, p. 1520. (in Chinese)
1984   Pan Guang, p. 959. (in English)
Type species: (without specific name)
Taxonomic status: "primitive angiosperms"
Distribution and Age: western Liaoning, China; Middle Jurassic.

Illicites sp. indet.
[Notes: Generic name was given only, but without specific name (or type species) in the original paper]
1983   *Illicites* sp. indet., Pan Guang, p. 1520; western Liaoning (about 45°58′N, 120°21′E); Middle Jurassic Haifanggou Formation. (in Chinese)
1984   *Illicites* sp. indet., Pan Guang, p. 959; western Liaoning (about 45°58′N, 120°21′E); Middle Jurassic Haifanggou Formation. (in English)

Genus *Jaenschea* Mathews, 1947—1948
1947—1948   Mathews G B, p. 239.
Type species: *Jaenschea sinensis* Mathews, 1947—1948
Taxonomic status: Osmundaceae, Filicopsida?
Distribution and Age: West Hill of Beijing, China; Permian (?) or Triassic (?).

Jaenschea sinensis Mathews, 1947—1948
1947—1948   Mathews G B, p. 239, fig. 2; fertile pinna; West Hill, Beijing; Permian (?) or Triassic (?) Shuantsuang Series.

Genus *Jiangxifolium* Zhou, 1988
1988   Zhou Xianding, p. 126.
Type species: *Jiangxifolium mucronatum* Zhou, 1988
Taxonomic status: Filicopsida
Distribution and Age: Youluo in Fengcheng of Jiangxi, China; Late Triassic.

Jiangxifolium mucronatum Zhou, 1988
1988   Zhou Xianding, p. 126, pl. 1, figs. 1, 2, 5, 6; text-fig. 1; frond; Reg. No.: No. 2228, No. 1862, No. 1348, No. 2867; Holotype: No. 2228 (pl. 1, fig. 1); Repository: 195th Coal-

geological Exploration Team of Jiangxi Province; Youluo of Fengcheng, Jiangxi; Late Triassic Anyuan Formation.

The other species:
Jiangxifolium denticulatum Zhou, 1988
1988   Zhou Xianding, p. 127, pl. 1, figs. 3, 4; frond; Reg. No. : No. 2135, No. 2867; Holotype: No. 2135 (pl. 1, fig. 3); Repository: 195ᵗʰ Coal-geological Exploration Team of Jiangxi Province; Youluo of Fengcheng, Jiangxi; Late Triassic Anyuan Formation.

Genus *Jiangxitheca* He, Liang et Shen, 1996 (in Chinese and English)
1996   He Xilin, Liang Dunshi and Shen Shuzhong, pp. 50, 160.
Type species: *Jiangxitheca xinanensis* He, Liang et Shen, 1996
Taxonomic status: Filices or Pteridospermopsida
Distribution and Age: Jiangxi, China; Late Permian.

Jiangxitheca xinanensis He, Liang et Shen, 1996 (in Chinese and English)
1996   He Xilin, Liang Dunshi and Shen Shuzhong, pp. 50, 160, pl. 36, fig. 5; pl. 37, figs. 1 — 4; fronds; Reg. No. : X88163 — X88167; Syntypes: X88163 (pl. 36, fig. 5), X88164 — X88167 (pl. 37, figs. 1 — 4); Repository: Department of Geology, China University of Mining and Technology; Xin'an Coal Mine of Qianshan, Jiangxi; Late Permian Tongjia Member of Shangrao Formation.

Genus *Jingmenophyllum* Feng, 1977
1977   Feng Shaonan and others, p. 250.
Type species: *Jingmenophyllum xiheense* Feng, 1977
Taxonomic status: Gymnospermae incertae sedis
Distribution and Age: Xihe in Jingmen of Hubei, China; Late Triassic.

Jingmenophyllum xiheense Feng, 1977
1977   Feng Shaonan and others, p. 250, pl. 94, fig. 9; cycadophyte leaf; Reg. No. : P25280; Holotype: P25280 (pl. 94, fig. 9); Repository: Hubei Institute of Geological Sciences; Xihe of Jingmen, Hubei; Late Triassic Lower Coal Formation of Hsiangchi Group. [Notes: This specimen lately was referred as *Compsopteris xiheensis* (Feng) Zhu, Hu et Meng (Zhu Jianan and others, 1984)]

Genus *Jixia* Guo et Sun, 1992
1992   Guo Shuangxing and Sun Ge, in Sun Ge and others, p. 547. (in Chinese)
1993   Guo Shuangxing and Sun Ge, in Sun Ge and others, p. 254. (in English)
Type species: *Jixia pinnatipartita* Guo et Sun, 1992

Taxonomic status: Dicotyledoneae

Distribution and Age: Chengzihe in Jixi of Heilongjiang, China; Early Cretaceous.

Jixia pinnatipartita Guo et Sun, 1992

1992 Guo Shuangxing and Sun Ge, in Sun Ge and others, p. 547, pl. 1, figs. 10 — 12; pl. 2, fig. 7; leaves; Reg. No. : PB16773 — PB16775, PB16775, PB16773A; Holotype: PB16774 (pl. 1, fig. 10); Repository: Nanjing Institute of Geology and Palaeontology, Chinese Academy of Sciences; Chengzihe of Jixi, Heilongjiang; Early Cretaceous upper part of Chengzihe Formation. (in Chinese)

1993 Guo Shuangxing and Sun Ge, in Sun Ge and others, p. 254, pl. 1, figs. 10 — 12; pl. 2, fig. 7; leaves; Reg. No. : PB16773 — PB16775, PB16775, PB16773A; Holotype: PB16774 (pl. 1, fig. 10); Repository: Nanjing Institute of Geology and Palaeontology, Chinese Academy of Sciences; Chengzihe of Jixi, Heilongjiang; Early Cretaceous upper part of Chengzihe Formation. (in English)

Genus *Junggaria* Dou, 1983

1983 Dou Yawei and others, p. 562.

Type species: *Junggaria spinosa* Dou, 1983

Taxonomic status: Psilophytes incertae sedis

Distribution and Age: Xinjiang, China; Early Devonian.

Junggaria spinosa Dou, 1983

1983 Dou Yawei nd others, p. 562, pl. 189, figs. 1 — 4; plants; Col. No. : 730H-1-44; Reg. No. : XPA001 — XPA003, XPA004; Syntypes: XPA001 — XPA003 (pl. 189, figs. 1 — 3); Hebukesaier County of West Junggar Basin, Xinjiang; Early Devonian Wutubulake Formation.

Genus *Juradicotis* Pan, 1983 (nom. nud.)

1983 Pan Guang, p. 1520. (in Chinese)
1984 Pan Guang, p. 958. (in English)

Type species: (without specific name)

Taxonomic status: "hemiangiosperms"

Distribution and Age: western Liaoning, China; Middle Jurassic.

Juradicotis sp. indet.

[Notes: Generic name was given only, but without specific name (or type species) in the original paper]

1983 *Juradicotis* sp. indet. , Pan Guang, p. 1520; western Liaoning (about 45°58′N, 120°21′E); Middle Jurassic Haifanggou Formation. (in Chinese)

1984 *Juradicotis* sp. indet. , Pan Guang, p. 958; western Liaoning (about 45°58′N, 120°21′E);

Middle Jurassic Haifanggou Formation. (in English)

Genus *Juramagnolia* Pan, 1983 (nom. nud.)

1983　Pan Guang, p. 1520. (in Chinese)
1984　Pan Guang, p. 959. (in English)
Type species: (without specific name)
Taxonomic status: "primitive angiosperms"
Distribution and Age: western Liaoning, China; Middle Jurassic.

Juramagnolia sp. indet.

[Notes: Generic name was given only, but without specific name (or type species) in the original paper]

1983　*Juramagnolia* sp. indet., Pan Guang, p. 1520; western Liaoning (about 45°58′N, 120°21′E); Middle Jurassic Haifanggou Formation. (in Chinese)
1984　*Juramagnolia* sp. indet., Pan Guang, p. 959; western Liaoning (about 45°58′N, 120°21′E); Middle Jurassic Haifanggou Formation. (in English)

Genus *Jurastrobus* Wang, Li et Cui, 2006 (in English)

2006　Wang Xin, Li Nan and Cui Jinzhong, p. 214.
Type species: *Jurastrobus chenii* Wang, Li et Cui, 2006
Taxonomic status: Cycadales
Distribution and Age: Xilinhaote of Inner Mongolia, China; Early Jurassic.

Jurastrobus chenii Wang, Li et Cui, 2006 (in English)

2006　Wang Xin, Li Nan and Cui Jinzhong, p. 214, figs. 2—6; pollen cone situated on the apex of the stem; No.: No. 9221; Holotype: No. 9221 (fig. 2); Repository: Palaeobotanica Collection, Laboratory of Palaeobotany, Beijing Institute of Botany, Chinese Academy of Sciences; Coal Well 795 of Xilinhaote, Inner Mongolia; late Early Jurassic (? Toarcian) K-1 seam of Manitemiao Group.

Genus *Kadsurrites* Pan, 1983 (nom. nud.)

1983　Pan Guang, p. 1520. (in Chinese)
1984　Pan Guang, p. 959. (in English)
Type species: (without specific name)
Taxonomic status: "primitive angiosperms"
Distribution and Age: western Liaoning, China; Middle Jurassic.

Kadsurrites sp. indet.

[Notes: Generic name was given only, but without specific name (or type species) in the

original paper]
1983 *Kadsurrites* sp. indet., Pan Guang, p. 1520; western Liaoning (about 45°58′N, 120°21′E); Middle Jurassic Haifanggou Formation. (in Chinese)
1984 *Kadsurrites* sp. indet., Pan Guang, p. 959; western Liaoning (about 45°58′N, 120°21′E); Middle Jurassic Haifanggou Formation. (in English)

Genus *Kaipingia* Stockmans et Mathieu, 1957
1957 Stockmans F and Mathieu F F, p. 62.
Type species: *Kaipingia sinica* Stockmans et Mathieu, 1957
Taxonomic status: Lycopodiales
Distribution and Age: Kaiping of Hebei, China; Late Carboniferous.

Kaipingia sinica Stockmans et Mathieu, 1957
1957 Stockmans F and Mathieu F F, p. 62, pl. 1, figs. 1, 1a; stem with leaf cushions; Kaiping, Hebei; Late Carboniferous.

Genus *Khitania* Guo, Sha, Bian et Qiu, 2009 (in English)
2009 Guo Shuangxing, Sha Jingeng, Bian Lizeng and Qiu Yinlong, p. 94.
Type species: *Khitania columnispicata* Guo, Sha, Bian et Qiu, 2009
Taxonomic status: Gnetaceae, Gnetales, Gnetopsida
Distribution and Age: Beipiao of Liaoning, China; Early Cretaceous.

Khitania columnispicata Guo, Sha, Bian et Qiu, 2009 (in English)
2009 Guo Shuangxing, Sha Jingeng, Bian Lizeng and Qiu Yinlong, p. 94, figs. 1 (left), 2A—2F; male spike strobili; Reg. No.: PB20189; Holotype: PB20189 [fig. 1 (left)]; Repository: Nanjing Institute of Geology and Palaeontology, Chinese Academy of Sciences; Eastern hillside of Huangbanjigou to western hillside of Jianshangou Villages (41°12′N, 119°22′E) in Beipiao, Liaoning; Early Cretaceous (Barremian) Jianshangou Member in Yixian Formation of Jehol Group.

Genus *Klukiopsis* Deng et Wang, 1999 (2000) (in Chinese and English)
1999 Deng Shenghui and Wang Shijun, p. 552. (in Chinese)
2000 Deng Shenghui and Wang Shijun, p. 356. (in English)
Type species: *Klukiopsis jurassica* Deng et Wang, 1999 (2000)
Taxonomic status: Schzaeaceae, Filicopsida
Distribution and Age: Yima of Henan, China; Middle Jurassic.

Klukiopsis jurassica Deng et Wang, 1999 (2000) (in Chinese and English)
1999 Deng Shenghui and Wang Shijun, p. 552, fig. 1; frond, fertile pinnae, sporangia and

spora; No. : YM98-303; Holotype: YM98-303 (fig. 1a); Yima, Henan; Middle Jurassic. (in Chinese)

2000 Deng Shenghui and Wang Shijun, p. 356, fig. 1; frond, fertile pinnae, sporangia and spora; No. : YM98-303; Holotype: YM98-303 (fig. 1a); Yima, Henan; Middle Jurassic. (in English)

Genus *Koilosphenus* Bohlin, 1971

1971 Bohlin B, p. 47.

Type species: *Koilosphenus cuneifolius* Bohlin, 1971

Taxonomic status: Conifers

Distribution and Age: Kansu, China; Late Palaeozoic.

Koilosphenus cuneifolius Bohlin, 1971

1971 Bohlin B, p. 47, pl. 7, figs. 6, 7; text-figs. 91A-D; shoots with apex and foliage; Yuerhhung, Kansu; Late Palaeozoic.

The other species:

? *Koilosphenus* sp.

1971 ? *Koilosphenus* sp., Bohlin B, p. 48, pls. 6, 7; text-figs. 90A-D; fragments of leaves; Yuerhhung, Kansu; Late Palaeozoic.

Genus *Kongshania* Wang, 2000 (in English)

2000 Wang Yi, p. 47.

Type species: *Kongshania synangioides* Wang, 2000

Taxonomic status: Sphenopterides

Distribution and Age: Jiangning of Jiangsu, China; Late Devonian.

Kongshania synangioides Wang, 2000 (in English)

2000 Wang Yi, p. 47, pls. 1 — 4; text-figs. 1 — 6; plant with rachies, mainly dichosympodial branching system; Reg. No. : PB17201 — PB17204, PB17205a, PB17205b, PB17206-PB17210, PB17211a, PB17211b, PB17212, PB17213, PB17214a, PB17214b; Holotype: PB17203 (pl. 1, fig. 3); Paratypes: PB17205b (pl. 1, fig. 10), PB17211a (pl. 2, fig. 1), PB17211b (pl. 2, fig. 2), PB17212 (pl. 2, fig. 3); Repository: Nanjing Institute of Geology and Palaeontology, Chinese Academy of Sciences; Kongshan of Jiangning, Jiangsu; Late Devonian Wutong Formation.

Genus *Konnoa* Asama, 1959

1959 Asama K, p. 63.

Type species: *Konnoa koraiensis* (Tokunaga) Asama, 1959

Taxonomic status: Callipterides

Distribution and Age: Liaoning and Shanxi of China, Korea; Late Palaeozoic.

Konnoa koraiensis (Tokunaga) Asama, 1959

1915 *Alethopteris koraiensis* Tokunaga, Tokunaga S, p. 52; text-figure.

1959 Asama K, p. 64, pl. 14, figs. 1 — 3; pl. 15, figs. 1 — 3; pl. 16, figs. 1 — 4; pl. 17, fig. 5; frond; Korea (Jido Series), Liaoning (Huangchi and Lintang Series) and Shanxi (Taiyuan and Shansi Series) of China; Late Palaeozoic.

The other species:

Konnoa penchihuensis Asama, 1959

1959 Asama K, p. 65, pl. 17, figs. 1 — 4; pl. 18, figs. 1 — 3; pl. 19, figs. 1, 2; frond; Benxi (Penchi), Liaoning; Late Palaeozoic (Huangchi Series).

Genus *Kuandiania* Zheng et Zhang, 1980

1980 Zheng Shaolin and Zhang Wu, in Zhang Wu and others, p. 279.

Type species: *Kuandiania crassicaulis* Zheng et Zhang, 1980

Taxonomic status: Cycadopsida

Distribution and Age: Kuandian in Benxi of Liaoning, China; Middle Jurassic.

Kuandiania crassicaulis Zheng et Zhang, 1980

1980 Zheng Shaolin and Zhang Wu, in Zhang Wu and others, p. 279, pl. 144, fig. 5; cycadophyte leaf; Reg. No.: D423; Repository: Shenyang Institute of Geology and Mineral Resources; Kuandian of Benxi, Liaoning; Middle Jurassic Zhuanshanzi Formation.

Genus *Leeites* Zodrow et Gao, 1991

1991 Zodrow E L and Gao Zhifeng, p. 63.

Type species: *Leeites oblongifolis* Zodrow et Gao, 1991

Taxonomic status: Sphenophyllales, Sphenopsida

Distribution and Age: Canada; Late Carboniferous.

Leeites oblongifolis Zodrow et Gao, 1991

1991 Zodrow E L and Gao Zhifeng, p. 64, pls. 1 — 8; text-figs. 1B — 1F; text-figs. 2 — 7; text-figs. 8A, 8B; text-fig. 9; foliage and strobili; Holotype: 982FG-315"5" (pl. 1, fig. 1; text-fig. 3D); Paratype: 982FG-315"1" (pl. 2, fig. 1); Cape Breton Island, Nova Scotia, Canada; Late Carboniferous.

Genus *Lepingia* Liu et Yao, 2002 (in English)

2002 Liu Lujun and Yao Zhaoqi, p. 177.

Type species: *Lepingia emarginata* Liu et Yao, 2002

Taxonomic status: Cycadalean?

Distribution and Age: Leping of Jiangxi, China; Late Permian.

Lepingia emarginata Liu et Yao, 2002 (in English)

2002 Liu Lujun and Yao Zhaoqi, p. 177, figs. 3, 5 — 7; plant with taeniopterid foliage to the cycadophytes; Holotype: PB18843 (fig. 5A); Repository: Nanjing Institute of Geology and Palaeontology, Chinese Academy of Sciences; Leping, Jiangxi; early Late Permian lower part in Laoshan Member of Leping Formation.

Genus *Lhassoxylon* Vozenin-Serra et Pons, 1990

1990 Voznin-Serra C and Pons D, p. 110.

Type species: *Lhassoxylon aptianum* Vozenin-Serra et Pons, 1990

Taxonomic status: Coniferopsida?

Distribution and Age: Lamba of Xizang (Tibet), China; Early Cretaceous (Aptian).

Lhassoxylon aptianum Vozenin-Serra et Pons, 1990

1990 Voznin-Serra C and Pons D, p. 110, pl. 1, figs. 1 — 7; pl. 2, figs. 1 — 8; pl. 3, figs. 1 — 7; pl. 4, figs. 1 — 3; text-figs. 2, 3; fossil wood; Col. No. : X/2 Pj/2 coll. J. J. Jaeger; Reg. No. : n° 10468; Holotype: n° 10468; Repository: Laboratoire de Paleobotanique et Palynologie evolutives, Universite Pierre et Marie Curie, Paris; Lamba, Xizang (Tibet); Early Cretaceous (Aptian).

Genus *Lianshanus* Pan, 1983 (nom. nud.)

1983 Pan Guang, p. 1520. (in Chinese)

1984 Pan Guang, p. 959. (in English)

Type species: (without specific name)

Taxonomic status: "primitive angiosperms"

Distribution and Age: western Liaoning, China; Middle Jurassic.

Lianshanus sp. indet.

[Notes: Generic name was given only, but without specific name (or type species) in the original paper]

1983 *Lianshanus* sp. indet., Pan Guang, p. 1520; western Liaoning (about 45°58′N, 120°21′E); Middle Jurassic Haifanggou Formation. (in Chinese)

1984 *Lianshanus* sp. indet., Pan Guang, p. 959; western Liaoning (about 45°58′N, 120°21′E); Middle Jurassic Haifanggou Formation. (in English)

Genus *Liaoningdicotis* Pan, 1983 (nom. nud.)

1983 Pan Guang, p. 1520. (in Chinese)

1984　Pan Guang, p. 958. (in English)

Type species: (without specific name)

Taxonomic status: "hemiangiosperms"

Distribution and Age: western Liaoning, China; Middle Jurassic.

Liaoningdicotis sp. indet.

[Notes: Generic name was given only, but without specific name (or type species) in the original paper]

1983　*Liaoningdicotis* sp. indet., Pan Guang, p. 1520; western Liaoning (about 45°58′N, 120°21′E); Middle Jurassic Haifanggou Formation. (in Chinese)

1984　*Liaoningdicotis* sp. indet., Pan Guang, p. 958; western Liaoning (about 45°58′N, 120°21′E); Middle Jurassic Haifanggou Formation. (in English)

Genus *Liaoningocladus* Sun, Zheng et Mei, 2000 (in English)

2000　Sun Ge, Zheng Shaolin and Mei Shengwu, p. 202.

Type species: *Liaoningocladus boii* Sun, Zheng et Mei, 2000

Taxonomic status: Conifers

Distribution and Age: Beipiao of Liaoning, China; Late Jurassic.

Liaoningocladus boii Sun, Zheng et Mei, 2000 (in English)

2000　Sun Ge, Zheng Shaolin and Mei Shengwu, p. 202, pl. 1, figs. 1 — 5; pl. 2, figs. 1 — 7; pl. 3, figs. 1 — 5; pl. 4, figs. 1 — 5; long and dwarf shoots, leaves and cuticles; Holotype: YB001 (pl. 1, fig. 1); Repository: Nanjing Institute of Geology and Palaeontology, Chinese Academy of Sciences; Huangbanjigou of Beipiao, Liaoning; Late Jurassic upper part of Yixian Formation.

Genus *Liaoningoxylon* Zhang et Zheng, 2006 (2008) (in Chinese and English)

2006　Zhang Wu and Zheng Shaolin, in Zhang Wu and others, p. 110. (in Chinese)

2008　Zhang Wu and Zheng Shaolin, in Zhang Wu and others, p. 110. (in English)

Type species: *Liaoningoxylon chaoyangehse* Zhang et Zheng, 2006 (2008)

Taxonomic status: Conifers incertae sedis

Distribution and Age: Chaoyang of Liaoning, China; Early Triassic.

Liaoningoxylon chaoyangehse Zhang et Zheng, 2006 (2008) (in Chinese and English)

2006　Zhang Wu and Zheng Shaolin, in Zhang Wu and others, p. 110, pls. 4-8 — 4-10; fossil woods; Holotype: GJ6-49; Paratype: GJ11-1; Repository: Shenyang Institute of Geology and Mineral Resources; Duanmutougou near Chaoyang, Liaoning; Early Triassic Hongla Formation. (in Chinese)

2008　Zhang Wu and Zheng Shaolin, in Zhang Wu and others, p. 110, pls. 4-8 — 4-10; fossil woods; Holotype: GJ6-49; Paratype: GJ11-1; Repository: Shenyang Institute of Geology and Mineral Resources; Duanmutougou near Chaoyang, Liaoning; Early Triassic Hongla Formation. (in English)

Genus *Liaoxia* Cao et Wu S Q, 1997 (1998) (in Chinese and English)

[Notes: The type species of the genus lately was referred by Guo Shuangxing and Wu Xiangwu into *Ephedrites* (Chlamydospermopsida or Gnetales) and named as *Ephedrites chenii* (Cao et Wu S Q) Guo et Wu X W (Guo Shuangxing, Wu Xiangwu, 2000); or this generic name was also referred by Wu Shunqing (1999) into Gnetales]

1997　Cao Zhengyao and Wu Shunqing, in Cao Zhengyao and others, p. 1765. (in Chinese)
1998　Cao Zhengyao and Wu Shunqing, in Cao Zhengyao and others, p. 231. (in English)

Type species: *Liaoxia chenii* Cao et Wu S Q, 1997 (1998)

Taxonomic status: Cyperaceae, Monocotyledoneae

Distribution and Age: Beipiao of Liaoning, China; Late Jurassic.

Liaoxia chenii Cao et Wu S Q, 1997 (1998) (in Chinese and English)

1997　Cao Zhengyao and Wu Shunqing, in Cao Zhengyao and others, p. 1765, pl. 1, figs. 1—2, 2a, 2b, 2c; herbaceous plant; Reg. No.: PB17800, PB17801; Holotype: PB17800 (pl. 1, fig. 1); Repository: Nanjing Institute of Geology and Palaeontology, Chinese Academy of Sciences; Shangyuan of Beipiao, western Liaoning; Late Jurassic Jianshangou Bed of Yixian Formation. (in Chinese)

1998　Cao Zhengyao and Wu Shunqing, in Cao Zhengyao and others, p. 231, pl. 1, figs. 1—2, 2a, 2b, 2c; herbaceous plant; Reg. No.: PB17800, PB17801; Holotype: PB17800 (pl. 1, fig. 1); Repository: Nanjing Institute of Geology and Palaeontology, Chinese Academy of Sciences; Shangyuan of Beipiao, western Liaoning; Late Jurassic Jianshangou Bed of Yixian Formation. (in English)

Genus *Liella* Yang et Zhao, 2006 (in Chinese and English)

2006　Yang Guanxiu and Zhao Jiming, in Yang Guanxiu and others, pp. 144, 288.

Type species: *Liella mirabilis* Yang et Zhao, 2006

Taxonomic status: Cycadales, Cycadophyta

Distribution and Age: Linru and Yiyang of western Henan, China; Middle Permian.

Liella mirabilis Yang et Zhao, 2006 (in Chinese and English)

2006　Yang Guanxiu and Zhao Jiming, in Yang Guanxiu and others, pp. 144, 288, pl. 40, fig. 3; pertile pinna Taeniopteris-type; Holotype: HEP3931 (pl. 40, fig. 3); Repository: China University of Geosciences, Beijing; Pochi of Linru and Ligou of Yiyang, western Henan; Middle Permian Xiaofengkou Formation.

Genus *Lilites* Wu S Q, 1999 (in Chinese)

1999 Wu Shunqing, p. 23.

Type species: *Lilites reheensis* Wu S Q, 1999

Taxonomic status: Liliaceae, Monocotyledoneae

Distribution and Age: Beipiao of Liaoning, China; Late Jurassic.

Lilites reheensis Wu S Q, 1999 (in Chinese)

1999 Wu Shunqing, p. 23, pl. 18, figs. 1, 1a, 2, 4, 5, 7, 7a, 8A; leaves and fruit; Col. No. : AEO-11, 219, 245, 134, 158, 246; Reg. No. : PB18327 — PB18332; Syntype 1: PB18327 (pl. 18, fig. 1); Syntype 2: PB18330 (pl. 18, fig. 5); Repository: Nanjing Institute of Geology and Palaeontology, Chinese Academy of Sciences; Huangbanjigou in Shangyuan of Beipiao, western Liaoning; Late Jurassic Jianshangou Bed in lower part of Yixian Formation. [Notes: The type species of the genus lately was referred by Sun Ge and Zheng Shaolin into *Podocarpites* (Coniferophytes) and named as *Podocarpites reheensis* (Wu) Sun et Zheng W (Sun Ge and others, 2001)]

Genus *Lingxiangphyllum* Meng, 1981

1981 Meng Fansong, p. 100.

Type species: *Lingxiangphyllum princeps* Meng, 1981

Taxonomic status: Plantae incertae sedis

Distribution and Age: Daye of Hubei, China; Early Cretaceous.

Lingxiangphyllum princeps Meng, 1981

1981 Meng Fansong, p. 100, pl. 1, figs. 12 — 13; text-fig. 1; single leaf; Reg. No. : CHP7901, CHP7902; Holotype: CHP7901 (pl. 1, fig. 12); Repository: Yichang Institute of Geology and Mineral Resources; Changpinghu in Lingxiang of Daye, Hubei; Early Cretaceous Lingxiang Group.

Genus *Linophyllum* Zhao, 1980

1980 Zhao Xiuhu and others, p. 83.

Type species: *Linophyllum xuanweiense* Zhao, 1980

Taxonomic status: Gigantopterides

Distribution and Age: Xuanwei of Yunnan, China; Late Permian.

Linophyllum xuanweiense Zhao, 1980

1980 Zhao Xiuhu and others, p. 83, pl. 22, figs. 5, 5a; leaves; Col. No. : XL-16; Reg. No. : PB7103; Repository: Nanjing Institute of Geology and Palaeontology, Chinese Academy of Sciences; Xuanwei, Yunnan; early Late Permian lower member of Xuanwei Formation.

Genus *Lioxylon* Zhang, Wang, Saiki, Li et Zheng, 2006 (in English)

2006　Zhang Wu, Wang Yongdong, Ken'ichi Saiki, Li Nan and Zheng Shaolin, in Zhang Wu and others, p. 237. (in Chinese)
2006　Wang Yongdong and others, p. 125. (in Chinese)
2008　Wang Yongdong and others, p. 125. (in English)

Type species: *Lioxylon liaoningense* Zhang, Wang, Saiki, Li et Zheng, 2006
Taxonomic status: Cycadales
Distribution and Age: Beipiao of Liaoning, China; Middle Jurassic (Bathonian—Callovian).

Lioxylon liaoningense Zhang, Wang, Saiki, Li et Zheng, 2006 (in English)

2006　Zhang Wu, Wang Yongdong, Ken'ichi Saiki, Li Nan and Zheng Shaolin, in Zhang Wu and others, p. 237, figs. 1—6; structurally preserved cycad-like stem; No.: DMG-1, DMG-2, DMG-3, DMG-6, LMY-133, LMY-138, XCG-19; Holotype: DMG-1; Paratypes: DMG-2, DMG-3, DMG-6, LMY-133, LMY-138, XCG-19; Repository: Shenyang Institute of Geology and Mineral Resources, Changgao of Beipiao, Liaoning; Middle Jurassic (Bathonian—Callovian) Tiaojishan Formation.
2006　Wang Yongdong and others, p. 125, pls. 5-1—5-6; structurally preserved cycad-like stem; Changgao of Beipiao, Liaoning; Middle Jurassic (Bathonian—Callovian) Tiaojishan Formation. (in Chinese)
2008　Wang Yongdong and others, p. 125, pls. 5-1—5-6; structurally preserved cycad-like stem; Changgao of Beipiao, Liaoning; Middle Jurassic (Bathonian—Callovian) Tiaojishan Formation. (in English)

Genus *Liulinia* Wang, 1986

1986　Wang Ziqiang, pp. 611, 615.

Type species: *Liulinia lacinulata* Wang, 1986
Taxonomic status: Cycadales, Cycaopsida
Distribution and Age: Liulin of Shanxi, China; Late Permian.

Liulinia lacinulata Wang, 1986

1986　Wang Ziqiang, pp. 611, 615, pl. 1, figs. 1—7; pl. 2, figs. 1—11; text-fig. 1; male cone; Col. No.: 8402-208, 8402-209, 8309-157, 8309-158; Holotype: 8402-209 (pl. 1, fig. 1); Repository: Nanjing Institute of Geology and Palaeontology, Chinese Academy of Sciences; Liulin, Shanxi; Late Permian middle member of Sunjiagou Formation.

Genus *Lixotheca* Yao, Liu et Zhang, 1993

1993　Yao Zhaoqi, Liu Lujun and Zhang Shi, pp. 526, 535.

Type species: *Lixotheca* (*Cladophlebis*) *permica* (Lee et Wang) Yao, Liu et Zhang, 1993
Taxonomic status: Hymenophyllaceae, Filicopsida
Distribution and Age: Shanxi, Henan, Fujian and Jiangsu, China; Permian.

Lixotheca (*Cladophlebis*) *permica* (Lee et Wang) Yao, Liu et Zhang, 1993

1956   *Cladophlebis permica* Lee et Wang, Lee H H and Wang S, p. 346, pls. 1, 2; pl. 3, fig. 1 (no figs. 2 — 4); fronds; Reg. No.: PB2538 — PB2545; Repository: Nanjing Institute of Geology and Palaeontology, Chinese Academy of Sciences; Wuxiang, Shanxi; Late Permian Upper Shihhotse Formation; Longtan of Nanjing, Jiangsu; Late Permian.

1993   Yao Zhaoqi, Liu Lujun and Zhang Shi, pp. 526, 535, pls. 1 — 3; text-figs. 1, 2; sterile frond and fertile frond; Dengfeng and Yuxian of Henan, Wuxiang and Anze of Shanxi; Late Permian Upper Shihhotse Formation; Longyan and Yongchun, Fujian; Late Permian Tongziyan Formation.

Genus *Lobatannulariopsis* Yang, 1978

1978   Yang Xianhe, p. 472.

Type species: *Lobatannulariopsis yunnanensis* Yang, 1978
Taxonomic status: Equisetales, Sphenopsida
Distribution and Age: Guangtong of Yunnan, China; Late Triassic.

Lobatannulariopsis yunnanensis Yang, 1978

1978   Yang Xianhe, p. 472, pl. 158, fig. 6; vegetal shoot with leaves; Reg. No.: Sp0009; Holotype: Sp0009 (pl. 158, fig. 6); Repository: Chengdu Institute of Geology and Mineral Resources; Yipinglang of Guangtong, Yunnan; Late Triassic Ganhaizi Formation.

Genus *Longjingia* Sun et Zheng, 2000 (nom. nud.)

2000   Sun Ge and Zheng Shaolin, in Sun Ge and others, pl. 4, figs. 5, 6.

Type species: *Longjingia gracilifolia* Sun et Zheng, 2000
Taxonomic status: Dicotyledoneae
Distribution and Age: Longjing of Jilin, China; Early Cretaceous.

Longjingia gracilifolia Sun et Zheng, 2000 (nom. nud.)

2000   Sun Ge and Zheng Shaolin, in Sun Ge and others, pl. 4, figs. 5, 6; leaf; Zhixin (Dalazi) of Longjing, Jilin; Early Cretaceous Dalazi Formation.

Genus *Longostachys* Zhu, Hu et Feng, 1983

1983   Zhu Jianan, Hu Yufan and Feng Shaonan, p. 79.

Type species: *Longostachys latisporophyllus* Zhu, Hu et Feng, 1983

Taxonomic status: Lycopods incertae sedis
Distribution and Age: Lixian of Hunan, China; Middle Devonian.

Longostachys latisporophyllus Zhu, Hu et Feng, 1983
1983   Zhu Jianan, Hu Yufan and Feng Shaonan, p. 79, pl. 1, figs. 1a, 1b; fertile branch; No. : PB25604; Holotype: PB25604 (pl. 1, figs. 1a, 1b); Lixian, Hunan; Middle Devonian Yuntaiguan Formation.

Genus *Lopadiangium* Zhao, 1980
1980   Zhao Xiuhu and others, p. 90.

Type species: *Lopadiangium acmodontum* Zhao, 1980

Taxonomic status: Plantae incertae sedis

Distribution and Age: Panxian and Qinglong of Guizhou, China; early Late Permian.

Lopadiangium acmodontum Zhao, 1980
1980   Zhao Xiuhu and others, p. 90, pl. 23, figs. 9, 9a, 10, 10a, 11 — 14; fertile frond and sporangia; Col. No. : H11-6; Reg. No. : PB7115 — PB7119; Repository: Nanjing Institute of Geology and Palaeontology, Chinese Academy of Sciences; Panxian, Guizhou; early Late Permian lower member of Xuanwei Formation; Qinglong, Guizhou; early Late Permian.

Genus *Lophotheca* Zhao, 1980
1980   Zhao Xiuhu and others, p. 89.

Type species: *Lophotheca panxianensis* Zhao, 1980

Taxonomic status: Plantae incertae sedis

Distribution and Age: Panxian, Guizhou; early Late Permian.

Lophotheca panxianensis Zhao, 1980
1980   Zhao Xiuhu and others, p. 89, pl. 23, figs. 5, 5a, 6 — 8; fertile frond and sporangia; Col. No. : PL18-15; Reg. No. : PB7111 — PB7114; Repository: Nanjing Institute of Geology and Palaeontology, Chinese Academy of Sciences; Panxian, Guizhou; early Late Permian lower member of Xuanwei Formation.

Genus *Lopinopteris* Sze, 1958
1958   Sze H C, p. 383. (in Chinese)
1959   Sze H C, p. 322. (in English)

Type species: *Lopinopteris intercalata* Sze, 1958

Taxonomic status: Filices or Pteridospermopsida

Distribution and Age: Leping of Jiangxi, China; Middle Carboniferous (Westphalian).

Lopinopteris intercalata Sze, 1958

1958　Sze H C, p. 383, pl. 2, figs. 1 — 4; pl. 3, figs. 4 — 6; fern-like fronds; Reg. No. : PB2626 — PB2629, PB2632, PB2633; Repository: Nanjing Institute of Geology and Palaeontology, Chinese Academy of Sciences; Leping, Jiangxi; Middle Carboniferous (Westphalian) Tzushan Coal Series. (in Chinese)

1959　Sze H C, p. 322, pl. 2, figs. 1 — 4; pl. 3, figs. 4, 5, 5a; fern-like fronds; Reg. No. : PB2626 — PB2629, PB2632, PB2633; Repository: Nanjing Institute of Geology and Palaeontology, Chinese Academy of Sciences; Leping, Jiangxi; Middle Carboniferous (Westphalian) Tzushan Coal Series. (in English)

Genus *Loroderma* Geng et Hilton, 1999 (in English)

1999　Geng Baoyin and Hilton J, p. 127.

Type species: *Loroderma henania* Geng et Hilton, 1999

Taxonomic status: Coniferophytes

Distribution and Age: Yima of Henan, China; Early Permian (Sakmarian).

Loroderma henania Geng et Hilton, 1999 (in English)

1999　Geng Baoyin and Hilton J, p. 127, figs. 6 — 22; coniferophyte ovulate structures; No. : CBP9191 — CBP9197; Holotype: CBP9191 (fig. 14); Repository: Institute of Botany, Chinese Academy of Sciences; Yima, Henan; Early Permian (Sakmarian) Shanxi Formation.

Genus *Lorophyton* Fairon-Demaret et Li, 1993

1993　Fairon-Demaret M and Li Chengsen, p. 15.

Type species: *Lorophyton goense* Fairon-Demaret et Li, 1993

Taxonomic status: Cladoxylopsida

Distribution and Age: Belgium; Middle Devonian (Lower Givetian).

Lorophyton goense Fairon-Demaret et Li, 1993

1993　Fairon-Demaret M and Li Chengsen, p. 15, pls. 1 — 6; text-figs. 1 — 6; Holotypes: ULg2056a, ULg2056b (pl. 1, fig. 3; pl. 4, figs. 1 — 5; pl. 6, figs. 1, 4 — 7); Paratypes: ULg2057a (pl. 1, fig. 1; pl. 2, fig. 2), ULg2057b (pl. 1, fig. 2), ULg2065 (pl. 3, fig. 1), ULg2066 (pl. 2, fig. 1), ULg2066 (pl. 5, figs. 2, 3); Repository: Paleobotanical collections of the University of Liege (Belgium); Vestre Synclinorium, Belgium; Middle Devonian (Lower Givetian).

Genus *Luereticopteris* Hsu et Chu, 1974

1974　Hsu J and Chu Chinan, in Hsu J and others, p. 270.

Type species: *Luereticopteris megaphylla* Hsu et Chu, 1974

Taxonomic status: Filicopsida

Distribution and Age: Yongren of Yunnan, China; Late Triassic.

Luereticopteris megaphylla Hsu et Chu, 1974

1974   Hsu J and Chu Chinan, in Hsu J and others, p. 270, pl. 2, figs. 5 — 11; pl. 3, figs. 2 — 3; text-fig. 2; frond; No. ; No. 742a-c, 2515; Syntypes: No. 742a-c, 2515 (pl. 2, figs. 5 — 11; pl. 3, figs. 2 — 3); Repository: Institute of Botany, Chinese Academy of Sciences; Huashan of Yongren, Yunnan; Late Triassic middle part of Daqiaodi Formation.

Genus *Macroglossopteris* Sze, 1931

1931   Sze H C, p. 5.

Type species: *Macroglossopteris leeiana* Sze, 1931

Taxonomic status: Pteridospermopsida

Distribution and Age: Pingxiang of Jiangxi, China; Early Jurassic (Lias).

Macroglossopteris leeiana Sze, 1931

1931   Sze H C, p. 5, pl. 3, fig. 1; pl. 4, fig. 1; fern-like leaf; Pingxiang, Jiangxi; Early Jurassic (Lias). [Notes: This type species lately was referred as *Anthrophyopsis leeiana* (Sze) Florin (Florin, 1933)]

Genus *Manchurostachys* Kon'no, 1960

1960   Kon'no E, p. 164.

Type species: *Manchurostachys manchuriensis* Kon'no, 1960

Taxonomic status: Equisetales

Distribution and Age: Benxi (Penchihu) of Liaoning, China; Permian.

Manchurostachys manchuriensis Kon'no, 1960

1960   Kon'no E, p. 164, pl. 20; fructification of *Schizoneura manchuriensis*; Benxi (Penchihu Coal-field), Liaoning; Permian (Tsaichia Formation).

Genus *Manica* Watson, 1974

1974   Watson J, p. 428.

Type species: *Manica parceramosa* (Fontaine) Watson, 1974

Taxonomic status: Cheirolepidiaceae, Coniferopsida

Distribution and Age: Virginia, USA; Early Cretaceous.

Manica parceramosa (Fontaine) Watson, 1974

1889   *Frenilopsis parceramosa* Fontaine, p. 218, pls. 111, 112, 158; leafy shoots; Virginia,

    USA;Early Cretaceous.
1974 Watson J,p. 428;Virginia,USA;Early Cretaceous.

Subgenus *Manica* (*Chanlingia*) Chow et Tsao,1977
1977 Chow Tseyen and Tsao Chengyao,p. 172.
Type species:*Manica* (*Chanlingia*) *tholistoma* Chow et Tsao,1977
Taxonomic status:Cheirolepidiaceae,Coniferopsida
Distribution and Age:Changling and Fuyu of Jilin,Lanxi of Zhejiang,China;Cretaceous.

Manica (*Chanlingia*) *tholistoma* Chow et Tsao,1977
[Notes:The species was later referred as *Pseudofrenelopsis tholistoma* (Chow et Tsao) (Cao Zhengyao,1989)]
1977 Chow Tseyen and Tsao Chengyao,p. 172,pl. 2,figs. 16,17;pl. 5,figs. 1 — 10;text-fig. 4; leafy shoots and cuticles;Reg. No. :PB6265,PB6272;Holotype:PB6272 (pl. 5,figs. 1, 2);Repository:Nanjing Institute of Geology and Palaeontology, Chinese Academy of Sciences;Changling,Jilin;Early Cretaceous Qingshankou Formation;Fuyu,Jilin;Early Cretaceous Quantou Formation;Lanxi,Zhejiang;Late Cretaceous Qujiang Group.

Subgenus *Manica* (*Manica*) Chow et Tsao,1977
1977 Chow Tseyen and Tsao Chengyao,p. 169.
Type species:*Manica* (*Manica*) *parceramosa* (Fontaine) Chow et Tsao,1977
Taxonomic status:Cheirolepidiaceae,Coniferopsida
Distribution and Age:Jilin,Ningxia and Zhejiang of China,Virginia of USA;Early Cretaceous.

Manica (*Manica*) *parceramosa* (Fontaine) Chow et Tsao,1977
1889 *Frenilopsis parceramosa* Fontaine,Fontaine O,p. 218,pls. 111,112,158;leafy shoot; Virginia,USA;Early Cretaceous.
1977 Chow Tseyen and Tsao Chengyao,p. 169.

The other species:
Manica (*Manica*) *dalatzensis* Chow et Tsao,1977
[Notes:The species was later referred as *Pseudofrenelopsis dalatzensis* (Chow et Tsao) Cao ex Zhou (Zhou Zhiyan,1995)]
1977 Chow Tseyen and Tsao Chengyao,p. 171,pl. 3,figs. 5 — 11;pl. 4,fig. 13;text-fig. 3;leafy shoots and cuticles;Reg. No. :PB6267,PB6268;Holotype:PB6267 (pl. 3, fig. 5); Repository:Nanjing Institute of Geology and Palaeontology, Chinese Academy of Sciences;Zhixin (Dalatze) of Yanji,Jilin;Early Cretaceous Dalatze Formation.

Manica (*Manica*) *foveolata* Chow et Tsao,1977
[Notes:The species was later referred as *Pseudofrenelopsis foveolata* (Chow et Tsao) (Tsao Chengyao,1989) and as *Pseudofrenelopsis papillosa* (Chow et Tsao) Cao ex Zhou (Chow Tseyen,1995)]
1977 Chow Tseyen and Tsao Chengyao,p. 171,pl. 4,figs. 1 — 7,14;leafy shoots and cuticles;

Reg. No. : PB6269, PB6270; Holotype: PB6269 (pl. 4, figs. 1, 2); Repository: Nanjing Institute of Geology and Palaeontology, Chinese Academy of Sciences; Gaodian of Guyuan and Qianyanghe of Xiji, Ningxia; Early Cretaceous Liupanshan Group.

Manica (*Manica*) *papillosa* Chow et Tsao, 1977

[Notes: The species was later referred as *Pseudofrenelopsis papillosa* (Chow et Tsao) Cao ex Zhou (Zhou Zhiyan, 1995)]

1977   Chow Tseyen and Tsao Chengyao, p. 169, pl. 2, fig. 15; pl. 3, figs. 1 — 4; pl. 4, fig. 12; text-fig. 2; leafy shoots, cuticles and cones; Reg. No. : PB6264, PB6266; Holotype: PB6266 (pl. 3, fig. 1); Repository: Nanjing Institute of Geology and Palaeontology, Chinese Academy of Sciences; Xinchang, Zhejiang; Early Cretaceous Guantou Formation; Guyuan, Ningxia; Early Cretaceous Liupanshan Group.

Genus *Mediocycas* Li et Zheng, 2005 (in Chinese and English)

2005   Li Nan and Zheng Shaolin, in Li Nan and others, pp. 425, 433.

Type species: *Mediocycas kazuoensis* Li et Zheng, 2005

Taxonomic status: Cycadales, Cycadopsida

Distribution and Age: western Liaoning, China; Early Triassic.

Mediocycas kazuoensis Li et Zheng, 2005 (in Chinese and English)

1986   Problematicum 1, Zheng Shaolin and Zhang Wu, p. 181, pl. 1, figs. 10 — 11; Yangshugou of Kazuo, western Liaoning; Early Triassic Hongla Formation.

1986   *Carpolithus* sp. , Zheng Shaolin and Zhang Wu, pl. 3, fig. 11; seed; Yangshugou of Kazuo, western Liaoning; Early Triassic Hongla Formation.

1986   *Carpolithus*? sp. , Zheng Shaolin and Zhang Wu, p. 14, pl. 3, figs. 12 — 14; seeds; Yangshugou of Kazuo, western Liaoning; Early Triassic Hongla Formation.

2005   Li Nan and Zheng Shaolin, in Li Nan and others, pp. 425, 433; text-figs. 3A — 3F, 5E; megasporophylls; No. : SG110280 — SG110283 (couterpart), SG11026 — SG11028; Holotypes: SG110280 — SG110283 (text-fig. 3A); Paratypes: SG110280 — SG110283 (text-fig. 3B); Repository: Shenyang Institute of Geology and Mineral Resources; Yangshugou of Kazuo, western Liaoning; Early Triassic Hongla Formation.

Genus *Megalopteris* Schenk, 1883 (non Andrews E B, 1875)

[Notes: This generic name *Megalopteris* Schenk, 1883 is a late homomum (homomum junius) of *Megalopteris* Andrews E B, 1875, its type species *Megalopteris dawsoni* (Hartt) Andrews 1875 (Andrews E B, 1875, p. 415; fern or pteridosperm foliage; New Brunswick, Canada; Devonian?. For *Neuropteris dawsoni* Hartt, in Dawson, 1868, p. 551, fig. 193)]

1883   Schenk A, p. 238.

Type species: *Megalopteris nicotianaefolia* Schenk, 1883

Taxonomic status: Pteridospermopsida, Gigantopterides

Distribution and Age: Hunan, China; Late Carboniferous.

Megalopteris nicotianaefolia Schenk, 1883
{Notes: The name *Gigantopteris nicotianaefolia* Schenk, for *Megalopteris nicotianaefolia* Schenk [Whit D, 1912; Nanjing Institute of Geology and Palaeontology, Institute of Botany, Chinese Academy of Sciences (Gu et Zhi), in *Palaeozoic Plants from China*, 1974]}
1883   Schenk A, p. 238, pl. 32, figs. 6 — 8; pl. 33, figs. 1 — 3; pl. 35, fig. 6; foliage, affinity uncertain; Lui-pa-Kou, Hunan; Late Carboniferous.

Genus *Meia* He, Liang et Shen, 1996 (in Chinese and English)
1996   He Xilin, Liang Dunshi and Shen Shuzhong, pp. 77, 163.
Type species: *Meia mingshanensis* He, Liang et Shen, 1996
Taxonomic status: Filices or Pteridospermopsida
Distribution and Age: Jiangxi and Guizhou, China; Late Permian.

Meia mingshanensis He, Liang et Shen, 1996 (in Chinese and English)
1996   He Xilin, Liang Dunshi and Shen Shuzhong, pp. 77, 164, pl. 75; pl. 76, fig. 1; pl. 77, figs. 1, 2; pl. 78, fig. 1; pl. 79, fig. 3; pl. 92; shoot frond-like and cuticle; Reg. No. : X88313 — X88317; Syntypes: X88313 — X88315 (pl. 75, figs. 1 — 3), X88316 (pl. 78, fig. 1), X88317 (pl. 79, fig. 3); Repository: Department of Geology, China University of Mining and Technology; Mingshan Coal Mine of Leping, Jiangxi; Late Permian Lower Laoshan Submember of Leping Formation.

The other species:
Meia magnifolia He, Liang et Shen, 1996 (in Chinese and English)
1996   He Xilin, Liang Dunshi and Shen Shuzhong, pp. 78, 165, pl. 78, figs. 2 — 4; pls. 90, 91; shoot frond-like and cuticle; Reg. No. : X88318 — X88320; Syntypes: X88318 — X88320 (pl. 78, figs. 2 — 4); Repository: Department of Geology, China University of Mining and Technology; Mingshan Coal Mine of Leping, Jiangxi; Late Permian Lower Laoshan Submember of Leping Formation.

Genus *Membranifolia* Sun et Zheng, 2001 (in Chinese and English)
2001   Sun Ge and Zheng Shaolin, in Sun Ge and others, pp. 108, 208.
Type species: *Membranifolia admirabilis* Sun et Zheng, 2001
Taxonomic status: Plantae incertae sedis
Distribution and Age: Lingyuan of Liaoning, China; Late Jurassic.

Membranifolia admirabilis Sun et Zheng, 2001 (in Chinese and English)
2001   Sun Ge and Zheng Shaolin, in Sun Ge and others, pp. 108, 208, pl. 26, figs. 1 — 2; pl. 67, figs. 3 — 6; leaves; No. : PB19184 — PB19185, PB19187, PB19196; Holotype: PB19184 (pl. 26, fig. 1); Repository: Nanjing Institute of Geology and Palaeontology, Chinese

Academy of Sciences; Lingyuan, western Liaoning; Late Jurassic Jianshangou Formation.

Genus *Metacladophyton* Wang et Geng, 1997 (in English)
1997　Wang Zhong and Geng Baoyin, p. 93.
Type species: *Metacladophyton tetraxylum* Wang et Geng, 1997
Taxonomic status: Cladoxylates, Pteridophyta
Distribution and Age: Hubei, China; late Middle Devonian.

Metacladophyton tetraxylum Wang et Geng, 1997 (in English)
1997　Wang Zhong and Geng Baoyin, p. 93, pls. 1 — 11; text-figs. 1 — 7; plant; Holotype: No. 9167 (pl. 3, fig. 1); Paratypes: No. 9168 — No. 9190; Repository: Institute of Botany, Chinese Academy of Sciences; Changyang, Hubei; late Middle Devonian Yuntaiguan Formation.

Genus *Metalepidodendron* Shen (MS) ex Wang X F, 1984 (nom. nud.)
1984　Shen Guanglong, in Wang Xifu, p. 297.
Type species: *Metalepidodendron sinensis* Shen (MS) ex Wang X F, 1984
Taxonomic status: Lycopodiales, Lycoposida
Distribution and Age: Gansu and Chengde of Hebei, China; Early — Middle Triassic.

Metalepidodendron sinensis Shen (MS) ex Wang X F, 1984 (nom. nud.)
1984　Shen Guanglong, in Wang Xifu, p. 297; Gansu; Middle Triassic.

The other species:
Metalepidodendron xiabanchengensis Wang X F et Cui, 1984
1984　Wang Xifu, p. 297, pl. 175, figs. 8 — 11; stem; Reg. No. : HB-57, HB-58; Xiabancheng of Chengde, Hebei; Early Triassic upper part of Heshanggou Formation.

Genus *Metasequoia* Miki, 1941 ex Hu et Cheng, 1948
1941　Miki S, p. 262. (fossil species)
1948　Hu Hsenhsu and Cheng Wanchun, p. 153. (living species)
Type species: *Metasequoia disticha* Miki, 1941 (fossil species)
　　　Metasequoia glyptostroboides Hu et Cheng, 1948 (living species)
Taxonomic status: Taxodiaceae, Coniferopsida
Distribution and Age: Northern Hemisphere; Cretaceous, Paleocene and Neocene.

living species:
Metasequoia glyptostroboides Hu et Cheng, 1948
1948　Hu Hsenhsu and Cheng Wanchun, p. 153; text-figs. 1 — 2; Modaoxi of Wanxian,

Sichuan; a living species of the genus *Metasequoia*.

fossil species:
Metasequoia disticha Miki, 1941
1876   *Sequoia disticha* Heer, Heer O, p. 63, pl. 12, fig. 2a; pl. 13, figs. 9 — 11; twig and cone; Northern Hemisphere; Cretaceous, Paleocene and Neocene.
1941   Miki S, p. 262, pl. 5, figs. A — Ca; text-figs. 8, A — G; twig and cone; Northern Hemisphere; Cretaceous, Paleocene and Neocene.

Genus *Metzgerites* Wu et Li, 1992
1992   Wu Xiangwu and Li Baoxian, pp. 268, 276.
Type species: *Metzgerites yuxinanensis* Wu et Li, 1992
Taxonomic status: Hepaticae
Distribution and Age: Yuxian of Hebei, China; Middle Jurassic.

Metzgerites yuxinanensis Wu et Li, 1992
1992   Wu Xiangwu and Li Baoxian, pp. 268, 276, pl. 3, figs. 3 — 5a; pl. 6, figs. 1, 2; text-fig. 6; thallus; Col. No.: ADN41-01, ADN41-02; Reg. No.: PB15480 — PB15483; Holotype: PB15481 (pl. 3, fig. 4); Repository: Nanjing Institute of Geology and Palaeontology, Chinese Academy of Sciences; Yuxian, Hebei; Middle Jurassic Qiaoerjian Formation.

Genus *Minarodendron* Li, 1990
1990   Li Chengsen, p. 105.
Type species: *Minarodendron cathaysiense* (Schweitzer et Cai) Li, 1990
Taxonomic status: Protolepidodendraceae, Protolepidophytales, Lycophyta
Distribution and Age: Yunnan, Hunan and Guangxi, China; late Middle Devonian (Givetium).

Minarodendron cathaysiense (Schweitzer et Cai) Li, 1990
1987   *Protolepidodendron cathaysiense* Schweitzer et Cai, Schweitzer H J and Cai Chongyang, p. 33 (5), pls. 1, 2; text-fig. 4; herbaceous lycopod; Longhuashan of Zhanyi, Yunnan; late Middle Devonian (Givetium) Haikou (Xichong) Formation; Liuyang, Hunan; late Middle Devonian Tiaomajian Formation; Luzhai and Luocheng, Guangxi; late Middle Devonian Tungkangling Formation.
1990   Li Chengsen, p. 105, pls. 1 — 11; text-figs. 1 — 7; herbaceous lycopod; Longhuashan of Zhanyi, Yunnan; late Middle Devonian (Givetium) Haikou Formation; Liuyang, Hunan; late Middle Devonian Tiaomajian Formation; Luzhai and Luocheng, Guangxi; late Middle Devonian Tungkangling Formation.

Genus *Mirabopteris* Mi et Liu, 1993
1993   Mi Jiarong and Liu Maoqiang, in Mi Jiarong and others, p. 102.

Type species: *Mirabopteris hunjiangensis* (Mi et Liu) Mi et Liu, 1993

Taxonomic status: Pteridospermopsida

Distribution and Age: Hunjiang of Jilin, China; Late Triassic.

Mirabopteris hunjiangensis (Mi et Liu) Mi et Liu, 1993

1977   *Paradoxopteris hunjiangensis* Mi et Liu, Mi Jiarong and Liu Maoqiang, in Surveying Group of Department of Geological Exploration of Changchun Institute of Geology, Regional Geological Surveying Team of Geological Bureau of Kirin Province, 102 Surveying Team of Coal Geology Exploration Company of Kirin Province, p. 8, pl. 3, fig. 1; text-fig. 1; fern-like leaves; Reg. No.: X-008; Repository: Department of Geology, Changchun University of Science and Technology; Shiren of Hunjiang, Jilin; Late Triassic "Beishan Formation".

1993   Mi Jiarong and Liu Maoqiang, in Mi Jiarong and others, p. 102, pl. 18, fig. 3; pl. 53, figs. 1, 2, 6; text-fig. 21; fern-like leaves and cuticles; Shiren of Hunjiang, Jilin; Late Triassic Beishan Formation (Xiaohekou Formation).

Genus *Mironeura* Zhou, 1978

1978   Zhou Tongshun, p. 114.

Type species: *Mironeura dakengensis* Zhou, 1978

Taxonomic status: Nilssoniales or Cycadales, Cycadopsida

Distribution and Age: Zhangping of Fujian, China; Late Triassic.

Mironeura dakengensis Zhou, 1978

1978   Zhou Tongshun, p. 114, pl. 25, figs. 1, 2, 2a; text-fig. 4; cycadophyte leaves; Col. No.: $WFT_3W_1^1$-9; Reg. No.: FKP135; Repository: Institute of Geology, Chinese Academy of Geological Sciences; Dakeng of Zhangping (Wenbinshan), Fujian; Late Triassic lower member of Wenbinshan Formation.

Genus *Mixophylum* Meng, 1983

1983   Meng Fansong, p. 228.

Type species: *Mixophylum simplex* Meng, 1983

Taxonomic status: Plantae incertae sedis

Distribution and Age: Nanzhang of Hubei, China; Late Triassic.

Mixophylum simplex Meng, 1983

1983   Meng Fansong, p. 228, pl. 3, fig. 1; leaf; Reg. No.: D76018; Holotype: D76018 (pl. 3, fig. 1); Repository: Yichang Institute of Geology and Mineral Resources; Donggong of Nanzhang, Hubei; Late Triassic Jiuligang Formation.

Genus *Mixopteris* Hsu et Chu C N,1974
1974   Hsu J and Chu Chinan,in Hsu J and others,p. 271.
Type species:*Mixopteris intercalaris* Hsu et Chu C N,1974
Taxonomic status:Filicopsida?
Distribution and Age:Yongren of Yunnan,China;Late Triassic.

Mixopteris intercalaris Hsu et Chu C N,1974
1974   Hsu J and Chu Chinan,in Hsu J and others,p. 271,pl. 3,figs. 4 — 7;text-fig. 4;frond;
        No. :No. 2610;Repository:Institute of Botany,Chinese Academy of Sciences;Nalajing of
        Yongren,Yunnan;Late Triassic bottom part of Daqiaodi Formation.

Genus *Mnioites* Wu X W,Wu X Y et Wang,2000 (in English)
2000   Wu Xiangwu,Wu Xiuyuan and Wang Yongdong,p. 170.
Type species:*Mnioites brachyphylloides* Wu X W,Wu X Y et Wang,2000
Taxonomic status:Btyiidae
Distribution and Age:Karamay of Xinjiang,China;Middle Jurassic.

Mnioites brachyphylloides Wu X W,Wu X Y et Wang,2000 (in English)
2000   Wu Xiangwu,Wu Xiuyuan and Wang Yongdong,p. 170,pl. 2,fig. 5;pl. 3,figs. 1 — 2d;
        caulidium;Col. No. :92-T61,Reg. No. :PB17797 — PB17799;Holotype:PB17798 (pl. 3,
        figs. 1 — 1c);Paratypes:PB17797 (pl. 3,figs. 2 — 2d),PB17799 (pl. 2,fig. 5);
        Repository: Nanjing Institute of Geology and Palaeontology, Chinese Academy of
        Sciences;Tuzi'Arkneigou of Karamay,Xinjiang;Middle Jurassic Xishanyao Formation.

Genus *Monogigantonoclea* Yang,2006 (in Chinese and English)
2006   Yang Guanxiu and others,pp. 189,311.
Type species:*Monogigantonoclea colocasifolia* (Yang) Yang,2006
Taxonomic status:Gigantopteridales
Distribution and Age:Yuzhou of western Henan,China;Permian.

Monogigantonoclea colocasifolia (Yang) Yang,2006 (in Chinese and English)
1987b   *Gigantonoclea colocasifolia* Yang,Yang Guanxiu and others,p. 185,pl. 2,figs. 1 — 6.
2006   Yang Guanxiu and others,pp. 190,311,pl. 65,figs. 1 — 3;pl. 66,figs. 1 — 4;pl. 67,fig. 6;
        text-fig. 7-10; simple leaf; Syntypes: HEP0152, 0149, 0124 (pl. 65, figs. 1 — 3),
        HEP0147,0150,0153,3621 (pl. 66,figs. 1 — 4),HEP0148 (pl. 67,fig. 6);Repository:
        China University of Geosciences, Beijing; Dafengkou, Yungaishan and Dishuitan in
        Fanshan of Yuzhou, western Henan; Late Permian coal member 6 of Yungaishan
        Formation.

The other species:
Monogigantonoclea rotundifolia (Yang) Yang, 2006 (in Chinese and English)
1987b *Gigantonoclea rotundifolia* Yang, Yang Guanxiu and others, p. 186, pl. 2, figs. 8, 9.
2006   Yang Guanxiu and others, pp. 191, 312, pl. 67, figs. 1—5; text-fig. 7-11; simple leaf; Syntypes: HEP0172 (pl. 67, fig. 1), HEP0168 (pl. 67, fig. 2), HEP0170 (pl. 67, fig. 3), HEP0171 (pl. 67, fig. 4), HEP0169 (pl. 67, fig. 5); Repository: China University of Geosciences, Beijing; Dafengkou and Yungaishan of Yuzhou, western Henan; Late Permian coal member 6 of Yungaishan Formation.

Monogigantonoclea latiovata Yang et Wu, 2006 (in Chinese and English)
2006   Yang Guanxiu and Wu Yuehui, in Yang Guanxiu and others, pp. 192, 313, pl. 68, figs. 1—3; text-fig. 7-12; simple leaf; Syntypes: HEP0157 (pl. 68, fig. 1), HEP0159 (pl. 68, fig. 2), HEP3149 (pl. 68, fig. 3); Repository: China University of Geosciences, Beijing; Dengcao of Dengfeng and Yungaishan of Yuzhou, western Henan; Late Permian coal members 7 and 8 of Yungaishan Formation.

Monogigantonoclea grandidenia Yang et Sheng, 2006 (in Chinese and English)
2006   Yang Guanxiu and Sheng Axing, in Yang Guanxiu and others, pp. 193, 313, pl. 55, figs. 3, 4; simple leaf; Syntypes: HEP5106 (pl. 55, fig. 3), HEP0203 (pl. 55, fig. 4); Repository: China University of Geosciences, Beijing; Dafengkou of Yuzhou and Pingdingshan Coal Field, western Henan; Middle Permian Xiaofengkou Formation.

Monogigantonoclea aceroides Yang, 2006 (in Chinese and English)
2006   Yang Guanxiu and others, pp. 194, 314, pl. 65, fig. 4; simple leaf; Holotype: HEP0154 (pl. 65, fig. 4); Repository: China University of Geosciences, Beijing; Yungaishan of Yuzhou, Henan; Late Permian Yungaishan Formation.

Genus *Monogigantopteris* Yang, 2006 (in Chinese and English)
2006   Yang Guanxiu and others, pp. 199, 319.
Type species: *Monogigantopteris clathroreticulatus* Yang, 2006
Taxonomic status: Gigantopteridales
Distribution and Age: Yuzhou and Dengfeng of western Henan, China; Permian.

Monogigantopteris clathroreticulatus Yang, 2006 (in Chinese and English)
2006   Yang Guanxiu and others, pp. 199, 319, pl. 70, figs. 1—7; large simple leaf; Syntypes: HEP0207 (pl. 70, fig. 1), HEP0205 (pl. 70, fig. 2), HEP0209 (pl. 70, fig. 3), HEP0208 (pl. 70, fig. 4), HEP0206 (pl. 70, fig. 5), HEP0210 (pl. 70, fig. 6), HEP0211 (pl. 70, fig. 7); Repository: China University of Geosciences, Beijing; Dafengkou of Yuzhou, western Henan; Middle Permian coal member 4 of Xiaofengkou Formation.

The other species:
Monogigantopteris densireticulatus Yang, 2006 (in Chinese and English)
2006   Yang Guanxiu and others, pp. 200, 320, pl. 72, figs. 1—6; text-fig. 7-15; large simple

leaf; Syntypes: HEP0123 (pl. 72, fig. 1), HEP0124 (pl. 72, fig. 3), HEP0125 (pl. 72, fig. 4), HEP0127 (pl. 72, fig. 2), HEP3121 (pl. 72, fig. 6), HEP3140 (pl. 72, fig. 5); Repository: China University of Geosciences, Beijing; Dafengkou and Yungaishan of Yuzhou and Dengcao of Dengfeng, western Henan; Late Permian coal member 8 of Yungaishan Formation.

Genus *Muricosperma* Seyfullah et Hilton, 2010 (in English)
2010   Seyfullah L J and Hilton J, in Seyfullah L J and others, p. 99.
Type species: *Muricosperma guizhouensis* Seyfullah et Hilton, 2010
Taxonomic status: Spermatophyta
Distribution and Age: Panxian of Guizhou, China; Late Permian (Wuchiapigian).

Muricosperma guizhouensis Seyfullah et Hilton, 2010 (in English)
2010   Seyfullah L J and Hilton J, in Seyfullah L J and others, p. 99, figs. 1 — 20; ovule; Holotype: GPP 2-001 (figs. 1 — 7); Repository: Institute of Botany, Chinese Academy of Sciences; Panxian, Guizhou; Late Permian (Wuchiapigian) Xuanwei Formation.

Genus *Myriophyllum* Xiao, 1985
1985   Xiao Suzhen and Zhang Enpeng, p. 584.
Type species: *Myriophyllum shanxiense* Xiao, 1985
Taxonomic status: Plantae incertae sedis
Distribution and Age: Shanyin of Shanxi, China; Late Carboniferous.

Myriophyllum shanxiense Xiao, 1985
1985   Xiao Suzhen and Zhang Enpeng, p. 584, pl. 203, fig. 13; pl. 204, figs. 1 — 3; cycadophyte leaves; Reg. No.: SH430, SH390, SH388, SH389; Holotype: SH390 (pl. 204, fig. 1); Paratypes: SH430 (pl. 203, fig. 13), SH388 (pl. 204, fig. 2), SH389 (pl. 204, fig. 3); Shanyin, Shanxi; late Late Carboniferous Shanxi Formation.

Genus *Nanpiaophyllum* Zhang et Zheng, 1984
1984   Zhang Wu and Zheng Shaolin, p. 389.
Type species: *Nanpiaophyllum cordatum* Zhang et Zheng, 1984
Taxonomic status: Plantae incertae sedis
Distribution and Age: Nanpiao of Liaoning, China; Late Triassic.

Nanpiaophyllum cordatum Zhang et Zheng, 1984
1984   Zhang Wu and Zheng Shaolin, p. 389, pl. 3, figs. 4 — 9; text-fig. 8; fern-like leaves; Reg. No.: J005-1 — J005-6; Repository: Shenyang Institute of Geology and Mineral Resources; Nanpiao, western Liaoning; Late Triassic Laohugou Formation.

Genus *Nanzhangophyllum* Chen, 1977

1977　Chen Gongxin, in Feng Shaonan and others, p. 246.

Type species: *Nanzhangophyllum donggongense* Chen, 1977

Taxonomic status: Gymnospermae incertae sedis

Distribution and Age: Nanzhang of Hubei, China; Late Triassic.

Nanzhangophyllum donggongense Chen, 1977

1977　Chen Gongxin, in Feng Shaonan and others, p. 246, pl. 99, figs. 6 — 7; text-fig. 82; cycadophyte leaf; No. : P5014, P5015; Syntype 1: P5014 (pl. 99, fig. 6); Syntype 2: P5015 (pl. 99, fig. 7); Repository: Geological Bureau of Hubei Province; Donggong of Nanzhang, Hubei; Late Triassic Lower Coal Formation of Hsiangchi Group.

Genus *Neoannularia* Wang, 1977

1977　Wang Xifu, p. 186.

Type species: *Neoannularia shanxiensis* Wang, 1977

Taxonomic status: Equisetales, Sphenopsida

Distribution and Age: Yijun of Shaanxi and Dukou of Sichuan, China; Late Triassic.

Neoannularia shanxiensis Wang, 1977

1977　Wang Xifu, p. 186, pl. 1, figs. 1 — 9; articulatean shoot with whorled leaves; Col. No. : JP672001 — JP672009; Reg. No. : 76003 — 76011; Jiaoping of Yijun, Shaanxi; Late Triassic upper part of Yenchang Group.

The other species:

Neoannularia chuandianensis Wang, 1977

1977　Wang Xifu, p. 187, pl. 1, fig. 10; text-fig. 1; articulatean shoot with whorled leaves; Col. No. : DK70502; Reg. No. : 76002; Moshahe of Dukou, Sichuan; Late Triassic Daqing Formation.

Genus *Neocordaites* Yang et Wang, 2006 (in Chinese and English)

2006　Yang Guanxiu and Wang Hongshan, in Yang Guanxiu and others, pp. 159, 297.

Type species: *Neocordaites lanceolatus* Yang et Wang, 2006

Taxonomic status: Cordaitopsida

Distribution and Age: Pingdingshan of Henan, China; Middle Permian.

Neocordaites lanceolatus Yang et Wang, 2006 (in Chinese and English)

2006　Yang Guanxiu and Wang Hongshan, in Yang Guanxiu and others, pp. 159, 297, pl. 48, figs. 9, 9a; leaf lanceolate; Holotype: HEP4362 (pl. 48, fig. 9); Repository: China

University of Geosciences, Beijing; Pingdingshan Coal Field, western Henan; Middle Permian Shenhou Formation.

Genus *Neogigantopteridium* Yang, 1987
1987b Yang Guanxiu, pp. 188, 194.
Type species: *Neogigantopteridium spiniferum* Yang, 1987
Taxonomic status: Gigantopterides
Distribution and Age: Yuxian of Henan, China; Late Permian.

Neogigantopteridium spiniferum Yang, 1987
1987b Yang Guanxiu, pp. 188, 194, pl. 3, figs. 3 — 5; text-fig. 9; frond leaves; Reg. No. : HEP 223 — HEP 225; Syntype 1: HEP 223 (pl. 2, figs. 4, 4a); Syntype 2: HEP 224 (pl. 2, fig. 5); Syntype 3: HEP 225; Repository: China University of Geosciences, Beijing; Chenzhuang of Yuxian, Henan; late Late Permian coal member 7 of Yungaishan Formation.

Genus *Neostachya* Wang, 1977
1977 Wang Xifu, p. 188.
Type species: *Neostachya shanxiensis* Wang, 1977
Taxonomic status: Equisetales, Sphenopsida
Distribution and Age: Yijun of Shaanxi, China; Late Triassic.

Neostachya shanxiensis Wang, 1977
1977 Wang Xifu, p. 188, pl. 2, figs. 1 — 10; articulatean fertile shoot; Col. No. : JP672010 — JP672017; Reg. No. : 76012 — 76019; Jiaoping of Yijun, Shaanxi; Late Triassic upper part of Yenchang Group.

Genus *Neurophyllites* Zhang, 1980
1980 Zhang Shanzhen, in Zhao Xiuhu and others, p. 85.
Type species: *Neurophyllites pecopteroides* Zhang, 1980
Taxonomic status: Plantae incertae sedis
Distribution and Age: Fuyuan, Yunnan; early Late Permian.

Neurophyllites pecopteroides Zhang, 1980
1980 Zhang Shanzhen, in Zhao Xiuhu and others, p. 85, pl. 14, figs. 1, 1a, 2, 2a, 2b; frond leaves; Col. No. : PQ-118; Reg. No. : PB7045, PB7046; Repository: Nanjing Institute of Geology and Palaeontology, Chinese Academy of Sciences; Fuyuan, Yunnan; early Late Permian lower member of Xuanwei Formation.

Genus *Ningxiaphyllum* Zhao, Wu et Gu, 1986

1986   Zhao Xiuhu, Wu Xiuyuan and Gu Qichang, pp. 554, 558.

Type species: *Ningxiaphyllum trilobatum* Zhao, Wu et Gu, 1986

Taxonomic status: Plantae incertae sedis

Distribution and Age: Qingtongxia of Ningxia, China; Late Devonian.

Ningxiaphyllum trilobatum Zhao, Wu et Gu, 1986

1986   Zhao Xiuhu, Wu Xiuyuan and Gu Qichang, pp. 555, 558, pl. 4, fig. 3; pl. 5, figs. 1, 2; text-fig. 9; large petiolate flabelliform leaf; Col. No.: QD-H-60, QD-H-84, QD-H-85; Reg. No.: PB11493, PB11495, PB11496; Holotype: PB11495 (pl. 5, fig. 1); Repository: Nanjing Institute of Geology and Palaeontology, Chinese Academy of Sciences; Yingshengtai of Qingtongxia, Ningxia; Late Devonian Zhongning Formation.

Genus *Norinia* Halle, 1927

1927   Halle T G, p. 218.

Type species: *Norinia cucullata* Halle, 1927

Taxonomic status: Plantae incertae sedis

Distribution and Age: Ch'en-chia-yu of central Shanxi, China; early Late Permian.

Norinia cucullata Halle, 1927

1927   Halle T G, p. 218, pl. 56, figs. 8 — 12; leaves; Ch'en-chia-yu, central Shanxi; early Late Permian Upper Shihhotse Series.

Genus *Nudasporestrobus* Feng, Wang et Bek, 2008 (in English)

2008   Feng Zuo, Wang Jun and Bek J, p. 152.

Type species: *Nudasporestrobus ningxicus* Feng, Wang et Bek, 2008

Taxonomic status: Sigillariaceae, Lepidodendrales

Distribution and Age: Zhongwei of Ningxia, China; Late Carboniferous.

Nudasporestrobus ningxicus Feng, Wang et Bek, 2008 (in English)

2008   Feng Zuo, Wang Jun and Bek J, p. 152, pls. 1 — 4; sigillarian strobili; Reg. No.: PB20826 — PB20832; Holotype: PB20826 (pl. 1, figs. 1 — 4); Repository: Nanjing Institute of Geology and Palaeontology, Chinese Academy of Sciences; Xiaheyan of Zhongwei, Ningxia; Late Carboniferous [Early Pennsylvanian (Bashkirian stage)] Yanghugou Formation.

Genus *Nystroemia* Halle, 1927
1927   Halle T G, p. 221.

Type species: *Nystroemia pectiniformis* Halle, 1927

Taxonomic status: Pteridospermopsida?

Distribution and Age: Ch'en-chia-yu of central Shanxi, China; early Late Permian.

Nystroemia pectiniformis Halle, 1927
1927   Halle T G, p. 221, pl. 59; seed-bearing organ and microsporangia; Ch'en-chia-yu, central Shanxi; early Late Permian Upper Shihhotse Series.

Genus *Odontosorites* Kobayashi et Yosida, 1944
1944   Kobayashi T and Yosida T, p. 257.

Type species: *Odontosorites heerianus* (Yokoyama) Kobayashi et Yosida, 1944

Taxonomic status: Filicopsida

Distribution and Age: Heihe (Heiho) of Heilongjiang, China; Jurassic. Japan; Early Cretaceous.

Odontosorites heerianus (Yokoyama) Kobayashi et Yosida, 1944
1889   *Adiantites heerianus* Yokoyama, Yokoyama M, p. 28, pl. 12, figs. 1, 1a, 1b, 2; Japan; Early Cretaceous (Tetori Series).

1944   Kobayashi T and Yosida T, p. 257, pl. 28, figs. 6 — 7; text-figs. a — c; sterile pinnae and fertile pinnae; Ryokusin or Lushen of Heihe (Heiho), Heilongjiang; Jurassic. [Notes: This specimen lately was referred as ? *Coniopteris burejensis* (Zalessky) Seward (Sze H C, Lee H H and others, 1963)]

Genus *Orchidites* Wu S Q, 1999 (in Chinese)
1999   Wu Shunqing, p. 23.

Type species: *Orchidites linearifolius* Wu S Q, 1999 (Notes: The type species was not designated in the original paper)

Taxonomic status: Orchidaceae, Monocotyledoneae

Distribution and Age: Beipiao of western Liaoning, China; Late Jurassic.

Orchidites linearifolius Wu S Q, 1999 (in Chinese)
1999   Wu Shunqing, p. 23, pl. 16, fig. 7; pl. 17, figs. 1 — 3; herbaceous plants; Col. No. : AEO-123, 104, 29; Reg. No. : PB18321, PB18324, PB18325; Repository: Nanjing Institute of Geology and Palaeontology, Chinese Academy of Sciences; Huangbanjigou in Shangyuan of Beipiao, western Liaoning; Late Jurassic Jianshangou Bed in lower part of Yixian Formation.

The other species:
Orchidites lancifolius Wu S Q, 1999 (in Chinese)
1999   Wu Shunqing, p. 23, pl. 17, figs. 4, 4a; herbaceous plants; Col. No. : AEO196; Reg. No. : PB18326; Repository: Nanjing Institute of Geology and Palaeontology, Chinese Academy of Sciences; Huangbanjigou in Shangyuan of Beipiao, western Liaoning; Late Jurassic Jianshangou Bed in lower part of Yixian Formation.

Genus *Otofolium* Gu et Zhi, 1974
1974   Nanjing Institute of Geology and Palaeontology, Institute of Botany, Chinese Academy of Sciences, in *Palaeozoic Plants from China*, p. 164.

Type species: *Otofolium polymorphum* Gu et Zhi, 1974

Taxonomic status: Plantae incertae sedis

Distribution and Age: Jiangsu, Anhui, Guangdong, Jiangxi and Guizhou, China; Late Permian.

Otofolium polymorphum Gu et Zhi, 1974
1974   Nanjing Institute of Geology and Palaeontology, Institute of Botany, Chinese Academy of Sciences, in *Palaeozoic Plants from China*, p. 165, pl. 127, figs. 2 — 6; text-fig. 134; pinnae; Reg. No. : PB3772 (pl. 127, fig. 2), PB3771 (pl. 127, fig. 3), PB3715 (pl. 127, fig. 5), PB3713; Holotype: PB3771; Repository: Nanjing Institute of Geology and Palaeontology, Chinese Academy of Sciences; Jiangning of Jiangsu, Jingxian of Anhui, Qujiang of Guangdong; early Late Permian Longtan Formation; Jinxian, Jiangxi; Late Permian Leping Formation.

The other species:
Otofolium ovatum Gu et Zhi, 1974
1974   Nanjing Institute of Geology and Palaeontology, Institute of Botany, Chinese Academy of Sciences, in *Palaeozoic Plants from China*, p. 165, pl. 127, figs. 7 — 9; pinnae; Reg. No. : PB4993 (pl. 127, fig. 7), PB4994 (pl. 127, fig. 8); Holotype: PB4994; Repository: Nanjing Institute of Geology and Palaeontology, Chinese Academy of Sciences; Panxian, Guizhou; early Late Permian Xuanwei Formation.

Genus *Palaeoginkgoxylon* Feng, Wang et Roessler, 2010 (in English)
2010   Feng Zhuo, Wang Jun and Roessler R, p. 149.

Type species: *Palaeoginkgoxylon zhoui* Feng, Wang et Roessler, 2010

Taxonomic status: Ginkgophytes

Distribution and Age: Alxa Left Banner of Inner Mongolia, China; Middle Permian (Guadalupian).

Palaeoginkgoxylon zhoui Feng, Wang et Roessler, 2010 (in English)
2010   Feng Zhuo, Wang Jun and Roessler R, p. 149, pls. 1 — 4; ginkgophyte wood; Reg. No. :

YKLP20006; Holotype: YKLP20006 (pls. 1 — 4); Repository: Yunnan Key Laboratory for Palaeobiology, Yunnan University; Hulstai of Alxa Left Banner, Inner Mongolia; Middle Permian (Guadalupian) Lower Shihhotse Formation.

Genus *Palaeognetaleaana* Wang, 2004 (in English)

2004　Wang Ziqiang, p. 282.

Type species: *Palaeognetaleaana auspicia* Wang, 2004

Taxonomic status: Gnetales

Distribution and Age: northwestern Shanxi, China; Late Permian.

Palaeognetaleaana auspicia Wang, 2004 (in English)

2004　Wang Ziqiang, p. 282, figs. 2 — 6; cones; No. : 9107-1, 9107-8, 9705-0, 9705-37, 9705-39, 9705-39a, 9705-41, 9705-45, 9805-045, 9805-048; Holotype: 9107-1 (figs. 2 — 3); Repository: Nanjing Institute of Geology and Palaeontology, Chinese Academy of Sciences; Baode section, northwestern Shanxi; Late Permian.

Genus *Palaeoskapha* Jacques et Guo, 2007 (in English)

2007　Jacques Frédérrie M B and Guo Shuangxing, p. 578.

Type species: *Palaeoskapha sichuanensis* Jacques et Guo, 2007

Taxonomic position: Menispermaceae, Dicotyledoneae

Distribution and Age: Litang of western Sichuan, China; Eocene.

Palaeoskapha sichuanensis Jacques et Guo, 2007 (in English)

2007　Jacques Frédérrie M B and Guo Shuangxing, p. 578, figs. 2A, 2B; fruit; Reg. No. : PB12702, PB12703; Syntypes: PB12703 (fig. 2A), PB12702 (fig. 2B); Repository: Nanjing Institute of Geology and Palaeontology, Chinese Academy of Sciences; Relucun (ca. 30°N, 100°32′E) of Litang, western Sichuan; Eocene Relu Formation.

Genus *Pania* Yang, 2006 (in Chinese and English)

2006　Yang Guanxiu and others, pp. 144, 289.

Type species: *Pania cycadina* Yang, 2006

Taxonomic status: Cycadales, Cycadophyta

Distribution and Age: Yuzhou of western Henan, China; Late Permian.

Pania cycadina Yang, 2006 (in Chinese and English)

2006　Yang Guanxiu and others, pp. 144, 289, pl. 40, figs. 4 — 6; male cone; Syntypes: HEP0885 (pl. 40, fig. 4), HEP0888 (pl. 40, fig. 5), HEP0887 (pl. 40, fig. 6); Repository: China University of Geosciences, Beijing; Dafengkou of Yuzhou, western Henan; Late Permian member 7 of Yungaishan Formation.

Genus *Pankuangia* Kimura, Ohana, Zhao et Geng, 1994

1994 Kimura T, Ohana T, Zhao Liming and Geng Baoyin, p. 256.

Type species: *Pankuangia haifanggouensis* Kimura, Ohana, Zhao et Geng, 1994

Taxonomic status: Cycadales, Cycadopsida

Distribution and Age: Jinxi of Liaoning, China; Middle Jurassic.

Pankuangia haifanggouensis Kimura, Ohana, Zhao et Geng, 1994

1994 Kimura T, Ohana T, Zhao Liming and Geng Baoyin, p. 257, figs. 2 — 4, 8; cycadophyte leaf; No.: LJS-8690, 8555, 8554, 8807, L0407A [regarded by Pan Kuang as *Juradicotes crecta* Pan (MS)]; Holotype: LJS-8690 (fig. 2A); Repository: Institute of Botany, Chinese Academy of Sciences; Sanjiaochengcun (roughly 40°58′N, 120°21′E) of Jinxi, western Liaoning; Middle Jurassic Haifanggou Formation. [Notes: This specimen lately was referred as *Anomozamites haifanggouensis* (Kimura, Ohana, Zhao et Geng) Zheng et Zhang (Zheng Shaolin and others, 2003)]

Genus *Papilionifolium* Cao, 1999 (in Chinese and English)

1999 Cao Zhengyao, pp. 102, 160.

Type species: *Papilionifolium hsui* Cao, 1999

Taxonomic status: Plantae incertae sedis

Distribution and Age: Wencheng of Zhejiang, China; Early Cretaceous.

Papilionifolium hsui Cao, 1999 (in Chinese and English)

1999 Cao Zhengyao, pp. 102, 160, pl. 21, figs. 12 — 15; text-fig. 35; leaf-bearing stem; Col. No.: Zh301; Reg. No.: PB14467 — PB14470; Holotype: PB14469 (pl. 21, fig. 14); Repository: Nanjing Institute of Geology and Palaeontology, Chinese Academy of Sciences; Konglong of Wencheng, Zhejiang; Early Cretaceous Guantou Formation.

Genus *Paracaytonia* Wang, 2010 (in English)

2010 Wang Xin, p. 208.

Type species: *Paracaytonia hongtaoi* Wang, 2010

Taxonomic status: Caytoniales

Distribution and Age: Jianchang of Liaoning, China; Early Cretaceous (Barremian).

Paracaytonia hongtaoi Wang, 2010 (in English)

2010 Wang Xin, p. 208, figs. 1 — 3; female reproductive organ; No.: GBM1A, B; Holotype: GBM1 (fig. 1); Repository: Palaeontological Museum, Fairy Lake Botanical Garden, Shenzhen; Langjiagou and Yaolugou of Jianchang, Liaoning; Early Cretaceous (Barremian) Yixian Formation.

Genus *Paraconites* Hu, 1984 (nom. nud.)
1984　Hu Yufan, p. 571.

Type species: *Paraconites longifolius* Hu, 1984

Taxonomic status: Taxodiaceae, Coniferopsida

Distribution and Age: Datong of Shanxi, China; Early Jurassic.

Paraconites longifolius Hu, 1984 (nom. nud.)
1984　Hu Yufan, p. 571; cones; Meiyukou of Datong, Shanxi; Early Jurassic Yongdingzhuang Formation.

Genus *Paradoxopteris* Mi et Liu, 1977 (non Hirmer, 1927)
{Notes: this generic name is a late synonym (homonymum junius) of *Paradoxopteris* Hirmer, 1927 (Wu Xiangwu, 1993a, 1993b), its type species *Paradoxopteris strumeri* Hirmer, 1927 [Hirmer, 1927, p. 609; text-figs. 733 — 736; Bahariji Oasis, Egypt; Cretaceous (Cenomanian)]; and lately was referred as *Mirabopteris* (Mi et Liu) Mi et Liu (Mi Jiarong and others, 1993)}

1977　Mi Jiarong and Liu Maoqiang, in Department of Geological Exploration, Changchun College of Geology and others, p. 8.

Type species: *Paradoxopteris hunjiangensis* Mi et Liu, 1977

Taxonomic status: Pteridospermopsida

Distribution and Age: Hunjiang of Jilin, China; Late Triassic.

Paradoxopteris hunjiangensis Mi et Liu, 1977
1977　Mi Jiarong and Liu Maoqiang, in Department of Geological Exploration, Changchun College of Geology and others, p. 8, pl. 3, fig. 1; text-fig. 1; fern-like leaves; No.: X-08; Repository: Department of Geological Exploration, Changchun Institute of Geology; Shiren of Hunjiang, Jilin; Late Triassic Xiaohekou Formation. [Notes: This species lately was referred as *Mirabopteris hunjiangensis* (Mi et Liu) Mi et Liu (Mi Jiarong and others, 1993)]

Genus *Paradrepanozamites* Chen, 1977
1977　Chen Gongxin, in Feng Shaonan and others, p. 236.

Type species: *Paradrepanozamites dadaochangensis* Chen, 1977

Taxonomic status: Cycadopsida

Distribution and Age: Nanzhang of Hubei, China; Late Triassic.

Paradrepanozamites dadaochangensis Chen, 1977
1977　Chen Gongxin, in Feng Shaonan and others, p. 236, pl. 99, figs. 1 — 2; text-fig. 81; cycadophyte leaf; Reg. No.: P5107, P25269; Syntype 1: P5107 (pl. 99, fig. 1);

Repository: Geological Bureau of Hubei Province; Syntype 2: P25269 (pl. 99, fig. 2); Repository: Hupei Institute of Geological Sciences; Donggong of Nanzhang, Hubei; Late Triassic Lower Coal Formation of Hsiangchi Group.

Genus *Parasphenophyllum* Asama, 1970

1970   Asama K, p. 301.

Type species: *Parasphenophyllum shansiense* (Asama) Asama, 1970

Taxonomic status: Sphenophyllales

Distribution and Age: China and Korea; Permian.

Parasphenophyllum shansiense (Asama) Asama, 1970

1970   Asama K, p. 301, pl. 3, fig. 1; leaves (Sphenophyllales); China and Korea; Permian.

Genus *Parasphenopteris* Sun et Deng, 2006 (in English)

2006   Sun Keqin and Deng Shenghui, p. 161.

Type species: *Parasphenopteris orientalis* Sun et Deng, 2006

Taxonomic status: Sphenopterides of Filicopsida or Pteridospermopsia

Distribution and Age: Wuda area of Inner Mongolia, China; Early Permian.

Parasphenopteris orientalis Sun et Deng, 2006 (in English)

2006   Sun Keqin and Deng Shenghui, p. 162, fig. 1; frond; Holotype: WD20026 (fig. 1); Repository: China University of Geosciences, Beijing; Wuda area, Inner Mongolia; Early Permian Shanxi Formation.

Genus *Parastorgaardis* Zeng, Shen et Fan, 1995

1995   Zeng Yong, Shen Shuzhong and Fan Bingheng, p. 67.

Type species: *Parastorgaardis mentoukouensis* (Stockmans et Mathieu) Zeng, Shen et Fan, 1995

Taxonomic status: Taxodiaceae, Coniferopsida

Distribution and Age: Mentougou (Mentoukou) of Beijing and Yima of Henan, China; Jurassic.

Parastorgaardis mentoukouensis (Stockmans et Mathieu) Zeng, Shen et Fan, 1995

1941   *Podocarpites mentoukouensis* Stockmans et Mathieu, Stockmans F and Mathieu F F, p. 53, pl. 7, figs. 5, 6; leafy shoot; Mentougou (Mentoukou), Beijing; Jurassic.

1995   Zeng Yong, Shen Shuzhong and Fan Bingheng, p. 67, pl. 20, fig. 3; pl. 23, fig. 3; pl. 29, figs. 6 — 8; leafy shoot and cuticle; Yima, Henan; Middle Jurassic lower coal-bearing member of Yima Formation.

Genus *Parataxospermum* Li, 1993

1993b Li Zhongming, p. 66.

Type species: *Parataxospermum taiyuanesis* Li, 1993

Taxonomic status: Cardiocarpales

Distribution and Age: Taiyuan of Shanxi, China; Late Carboniferous.

Parataxospermum taiyuanesis Li, 1993

1993b Li Zhongming, p. 66, pls. 1 — 4; text-fig. 1; seeds (coal balls); Holotype: BSC3. 363 (pl. 1, figs. 1 — 4; pl. 2; pl. 4, figs. 1 — 3); Paratype: BSC3. 102 (pl. 1, fig. 5; pl. 3; pl. 4, fig. 2); Repository: Institute of Botany, Chinese Academy of Sciences; West Mountain of Taiyuan, Shanxi; Late Carboniferous Taiyuan Formation.

Genus *Paratingia* Zhang, 1987

1987 Zhang Hong, p. 200.

Type species: *Paratingia datongensis* Zhang, 1987

Taxonomic status: Noeggerathopsida

Distribution and Age: Datong of Shanxi, China; Early Permian.

Paratingia datongensis Zhang, 1987

1987 Zhang Hong, p. 200, pl. 10, figs. 1 — 4; pl. 11, figs. 3, 4; fornd; Reg. No. : Mp-84061, Mp-84062, Mp-84063, Mp-84064; Datong, Shanxi; Early Permian.

Genus *Paratingiostachya* Sun, Deng, Cui et Shang, 1999 (in Chinese and English)

1999 Sun Keqin, Deng Shenghui, Cui Jinzhong and Shang Ping, pp. 1024, 1025.

Type species: *Paratingiostachya cathaysiana* Sun, Deng, Cui et Shang, 1999

Taxonomic status: Noeggerathopsida

Distribution and Age: Wuda area of Inner Mongolia, China; Early Permian.

Paratingiostachya cathaysiana Sun, Deng, Cui et Shang, 1999 (in Chinese and English)

1999 Sun Keqin, Deng Shenghui, Cui Jinzhong and Shang Ping, pp. 1024, 1025, pl. 3, figs. 3, 3a, 4, 4a; cone; Repository: China University of Geosciences, Beijing; Wuda area, Inner Mongolia; Early Permian Shanxi Formation.

Genus *Pavoniopteris* Li et He, 1986

1986 Li Peijuan and He Yuanliang, p. 279.

Type species: *Pavoniopteris matonioides* Li et He, 1986

Taxonomic status: Filicopsida
Distribution and Age: Dulan of Qinghai, China; Late Triassic.

Pavoniopteris matonioides Li et He, 1986
1986　Li Peijuan and He Yuanliang, p. 279, pl. 2, fig. 1; pl. 3, figs. 3 — 4; pl. 4, figs. 1 — 1d; text-figs. 1, 2; sterile frond and fertile frond; Col. No.: 79PIVF22-3; Reg. No.: PB10866, PB10869 — PB10871; Holotype: PB10871 (pl. 4, figs. 1 — 1d); Repository: Nanjing Institute of Geology and Palaeontology, Chinese Academy of Sciences; Babaoshan of Dulan, Qinghai; Late Triassic Lower Rock Formation of Babaoshan Group.

Genus *Pectinangium* Gu et Zhi, 1974
1974　Nanjing Institute of Geology and Palaeontology, Institute of Botany, Chinese Academy of Sciences, in *Palaeozoic Plants from China*, p. 166.

Type species: *Pectinangium lanceolatum* Gu et Zhi, 1974

Taxonomic status: Plantae incertae sedis

Distribution and Age: Jiangning of Jiangsu, China; early Late Permian.

Pectinangium lanceolatum Gu et Zhi, 1974
1974　Nanjing Institute of Geology and Palaeontology, Institute of Botany, Chinese Academy of Sciences, in *Palaeozoic Plants from China*, p. 166, pl. 128, figs. 9 — 12; text-fig. 135; pinnae; Reg. No.: PB3807 (pl. 128, fig. 9), PB3804 (pl. 128, fig. 10); Holotype: PB3804; Repository: Nanjing Institute of Geology and Palaeontology, Chinese Academy of Sciences; Jiangning, Jiangsu; early Late Permian Longtan Formation.

Genus *Perisemoxylon* He et Zhang, 1993
1993　He Dechang and Zhang Xiuyi, pp. 262, 264.

Type species: *Perisemoxylon bispirale* He et Zhang, 1993

Taxonomic status: Cycadales, Cycadopsida

Distribution and Age: Yima of Henan, China; Middle Jurassic.

Perisemoxylon bispirale He et Zhang, 1993
1993　He Dechang and Zhang Xiuyi, pp. 262, 264, pl. 1, figs. 1, 2; pl. 2, fig. 5; pl. 4, fig. 3; fusain woods; Col. No.: 9001, 9002; Reg. No.: S006, S007; Holotype: S006 (pl. 1, fig. 1); Paratype: S007 (pl. 1, fig. 2); Repository: Xi'an Branch, China Coal Research Institute; Yima, Henan; Middle Jurassic.

The other species:
Perisemoxylon sp.
1993　*Perisemoxylon* sp., He Dechang and Zhan Xiuyi, p. 263, pl. 2, figs. 1 — 4; fusain woods; Yima, Henan; Middle Jurassic.

Genus *Phoenicopsis* Heer, 1876

1876   Heer O, p. 51.

Type species: *Phoenicopsis angustifolia* Heer, 1876

Taxonomic status: Czekanowskiales

Phoenicopsis angustifolia Heer, 1876

1876   Heer O, p. 51, pl. 1, fig. 1d; pl. 2, fig. 3b; p. 113, pl. 31, figs. 3, 8; leaves; Irkutsk of upper reaches of Heilongjiang River Basin; Jurassic.

Subgenus *Phoenicopsis* (*Stephenophyllum*) ex Li, 1988

[Notes: This subgeneric name was proposed by Li Peijuan and others (1988), but not mentioned clearly as a nomen novum and type species]

1936   *Stephanophyllum* Florin, Florin R, p. 82.

1988   Li Peijuan and others, p. 106.

Type species: *Phoenicopsis* (*Stephenophyllum*) *solmsi* (Seward) [Notes: *Stephenophyllum solmsi* (Seward) Florin, is the type species of *Stephenophyllum* (Florin, 1936)]

Taxonomic status: Czekanowskiales

Distribution and Age: Qinghai and Xinjiang of China and Franz Josef Land; Jurassic.

Phoenicopsis (*Stephenophyllum*) *solmsi* (Seward)

1919   *Desmiophllum solmsi* Seward, Seward A C, p. 71, fig. 662; transverse sections of leaves and stomata; Franz Josef Land; Jurassic.

1936   *Stephenophyllum solmsi* (Seward) Florin, Florin R, p. 82, pl. 11, figs. 7 — 10; pls. 12 — 16; text-figs. 3, 4; leaves and cuticles; Franz Josef Land; Jurassic.

The other species:

Phoenicopsis (*Stephenophyllum*) *decorata* Li, 1988

1988   Li Peijuan and others, p. 106, pl. 68, fig. 5B; pl. 79, figs. 4, 4a; pl. 120, figs. 1 — 6; leaves and cuticles; Col. No. : 80LFu; Reg. No. : PB13630, PB13631; Holotype: PB13631 (pl. 79, figs. 4, 4a); Repository: Nanjing Institute of Geology and Palaeontology, Chinese Academy of Sciences; Lvcaogou in Lvcaoshan of Qaidam, Qinghai; Middle Jurassic *Nilssonia* Bed of Shimengou Formation.

Phoenicopsis (*Stephenophyllum*) *enissejensis* (Samylina) ex Li, 1988

[Notes: This species name was proposed by Li Peijuan (1988), but not mentioned clearly as a nomen novum]

1972   *Phoenicopsis* (*Phoenicopsis*) *enissejensis* Samylina, p. 63, pl. 2, figs. 1, 2; pl. 3, figs. 1 — 4; pl. 4, figs. 1 — 5; leaves and cuticles; West Siberia; Middle Jurassic.

1988   Li Peijuan and others, p. 106, pl. 85, figs. 2 2a; pl. 86, fig. 1; pl. 87, fig. 1; pl. 121, figs. 1 — 6; leaves and cuticles; Lvcaogou in Lvcaoshan of Qaidam, Qinghai; Middle Jurassic *Nilssonia* Bed of Shimengou Formation.

Phoenicopsis (*Stephenophyllum*) *mira* Li, 1988

1988 Li Peijuan and others, p. 107, pl. 80, figs. 2 — 4a; pl. 81, fig. 2; pl. 122, figs. 5, 6; pl. 123, figs. 1 — 4; pl. 136, fig. 5; pl. 138, fig. 4; leaves and cuticles; Col. No. : $80DP_1F_{89}$, $80DJ_{2d}$ Fu; Reg. No. : PB13635 — PB13637; Holotype: PB13635 (pl. 81, fig. 2); Repository: Nanjing Institute of Geology and Palaeontology, Chinese Academy of Sciences; Dameigou of Qaidam, Qinghai; Middle Jurassic *Coniopteri murrayana* Bed of Yinmagou Formation and *Tyrmia-Sphenobaiera* Bed of Dameigou Formation.

Phoenicopsis (*Stephenophyllum*) *taschkessiensis* (**Krasser**) ex Li, 1988

[Notes: The species name was proposed by Li Peijuan (1988), but not mentioned clearly as a nomen novum]

1901 *Phoenicopsis taschkessiensis* Krasser, Krasser F, p. 150, pl. 4, fig. 2; pl. 3, fig. 4t; leaves; Sandaoling (Santoling) between Hami and Turfan, Xinjiang; Jurassic.

1988 Li Peijuan and others, p. 3.

Genus *Phoroxylon* Sze, 1951

1951b Sze H C, pp. 443, 451.

Type species: *Phoroxylon scalariforme* Sze, 1951

Taxonomic status: Bennetittales

Distribution and Age: Jixi of Heilongjiang, China; Late Cretaceous.

Phoroxylon scalariforme Sze, 1951

1951b Sze H C, pp. 443, 451, pl. 5, figs. 2, 3; pl. 6, figs. 1 — 4; pl. 7, figs. 1 — 4; text-figs. 3A — 3E; petrified woods; Chengzihe of Jixi, Heilongjiang; Late Cretaceous.

Genus *Phylladendroid* He, Liang et Shen, 1996 (in Chinese and English)

1996 He Xilin, Liang Dunshi and Shen Shuzhong, pp. 92, 167.

Type species: *Phylladendroid jiangxiensis* He, Liang et Shen, 1996

Taxonomic status: Plantae incertae sedis

Distribution and Age: Fengcheng of Jiangxi, China; Late Permian.

Phylladendroid jiangxiensis He, Liang et Shen, 1996 (in Chinese and English)

1996 He Xilin, Liang Dunshi and Shen Shuzhong, pp. 92, 167, pl. 74, figs. 3, 4; stem; Reg. No. : X88311, X88312; Syntypes: X88311 (pl. 74, fig. 3), X88312 (pl. 74, fig. 4); Repository: Department of Geology, China University of Mining and Technology; Fengcheng Coal Mine of Leping, Jiangxi; Late Permian Lower Laoshan Submember of Leping Formation.

Genus *Pinnagigantonoclea* Yang, 2006 (in Chinese and English)

2006　Yang Guanxiu and others, pp. 184, 306.

Type species: *Pinnagigantonoclea zelkovoides* Yang, 2006

Taxonomic status: Gigantopteridales

Distribution and Age: Yuzhou of western Henan, Panxian of Guizhou and Kaiping of Hebei, China; Early Permian.

Pinnagigantonoclea zelkovoides Yang, 2006 (in Chinese and English)

2006　Yang Guanxiu and others, pp. 184, 307, pl. 62, figs. 5, 6; pinnate compound leaf; Syntypes: HEP3630 (pl. 62, fig. 5), HEP3601 (pl. 62, fig. 6); Repository: China University of Geosciences, Beijing; Dishuitan in Fangshan of Yuzhou, western Henan; Late Permian coal member 6 of Yungaishan Formation.

The other species:

Pinnagigantonoclea heteroeura Yang et Wang, 2006 (in Chinese and English)

2006　Yang Guanxiu and Wang Hongshan, in Yang Guanxiu and others, pp. 185, 307, pl. 57, figs. 1, 1a; pinnate compound leaf; Holotype: HEP4236 (pl. 57, figs. 1, 1a); Repository: China University of Geosciences, Beijing; Dafengkou of Pingdingshan Coal Field, western Henan; Middle Permian Xiaofengkou Formation.

Pinnagigantonoclea mira (Gu et Zhi) Yang, 2006 (in Chinese and English)

1974　*Gigantonoclea mira* Gu et Zhi, Nanjing Institute of Geology and Palaeontology, Institute of Botany, Chinese Academy of Sciences, in *Palaeozoic Plants from China*, p. 128, pl. 97, figs. 1—5; fertile pinnae; Kaiping, Hebei; Early Permian — Late Carboniferous upper part of Zhaogezhuang Group; Benxi, Liaoning; Early — Middle Permian Caijia Group.

2006　Yang Guanxiu and others, pp. 185, 308, pl. 55, figs. 1, 2; pinnate compound leaf; Dafengkou of Yuzhou, western Henan; Middle Permian Xiaofengkou Formation.

Pinnagigantonoclea guizhouensis (Gu et Zhi) Yang, 2006 (in Chinese and English)

1974　*Gigantonoclea guizhouensis* Gu et Zhi, Nanjing Institute of Geology and Palaeontology, Institute of Botany, Chinese Academy of Sciences, in *Palaeozoic Plants from China*, p. 127, pl. 96, figs. 7 — 10; fertile pinnae; Panxian, Guizhou; Late Permian Xuanwei Formation.

2006　Yang Guanxiu and others, pp. 186, 308, pl. 61, figs. 4, 4a; pinnate compound leaf; Dishuitan in Fangshan of Yuzhou, western Henan; Late Permian coal member 6 of Yungaishan Formation.

Pinnagigantonoclea mucronata Yang, 2006 (in Chinese and English)

2006　Yang Guanxiu and others, pp. 186, 308, pl. 63, fig. 7; pl. 64, figs. 1, 2; pinnate compound leaf; Syntypes: HEP0137 (pl. 63, fig. 7), HEP3608 (pl. 64, fig. 1), HEP3607 (pl. 64, fig. 2); Repository: China University of Geosciences, Beijing; Yungaishan and Fangshan of Yuzhou and Magou of Xinmi, western Henan; Late Permian coal member 6 of

Yungaishan Formation.

Pinnagigantonoclea spatulata (Yang) Yang, 2006 (in Chinese and English)

1987 *Gigantonoclea spatulata* Yang, Yang Guanxiu and others, p. 187, pl. 58; pl. 3, figs. 1, 2, 7; text-fig. 8.

2006 Yang Guanxiu and others, pp. 187, 309, pl. 58; pl. 59, fig. 1; pl. 60, figs. 3, 4; text-fig. 7-9; pinnate compound leaf; Syntypes: HEP0185 (pl. 58), HEP0183 (pl. 59, fig. 1), HEP0184 (pl. 60, fig. 3), HEP0186 (pl. 60, fig. 4; text-fig. 7-9. 3); Repository: China University of Geosciences, Beijing; Dafengkou and Chenzhuang of Yuzhou, western Henan; Late Permian coal member 8 of Yungaishan Formation.

Pinnagigantonoclea rosulata (Gu et Zhi) Yang et Xie, 2006 (in Chinese and English)

1974 *Gigantonoclea rosulata* Gu et Zhi, Nanjing Institute of Geology and Palaeontology, Institute of Botany, Chinese Academy of Sciences, in *Palaeozoic Plants from China*, p. 126, pl. 96, figs. 1 — 6; text-fig. 104; fertile pinnae; Reg. No.: PB4969 — PB4972; Syntypes: PB4969 — PB4972 (pl. 96, figs. 1 — 6); Repository: Nanjing Institute of Geology and Palaeontology, Chinese Academy of Sciences; Yuzhou, Henan; Late Permian Upper Shihhotse Formation.

2006 Yang Guanxiu and Xie Jianhua, in Yang Guanxiu and others, pp. 188, 309, pl. 60, figs. 1, 2; pl. 75, fig. 4; pinnate compound leaf; Dafengkou of Yuzhou, western Henan; Late Permian coal member 8 of Yungaishan Formation; Dishuitan in Fangshan of Yuzhou, western Henan; Late Permian coal member 6 of Yungaishan Formation.

Pinnagigantonoclea dryophylloides Yang et Xie, 2006 (in Chinese and English)

2006 Yang Guanxiu and Xie Jianhua, in Yang Guanxiu and others, pp. 188, 310, pl. 61, figs. 5 — 7; pinnate compound leaf; Syntypes: HEP3377 (pl. 61, fig. 5), HEP3376 (pl. 61, fig. 6), HEP3378 (pl. 61, fig. 7); Repository: China University of Geosciences, Beijing; Pochi of Linru, western Henan; Late Permian coal member 8 of Yungaishan Formation.

Pinnagigantonoclea polymorpha Xie, 2006 (in Chinese and English)

2006 Xie Jianhua, in Yang Guanxiu and others, pp. 189, 311, pl. 62, figs. 1 — 4; pinnate compound leaf; Syntypes: HEP3373 (pl. 62, fig. 1), HEP3374 (pl. 62, fig. 2), HEP3622 (pl. 62, fig. 3), HEP3375 (pl. 62, fig. 4); Repository: China University of Geosciences, Beijing; Pochi of Linru, western Henan; Late Permian coal member 8 of Yungaishan Formation.

Genus *Pinnagigantopteris* Yang et Xie, 2006 (in Chinese and English)

2006 Yang Guanxiu and Xie Jianhua, in Yang Guanxiu and others, pp. 197, 317.

Type species: *Pinnagigantopteris nicotianaefolia* (Gu et Zhi) Yang, 2006

Taxonomic status: Gigantopteridales

Distribution and Age: Leiyang of Hunan, Jurong and Longtan of Jiangsu, Changxing of Zhejiang, Xuanwei of Yunnan and Yuzhou of western Henan, China; Permian.

Pinnagigantopteris nicotianaefolia (Gu et Zhi) **Yang, 2006** (in Chinese and English)

1974 *Gigantopteris nicotianaefolia* Schenk, Nanjing Institute of Geology and Palaeontology, Institute of Botany, Chinese Academy of Sciences, in *Palaeozoic Plants from China*, p. 130, pl. 100, figs. 2 — 4; pl. 101, fig. 1; pl. 102, fig. 7; text-figs. 103 — 105, 108; Leiyang of Hunan, Jurong and Longtan of Jiangsu, Changxing of Zhejiang; Late Permian Longtan Formation; Xuanwei, Yunnan; Late Permian Xuanwei Formation; Pingdingshan, western Henan (?); Late Permian Upper Shihhotse Formation.

2006 Yang Guanxiu and others, pp. 197, 317, pl. 71, figs. 4 — 7; pinnate compound leaf; Dafengkou of Yuzhou and Pochi of Linru, western Henan; Late Permian Yungaishan Formation.

The other species:

Pinnagigantopteris lanceolatus **Yang et Xie, 2006** (in Chinese and English)

2006 Yang Guanxiu and Xie Jianhua, in Yang Guanxiu and others, pp. 198, 317, pl. 64, figs. 3, 3a; text-figs. 7 — 14; pinnate compound leaf; Holotype: HEP3613 (pl. 64, figs. 3, 3a); Repository: China University of Geosciences, Beijing; Dishuitan in Fangshan of Yuzhou, western Henan; Late Permian coal member 6 of Yungaishan Formation.

Pinnagigantopteris oblongus **Chen, 2006** (in Chinese and English)

2006 Chen Yao, in Yang Guanxiu and others, pp. 198, 318, pl. 71, figs. 1 — 3a; possibly pinnate compound leaf; Syntypes: HEP3250 (pl. 71, fig. 1), HEP3259 (pl. 71, fig. 2), HEP3257 (pl. 71, fig. 3); Repository: China University of Geosciences, Beijing; Pochi of Linru, Henan; Middle Permian coal member 3 of Xiaofengkou Formation.

Genus *Pinnatiramosus* Geng, 1986

1986 Geng Baoyin, pp. 665, 671.

Type species: *Pinnatiramosus qianensis* Geng, 1986

Taxonomic status: Plantae incertae sedis

Distribution and Age: Fenggang of Guizhou, China; Middle Silurian.

Pinnatiramosus qianensis **Geng, 1986**

1986 Geng Baoyin, pp. 665, 671, pls. 1 — 6; plant consists pinnate branching system; No.: 8196 — 8203; Holotype: 8196 (pl. 1); Repository: Institute of Botany, Chinese Academy of Sciences; Fenggang, Guizhou; Middle Silurian Xiushan Formation.

Genus *Plagiozamiopsis* Sze, 1943

1943 Sze H C, p. 511.

Type species: *Plagiozamiopsis podozamoides* Sze, 1943

Taxonomic status: Cycadophytes

Distribution and Age: Guangdong and Jiangxi, China; early Late Permian.

Plagiozamiopsis podozamoides Sze,1943
1943　Sze H C, p. 511, pl. 1, figs. 1 — 10; leaves; Guangdong, Jiangxi; early Late Permian *Gigantoperis* Coal Series.

Genus *Polygatites* Pan,1983 (nom. nud.)
1983　Pan Guang, p. 1520. (in Chinese)
1984　Pan Guang, p. 959. (in English)
Type species: (without specific name)
Taxonomic status: "primitive angiosperms"
Distribution and Age: western Liaoning, China; Middle Jurassic.

Polygatites sp. indet.
[Notes: Generic name was given only, but without specific name (or type species) in the original paper]
1983　*Polygatites* sp. indet., Pan Guang, p. 1520; western Liaoning (about 45°58′N, 120°21′E); Middle Jurassic Haifanggou Formation. (in Chinese)
1984　*Polygatites* sp. indet., Pan Guang, p. 959; western Liaoning (about 45°58′N, 120°21′E); Middle Jurassic Haifanggou Formation. (in English)

Genus *Polygonites* Wu S Q,1999 (non Saporta,1865) (in Chinese)
[Notes: This generic name is a late homonym (homonymum junius) of *Polygonites* Saporta, 1865, its type species is *Polygonites ulmaceus* Saporta,1865 (Saporta,1865, p. 92, pl. 3, fig. 14; winged fruit; St.-Jean-Garguier, France; Tertiary)]
1999　Wu Shunqing, p. 23.
Type species: *Polygonites polyclonus* Wu S Q,1999
Taxonomic status: Polygonaceae, Monocotyledoneae
Distribution and Age: Beipiao of western Liaoning, China; Late Jurassic.

Polygonites polyclonus Wu S Q,1999 (in Chinese)
1999　Wu Shunqing, p. 23, pl. 16, figs. 4, 4a; pl. 19, figs. 1, 1a, 3A — 4a; stem and shoot; Col. No.: AEO-170, 211, 171, 169; Reg. No.: PB18319, PB18335 — PB18337; Holotype: PB18337 (pl. 19, fig. 4); Repository: Nanjing Institute of Geology and Palaeontology, Chinese Academy of Sciences; Huangbanjigou in Shangyuan of Beipiao, western Liaoning; Late Jurassic Jianshangou Bed in lower part of Yixian Formation. (Notes: The type species was not designated in the original paper)

The other species:
Polygonites planus Wu S Q,1999 (in Chinese)
1999　Wu Shunqing, p. 24, pl. 19, fig. 2; shoot; Col. No.: AEO-122; Holotype: PB18338; Repository: Nanjing Institute of Geology and Palaeontology, Chinese Academy of

Sciences; Huangbanjigou in Shangyuan of Beipiao, western Liaoning; Late Jurassic Jianshangou Bed in lower part of Yixian Formation.

Genus *Polypetalophyton* Geng, 2003 (in English)
1995 Geng Baoyin, in Li Chengsen and Cui Jinzhong, p. 5. (nom. nud.)
2003 Geng Baoyin, in Hilton J and others, p. 795.
Type species: *Polypetalophyton wufengensis* Geng, 2003
Taxonomic status: Cladoxylopsida
Distribution and Age: Wufeng of Hubei, China; Late Devonian (Frasnian).

Polypetalophyton wufengensis Geng, 2003 (in English)
1995 Geng Baoyin, in Li Chengsen and Cui Jinzhong, pp. 29 — 35 (only figure). (nom. nud.)
2003 Geng Baoyin, in Hilton J and others, p. 795, figs. 2 — 10; plant with at least fout orders of branching; No. : CBP9796 — CBP9798; Holotype: CBP9796 (fig. 2); Repository: Institute of Botany, Chinese Academy of Sciences; Lijianwan of Wufeng, Hubei; Late Devonian (Frasnian).

Genus *Polythecophyton* Hao, Gensel et Wang, 2001 (in English)
2001 Hao Shougang, Gensel P G and Wang Deming, p. 57.
Type species: *Polythecophyton demissum* Hao, Gensel et Wang, 2001
Taxonomic status: Euphyllophytes
Distribution and Age: Wenshan of Yunnan, China; Early Devonian (Pragian).

Polythecophyton demissum Hao, Gensel et Wang, 2001 (in English)
2001 Hao Shougang, Gensel P G and Wang Deming, p. 57, pls. 1 — 3; text-figs. 1 — 3; fertile branches; Holotype: BUPB-801 (pl. 1, fig. 4); Paratypes: BUPB-802, BUPB-806, BUPB-802′ (pl. 1, fig. 2; pl. 2, fig. 1; pl. 3); Repository: Department of Geology, Peking University; Wenshan, Yunnan; Early Devonian Posongchong Formation (Pragian).

Genus *Primocycas* Zhu et Du, 1981
1981 Zhu Jianan and Du Xianming, p. 402.
Type species: *Primocycas chinensis* Zhu et Du, 1981
Taxonomic status: Cycadophytes
Distribution and Age: Taiyuan of Shanxi, China; Early Permian.

Primocycas chinensis Zhu et Du, 1981
1981 Zhu Jianan and Du Xianming, p. 402, pl. 1, figs. 1 — 6; pl. 2, figs. 1 — 4; text-figs. 1 — 4; macrosporophyll with ovules and seeds; No. : G001 — G010; Holotypes: G001 — G010; Dongshan of Taiyuan, Shanxi; Early Permian Lower Shihezi (Shihhotse) Formation.

The other species:
Primocycas muscariformis **Zhu et Du,1981**
1927   *Norinia cucullata* Halle, Halle T G, p. 218, pl. 56, figs. 8 — 12; leaves; Ch'en-chia-yu, central Shanxi; early Late Permian Upper Shihhotse Series.
1981   Zhu Jianan and Du Xianming, pp. 401, 403, 404.

Genus *Primoginkgo* **Ma et Du, 1989**
1989   Ma Jie and Du Xianming, pp. 1, 2.
Type species: *Primoginkgo dissecta* Ma et Du, 1989
Taxonomic status: Ginkgophytes?
Distribution and Age: Taiyuan of Shanxi, China; Early Permian.

Primoginkgo dissecta **Ma et Du, 1989**
1989   Ma Jie and Du Xianming, pp. 1, 2, pl. 1; pl. 2, figs. 1 — 3; leaves; Dongshan of Taiyuan, Shanxi; Early Permian Lower Shihezi (Shihhotse) Formation.

Genus *Primozamia* **Yang, 2006** (in Chinese and English)
2006   Yang Guanxiu and others, pp. 142, 285.
Type species: *Primozamia sinensis* Yang, 2006
Taxonomic status: Cycadales, Cycadophyta
Distribution and Age: Yuzhou of Henan, China; Middle Permian.

Primozamia sinensis **Yang, 2006** (in Chinese and English)
2006   Yang Guanxiu and others, pp. 142, 285, pl. 49, fig. 14; megasporophyll shield-like; Holotype: HEP0593 (pl. 49, fig. 14); Repository: China University of Geosciences, Beijing; Dafengkou of Yuzhou, western Henan; Middle Permian coal-bearing member 4 of Xiaofengkou Formation.

Genus *Prionophyllopteris* **Mo, 1980**
1980   Mo Zhuangguan, in Zhao Xiuhu and others, p. 86.
Type species: *Prionophyllopteris spiniformis* Mo, 1980
Taxonomic status: Plantae incertae sedis
Distribution and Age: Guizhou, China; early Late Permian.

Prionophyllopteris spiniformis **Mo, 1980**
1980   Mo Zhuangguan, in Zhao Xiuhu and others, p. 86, pl. 19, figs. 9, 10; leaves; Reg. No.: PB7081, PB7082; Repository: Nanjing Institute of Geology and Palaeontology, Chinese

Academy of Sciences; Panxian, Guizhou; early Late Permian lower member of Xuanwei Formation.

Genus *Procycas* Zhang et Mo, 1981
1981　Zhang Shanzhen and Mo Zhuangguan, p. 238.

Type species: *Procycas densinervioides* Zhang et Mo, 1981

Taxonomic status: Cycadophytes

Distribution and Age: Henan, China; Early Permian.

Procycas densinervioides Zhang et Mo, 1981
1981　Zhang Shanzhen and Mo Zhuangguan, p. 238, pl. 1, figs. 1 — 6; pl. 2, figs. 1 — 6; text-fig. 1; *Nilssonia* like leaves and aslender axis; Reg. No.: PB8766 — PB8772; Syntypes: PB8766 (pl. 1, fig. 1), PB8767 (pl. 1, figs. 2, 3), PB8768 (pl. 1, fig. 4), PB8769 (pl. 1, fig. 6), PB8770 (pl. 2, figs. 1, 3), PB8771 (pl. 2, fig. 2), PB8772 (pl. 2, fig. 4); Repository: Nanjing Institute of Geology and Palaeontology, Chinese Academy of Sciences; Henan; Early Permian upper part of Shanxi Formation.

Genus *Progigantonoclea* Yang, 2006 (in Chinese and English)
2006　Yang Guanxiu and others, pp. 178, 302.

Type species: *Progigantonoclea henanensis* (Yang) Yang, 2006

Taxonomic status: Gigantopteridales

Distribution and Age: Yuzhou of western Henan, China; Middle Permian.

Progigantonoclea henanensis (Yang) Yang, 2006 (in Chinese and English)
1987b　*Emplecotopteris henanensis* Yang, Yang Guanxiu, p. 183, pl. 1, figs. 1 — 4.
2006　Yang Guanxiu and others, pp. 178, 302, pl. 52, figs. 2 — 8; pl. 19, fig. 4; text-fig. 7.5; fern-like pinnate compound leaf; No.: HEP0114, HEP0116 — HEP0119; Syntypes: HEP0119 (pl. 52, fig. 2), HEP0118 (pl. 52, fig. 3), HEP0117 (pl. 52, fig. 4), HEP0114 (pl. 52, fig. 6), HEP0116 (pl. 52, fig. 8); Repository: China University of Geosciences, Beijing; Dajiancun of Yuzhou, Henan; Middle Permian Shenhou Formation.

Genus *Progigantopteris* Yang, 1987
1987b　Yang Guanxiu, pp. 189, 194.

Type species: *Progigantopteris brevireticulatus* Yang, 1987

Taxonomic status: Gigantopterides

Distribution and Age: Yuxian of Henan, China; Early Permian.

Progigantopteris brevireticulatus Yang, 1987
1987b　Yang Guanxiu, pp. 190, 194, pl. 1, figs. 10 — 12; text-fig. 10; frond leaves; Reg. No.: HEP

226 — HEP 228; Syntypes: HEP 226 — HEP 228 (pl. 1, figs. 10 — 12); Repository: China University of Geosciences, Beijing; Yuxian, Henan; Early Permian coal members 3 — 4 of Xiaofengkou Formation.

Genus *Proginkgoxylon* Zheng et Zhang, 2008 (in English)

2008　Zheng Shaolin and Zhang Wu, in Zheng Shaolin and others, p. 43.

Type species: *Proginkgoxylon benxiense* (Zheng et Zhang) Zheng et Zhang, 2008

Taxonomic status: Ginkgophytes

Distribution and Age: Liaoning and Inner Mongolia, China; Permian.

Proginkgoxylon benxiense (Zheng et Zhang) Zheng et Zhang, 2008 (in English)

2000　*Protoginkgoxylon benxiense* Zheng et Zhang, Zheng Shaolin and Zhang Wu, p. 121, pl. 1, figs. 1 — 6; pl. 2, figs. 1 — 5; ginkgophyte wood; No. : GJ6-21; Holotype: GJ6-21 (pl. 1, figs. 1 — 6; pl. 2, figs. 1 — 5); Repository: Shenyang Institute of Geology and Mineral Resources; Tianshifu of Benxi, Liaoning; Early Permian Shanxi (Shansi) Formation.

2006　*Protoginkgoxylon benxiense* Zheng et Zhang, Zheng Shaolin and Zhang Wu, in Zhang Wu and others, p. 43, pl. 3-8, figs. A — F; pl. 3-9, figs. A — E; Tianshifu of Benxi, Liaoning; Early Permian Shanxi (Shansi) Formation.

2008　Zheng Shaolin and Zhang Wu, in Zheng Shaolin and others, p. 47, pl. 3-8, figs. A — F; pl. 3-9, figs. A — E; ginkgophyte wood; Reg. No. : GJ6-21; Holotype: GJ6-21 (pl. 3-8, figs. A — F; pl. 3-9, figs. A — E); Repository: Shenyang Institute of Geology and Mineral Resources; Tianshifu of Benxi, Liaoning; Early Permian Shanxi (Shansi) Formation.

The other species:

Proginkgoxylon daqingshanense (Zheng et Zhang) Zheng et Zhang, 2008 (in English)

2000　*Protoginkgoxylon daqingshanense* Zheng et Zhang, Zheng Shaolin and Zhang Wu, p. 121, pl. 2, fig. 6; pl. 3, figs. 1 — 6; ginkgophyte wood; Reg. No. : M56-114; Holotype: M56-114 (pl. 2, fig. 6; pl. 3, figs. 1 — 6); Repository: Shenyang Institute of Geology and Mineral Resources; Shiguaizi of Daqingshan, Inner Mongolia; Early Permian "Daqingshan Formation".

2006　*Protoginkgoxylon daqingshanense* Zheng et Zhang, Zheng Shaolin and Zhang Wu, in Zhang Wu and others, p. 47, pl. 3-9, fig. F; pl. 3-10, figs. A — F; ginkgophyte wood; Shiguaizi of Daqingshan, Inner Mongolia; Early Permian "Daqingshan Formation".

2008　Zheng Shaolin and Zhang Wu, in Zheng Shaolin and others, p. 47, pl. 3-9, fig. F; pl. 3-10, figs. A — F; ginkgophyte wood; Reg. No. : M56-114; Holotype: M56-114 (pl. 3-9, fig. F; pl. 3-10, figs. A — F); Repository: Shenyang Institute of Geology and Mineral Resources; Shiguaizi of Daqingshan, Inner Mongolia; Early Permian "Daqingshan Formation".

Genus *Protoglyptostroboxylon* He,1995
1995　He Dechang,pp. 8 (in Chinese),10 (in English).

Type species:*Protoglyptostroboxylon giganteum* He,1995

Taxonomic status:Coniferopsida (fusainized wood)

Distribution and Age:Ewenki Banner of Inner Mongolia,China;Early Cretaceous.

Protoglyptostroboxylon giganteum He,1995
1995　He Dechang,pp. 8 (in Chinese),10 (in English),pl. 5,figs. 2 — 2c;pl. 6,figs. 1 — 1e,2; pl. 8, figs. 1 — 1d; fusainized woods; Reg. No. : 91363, 91370; Holotype: 91363; Repository: Xi'an Branch, Central Coal Research Institute; Yimin Mine of Ewenki Banner,Inner Mongolia;Early Cretaceous 16th seam of Yimin Formation.

The other species:

Protoglyptostroboxylon yimiense He,1995
1995　He Dechang,pp. 9 (in Chinese),11 (in English),pl. 1,fig. 3;pl. 2,fig. 5;pl. 7,figs. 1 — 1f;pl. 8, figs. 2 — 2a, 4 — 4a; pl. 9, fig. 2; fusainized woods; Reg. No. : 9114, 91403; Holotype:91403;Repository:Xi'an Branch,Central Coal Research Institute;Yimin Mine of Ewenki Banner,Inner Mongolia;Early Cretaceous 16th seam of Yimin Formation.

Genus *Protopteridophyton* Li et Hsu,1987
1987　Li Chengsen and Hsu J,p. 120.

Type species:*Protopteridophyton devonicum* Li et Hsu,1987

Taxonomic status:Primitive ferns

Distribution and Age:Hunan and Hubei,China;Middle Devonian — Late Devonian (Givetian — Frasnian).

Protopteridophyton devonicum Li et Hsu,1987
1987　Li Chengsen and Hsu J,p. 120,pls. 1 — 16;blants herbaceous;Holotype:No. 8150a (pl. 1,fig. 1), No. 8150b (pl. 2, fig. 1) (part and counterpart); Repository: Institute of Botany, Chinese Academy of Sciences; Changsha of Hunan and Hanyang of Hubei; Middle Devonian — Late Devonian (Givetian — Frasnian) Tiaomachien Formation — lower part of Luojiashan Formation.

Genus *Protosciadopityoxylon* Zhang,Zheng et Ding,1999 (in English)
1999　Zhang Wu,Zheng Shaolin and Ding Qiuhong,p. 1314.

Type species:*Protosciadopityoxylon liaoningensis* Zhang,Zheng et Ding,1999

Taxonomic status:Taxodiaceae,Coniferopsida (fossil wood)

Distribution and Age:Yixian of Liaoning,China;Early Cretaceous.

***Protosciadopityoxylon liaoningense* Zhang, Zheng et Ding, 1999** (in English)
1999 Zhang Wu, Zheng Shaolin and Ding Qiuhong, p. 1314, pls. 1—3; text-fig. 2; fossil wood; No. : Sha. 30; Holotype: Sha. 30 (pls. 1—3); Repository: Shenyang Institute of Geology and Mineral Resources; Bijiagou of Yixian, Liaoning; Early Cretaceous Shahai Formation.

Genus *Pseudopolystichum* Deng et Chen, 2001 (in Chinese and English)
2001 Deng Shenghui and Chen Fen, pp. 153, 229.
Type species: *Pseudopolystichum cretaceum* Deng et Chen, 2001
Taxonomic status: Filicopsida
Distribution and Age: Tiefa Basin of Liaoning, China; Early Cretaceous.

***Pseudopolystichum cretaceum* Deng et Chen, 2001** (in Chinese and English)
2001 Deng Shenghui and Chen Fen, pp. 153, 229, pl. 115, figs. 1—4; pl. 116, figs. 1—6; pl. 117, figs. 1—9; pl. 118, figs. 1—7; fertile pinnae; No. : TXQ-2520; Repository: Research Institute of Petroleum Exploration and Development; Tiefa Basin, Liaoning; Early Cretaceous Xiaoming'anbei Formation.

Genus *Pseudorhipidopsis* P'an, 1937
1937 P'an C H, p. 263.
Type species: *Pseudorhipidopsis brevicaulis* (Kawasaki et Kon'no) P'an, 1937
Taxonomic status: Ginkgophytes?
Distribution and Age: Henan of China and Korea; early Late Permian.

***Pseudorhipidopsis brevicaulis* (Kawasaki et Kon'no) P'an, 1937**
1932 *Rhipidopsis brevicaulis* Kawasaki et Kon'no, Kawasaki S and Kon'no E, p. 39, pl. 51, figs. 7, 8; shoots and leaves; Koto District, Korea; Permian Heian System.
1937 P'an C H, p. 265, pl. 1; pl. 2; pl. 3, figs. 4, 5; shoots and leaves; Yuxian, Henan; early Late Permian Upper Shihhotse Formation.

Genus *Pseudotaeniopteris* Sze, 1951
1951a Sze H C, p. 83.
Type species: *Pseudotaeniopteris piscatorius* Sze, 1951
Taxonomic status: Problematicum
Distribution and Age: Benxi of Liaoning, China; Early Cretaceous.

***Pseudotaeniopteris piscatorius* Sze, 1951**
1951a Sze H C, p. 83, pl. 1, figs. 1, 2; problematicum; Benxi, Liaoning; Early Cretaceous.

Genus *Pseudotaxoxylon* Prakash, Du et Tripathi, 1995
1995　Prakash U, Du Naizheng and Tripathi P P, p. 345.

Type species: *Pseudotaxoxylon chinensis* Prakash, Du et Tripathi, 1995

Taxonomic status: Taxaceae

Distribution and Age: Zibo of Shandong, China; Late Miocene.

Pseudotaxoxylon chinensis Prakash, Du et Tripathi, 1995
1995　Prakash U, Du Naizheng and Tripathi P P, p. 345, figs. 7 — 12; fossil woods; Repository: Institute of Botany, Chinese Academy of Sciences; Zibo, Shandong; Late Miocene.

Genus *Pseudotsugxylon* Yang, 1994
1994　Yang Jiaju, in Tao Junrong and others, pp. 111, 112.

Type species: *Pseudotsugxylon pingzhangensis* Yang, 1994

Taxonomic status: Pinales

Distribution and Age: Chifeng of Inner Mongolia, China; Miocene.

Pseudotsugxylon pingzhangensis Yang, 1994
1994　Yang Jiaju, in Tao Junrong and others, pp. 111, 113, pl. 1, figs. 1 — 8; fossil woods; No.: 48; Pingzhuang Coal Mine of Chifeng, Inner Mongolia; Miocene.

Genus *Pseudoullmannia* He, Liang et Shen, 1996 (in Chinese and English)
1996　He Xilin, Liang Dunshi and Shen Shuzhong, pp. 86, 168.

Type species: *Pseudoullmannia frumentarioides* He, Liang et Shen, 1996

Taxonomic status: Coniferopsida

Distribution and Age: Jiangxi, China; Late Permian.

Pseudoullmannia frumentarioides He, Liang et Shen, 1996 (in Chinese and English)
1996　He Xilin, Liang Dunshi and Shen Shuzhong, pp. 87, 168 pl. 69, figs. 1 — 5; pl. 70, fig. 2; pl. 96; leafy shoot; Reg. No.: X88287 — X88291; Syntypes: X88287 — X88289 (pl. 69, figs. 1, 4, 5); Repository: Department of Geology, China University of Mining and Technology; Leping, Fengcheng, Pinghu and Gaoan, Jiangxi; Late Permian Lower Laoshan Submember and Wangpanli Member of Leping Formation.

The other species:
Pseudoullmannia bronnioides He, Liang et Shen, 1996 (in Chinese and English)
1996　He Xilin, Liang Dunshi and Shen Shuzhong, pp. 88, 169, pl. 69, fig. 6; leafy shoot; Reg. No.: X88292; Holotype: X88292 (pl. 69, fig. 6); Repository: Department of Geology, China University of Mining and Technology; Fengcheng, Jiangxi; Late Permian Lower

Laoshan Submember of Leping Formation.

Genus *Pteridiopsis* Zheng et Zhang, 1983
1983　Zheng Shaolin and Zhang Wu, p. 381.
Type species: *Pteridiopsis didaoensis* Zheng et Zhang, 1983
Taxonomic status: Pteridiaceae, Filicopsida
Distribution and Age: Heilongjiang, China; Late Jurassic.

Pteridiopsis didaoensis Zheng et Zhang, 1983
1983　Zheng Shaolin and Zhang Wu, p. 381, pl. 1, figs. 1 — 3; text-figs. 1a — 1c; sterile pinnae and fertile pinnae; No. : HDN021 — HDN023; Holotype: HDN021 (pl. 1, figs. 1 — 1d); Didao of Jixi, Heilongjiang; Late Jurassic Didao Formation.

The other species:
Pteridiopsis tenera Zheng et Zhang, 1983
1983　Zheng Shaolin and Zhang Wu, p. 382, pl. 2, figs. 1 — 3; text-figs. 2c — 2f; sterile pinnae and fertile pinnae; No. : HDN036 — HDN038; Holotype: HDN036 (pl. 2, figs. 3 — 3c); Didao of Jixi, Heilongjiang; Late Jurassic Didao Formation.

Genus *Qinlingopteris* Wu et Wang, 2004 (in Chinese and English)
2004　Wu Xiuyuan and Wang Jun, pp. 494, 498.
Type species: *Qinlingopteris orientalis* Wu et Wang, 2004
Taxonomic status: Archaeopterides
Distribution and Age: Zhen'an of Shaanxi, China; Early Carboniferous (Visean).

Qinlingopteris orientalis Wu et Wang, 2004 (in Chinese and English)
2004　Wu Xiuyuan and Wang Jun, pp. 494, 498, pl. 2, figs. 5, 5a; fronds; Reg. No. : PB20205; Holotype: PB20205 (pl. 2, fig. 5); Repository: Nanjing Institute of Geology and Palaeontology, Chinese Academy of Sciences; Maoping of Zhen'an, Shaanxi; Early Carboniferous (Visean) Eryuhe Formation.

The other species:
Qinlingopteris sp.
2004　*Qinlingopteris* sp., Wu Xiuyuan and Wang Jun, p. 495, pl. 2, fig. 6; pinnae; Reg. No. : PB20206; Repository: Nanjing Institute of Geology and Palaeontology, Chinese Academy of Sciences; Maoping of Zhen'an, Shaanxi; Early Carboniferous (Visean) Eryuhe Formation.

Genus *Qionghaia* Zhou et Li, 1979
1979　Zhou Zhiyan and Li Baoxian, p. 454.

Type species: *Qionghaia carnosa* Zhou et Li, 1979

Taxonomic status: Incertae sedis or Bennettitales?

Distribution and Age: Qionghai of Hainan, China; Early Triassic.

Qionghaia carnosa Zhou et Li, 1979

1979   Zhou Zhiyan and Li Baoxian, p. 454, pl. 2, figs. 21, 21a; sporophylls; Reg. No.: PB7618; Repository: Nanjing Institute of Geology and Palaeontology, Chinese Academy of Sciences; Xinhua near Jiuqujiang of Qionghai, Hainan; Early Triassic Lingwen Group (Jiuqujiang Formation).

Genus *Radiatifolium* Meng, 1992

1992   Meng Fansong, pp. 705, 707.

Type species: *Radiatifolium magnusum* Meng, 1992

Taxonomic status: Ginkgophytes?

Distribution and Age: Nanzhang of Hubei, China; Late Triassic.

Radiatifolium magnusum Meng, 1992

1992   Meng Fansong, pp. 705, 707, pl. 1, figs. 1, 2; pl. 2, figs. 1, 2; leaves; Reg. No.: P86020 — P86024; Holotype: P86020 (pl. 1, fig. 1); Repository: Yichang Institute of Geology and Mineral Resources; Donggong of Nanzhang, Hubei; Late Triassic Jiuligang Formation.

Genus *Ramophyton* Wang, 2008 (in English)

2008   Wang Deming, p. 1101.

Type species: *Ramophyton givetianum* Wang, 2008

Taxonomic status: Cladoxylates, Pteridophyta

Distribution and Age: West Junggar Basin of Xinjiang, China; Middle Devonian (Givetian).

Ramophyton givetianum Wang, 2008 (in English)

2008   Wang Deming, p. 1101, figs. 2—13; plants; Holotypes: XZH25a (fig. 3a), XZH25b (fig. 3b); Paratypes: XZH21b, XZH21a, XZH10b, XZH10a, XZH09b, XZH09a (figs. 2a — 2f), XZH06a, XZH06b, XZH5a, XZH5b (figs. 5c, 5e, 5g, 5i), XZH26a, XZH26b, XZH13a, XZH18a (figs. 6b, 6d — 6f), XZH14a (fig. 7a); Repository: Department of Geology, Peking University; Hebukesaier of West Junggar Basin, Xinjiang; Middle Devonian (Givetian) Hujiersite Formation.

Genus *Ranunculophyllum* ex Tao et Zhang, 1990, emend Wu, 1993

[Notes: The generic name was originally not mentioned clearly as a new generic name (Wu Xiangwu, 1993a, b)]

1990   Tao Junrong and Zhang Chuanbo, pp. 221, 226.

1993a Wu Xiangwu, pp. 31, 232.
1993b Wu Xiangwu, pp. 508, 517.
Type species: *Ranunculophyllum pinnatisctum* Tao et Zhang, 1990
Taxonomic status: Ranunculaceae, Dicotyledoneae
Distribution and Age: Yanji of Jilin, China; Early Cretaceous.

Ranunculophyllum pinnatisctum Tao et Zhang, 1990

1990 Tao Junrong and Zhang Chuanbo, pp. 221, 226, pl. 2, fig. 4; text-fig. 3; leaf; No.: $K_1 d_{41-9}$; Repository: National Museum of Plant History of China, Institute of Botany, Chinese Academy of Sciences; Yanji, Jilin; Early Cretaceous Dalazi Formation.
1993a Wu Xiangwu, pp. 31, 232.
1993b Wu Xiangwu, pp. 508, 517.

Genus *Rastropteris* Galtier, Wang, Li et Hilton, 2001 (in English)

2001 Galtier Jean, Wang Shijun, Li Chengsen and Hilton Jason, p. 436.
Type species: *Rastropteris pingquanensis* Galtier, Wang, Li et Hilton, 2001
Taxonomic status: Filicales
Distribution and Age: Pingquan of Hebei, China; Early Permian (Sakmarian).

Rastropteris pingquanensis Galtier, Wang, Li et Hilton, 2001 (in English)

2001 Galtier Jean, Wang Shijun, Li Chengsen and Hilton Jason, p. 436, figs. 1 — 20; stem; Holotype: CBP-PQ42 (figs. 1 — 20); Repository: National Museum of Plant History of China, Institute of Botany, Chinese Academy of Sciences; Yangshuling Coal Mine near Pingquan, Hebei; Early Permian (Sakmarian) upper part of Taiyuan Formation.

Genus *Rehezamites* Wu S, 1999 (in Chinese)

1999 Wu Shunqing, p. 15.
Type species: *Rehezamites anisolobus* Wu S, 1999
Taxonomic status: Bennettitales?, Cycadopsida
Distribution and Age: Beipiao of western Liaoning, China; Late Jurassic.

Rehezamites anisolobus Wu S, 1999 (in Chinese)

1999 Wu Shunqing, p. 15, pl. 8, figs. 1, 1a; cycadophyte leaf; Col. No.: AEO-187; Reg. No.: PB18265; Repository: Nanjing Institute of Geology and Palaeontology, Chinese Academy of Sciences; Huangbanjigou near Shangyuan of Beipiao, western Liaoning; Late Jurassic Jianshangou Bed in lower part of Yixian Formation.

The other species:
Rehezamites sp.
1999 *Rehezamites* sp., Wu Shunqing, p. 15, pl. 7, figs. 1, 1a; cycadophyte leaf; Huangbanjigou near Shangyuan of Beipiao, western Liaoning; Late Jurassic Jianshangou Bed in lower

part of Yixian Formation.

Genus *Renifolium* Li H et Lan, 1982

1982   Li Hanmin and Lan Shanxian, in Li Hanmin and others, p. 375.

Type species: *Renifolium logipetiolatum* Li H et Lan, 1982

Taxonomic status: Plantae incertae sedis

Distribution and Age: Jiangsu, China; Late Permian.

Renifolium logipetiolatum Li H et Lan, 1982

1982   Li Hanmin and Lan Shanxian, in Li Hanmin and others, p. 375, pl. 157, figs. 3 — 10; sterile leaf and fertile leaf; No. : HP1594 — HP1598; Syntypes: HP1594 (pl. 157, fig. 3), HP1595 (pl. 157, fig. 4), HP1596 (pl. 157, fig. 6), HP1597 (pl. 157, fig. 7), HP1598 (pl. 157, fig. 8); Zhenjiang, Jiangsu; Late Permian Longtan Formation.

Genus *Reteophlebis* Lee et Tsao, 1976

1976   Lee P C and Tsao Chengyao, in Lee P C and others, p. 102.

Type species: *Reteophlebis simplex* Lee et Tsao, 1976

Taxonomic status: Osmundaceae, Filicopsida

Distribution and Age: Lufeng of Yunnan, China; Late Triassic.

Reteophlebis simplex Lee et Tsao, 1976

1976   Lee P C and Tsao Chengyao, in Lee P C and others, p. 102, pl. 10, figs. 3 — 8; pl. 11; pl. 12, figs. 4 — 5; text-fig. 3-2; sterile pinna and fertile pinna; Reg. No. : PB5203 — PB5214, PB5218 — PB5219; Holotype: PB5214 (pl. 11, fig. 8); Repository: Nanjing Institute of Geology and Palaeontology, Chinese Academy of Sciences; Yipinglang of Lufeng, Yunnan; Late Triassic Ganhaizi Member of Yipinglang Formation.

Genus *Reticalethopteris* Li, Shen et Wu, 1993

1993   Li Xingxue, Shen Guanglong and Wu Xiuyuan, pp. 542, 546.

Type species: *Reticalethopteris yuani* (Sze) Li, Shen et Wu, 1993

Taxonomic status: Plantae incertae sedis

Distribution and Age: Gansu, Ningxia and Inner Mongolia, China; Late Carboniferous.

Reticalethopteris yuani (Sze) Li, Shen et Wu, 1993

1933   *Palaeoweichselia yuani* Sze, Sze H C, p. 59, pl. 6, figs. 1 — 12; pl. 7, figs. 1 — 10; frond leaves; Jingtai, Gansu; Late Carboniferous Hongtuwa Formation.

1993   Li Xingxue, Shen Guanglong and Wu Xiuyuan, pp. 542, 546, pls. 1 — 4; frond leaves; Jingtai of Gansu, Zhongwei of Ningxia and Alxa Left Banner of Inner Mongolia; Late Carboniferous Hongtuwa Formation.

Genus *Rhizoma* Wu S Q, 1999 (in Chinese)

1999 Wu Shunqing, p. 24.

Type species: *Rhizoma elliptica* Wu S Q, 1999

Taxonomic status: Nymphaeaceae, Dicotyledoneae

Distribution and Age: Beipiao of western Liaoning, China; Late Jurassic.

Rhizoma elliptica Wu S Q, 1999 (in Chinese)

1999 Wu Shunqing, p. 24, pl. 16, figs. 9, 10; rhizome; Col. No.: AEO-110, AEO-197; Reg. No.: PB18322, PB18323; Repository: Nanjing Institute of Geology and Palaeontology, Chinese Academy of Sciences; Huangbanjigou near Shangyuan of Beipiao, western Liaoning; Late Jurassic Jianshangou Bed in lower part of Yixian Formation.

Genus *Rhizomopsis* Gothan et Sze, 1933

1933 Gothan W and Sze H C, p. 26.

Type species: *Rhizomopsis gemmifera* Gothan et Sze, 1933

Taxonomic status: Plantae incertae sedis

Distribution and Age: Longtan of Jiangsu, China; Carboniferous.

Rhizomopsis gemmifera Gothan et Sze, 1933

1933 Gothan W and Sze H C, p. 26, pl. 4, fig. 6; rhizome (?); Longtan, Jiangsu; Carboniferous.

Genus *Rhomboidopteris* Si, 1989

1989 Si Xingjian, pp. 51, 192.

Type species: *Rhomboidopteris yongwolensis* (Kawasaki) Si, 1989

Taxonomic status: Filices or Pteridospermopsida

Distribution and Age: Shanxi and Inner Mongolia of China and Korea; Permian.

Rhomboidopteris yongwolensis (Kawasaki) Si, 1989

1931 *Neuroptridium? yongwolensis* Kawasaki, Kawasaki S, pl. 50, figs. 129, 129a; Korea; Permian Heian System.

1934 *Neuroptridium? yongwolensis* Kawasaki, Kawasaki S, p. 143.

1989 Si Xingjian, pp. 51, 192, pl. 60, figs. 1 — 7; pl. 61, figs. 1 — 3; frond; Hequ district, Shanxi; Permian Shansi Formation; Jungar Banner, Inner Mongolia; Permian Shihhotse Group.

Genus *Riccardiopsis* Wu et Li, 1992

1992 Wu Xiangwu and Li Baoxian, pp. 268, 276.

Type species: *Riccardiopsis hsüi* Wu et Li, 1992

Taxonomic status: Hepaticae

Distribution and Age: Yuxian of Hebei, China; Middle Jurassic.

Riccardiopsis hsüi Wu et Li, 1992

1992   Wu Xiangwu and Li Baoxian, pp. 265, 275, pl. 4, figs. 5, 6; pl. 5, figs. 1 — 4A, 4a; pl. 6, figs. 4 — 6a; text-fig. 5; thallus; Col. No.: ADN41-03, ADN41-06, ADN41-07; Reg. No.: PB15472 — PB15479; Holotype: PB15475 (pl. 5, fig. 2); Repository: Nanjing Institute of Geology and Palaeontology, Chinese Academy of Sciences; Yuxian, Hebei; Middle Jurassic Qiaoerjian Formation.

Genus *Rireticopteris* Hsu et Chu, 1974

1974   Hsu J and Chu Chinan, in Hsu J and others, p. 269.

Type species: *Rireticopteris microphylla* Hsu et Chu, 1974

Taxonomic status: Osmundaceae, Filicopsida

Distribution and Age: Yongren of Yunnan and Dukou of Sichuan, China; Late Triassic.

Rireticopteris microphylla Hsu et Chu, 1974

1974   Hsu J and Chu Chinan, in Hsu J and others, p. 269, pl. 1, figs. 7 — 9; pl. 2, figs. 1 — 4; pl. 3, fig. 1; text-fig. 1; frond; No.: No. 2785, No. 2839, No. 825, No. 830; Syntype 1: No. 2785 (pl. 1, fig. 7); Syntype 2: No. 2839 (pl. 1, fig. 8); Repository: Institute of Botany, Chinese Academy of Sciences; Nalajing of Yongren, Yunnan; Late Triassic Daqiaodi Formation; Taipingchang of Dukou, Sichuan; Late Triassic bottom part of Daqiaodi Formation.

Genus *Rotafolia* Wang D M, Hao et Wang Q, 2005 (in English)

2005   Wang Deming, Hao Shougang and Wang Qi, p. 23.

Type species: *Rotafolia songziensis* (Feng) Wang D M, Hao et Wang Q, 2005

Taxonomic status: Sphenophyllales, Sphenopsida

Distribution and Age: Songzi of Hubei, China; Late Devonian (Famennian).

Rotafolia songziensis (Feng) Wang D M, Hao et Wang Q, 2005 (in English)

1984   *Boumanite songziensis* Feng, Feng Shaonan, p. 302, pl. 48, fig. 4.

1991   *Sphenophyllostachys songziensis* Feng et Ma, Feng Shaonan and Ma Jie, p. 142, pl. 1, fig. 1; pl. 2, fig. 1; text-fig. 2.

2005   Wang Deming, Hao Shougang and Wang Qi, p. 23, figs. 2 — 42; strobili; Holotype: A-076 (fig. 30; also see Feng Shaonan and Ma Jie, 1991, p. 142, pl. 1, fig. 1; pl. 2, fig. 1); Paratypes: Hu-27 (fig. 2), Hu-13 (fig. 4), Hu-20 (fig. 12), Hu-10 (fig. 14), Hu-06 (fig. 20), Hu-18 (fig. 23), Hu-B1 (fig. 26), Hu-19 (fig. 27), A-056 (fig. 32), Hu-22 (fig. 33), Hu-B2 (fig. 35), Hu-B1 (fig. 37), H-02 (figs. 43, 44); Repository: Yichang

Institute of Geology and Mineral Resources (A-076 and A-056) and Department of Geology, Peking University (other paratypes and figured specimens); Songzi, Hubei; Late Devonian (Famennian) Xiejingsi Formation.

Genus *Sabinites* Tan et Zhu, 1982
1982　Tan Lin and Zhu Jianan, p. 153.

Type species: *Sabinites neimonglica* Tan et Zhu, 1982

Taxonomic status: Cupressaceae, Coniferopsida

Distribution and Age: Guyang of Inner Mongolia, China; Early Cretaceous.

Sabinites neimonglica Tan et Zhu, 1982
1982　Tan Lin and Zhu Jianan, p. 153, pl. 39, figs. 2 — 6; leafy shoot and cone; Reg. No. : GR40, GR65, GR87, GR67, GR103; Holotype: GR87 (pl. 39, figs. 4, 4a); Paratype: GR65 (pl. 39, figs. 3, 3a); Guyang, Inner Mongolia; Early Cretaceous Guyang Formation.

The other species:
Sabinites gracilis Tan et Zhu, 1982
1982　Tan Lin and Zhu Jianan, p. 153, pl. 40, figs. 1, 2; leafy shoot and cone; Reg. No. : GR09, GR66; Holotype: GR09 (pl. 40, fig. 1); Paratype: GR66 (pl. 40, figs. 2); Guyang, Inner Mongolia; Early Cretaceous Guyang Formation.

Genus *Sagittopteris* Zhang E et Xiao, 1985 (non Zhang S et Xiao, 1987)
1985　Zhang Enpeng and Xiao Suzhen, in Xiao Suzhen and Zhang Enpeng, p. 584.

Type species: *Sagittopteris belemnopteroides* Zhang E et Xiao, 1985 (non Zhang S et Xiao, 1987)

Taxonomic status: Plantae incertae sedis

Distribution and Age: Taiyuan, Lingchuan and Qinshui of Shanxi, China; late Late Carboniferous — Early Permian.

Sagittopteris belemnopteroides Zhang E et Xiao, 1985 (non Zhang S et Xiao, 1987)
1985　Zhang Enpeng and Xiao Suzhen, in Xiao Suzhen and Zhang Enpeng, p. 584, pl. 202, figs. 1 — 4; pl. 203, figs. 1 — 2; leaves; Reg. No. : Sh361, Sh362, Sh363, Sh365, Sh366, Sh367; Holotype: Sh366 (pl. 202, fig. 1); Paratypes: Sh361 (pl. 202, fig. 3), Sh362 (pl. 203, fig. 1), Sh363 (pl. 203, fig. 3), Sh365 (pl. 202, fig. 4), Sh367 (pl. 202, fig. 2); Taiyuan, Lingchuan and Qinshui, Shanxi; late Late Carboniferous Shanxi Formation.

Genus *Sagittopteris* Zhang S et Xiao, 1987 (non Zhang E et Xiao, 1985)
[Notes: This generic name *Sagittopteris* Zhang S et Xiao, 1987 is a late homomum (homomum junius) of *Sagittopteris* Zhang E et Xiao, 1985]

1987 Zhang Shanzhen and Xiao Suzhen, pp. 181, 185.

Type species: *Sagittopteris belemnopteroides* Zhang S et Xiao, 1987 (non Zhang E et Xiao, 1985)

Taxonomic status: Plantae incertae sedis

Distribution and Age: Taiyuan and Lingchuan of Shanxi, China; Early Permian.

Sagittopteris belemnopteroides Zhang S et Xiao, 1987 (non Zhang E et Xiao, 1985)

[Notes: This specific name *Sagittopteris* Zhang S et Xiao, 1987 is a late homomum (homomum junius) of *Sagittopteris* Zhang E et Xiao, 1985]

1987 Zhang Shanzhen and Xiao Suzhen, pp. 181, 185, pl. 1, figs. 1 — 4; pl. 2, figs. 1 — 6; leaves; Reg. No.: PB11272, PB11273, PB11274, PB11275, PB11276, PB11277; Holotype: PB11273 (pl. 1, fig. 2); Repository: Nanjing Institute of Geology and Palaeontology, Chinese Academy of Sciences; Reg. No.: SH672; Repository: Regional Geological Survey Team, Geological Bureau of Shanxi; Taiyuan and Lingchuan, Shanxi; Early Permian Shanxi Formation.

Genus *Schizoneuropsis* Yabe et Shimakura, 1940

1940a Yabe H and Shimakura M, p. 177.

Type species: *Schizoneuropsis tokudae* Yabe et Shimakura, 1940

Taxonomic status: Sphenophytes

Distribution and Age: Anhui, China; Permian.

Schizoneuropsis tokudae Yabe et Shimakura, 1940

1940a Yabe H and Shimakura M, p. 177, pl. 15, figs. 1 — 4; two opposite leaves at each node; Huainan Coal Mine, Anhui; Permian.

Genus *Sciadocillus* Geng, 1992

1992a Geng Baoyin, pp. 197, 206.

Type species: *Sciadocillus cuneifidus* Geng, 1992

Taxonomic status: Marchantiales?

Distribution and Age: Jiangyou of Sichuan, China; Early Devonian.

Sciadocillus cuneifidus Geng, 1992

1992a Geng Baoyin, pp. 197, 206, pl. 7, figs. 53 — 57; thallus; No.: 8353, 8354, 8355; Holotype: 8353 [pl. 7, figs. 53, 54 (counterpart)]; Repository: Institute of Botany, Chinese Academy of Sciences; Yanmenba of Jiangyou, Sichuan; Early Devonian Pingyipu Formation.

Genus *Scoparia* Wang, 1993

1993 Wang Qingzhi, p. 223.

Type species: *Scoparia plumaria* Wang, 1993

Taxonomic status: Cycadophytes

Distribution and Age: Quyang of Hebei, China; Permian.

Scoparia plumaria Wang, 1993

1993　Wang Qingzhi, pp. 223, 225, pl. 3, figs. 3, 4; frond; Col. No. : 7DPH250; Reg. No. : 3882; Repository: Shanghai Museum of Natural History; Lingshan of Quyang, Hebei; Permian Lower Shihhotse Formation.

Genus *Semenalatum* Dilcher, Mei et Du, 1997 (in English)

1997　Dilcher D L, Mei Meitang and Du Meili, p. 248.

Type species: *Semenalatum paucum* Dilcher, Mei et Du, 1997

Taxonomic status: Plantae incertae sedis

Distribution and Age: Huaibei of Anhui, China; Early Permian.

Semenalatum paucum Dilcher, Mei et Du, 1997 (in English)

1997　Dilcher D L, Mei Meitang and Du Meili, p. 248, pl. 1; winged seed; Holotype: UF♯14983 (Specimen B) (pl. 1, figs. 4—8); Paratype: UF♯14982 (Specimen A) (pl. 1, figs. 1—3); Repository: Florida Museum of Natural History, Paleobotanical Collections; Huaibei, Anhui; Early Permian Lower Shihhotse Formation.

Genus *Setarites* Pan, 1983 (nom. nud.)

1983　Pan Guang, p. 1520. (in Chinese)

1984　Pan Guang, p. 959. (in English)

Type species: (without specific name)

Taxonomic status: "primitive angiosperms"

Distribution and Age: western Liaoning, China; Middle Jurassic.

Setarites sp. indet.

[Notes: Generic name was given only, but without specific name (or type species) in the original paper]

1983　*Setarites* sp. indet. , Pan Guang, p. 1520; western Liaoning (about 45°58′N, 120°21′E); Middle Jurassic Haifanggou Formation. (in Chinese)

1984　*Setarites* sp. indet. , Pan Guang, p. 959; western Liaoning (about 45°58′N, 120°21′E); Middle Jurassic Haifanggou Formation. (in English)

Genus *Shangyuania* Zheng, Gao et Bo, 2008 (in Chinese and English)

2008　Zheng Shaolin, Gao Jiajun and Bo Xue, pp. 329, 338.

Type species: *Shangyuania caii* Zheng, Gao et Bo, 2008

Taxonomic status: Monocots?

Distribution and Age: Beipiao of Liaoning, China; Early Cretaceous.

Shangyuania caii Zheng, Gao et Bo, 2008 (in Chinese and English)
2008 Zheng Shaolin, Gao Jiajun and Bo Xue, pp. 329, 339; text-figs. 1. 1A, 3 — 8; text-figs. 2. 1A, 4, 6; plants with flowers; Holotype: LBY2001 (text-figs. 1. 1A, 3 — 8; text-figs. 2. 1A, 4, 6); Repository: Mr. Cai Shuren, Jinzhou, Liaoning; Huangbanjigou near Shangyuan of Beipiao, Liaoning; Early Cretaceous Jianshangou Bed of Yixian Formation.

Genus *Shanxicladus* Wang Z et Wang L, 1990
1990b Wang Ziqiang and Wang Lixin, p. 308.

Type species: *Shanxicladus pastulosus* Wang Z et Wang L, 1990

Taxonomic status: Filicopsida? or Pteridospermae?

Distribution and Age: Wuxiang of Shanxi, China; Middle Triassic.

Shanxicladus pastulosus Wang Z et Wang L, 1990
1990b Wang Ziqiang and Wang Lixin, p. 308, pl. 5, figs. 1 — 2; rachis; No.: No. 8407-4; Holotype: No. 8407-4 (pl. 5, figs. 1, 2); Repository: Nanjing Institute of Geology and Palaeontology, Chinese Academy of Sciences; Sizhuang of Wuxiang, Shanxi; Middle Triassic base part of Ermaying Formation.

Genus *Shanxioxylon* Tian et Wang, 1987
1987 Tian Baolin and Wang Shijun, pp. 196, 202.

Type species: *Shanxioxylon sinense* Tian et Wang, 1987

Taxonomic status: Cordaiphytes insertae sedis

Distribution and Age: Taiyuan of Shanxi, China; Late Carboniferous.

Shanxioxylon sinense Tian et Wang, 1987
1987 Tian Baolin and Wang Shijun, pp. 196, 202, pl. 1, figs. 1 — 8; text-fig. 1; pith of stem; No.: T7-39A/1, T7-39A-R/1, T7-60B-L/5, TN-292; Holotype: T7-39A/1 (pl. 1, fig. 1); Paratype: T7-60B-L/5 (pl. 1, fig. 3); Repository: Beijing Graduate School, China University of Mining and Technology; Taiyuan, Shanxi; Late Carboniferous Taiyuan Formation.

The other species:
Shanxioxylon taiyuanense Tian et Wang, 1987
1987 Tian Baolin and Wang Shijun, pp. 198, 202, pl. 2, figs. 1 — 7; text-figs. 2 — 4; pith of stem; No.: T7-59A, T7-59A-L, T7-59-L, T7-59A, T7-59A/1, T7-59A/3; Holotype: T7-59A (pl. 2, fig. 1); Repository: Beijing Graduate School, China University of Mining and Technology; Taiyuan, Shanxi; Late Carboniferous Taiyuan Formation.

Genus *Shenea* Mathews, 1947 — 1948

1947 — 1948　Mathews G B, p. 240.

Type species: *Shenea hirschmeierii* Mathews, 1947 — 1948

Taxonomic status: Plantae incertae sedis (Filicopsida or Pteridospermopsida)

Distribution and Age: West Hill of Beijing, China; Permian (?) or Triassic (?).

Shenea hirschmeierii Mathews, 1947 — 1948

1947 — 1948　Mathews G B, p. 240, fig. 3; fertile frond; West Hill, Beijing; Permian (?) or Triassic (?) Shuantsuang Series.

Genus *Shenkuoia* Sun et Guo, 1992

1992　Sun Ge and Guo Shuangxing, in Sun Ge and others, p. 546. (in Chinese)

1993　Sun Ge and Guo Shuangxing, in Sun Ge and others, p. 254. (in English)

Type species: *Shenkuoia caloneura* Sun et Guo, 1992

Taxonomic status: Dicotyledoneae

Distribution and Age: Jixi of Heilongjiang, China; Early Cretaceous.

Shenkuoia caloneura Sun et Guo, 1992

1992　Sun Ge and Guo Shuangxing, in Sun Ge and others, p. 547, pl. 1, figs. 13, 14; pl. 2, figs. 1 — 6; leaves and cuticles; Reg. No.: PB16775, PB16777; Holotype: PB16775 (pl. 1, fig. 13); Repository: Nanjing Institute of Geology and Palaeontology, Chinese Academy of Sciences; Chengzihe of Jixi, Heilongjiang; Early Cretaceous upper part of Chengzihe Formation. (in Chinese)

1993　Sun Ge and Guo Shuangxing, in Sun Ge and others, p. 546, pl. 1, figs. 13, 14; pl. 2, figs. 1 — 6; leaves and cuticles; Reg. No.: PB16775, PB16777; Holotype: PB16775 (pl. 1, fig. 13); Repository: Nanjing Institute of Geology and Palaeontology, Chinese Academy of Sciences; Chengzihe of Jixi, Heilongjiang; Early Cretaceous upper part of Chengzihe Formation. (in English)

Genus *Shenzhouphyllum* Yang et Xie, 2006 (in Chinese and English)

2006　Yang Guanxiu and Xie Jianhua, in Yang Guanxiu and others, pp. 128, 275.

Type species: *Shenzhouphyllum undulatum* (Yang) Yang et Xie, 2006

Taxonomic status: Peltaspermaceae, Peltaspermales, Pteridospermophyta

Distribution and Age: Yuzhou, Dengfeng and Linru of western Henan, China; Late Permian.

Shenzhouphyllum undulatum (Yang) Yang et Xie, 2006 (in Chinese and English)

1987　*Psygmophyllum undulatum* Yang, Yang Guanxiu, p. 53, pl. 14, fig. 6.

2006　Yang Guanxiu and Xie Jianhua, in Yang Guanxiu and others, pp. 128, 276, pl. 32, figs.

2—6; simple leaves; Syntypes: HEP3417 (pl. 32, fig. 2), HEP3408 (pl. 32, fig. 3), HEP3436 (pl. 32, fig. 4), HEP0702 (pl. 32, fig. 5), HEP3428 (pl. 32, fig. 6); Repository: China University of Geosciences, Beijing; Dafengkou of Yuzhou, Dengcao of Dengfeng and Pochi of Linru, western Henan; Late Permian members 7 and 8 of Yungaishan Formation.

The other species:
Shenzhouphyllum rotundatum Xie, 2006 (in Chinese and English)
2006　Xie Jianhu, in Yang Guanxiu and others, pp. 129, 277, pl. 33, figs. 1—2; simple leaves; Syntypes: HEP3468 (pl. 33, fig. 1), HEP3430 (pl. 33, fig. 2); Repository: China University of Geosciences, Beijing; Pochi of Linru, western Henan; Late Permian member 8 of Yungaishan Formation.

Shenzhouphyllum spatulatum Xie et Wu, 2006 (in Chinese and English)
2006　Xie Jianhua and Wu Yuehui, in Yang Guanxiu and others, pp. 129, 278, pl. 33, figs. 3—8; pl. 32, fig. 1; simple leaves; Syntypes: HEP3149 (pl. 33, fig. 8), HEP3427 (pl. 33, fig. 5), HEP3388 (pl. 33, fig. 5); Repository: China University of Geosciences, Beijing; Dengcao of Dengfeng and Pochi of Linru, western Henan; Late Permian member 8 of Yungaishan Formation.

Genus *Shenzhouspermum* Yang, Xie et Wu, 2006 (in Chinese and English)
2006　Yang Guanxiu, Xie Jianhua and Wu Yuehui, in Yang Guanxiu and others, pp. 126, 273.
Type species: *Shenzhouspermum trichotomum* Yang, Xie et Wu, 2006
Taxonomic status: Peltaspermaceae, Peltaspermales, Pteridospermophyta
Distribution and Age: Dengfeng and Linru of western Henan, China; Late Permian.

Shenzhouspermum trichotomum Yang, Xie et Wu, 2006 (in Chinese and English)
2006　Yang Guanxiu, Xie Jianhua and Wu Yuehui, in Yang Guanxiu and others, pp. 126, 273, pl. 29, figs. 1—3; pl. 31, figs. 5—7; text-fig. 7-1; fertile reproduction; Syntypes: HEP3409, HEP3154, HEP3153 (pl. 29, figs. 1—3), HEP3425, HEP3156, HEP3135 (pl. 31, figs. 5—7); Repository: China University of Geosciences, Beijing; Dengcao of Dengfeng and Pochi of Linru, western Henan; Late Permian member 8 of Yungaishan Formation.

Genus *Shenzhoutheca* Yang et Wu, 2006 (in Chinese and English)
2006　Yang Guanxiu and Wu Yuehui, in Yang Guanxiu and others, pp. 127, 275.
Type species: *Shenzhoutheca aspergilliformis* Yang et Wu, 2006
Taxonomic status: Peltaspermaceae, Peltaspermales, Pteridospermophyta
Distribution and Age: Dengfeng and Linru of western Henan, China; Late Permian.

***Shenzhoutheca aspergilliformis* Yang et Wu, 2006** (in Chinese and English)

2006　　Yang Guanxiu and Wu Yuehui, in Yang Guanxiu and others, pp. 127, 275, pl. 29, figs. 4 — 6; pl. 31, fig. 4; pollen organ (or microsporoclad); Syntypes: HEP3129, HEP3130, HEP3152 (pl. 29, figs. 4 — 6); Repository: China University of Geosciences, Beijing; Dengcao of Dengfeng and Pochi of Linru, western Henan; Late Permian member 8 of Yungaishan Formation.

Genus *Shuangnangostachya* Gao et Thomas, 1991

1991　　Gao Zhifeng and Thomas B A, p. 198.

Type species: *Shuangnangostachya gracilis* Gao et Thomas, 1991

Taxonomic status: Sphenophytes

Distribution and Age: Taiyuan of Shanxi, China; Early Permian.

Shuangnangostachya gracilis Gao et Thomas, 1991

1991　　Gao Zhifeng and Thomas B A, p. 198, pl. 1, figs. 1 — 7; text-fig. 1; strobilus; No.: GP0108; Holotype: GP0108 (pl. 1, figs. 1 — 7); Repository: Department of Geology, Peking University; East Hill of Taiyuan, Shanxi; Early Permian Lower Shihhotse Formation.

Genus *Shuichengella* Li, 1993

1993a　　Li Zhongming, p. 53.

Type species: *Shuichengella primitiva* (Li) Li, 1993

Taxonomic status: Osmundales

Distribution and Age: Shuicheng of Guizhou, China; Late Permian.

Shuichengella primitiva (Li) Li, 1993

1983　　*Palaeosmunda primitiva* Li, Li Zhongming, p. 154, pls. 9 — 13; stems (coal balls); Syntypes: GP2. 377-3-2/4-1, GP2. 377-3-2/6-1, GP2. 377-3-2/6-10; Repository: Institute of Botany, Chinese Academy of Sciences; Wangjiazhai Coal Mine of Shuicheng, Guizhou; Late Permian Wangjiazhai Formation.

1993a　　Li Zhongming, p. 53, pls. 1 — 4; stems (coal balls); Wangjiazhai Coal Mine of Shuicheng, Guizhou; Late Permian Wangjiazhai Formation.

Genus *Siella* Yang, 2006 (in Chinese and English)

2006　　Yang Guanxiu and others, pp. 98, 250.

Type species: *Siella leptocostata* Yang, 2006

Taxonomic status: Equisetaceae, Sphenophylates

Distribution and Age: Yuzhou of western Henan, China; Late Permian.

Siella leptocostata **Yang, 2006** (in Chinese and English)

2006　Yang Guanxiu and others, pp. 98, 250, pl. 10, fig. 9; pl. 11, figs. 4 — 6, 6a; pl. 73, fig. 1; vegetative stem covered by sheath-like overlapping leaves; Syntypes: HEP0914 (pl. 10, fig. 9), HEP0915 (pl. 11, fig. 4), HEP0916 (pl. 11, fig. 5), HEP0917 (pl. 11, fig. 6); Repository: China University of Geosciences, Beijing; Dafengkou and Yungaishan of Yuzhou, western Henan; Late Permian Yungaishan Formation.

Genus *Sinocarpus* **Leng et Friis, 2003** (in English)

2003　Leng Qin and Friis E M, p. 79.

Type species: *Sinocarpus decussatus* Leng et Friis, 2003

Taxonomic status: Angiospermae incertae sedis

Distribution and Age: Chaoyang of Liaoning, China; Early Cretaceous (Barremian or Aptian).

Sinocarpus decussatus **Leng et Friis, 2003** (in English)

2003　Leng Qin and Friis E M, p. 79, figs. 2 — 22; fruits; Holotype: B0162 [fig. 2 left (B0162A part), fig. 2 right (B0162B counterpart) and figs. 11 — 22 SEM micrographs]; Repository: Institute of Vertebrate Paleontology and Paleoanthropology, Chinese Academy of Sciences; Dawangzhangzi in Lingyuan of Chaoyang, Liaoning (41°15′N, 119°15′E); Early Cretaceous (Barremian or Aptian) Dawangzhangzi Bed of Yixian Formation. [Notes: The specimen was later referred as *Hyrcantha decussata* (Leng et Friis) Dilcher, Sun, Ji et Li (David L Dilcher and others, 2007)]

Genus *Sinoctenis* **Sze, 1931**

1931　Sze H C, p. 14.

Type species: *Sinoctenis grabauiana* Sze, 1931

Taxonomic status: Cycadopsida

Distribution and Age: Pingxiang of Jiangxi, China; Early Jurassic (Lias).

Sinoctenis grabauiana **Sze, 1931**

1931　Sze H C, p. 14, pl. 2, fig. 1; pl. 4, fig. 2; cycadophyte leaf; Pingxiang, Jiangxi; Early Jurassic (Lias).

Genus *Sinodicotis* **Pan, 1983** (nom. nud.)

1983　Pan Guang, p. 1520. (in Chinese)

1984　Pan Guang, p. 958. (in English)

Type species: (without specific name)

Taxonomic status: "hemiangiosperms"

Distribution and Age: western Liaoning, China; Middle Jurassic.

Sinodicotis sp. indet.

[Notes: Generic name was given only, but without specific name (or type species) in the original paper]

1983 *Sinodicotis* sp. indet., Pan Guang, p. 1520; western Liaoning (about 45°58′N, 120°21′E); Middle Jurassic Haifanggou Formation. (in Chinese)

1984 *Sinodicotis* sp. indet., Pan Guang, p. 958; western Liaoning (about 45°58′N, 120°21′E); Middle Jurassic Haifanggou Formation. (in English)

Genus *Sinopalaeospiroxylon* Zhang, Wang, Zheng, Yang, Li, Fu et Li, 2007 (in English)

2006 Zhang Wu and others, p. 74. (nom. nud.) (in Chinese)

2007 Zhang Wu, Wang Yongdong, Zheng Shaolin, Yang Xiaoju, Li Yong, Fu Xiaoping and Li Nan, p. 266.

2008 Zhang Wu and others, p. 74. (nom. nud.) (in English)

Type species: *Sinopalaeospiroxylon baoligemiaoense* Zhang, Wang, Zheng, Yang, Li, Fu et Li, 2007

Taxonomic status: Coniferophytes

Distribution and Age: Inner Mongolia, Liaoning and Hebei, China; Late Carboniferous — Early and Middle Permian.

Sinopalaeospiroxylon baoligemiaoense Zhang, Wang, Zheng, Yang, Li, Fu et Li, 2007 (in English)

2006 Zhang Wu and others, p. 75, pl. 3-31; pl. 3-32; text-fig. 3. 2; woods; Sonid Left Banner, Inner Mongolia; Late Carboniferous Baoligemiao Formation. (nom. nud.) (in Chinese)

2007 Zhang Wu, Wang Yongdong, Zheng Shaolin, Yang Xiaoju, Li Yong, Fu Xiaoping and Li Nan, p. 266, fig. 4A-D; fig. 5A-G; fig. 6A-G; woods; Reg. No. : No. 3P2H4-2; Holotype: No. 3P2H4-2 (fig. 5A-G, fig. 6A-G); Repository: Shenyang Institute of Geology and Mineral Resources; Dalaihuduge of Sonid Left Banner, Inner Mongolia; Late Carboniferous Baoligemiao Formation. (in English)

2008 Zhang Wu and others, p. 75, pl. 3-31; pl. 3-32; text-fig. 3. 2; woods; Sonid Left Banner, Inner Mongolia; Late Carboniferous Baoligemiao Formation. (nom. nud.) (in English)

The other species:

Sinopalaeospiroxylon napiaoense Zhang et Zheng, 2006 (2008) (in Chinese and English)

2006 Zhang Wu and Zheng Shaolin, in Zhang Wu and others, p. 78, pl. 3-33; pl. 3-34; woods; No. : GJ6-22; Holotype: GJ6-22 (pl. 3-33; pl. 3-34); Repository: Shenyang Institute of Geology and Mineral Resources; Nanpiao of Jinxi, Liaoning; Early Permian Taiyuan Formation. (in Chinese)

2008 Zhang Wu and Zheng Shaolin, in Zhang Wu and others, p. 78, pl. 3-33; pl. 3-34; woods; No. : GJ6-22; Holotype: GJ6-22 (pl. 3-33; pl. 3-34); Repository: Shenyang Institute of Geology and Mineral Resources; Nanpiao of Jinxi, Liaoning; Early Permian Taiyuan Formation. (in English)

***Sinopalaeospiroxylon pingquanense* Zhang et Zheng, 2006 (2008)** (in Chinese and English)
2006　Zhang Wu and Zheng Shaolin, in Zhang Wu and others, p. 78, pl. 3-35; pl. 3-36; woods; No.: GJ6-3; Holotype: GJ6-3 (pl. 3-35; pl. 3-36); Repository: Shenyang Institute of Geology and Mineral Resources; Pingquan, Hebei; Middle Permian Shanxi Formation. (in Chinese)
2008　Zhang Wu and Zheng Shaolin, in Zhang Wu and others, p. 78, pl. 3-35; pl. 3-36; woods; No.: GJ6-3; Holotype: GJ6-3 (pl. 3-35; pl. 3-36); Repository: Shenyang Institute of Geology and Mineral Resources; Pingquan, Hebei; Middle Permian Shanxi Formation. (in English)

Genus *Sinophyllum* Sze et Lee, 1952
1952　Sze H C and Lee H H, pp. 12, 32.
Type species: *Sinophyllum suni* Sze et Lee, 1952
Taxonomic status: Ginkgophytes?
Distribution and Age: Baxian of Sichuan, China; Early Jurassic.

Sinophyllum suni Sze et Lee, 1952
1952　Sze H C and Lee H H, pp. 12, 32, pl. 5, fig. 1; pl. 6, fig. 1; text-fig. 2; leaf; Repository: Nanjing Institute of Geology and Palaeontology, Chinese Academy of Sciences; Yipinchang of Baxian, Sichuan; Early Jurassic Hsiangchi Group.

Genus *Sinozamites* Sze, 1956
1956a　Sze H C, pp. 46, 150.
Type species: *Sinozamites leeiana* Sze, 1956
Taxonomic status: Cycadopsida
Distribution and Age: Yijun of Shaanxi, China; Late Triassic.

Sinozamites leeiana Sze, 1956
1956a　Sze H C, pp. 47, 151, pl. 39, figs. 1—3; pl. 50, fig. 4; pl. 53, fig. 5; cycadophyte leaf; Reg. No.: PB2447—PB2450; Repository: Nanjing Institute of Geology and Palaeontology, Chinese Academy of Sciences; Huangcaowan near Xingshuping of Yijun, Shaanxi; Late Triassic upper part of Yenchang Formation.

Genus *Siphonospermum* Rydin et Friis, 2010 (in English)
2010　Rydin C and Friis E M, p. 5.
Type species: *Siphonospermum simplex* Rydin et Friis, 2010
Taxonomic status: Gnetales
Distribution and Age: Northeast China; Early Cretaceous.

Siphonospermum simplex **Rydin et Friis, 2010** (in English)
2010　　Rydin C and Friis E M, p. 5, figs. 1 — 3; erect stem with terminal units of reproductive structures; Holotypes: 9880A (fig. 1a), 9880B (fig. 1b); Repository: Institute of Botany, Chinese Academy of Sciences; Northeast China; Early Cretaceous Yixian Formation.

Genus *Solaranthus* **Zheng et Wang, 2010** (in English)
2010　　Zheng Shaolin and Wang Xin, p. 896.

Type species: *Solaranthus daohugouensis* Zheng et Wang, 2010

Taxonomic status: Angiosperms

Distribution and Age: Ningcheng of Inner Mongolia, China; Middle Jurassic.

Solaranthus daohugouensis **Zheng et Wang, 2010** (in English)
2010　　Zheng Shaolin and Wang Xin, p. 896, figs. 2 — 4; "inflorescence"; Reg. No. : PB21046, PB21107, B0179, B0201, No. 47 — 277, GBM3; Holotype: PB21046 (figs. 2c, f, l — r); Paratypes: B0179 (figs. 2a, d, e, i), B0201 (figs. 2b, j, k), PB21107 (figs. 2g — h), No. 47 — 277, GBM3; Repository: PB21046, PB21107 deposited in Nanjing Institute of Geology and Palaeontology, Chinese Academy of Sciences; B0179, B0201 deposited in Institute of Vertebrate Paleontology and Paleoanthropology (IVPP), Chinese Academy of Sciences; No. 47 — 277 deposited in Shandong Tianyu Museum of Natural History; GBM3 deposited in Palaeotological Museum, Shenzhen Pairy Lake Botanical Garden; Daohugou of Ningcheng, Inner Mongolia; Middle Jurassic Jiulongshan Formation.

Genus *Speirocarpites* **Yang, 1978**
1978　　Yang Xianhe, p. 479.

Type species: *Speirocarpites virginiensis* (Fontaine) Yang, 1978

Taxonomic status: Osmundaceae, Filicopsida

Distribution and Age: Dukou of Sichuan and Xiangyun of Yunnan, China and Virginia, USA; Late Triassic.

Speirocarpites virginiensis **(Fontaine) Yang, 1978**
[Notes: This species lately was referred as *Cynepteris lasiophora* Ash (Ye Meina and others, 1986)]

1883　　*Lonchopteris virginiensis* Fontaine, p. 53, pl. 28, figs. 1, 2; pl. 29, figs. 1 — 4; sterile frond; Virginia, USA; Late Triassic.

1978　　Yang Xianhe, p. 479; text-fig. 101; Virginia, USA; Late Triassic.

The other species:

Speirocarpites dukouensis **Yang, 1978**
[Notes: This species lately was referred as *Cynepteris lasiophora* Ash (Ye Meina and others, 1986)]

1978　Yang Xianhe, p. 480, pl. 164, figs. 1 — 2; sterile frond and fertile pinna; No. : Sp0044, Sp0045; Holotype: Sp0044 (pl. 164, fig. 1); Repository: Chengdu Institute of Geology and Mineral Resources; Moshahe of Dukou, Sichuan; Late Triassic Daqiaodi Formation; Xiangyun, Yunnan; Late Triassic Ganhaizi Formation.

Speirocarpites rireticopteroides Yang, 1978
[Notes: This species lately was referred as *Cynepteris lasiophora* Ash (Ye Meina and others, 1986)]

1978　Yang Xianhe, p. 480, pl. 164, fig. 3; sterile frond; No. : Sp0046; Holotype: Sp0046 (pl. 164, fig. 3); Repository: Chengdu Institute of Geology and Mineral Resources; Huijiasuo of Dukou, Sichuan; Late Triassic Daqiaodi Formation.

Speirocarpites zhonguoensis Yang, 1978
[Notes: This species lately was referred as *Cynepteris lasiophora* Ash (Ye Meina and others, 1986)]

1978　Yang Xianhe, p. 481, pl. 164, figs. 4 — 5; sterile frond and fertile pinnae; No. : Sp0047, Sp0048; Holotype: Sp0048 (pl. 164, fig. 5); Repository: Chengdu Institute of Geology and Mineral Resources; Moshahe of Dukou, Sichuan; Late Triassic Daqiaodi Formation.

Genus *Sphenobaieroanthus* Yang, 1986
1986　Yang Xianhe, p. 54.

Type species: *Sphenobaieroanthus sinensis* Yang, 1986

Taxonomic status: Ginkgopsida, Sphenobaierales Yang (1986), Sphenobaieraceae Yang (1986)

Distribution and Age: Dazu of Chongqing, China; Late Triassic.

Sphenobaieroanthus sinensis Yang, 1986
1986　Yang Xianhe, p. 54, pl. 1, figs. 1 — 2a; text-fig. 2; long shoots with leaves, short shoots and male flowers; Col. No. : H2-5; Reg. No. : SP301 (Syntype); Repository: Chengdu Institute of Geology and Mineral Resources; Ranjiawan in Xinglong of Dazu, Chongqing; Late Triassic Xujiahe (Hsuchiaho) Formation.

Genus *Sphenobaierocladus* Yang, 1986
1986　Yang Xianhe, p. 53.

Type species: *Sphenobaierocladus sinensis* Yang, 1986

Taxonomic status: Sphenobaieraceae, Sphenobaierales, Ginkgopsida

Distribution and Age: Dazu of Chongqing, China; Late Triassic.

Sphenobaierocladus sinensis Yang, 1986
1986　Yang Xianhe, p. 53, pl. 1, figs. 1 — 2a; text-fig. 2; long shoots with leaves, short shoots and male flowers; Col. No. : H2-5; Reg. No. : SP301; Repository: Chengdu Institute of Geology and Mineral Resources; Ranjiawan in Xinglong of Dazu, Chongqing; Late

Triassic Xujiahe (Hsuchiaho) Formation.

Genus *Sphenopecopteris* Zhang et Mo, 1985
1985   Zhang Shanzhen and Mo Zhuangguan, p. 173.
Type species: *Sphenopecopteris beaniata* Zhang et Mo, 1985
Taxonomic status: Pteridospermopsida
Distribution and Age: Henan, China; Late Permian.

Sphenopecopteris beaniata Zhang et Mo, 1985
1985   Zhang Shanzhen and Mo Zhuangguan, p. 173, pl. 1, figs. 1 — 5; pl. 2, fig. 1; pinna with seed; Reg. No. : PB8106 — PB8109; Holotype: PB8106 (pl. 1, fig. 1); Repository: Nanjing Institute of Geology and Palaeontology, Chinese Academy of Sciences; Henan; Late Permian Upper Shihhotse Series.

Genus *Sphinxia* Li, Hilton et Hemsley, 1997 (non Reid et Chandler, 1933) (in English)
[Notes: This generic name *Sphinxia* Li, Hilton et Hemsley, 1997 is a late homomum (homomum junius) of *Sphinxia* Reid et Chandler, 1933, its type species is *Sphinxia ovalis* Reid et Chandler (Reid E M and Chandler M E J, 1933, p. 397, pl. 20, figs. 12 — 23; fruit; Sheppey, Kent, England; London Clay, Eocene)]
1997   Li Chengsen, Hilton J and Hemsley A R, p. 139.
Type species: *Sphinxia wuhania* Li, Hilton et Hemsley, 1997
Taxonomic status: Tracheophyta incertae sedis
Distribution and Age: Wuhan of Hubei, China; Late Devonian (Frasnian).

Sphinxia wuhania Li, Hilton et Hemsley, 1997 (in English)
1997   Li Chengsen, Hilton J and Hemsley A R, p. 139, figs. 1 — 24; Morphological observations of the Seed-like Structure; No. : CBMh 101 — CBMh 147; Holotype: CBMh 105 (fig. 6); Repository: Institute of Botany, Chinese Academy of Sciences; Milianshan Quarry near Wuhan, Hubei; Late Devonian (Frasnian).

Genus *Sphinxiocarpon* Wang, Xue et Prestianni, 2007 (in English)
1997   *Sphinxia* Li, Hilton et Hemsley, Li Chengsen, Hilton J and Hemsley A R, p. 139.
2007   Wang Qi, Xue Jinzhuang and Prestianni C, p. 393.
Type species: *Sphinxiocarpon wuhania* (Li, Hilton et Hemsley) Wang, Xue et Prestianni, 2007
Taxonomic status: Tracheophyta incertae sedis
Distribution and Age: Wuhan of Hubei, China; Late Devonian (Frasnian).

Sphinxiocarpon wuhania (Li, Hilton et Hemsley) Wang, Xue et Prestianni, 2007 (in English)
1997   *Sphinxia wuhania* Li, Hilton et Hemsley, Li Chengsen, Hilton J and Hemsley A R, p.

139, figs. 1 — 24; Morphological observations of the Seed-like Structure; No. : CBMh 101 — CBMh 147; Holotype: CBMh 105 (fig. 6); Repository: Institute of Botany, Chinese Academy of Sciences; Milianshan Quarry near Wuhan, Hubei; Late Devonian (Frasnian).
2007　Wang Qi, Xue Jinzhuang and Prestianni C, p. 393.

Genus *Spinolepidodendron* Chen, 1999 (in Chinese and English)
1999　Chen Qishi, p. 17.
Type species: *Spinolepidodendron hangzhouense* Chen, 1999
Taxonomic status: Lycopods incertae sedis
Distribution and Age: Xiaoshan of Zhejiang, China; Late Devonian.

Spinolepidodendron hangzhouense Chen, 1999 (in Chinese and English)
1999　Chen Qishi, p. 17, pl. 2, figs. 1 — 7; pl. 3, figs. 1, 1a; the stems with leaf cushions; No. : M3587a, M3641, M3649c; Repository: Zhejiang Museum of Natural History; Hushan of Xiaoshan, Zhejiang; Late Devonian Xihu Formation.

Genus *Squamocarpus* Mo, 1980
1980　Mo Zhuangguan, in Zhao Xiuhu and others, p. 87.
Type species: *Squamocarpus papilioformis* Mo, 1980
Taxonomic status: Gymnospermae?
Distribution and Age: Fuyuan of Yunnan, China; Early Triassic.

Squamocarpus papilioformis Mo, 1980
1980　Mo Zhuangguan, in Zhao Xiuhu and others, p. 87, pl. 19, figs. 13, 14 (counterpart); cone-scale; Col. No. : FQ-36; Reg. No. : PB7085, PB7086; Repository: Nanjing Institute of Geology and Palaeontology, Chinese Academy of Sciences; Qingyun of Fuyuan, Yunnan; Early Triassic "Kayitou Bed".

Genus *Squarmacarpus* Wang Z et Wang L, 1986
1986　Wang Ziqiang and Wang Lixin, p. 43.
Type species: *Squarmacarpus cuneiformus* Wang Z et Wang L, 1986
Taxonomic status: Plantae incertae sedis
Distribution and Age: Liulin of Shanxi, China; Late Permian.

Squarmacarpus cuneiformus Wang Z et Wang L, 1986
1986　Wang Ziqiang and Wang Lixin, p. 43, pl. 16, figs. 1, 2 (counterpart of p. 16, fig. 1); pl. 1, fig. 13 (?); text-fig. 23; seed-scale; No. : 8402-222, 8402-223, 8309-46; Holotype: 8402-222 (pl. 16, fig. 1); Isotype: 8402-223 (pl. 16, fig. 2); Repository: Nanjing Institute of

Geology and Palaeontology, Chinese Academy of Sciences; Moshigou of Liulin, Shanxi; Late Permian middle number of Sunjiagou Formation.

Genus *Stachybryolites* Wu X W, Wu X Y et Wang, 2000 (in English)
2000   Wu Xiangwu, Wu Xiuyuan and Wang Yongdong, p. 168
Type species: *Stachybryolites zhoui* Wu X W, Wu X Y et Wang, 2000
Taxonomic status: Bryiidae
Distribution and Age: Karamay of Xinjiang, China; Early Jurassic.

Stachybryolites zhoui Wu X W, Wu X Y et Wang, 2000 (in English)
2000   Wu Xiangwu, Wu Xiuyuan and Wang Yongdong, p. 168, pl. 1, figs. 1 — 5; pl. 2, figs. 1 — 4; caulidium; Col. No.: 92-T-22; Reg. No.: PB17786 — PB17796; Syntype 1: PB17786 (pl. 1, figs. 1, 1a, 1b, 1c); Syntype 2: PB17791 (pl. 2, fig. 1); Syntype 3: PB17796 (pl. 2, fig. 4); Repository: Nanjing Institute of Geology and Palaeontology, Chinese Academy of Sciences; Tuzi'Arkneigou of Karamay, Xinjiang; Early Jurassic Badaowan Formation.

Genus *Stachyophyton* Geng, 1983
1983   Geng Baoyin, p. 574.
Type species: *Stachyophyton yunnanense* Geng, 1983
Taxonomic status: Psilophytes incertae sedis
Distribution and Age: Yunnan, China; Early Devonian.

Stachyophyton yunnanense Geng, 1983
1983   Geng Baoyin, p. 574, pl. 1, figs. 1 — 9; pl. 2, figs. 1 — 10; text-fig. 1; plants and strobili; Holotype: 8091 (pl. 1, fig. 1; text-fig. 1); Repository: National Museum of Plant History of China, Institute of Botany, Chinese Academy of Sciences; Wenshan, Yunnan; Early Devonian Posongchong Formation.

Genus *Stalagma* Zhou, 1983
1983   Zhou Zhiyan, p. 63.
Type species: *Stalagma samara* Zhou, 1983
Taxonomic status: Podocarpaceae, Coniferopsida
Distribution and Age: Hengyang of Hunan, China; Late Triassic.

Stalagma samara Zhou, 1983
1983   Zhou Zhiyan, p. 63, pl. 3, fig. 7; pls. 4 — 11; text-figs. 3 — 6, 7C, 7I, 7J; foliage leaves, fertile shoots, female cones, seeds, pollen grains and cuticles; Reg. No.: PB9586, PB9588, PB9592 — PB9605; Holotype: PB9605 (pl. 4, fig. 4; text-fig. 3B); Repository: Nanjing Institute of Geology and Palaeontology, Chinese Academy of Sciences; Shanqiao

Coal Mine of Hengyang, Hunan; Late Triassic Yangbaichong Formation.

Genus *Stephanofolium* Guo, 2000 (in English)
2000　Guo Shuangxing, p. 233.
Type species: *Stephanofolium ovatiphyllum* Guo, 2000
Taxonomic status: Menisspermaceae, Dicotyledoneae
Distribution and Age: Hunchun of Jilin, China; Late Cretaceous.

Stephanofolium ovatiphyllum Guo, 2000 (in English)
2000　Guo Shuangxing, p. 233, pl. 2, fig. 8; pl. 6, figs. 1 — 6; leaves; Reg. No. : PB18630 — PB18633; Holotype: PB18632 (pl. 6, fig. 1); Repository: Nanjing Institute of Geology and Palaeontology, Chinese Academy of Sciences; Hunchun, Jilin; Late Cretaceous Hunchun Formation.

Genus *Strigillotheca* Gu et Zhi, 1974
1974　Nanjing Institute of Geology and Palaeontology, Institute of Botany, Chinese Academy of Sciences, in *Palaeozoic Plants from China*, p. 167.
Type species: *Strigillotheca fasciculata* Gu et Zhi, 1974
Taxonomic status: Plantae incertae sedis
Distribution and Age: Panxian of Guizhou, China; early Late Permian.

Strigillotheca fasciculata Gu et Zhi, 1974
1974　Nanjing Institute of Geology and Palaeontology, Institute of Botany, Chinese Academy of Sciences, in *Palaeozoic Plants from China*, p. 167, pl. 129, figs. 5 — 7; fertile pinnae; Reg. No. : PB4995 — PB4996; Syntypes: PB4995 (pl. 129, fig. 5), PB4996 (pl. 129, fig. 6); Repository: Nanjing Institute of Geology and Palaeontology, Chinese Academy of Sciences; Panxian, Guizhou; early Late Permian Xuanwei Formation.

Genus *Suturovagina* Chow et Tsao, 1977
1977　Chow Tseyen and Tsao Chengyao, p. 167.
Type species: *Suturovagina intermedia* Chow et Tsao, 1977
Taxonomic status: Cheirolepidiaceae, Coniferopsida
Distribution and Age: Nanjing of Jiangsu, China; Early Cretaceous.

Suturovagina intermedia Chow et Tsao, 1977
1977　Chow Tseyen and Tsao Chengyao, p. 167, pl. 2, figs. 1 — 14; text-fig. 1; leafy shoots and cuticles; Reg. No. : PB6256 — PB6260; Holotype: PB6256 (pl. 2, figs. 1); Repository: Nanjing Institute of Geology and Palaeontology, Chinese Academy of Sciences; Yanziji of Nanjing, Jiangsu; Early Cretaceous Gecun Formation.

Genus *Symopteris* Hsu, 1979

1876   *Bernoullia* Heer, Heer O, p. 88

1979   Hsu J and others, p. 17.

Type species: *Symopteris helvetica* (Heer) Hsu, 1979

Taxonomic status: Marattiaceae, Filicopsida

Distribution and Age: Switzerland and China; Late Triassic.

Symopteris helvetica (Heer) Hsu, 1979

1876   *Bernoullia helvetica* Heer, Heer O, p. 88, pl. 38, figs. 1 — 6; fertile fern; Switzerland; Late Triassic.

1979   Hsu J and others, p. 17.

The other species:

Symopteris densinervis Hsu et Tuan, 1979

1979   Hsu J and Duan Shuying, in Hsu J and others, p. 18, pls. 6 — 7, fig. 4; pl. 10, figs. 4 — 6; pl. 58; pl. 59, fig. 6; frond; No. : 814, 829, 831, 839, 846, 885; Repository: Institute of Botany, Chinese Academy of Sciences; Taipingchang of Baoding, Sichuan; Late Triassic Daqing Formation.

Symopteris zeilleri (Pan) Hsu, 1979

1936   *Bernoullia zeilleri* P'an, P'an C H, p. 26, pl. 9, figs. 6, 7; pl. 11, figs. 3, 3a, 4, 4a; pl. 14, figs. 5, 6, 6a; sterile pinna and fertile pinna; Qingjian of Yanchuan, Shaanxi; Late Triassic middle part of Yenchang Formation.

1979   Hsu J and others, p. 17.

Genus *Szea* Yao et Taylor, 1988

1988   Yao Zhaoqi and Taylor T N, p. 123.

Type species: *Szea sinensis* Yao et Taylor, 1988

Taxonomic status: Geicheniaceous, Filicopsida

Distribution and Age: Nanjing and Zhenjiang of Jiangsu, China; Early Permian.

Szea sinensis Yao et Taylor, 1988

1988   Yao Zhaoqi and Taylor T N, p. 123, pls. 1 — 4; text-figs. 1 — 3; fronds, fertile pinnae, sori and sporangia; Reg. No. : PB9270, PB9271, PB9272, PB9273, PB9274, PB9275, PB9276, PB9277; Holotype: PB9270 (pl. 1, fig. 1; pl. 3, figs. 1 — 6; pl. 4, figs. 1 — 3, 7 — 9); Paratypes: PB9271 — PB9277 (pl. 1, figs. 2 — 6; pl. 2, figs. 1 — 5); Repository: Nanjing Institute of Geology and Palaeontology, Chinese Academy of Sciences; Funiushan Coal Mine between Nanjing and Zhenjiang, Jiangsu; late Early Permian lower part of Longtan Formation.

Genus *Szecladia* Yao, Liu, Rothwell et Mapes, 2000 (in English)

2000 Yao Zhaoqi, Liu Lujun, Rothwell G W and Mapes G, p. 525.

Type species: *Szecladia multinervia* Yao, Liu, Rothwell et Mapes, 2000

Taxonomic status: Coniferopsida

Distribution and Age: Anshun of Guizhou, China; Late Permian.

Szecladia multinervia Yao, Liu, Rothwell et Mapes, 2000 (in English)

2000 Yao Zhaoqi, Liu Lujun, Rothwell G W and Mapes G, p. 525, figs. 3 — 5; conifer shoots; Reg. No.: PB18129, PB18130; Holotype: PB18129 (figs. 3.1, 3.2); Paratype: PB18130 (figs. 3.7, 4.5, 4.6); Repository: Nanjing Institute of Geology and Palaeontology, Chinese Academy of Sciences; Anshun, Guizhou; late Late Permian Dalong Formation.

Genus *Szeioxylon* Wang, Jiang et Qin, 1994

1994 Wang Shijun, Jiang Yaofa and Qin Yong, pp. 194, 195.

Type species: *Szeioxylon xuzhouene* Wang, Jiang et Qin, 1994

Taxonomic status: Coniferopsida

Distribution and Age: Xuzhou of Jiangsu, China; Carboniferous.

Szeioxylon xuzhouene Wang, Jiang et Qin, 1994

1994 Wang Shijun, Jiang Yaofa and Qin Yong, p. 195, pl. 1, figs. 1 — 8; pl. 2, figs. 1 — 8; wood; No.: XT-3; Xuzhou, Jiangsu; Carboniferous Taiyuan Formation.

Genus *Tachingia* Hu, 1975

1975 Hu Yufan, in Hsu J and others, p. 75.

Type species: *Tachingia pinniformis* Hu, 1975

Taxonomic status: Gymnospermae incertae sedis or Cycadopsida?

Distribution and Age: Taipingchang in Dukou of Sichuan, China; Late Triassic.

Tachingia pinniformis Hu, 1975

1975 Hu Yufan, in Hsu J and others, p. 75, pl. 5, figs. 1 — 4; cycadophyte leaf; No.: No. 801; Repository: Institute of Botany, Chinese Academy of Sciences; Taipingchang of Dukou, Sichuan; Late Triassic base part of Daqing Formation.

Genus *Taeniocladopsis* Sze, 1956

1956a Sze H C, pp. 63, 168.

Type species: *Taeniocladopsis rhizomoides* Sze, 1956

Taxonomic status: Equisetales, Sphenopsida

Distribution and Age: Yanchang of Shaanxi, China; Late Triassic.

Taeniocladopsis rhizomoides Sze, 1956

1956a Sze H C, pp. 63, 168, pl. 54, figs. 1, 1a; pl. 55, figs. 1 — 4; root-remains (?); Reg. No.: PB2494, PB2495 — PB2499; Repository: Nanjing Institute of Geology and Palaeontology, Chinese Academy of Sciences; Zhoujiawan of Yanchang, Shaanxi; Late Triassic Yangcaogou Formation.

Genus *Taipingchangella* Yang, 1978

1978 Yang Xianhe, p. 489.

Type species: *Taipingchangella zhongguoensis* Yang, 1978

Taxonomic status: Taipingchangellaceae, Filicopsida [Notes: This family was established by Yang Xianhe (1978), including two genera *Taipingchangella* and *Goeppertella*]

Distribution and Age: Taipingchang in Dukou of Sichuan, China; Late Triassic.

Taipingchangella zhongguoensis Yang, 1978

1978 Yang Xianhe, p. 489, pl. 172, figs. 4 — 6; pl. 170, figs. 1b — 2; pl. 171, fig. 1; frond; No.: Sp0071 — Sp0073, Sp0078; Syntypes: Sp0071 — Sp0073, Sp0078; Repository: Chengdu Institute of Geology and Mineral Resources; Taipingchang of Dukou, Sichuan; Late Triassic Daqiaodi Formation.

Genus *Taiyuanitheca* Gao et Thomas, 1993

1993 Gao Zhifeng and Thomas B A, p. 82.

Type species: *Taiyuanitheca tetralinea* Gao et Thomas, 1993

Taxonomic status: Marattiaceae, Marattiales

Distribution and Age: Taiyuan of Shanxi, China; Early Permian.

Taiyuanitheca tetralinea Gao et Thomas, 1993

1993 Gao Zhifeng and Thomas B A, p. 82; text-figs. 1 — 2; Holotype: GP0112 (text-fig. 2); frond; Taiyuan, Shanxi; Early Permian Upper Shihezi (Shihhotse) Formation.

Genus *Tchiaohoella* Lee et Yeh ex Wang, 1984 (nom. nud.)

(Notes: The genus name is probably error in spelling for *chsiaohoella*)

1984 Wang Ziqiang, p. 269.

Type species: *Tchiaohoella mirabilis* Lee et Yeh, 1964 (MS) ex Wang, 1984

Taxonomic status: Cycadopsida

Distribution and Age: Jiaohe of Jilin and Pingquan of Hebei, China; Early Cretaceous.

***Tchiaohoella mirabilis* Lee et Yeh ex Wang, 1984** (nom. nud.)
1984   Wang Ziqiang, p. 269.

The other species:
***Tchiaohoella* sp.**
1984   *Tchiaohoella* sp., Wang Ziqiang, p. 270, pl. 149, fig. 7; cycadophyte leaf; Pingquan, Hebei; Early Cretaceous Jiufotang Formation.

Genus *Tenuisa* Wang, 2007 (in English)
2007   Wang Deming, p. 1342.
Type species: *Tenuisa frasniana* Wang, 2007
Taxonomic status: Euphyllophytes
Distribution and Age: Changsha of Hunan, China; Late Devonian (Frasnian).

***Tenuisa frasniana* Wang, 2007** (in English)
2007   Wang Deming, p. 1342, figs. 3—5; Holotypes: HNCS-01a (fig. 3a), HNCS-01b (fig. 3b), HNCS-02 (fig. 5a); fertile stem; Repository: Department of Geology, Peking University; Lianhua of Changsha, Hunan; Late Devonian (Frasnian) Yunligang Formation.

Genus *Tetrafolia* Chu ex Feng et al., 1977
1977   Feng Shaonan and others, p. 673.
Type species: *Tetrafolia changshaense* (Ngo) Chu ex Feng et al., 1977
Taxonomic status: Plantae incertae sedis
Distribution and Age: Changsha of Hunan, China; Early Carboniferous.

***Tetrafolia changshaense* (Ngo) Chu ex Feng et al., 1977**
1963   ?*Sphenophyllum changshaense* Ngo, Ngo C K, p. 610, fig. 1; leaves; Changsha, Hunan; Early Carboniferous.
1977   Feng Shaonan and others, p. 673, pl. 235, fig. 5; leaves; Changsha, Hunan; Early Carboniferous Ceshui Member of Datang Formation.

Genus *Tharrisia* Zhou, Wu et Zhang, 2001 (in English)
2001   Zhou Zhiyan, Wu Xiangwu and Zhang Bole, p. 99.
Type species: *Tharrisia dinosaurensis* (Harris) Zhou, Wu et Zhang, 2001
Taxonomic status: Gymnospermae incertae sedis
Distribution and Age: East Greenland and China; Early Jurassic.

***Tharrisia dinosaurensis* (Harris) Zhou, Wu et Zhang, 2001** (in English)
1932   *Stenopteris dinosaurensis* Harris, Harris T M, p. 75, pl. 8, fig. 4; text-fig. 31; fern-like

leaves and cuticles; Scoresby Sound, East Greenland; Early Jurassic (*Thaumatopteris* Zone).

1988 *Stenopteris dinosaurensis* Harris, Li Peijuan and others, p. 77, pl. 53, figs. 1 — 2a; pl. 102, figs. 3 — 5; pl. 105, figs. 1 — 2; fern-like leaves and cuticles; Dameigou of Da Qaidam, Qinghai; Early Jurassic *Ephedrites* Bed of Tianshuigou Formation.

2001 Zhou Zhiyan, Wu Xiangwu and Zhang Bole, p. 99, pl. 1, figs. 7 — 10; pl. 3, fig. 2; pl. 4, figs. 1 — 2; pl. 5, figs. 1 — 5; pl. 7, figs. 1 — 2; text-fig. 3; leaves and cuticles; East Greenland; Early Jurassic (*Thaumatopteris* Zone); Sweden (?); Early Jurassic; Dameigou of Da Qaidam, Qinghai; Early Jurassic *Ephedrites* Bed of Tianshuigou Formation; Dianerwan of Fugu, Shaanxi; Early Jurassic Fuxian Formation.

The other species:
Tharrisia lata Zhou et Zhang, 2001 (in English)
2001 Zhou Zhiyan and Zhang Bole, in Zhou Zhiyan and others, p. 103, pl. 1, figs. 1 — 6; pl. 3, figs. 1, 3 — 8; pl. 5, figs. 5 — 8; pl. 6, figs. 1 — 8; text-fig. 5; leaves and cuticles; Reg. No.: PB18124 — PB1828; Holotype: PH18124 (pl. 1, fig. 1); Paratypes: PB18125 — PB18128; Repository: Nanjing Institute of Geology and Palaeontology, Chinese Academy of Sciences; Yima (34°40′N, 111°55′E), Henan; Middle Jurassic bed 4 in lower part of Yima Formation.

Tharrisia spectabilis (Mi, Sun C, Sun Y, Cui, Ai et al.) Zhou, Wu et Zhang, 2001 (in English)
1996 *Stenopteris spectabilis* Mi, Sun C, Sun Y, Cui, Ai et al., Mi Jiarong, Sun Chunlin, Sun Yuewu, Cui Shangsen, Ai Yongliang and others, p. 101, pl. 12, figs. 1, 7 — 9; text-fig. 5; fern-like leaves and cuticles; Taiji of Beipiao, Liaoning; Early Jurassic lower member of Beipiao Formation.

2001 Zhou Zhiyan, Wu Xiangwu and Zhang Bole, p. 101, pl. 2, fig. 14; pl. 4, figs. 3 — 7; pl. 7, figs. 3 — 8; text-fig. 4; leaves and cuticles; Taiji of Beipiao, Liaoning; Early Jurassic lower member of Beipiao Formation.

Genus *Thaumatophyllum* Yang, 1978
1978 Yang Xianhe, p. 515.

Type species: *Thaumatophyllum ptilum* (Harris) Yang, 1978

Taxonomic status: Bennettiales, Cycadopsida

Distribution and Age: East Greenland and China; Early Jurassic.

Thaumatophyllum ptilum (Harris) Yang, 1978
1932 *Pterophyllum ptilum* Harris, Harris T M, p. 61, pl. 5, figs. 1 — 5, 11; text-figs. 30, 31; cycadophyte leaves; Scoresby Sound, East Greenland; Late Triassic.

1954 *Pterophyllum ptilum* Harris, Hsu J, p. 58, pl. 51, figs. 2 — 4; cycadophyte leaves; Yipinglang of Yunnan, Anyuan of Jiangxi, Shimenkou of Hunan and Sichuan; Late Triassic.

1978 Yang Xianhe, p. 515, pl. 163, fig. 14; cycadophyte leaf; Taiping of Dayi, Sichuan; Late

Triassic Hsuchiaho Formation.

Genus *Thelypterites* Tao et Xiong, 1986, emend Wu, 1993

[Notes: This name was originally used by Tao Junrong and Xiong Xianzheng (1986), but not mentioned clearly as a new genus. The genus *Thelypterites* recognized as new species and the type species was appointed by Wu Xiangwu (1993a) as *Thelypterites* sp. A, Tao et Xiong, 1986]

1986 Tao Junrong and Xiong Xianzheng, p. 122.
1993a Wu Xiangwu, pp. 41, 240.

Type species: *Thelypterites* sp. A, Tao et Xiong, 1986

Taxonomic status: Thelypteridaceae, Filicopsida

Distribution and Age: Jiayin of Heilongjiang, China; Late Cretaceous.

Thelypterites sp. A

1986 *Thelypterites* sp. A, Tao et Xiong, Tao Junrong and Xiong Xianzheng, p. 122, pl. 5, fig. 2b; fertile pinnae; No.: 52701; Repository: Institute of Botany, Chinese Academy of Sciences; Jiayin, Heilongjiang; Late Cretaceous Wuyun Formation.
1993a *Thelypterites* sp. A, Wu Xiangwu, pp. 41, 240.

Thelypterites sp. B

1986 *Thelypterites* sp. B, Tao et Xiong, Tao Junrong and Xiong Xianzheng, p. 122, pl. 6, fig. 1; fertile pinnae; No.: 52706; Repository: Institute of Botany, Chinese Academy of Sciences; Jiayin, Heilongjiang; Late Cretaceous Wuyun Formation.

Genus *Tianoxylon* Zhang et Zheng, 2006 (2008) (in Chinese and English)

2006 Zhang Wu and Zheng Shaolin, in Zhang Wu and others, p. 114. (in Chinese)
2008 Zhang Wu and Zheng Shaolin, in Zhang Wu and others, p. 114. (in English)

Type species: *Tianoxylon duanmutouense* Zhang et Zheng, 2006 (in Chinese), 2008 (in English)

Taxonomic status: Conifers incertae sedis

Distribution and Age: Chaoyang of Liaoning, China; Early Triassic.

Tianoxylon duanmutouense Zhang et Zheng, 2006 (2008) (in Chinese and English)

2006 Zhang Wu and Zheng Shaolin, in Zhang Wu and others, p. 114, pls. 4-11 — 4-13; fossil woods; Holotype: GJ6-46; Paratype: GJ6-47; Repository: Shenyang Institute of Geology and Mineral Resources; Duanmutougou near Chaoyang, Liaoning; Early Triassic Hongla Formation. (in Chinese)
2008 Zhang Wu and Zheng Shaolin, in Zhang Wu and others, p. 114, pls. 4-8 — 4-10; fossil woods; Holotype: GJ6-46; Paratype: GJ6-47; Repository: Shenyang Institute of Geology and Mineral Resources; Duanmutougou near Chaoyang, Liaoning; Early Triassic Hongla Formation. (in English)

Genus *Tianshanopteris* Wu, 1983

1983　Wu Shaozu, in Dou Yawei and others, p. 599.

Type species: *Tianshanopteris wensuensis* Wu, 1983

Taxonomic status: Sphenopterides

Distribution and Age: Wensu of Xinjiang, China; Early Permian.

Tianshanopteris wensuensis Wu, 1983

1983　Wu Shaozu, in Dou Yawei and others, p. 600, pl. 217, figs. 1 — 5; frond leaves; Col. No. : 75KH2-6-14; Reg. No. : XPB-014 — XPB-018; Syntypes: XPB-014 — XPB-018 (pl. 217, figs. 1 — 5); Wensu, Xinjiang; Early Permian Kurgan Formation.

Genus *Tianshia* Zhou et Zhang, 1998 (in English)

1998　Zhou Zhiyan and Zhang Bole, p. 173.

Type species: *Tianshia patens* Zhou et Zhang, 1998

Taxonomic status: Czekanowskiales

Distribution and Age: Yima of Henan, China; Middle Jurassic.

Tianshia patens Zhou et Zhang, 1998 (in English)

1998　Zhou Zhiyan and Zhang Bole, p. 173, pl. 2, figs. 1 — 6; pl. 4, figs. 3, 4, 11; text-fig. 3; shoots, leaves and cuticles; Reg. No. : PB17912, PB17913, PB17914; Holotype: PB17912 (pl. 2, figs. 1, 4, 5); Repository: Nanjing Institute of Geology and Palaeontology, Chinese Academy of Sciences; Yima, Henan; Middle Jurassic middle part of Yima Formation.

Genus *Tingia* Halle, 1925

1925　Halle T G, p. 5.

Type species: *Tingia carbonica* (Schenk) Halle, 1925

Taxonomic status: Noeggerathiales

Distribution and Age: Taiyuan of Shanxi, China; Permian.

Tingia carbonica (Schenk) Halle, 1925

1883　*Pterophyllum carbonicum* Schenk, Schenk A, p. 214, pl. 44, figs. 4, 5; Taiyuan, Shanxi; Permian.

1925　Halle T G, p. 5, pl. 1, figs. 1 — 4; Taiyuan, Shanxi; Permian.

Genus *Toksunopteris* Wu S Q et Zhou, ap Wu X W, 1993

1986　*Xinjiangopteris* Wu et Zhou (non Wu S Z, 1983), Wu Shunqing and Zhou Hanzhong, pp. 642,645.

1993b Wu Shunqing and Zhou Hanzhong,in Wu Xiangwu,pp. 507,521.

Type species: *Toksunopteris opposita* (Wu et Zhou) Wu S Q et Zhou,ap Wu X W,1993

Taxonomic status: Filicopsida? or Pteridospermopsida?

Distribution and Age: Turpan Depression of Xinjiang,China; Early Jurassic.

Toksunopteris opposita (Wu et Zhou) Wu S Q et Zhou,ap Wu X W,1993

1986 *Xinjiangopteris opposita* Wu et Zhou,Wu Shunqing and Zhou Hanzhong,pp. 642,645, pl. 5,figs. 1 — 8,10,10a; frond; Col. No.: K215 — K217,K219 — K223,K228,K229; Reg. No.: PB11780 — PB11786,PB11793,PB11794; Holotype: PB11785 (pl. 5,fig. 10); Repository: Nanjing Institute of Geology and Palaeontology, Chinese Academy of Sciences; Toksun of northwestern Turpan Depression, Xinjiang; Early Jurassic Badaowan Formation.

1993b Wu Shunqing and Zhou Hanzhong, in Wu Xiangwu, pp. 507, 521; Toksun of northwestern Turpan Depression, Xinjiang; Early Jurassic Badaowan Formation.

Genus *Tongchuanophyllum* Huang et Zhou,1980

1980 Huang Zhigao and Zhou Huiqin,p. 91.

Type species: *Tongchuanophyllum trigonus* Huang et Zhou,1980

Taxonomic status: Pteridospermopsida

Distribution and Age: Tongchuan and Shenmu of Shaanxi,China; Middle Triassic.

Tongchuanophyllum trigonus Huang et Zhou,1980

1980 Huang Zhigao and Zhou Huiqin,p. 91,pl. 17,fig. 2; pl. 21,figs. 2,2a; fern-like leaf; Reg. No.: OP3035,OP151; Jinsuoguan of Tongchuan and Zaojing of Shenmu,Shaanxi; Middle Triassic upper member of Tongchuan Formation.

The other species:

Tongchuanophyllum concinnum Huang et Zhou,1980

1980 Huang Zhigao and Zhou Huiqin,p. 91,pl. 16,fig. 4; pl. 18,figs. 1 — 2; fern-like leaf; Reg. No.: OP149,OP131; Jinsuoguan of Tongchuan and Zaojing of Shenmu,Shaanxi; Middle Triassic upper member of Tongchuan Formation.

Tongchuanophyllum shensiense Huang et Zhou,1980

1980 Huang Zhigao and Zhou Huiqin,p. 91,pl. 13,fig. 5; pl. 14,fig. 3; pl. 18,fig. 3; pl. 21, fig. 1; pl. 22,fig. 1; fern-like leaf; Reg. No.: OP39,OP59,OP49,OP60; Jinsuoguan of Tongchuan and Zaojing of Shenmu, Shaanxi; Middle Triassic lower member of Tongchuan Formation.

Genus *Tonglucarpus* Chen et Zhu,1994

1994 Chen Qishi and Zhu Deshou,pp. 6,8.

Type species: *Tonglucarpus spectabilis* Chen et Zhu, 1994

Taxonomic status: Plantae incertae sedis

Distribution and Age: Tonglu of Zhejiang, China; Late Permian.

Tonglucarpus spectabilis Chen et Zhu, 1994

1994　Chen Qishi and Zhu Deshou, pp. 6, 8, pl. 1, figs. 1 — 4; text-fig. 1; seeds; No.: 75, 76, 78, 79; Repository: Zhejiang Museum of Natural History; Tonglu, Zhejiang; Late Permian Longtan Formation.

Genus *Tongshania* Stockmans et Mathieu, 1957

1957　Stockmans F and Mathieu F F, p. 66.

Type species: *Tongshania dentate* Stockmans et Mathieu, 1957

Taxonomic status: Plantae incertae sedis

Distribution and Age: Kaiping of Hebei, China; Late Carboniferous.

Tongshania dentate Stockmans et Mathieu, 1957

1957　Stockmans F and Mathieu F F, p. 66, pl. 2, figs. 5 — 7a; pl. 5, figs. 4, 4a; sporangiate organ; Kaiping, Hebei; Late Carboniferous.

Genus *Torreyocladus* Li et Ye, 1980

1980　Li Xingxue and Ye Meina, p. 10.

Type species: *Torreyocladus spectabilis* Li et Ye, 1980

Taxonomic status: Coniferopsida

Distribution and Age: Jiaohe of Jilin, China; Early Cretaceous.

Torreyocladus spectabilis Li et Ye, 1980

1980　Li Xingxue and Ye Meina, p. 10, pl. 4, fig. 5; leafy shoot; Reg. No.: PB8973; Genotype: PB8973 (pl. 4, fig. 5); Repository: Nanjing Institute of Geology and Paleontology, Chinese Academy of Sciences; Shansong of Jiaohe, Jilin; Early Cretaceous Moshilazi Formation. [Notes: The specimen was later referred as *Rhipidiocladus flabellata* Prynada (Li Xingxue and others, 1986)]

Genus *Tricoemplectopteris* Asama, 1959

1959　Asama K, p. 59

Type species: *Tricoemplectopteris taiyuanensis* Asama, 1959

Taxonomic status: Gigantopterides

Distribution and Age: Taiyuan of Shanxi, China; Late Palaeozoic.

Tricoemplectopteris taiyuanensis Asama, 1959

1927　*Gigantopteris nicotianaefolia* Halle, Halle T G, p. 164, pls. 43 — 44, fig. 9.

1959　Asama K, p. 59, pl. 3, fig. 4; frond; Taiyuan, Shanxi; Late Palaeozoic (Shihhotse Series).

Genus *Tricrananthus* Wang Z Q et Wang L X, 1990
1990a　Wang Ziqiang and Wang Lixin, p. 137.
Type species: *Tricrananthus sagittatus* Wang Z Q et Wang L X, 1990
Taxonomic status: Coniferopsida
Distribution and Age: Yushe, Heshun and Puxian of Shanxi, China; Early Triassic.

Tricrananthus sagittatus Wang Z Q et Wang L X, 1990
1990a　Wang Ziqiang and Wang Lixin, p. 137, pl. 21, figs. 13 — 17; pl. 26, fig. 6; male cone scale; No.: Z16-418, Z16-422, Z16-17, Z16-426, Z16-422a, Iso19-29; Holotype: Z16-422 (pl. 21, fig. 15); Repository: Nanjing Institute of Geology and Palaeontology, Chinese Academy of Sciences; Tuncun of Yushe and Mafang of Heshun, Shanxi; Early Triassic base part of Heshanggou Formation.

The other species:
Tricrananthus lobatus Wang Z Q et Wang L X, 1990
1990a　Wang Ziqiang and Wang Lixin, p. 137, pl. 26, figs. 5, 10; male cone scale; No.: Iso15-11, 8304-3; Syntypes: Iso15-11, 8304-3 (pl. 26, figs. 5, 10); Repository: Nanjing Institute of Geology and Palaeontology, Chinese Academy of Sciences; Puxian, Shanxi; Early Triassic base part of Heshanggou Formation.

Genus *Trinerviopteris* Zhu, 1995
1995　Zhu Jianan and Zhang Xiusheng, pp. 316, 317.
Type species: *Trinerviopteris cardiophylla* (Zhu et Geng) Zhu, 1995
Taxonomic status: Gigantopterides
Distribution and Age: Jiangle of Fujian, China; late Early Permian — early Late Permian.

Trinerviopteris cardiophylla (Zhu et Geng) Zhu, 1995
1995　Zhu Jianan and Zhang Xiusheng, pp. 316, 317, pl. 1; pl. 2, figs. 1 — 7; text-figs. 1 — 4; leaf; Holotype: B. Y. Geng. B23 (pl. 2, figs. 6, 7); Paratype: B. Y. Geng. B23 (pl. 2, fig. 1); Repository: Institute of Botany, Chinese Academy of Sciences; Jiangle, Fujian; late Early Permian — early Late Permian Longtan Formation.

Genus *Triqueteria* Stockmans et Mathieu, 1957
1957　Stockmans F and Mathieu F F, p. 68.
Type species: *Triqueteria sinensis* Stockmans et Mathieu, 1957
Taxonomic status: Plantae incertae sedis
Distribution and Age: Kaiping of Hebei, China; Late Carboniferous.

Triqueteria sinensis **Stockmans et Mathieu, 1957**
1957　Stockmans F and Mathieu F F, p. 68, pl. 7, figs. 6, 6a; Kaiping, Hebei; Late Carboniferous.

Genus *Tsaia* Wang et Berry, 2001 (in English)
2001　Wang Yi and Berry C M, p. 82.
Type species: *Tsaia denticulata* Wang et Berry, 2001
Taxonomic status: Euphyllophytes
Distribution and Age: Wuding of Yunnan, China; Middle Devonian (Givetian).

Tsaia denticulata **Wang et Berry, 2001** (in English)
2001　Wang Yi and Berry C M, p. 82, pl. 1; pl. 2; text-figs. 2 — 5; herbaceous erect plants; Holotype: PB18358 (pl. 2, fig. 1; text-fig. 5a); Paratypes: PB18349 — PB18357, PB18359 — PB18370 (pl. 1; pl. 2, figs. 2 — 13; text-figs. 2 — 4, 5b — 5f); Repository: Nanjing Institute of Geology and Palaeontology, Chinese Academy of Sciences; Wuding, Yunnan; Middle Devonian Xichong Formation (Givetian).

Genus *Tsiaohoella* Lee et Yeh ex Zhang et al., 1980 (nom. nud.)
[Notes: This generic name *Tchiaohoella* is probably error in spelling for *Chiaohoella*; The taxonomic status is also referred as Adiantaceae, Filicopsida (Li Xingxue and others, 1986, p. 13)]
1980　Zhang Wu and others, p. 279.
Type species: *Tsiaohoella mirabilis* Lee et Yeh ex Zhang et al., 1980
Taxonomic status: Cycadopsida
Distribution and Age: Jiaohe of Jilin, China; Early Cretaceous.

Tsiaohoella mirabilis **Lee et Yeh ex Zhang et al., 1980** (nom. nud.)
1980　Zhang Wu and others, p. 279, pl. 177, figs. 4 — 5; pl. 179, figs. 2, 4; cycadophyte leaf; Shansong of Jiaohe, Jilin; Early Cretaceous Moshilazi Formation.

The other species:
Tsiaohoella neozamioides **Lee et Yeh ex Zhang et al., 1980** (nom. nud.)
1980　Zhang Wu and others, p. 79, pl. 179, figs. 1, 4; cycadophyte leaf; Shansong of Jiaohe, Jilin; Early Cretaceous Moshilazi Formation.

Genus *Vasovinea* Li et Taylor, 1999 (in English)
1999　Li Hongqi and Taylor D W, p. 1564.
Type species: *Vasovinea tianii* Li et Taylor, 1999

Taxonomic status: Gigantopteridales
Distribution and Age: Panxian of Guizhou, China; Late Permian.

Vasovinea tianii **Li et Taylor, 1999** (in English)
1999　Li Hongqi and Taylor D W, p. 1564, figs. 1 — 32; stems; No.: PLY02, PLY03 (figs. 5, 7 — 20, 29 — 30); Holotypes: Slides L9407-C-B2, L9407-C-B16, L9407- D-T2 (figs. 1 — 4, 6); Paratypes: Slides PLY02-C10-1-1, PLY02-E1, PLY03-01, PLY03-06, PLY03-07, PLY03-11, PLY03-34 and PLY04-B; Repository: National Museum of Plant History of China, Institute of Botany, Chinese Academy of Sciences; Yueliangtian Coal Mine of Panxian, Guizhou; Late Permian upper part of Xuanwei Formation.

Genus *Vittifoliolum* **Zhou, 1984**
1984　Zhou Zhiyan, p. 49.

Type species: *Vittifoliolum segregatum* Zhou, 1984

Taxonomic status: Ginkgopsida? or Czekanowskiales? [Notes: The author compared this genus with other genera, such as *Desmiophyllum*, *Cordaites*, *Yuccites*, *Bambusium*, *Phoenicopsis*, *Culgouweria*, *Windwardia*, *Pseudotorellia*, and considered that it belongs to Ginkgopsida (Zhou Zhiyan, 1984); Li Peijuan and others (1988) attributed it to Ginkgoales (?)]

Distribution and Age: Hunan and Qinghai, China; Early Cretaceous.

Vittifoliolum segregatum **Zhou, 1984**
1984　Zhou Zhiyan, p. 49, pl. 29, figs. 4 — 4d; pl. 30, figs. 1 — 2b; pl. 31, figs. 1 — 2a, 4; text-fig. 12; leaves and cuticles; Reg. No.: PB8938 — PB8941, PB8943; Holotype: PB8937 (pl. 30, fig. 1); Repository: Nanjing Institute of Geology and Palaeontology, Chinese Academy of Sciences; Qiyang, Lingling, Lanshan, Hengnan, Jiangyong and Yongxing, Hunan; Early Jurassic middle and lower parts of Guanyintan Formation.

The other species:
Vittifoliolum segregatum forma *costatum* **Zhou, 1984**
1984　Zhou Zhiyan, p. 50, pl. 31, figs. 3 — 3b; leaves and cuticles; Reg. No.: PB8942; Repository: Nanjing Institute of Geology and Palaeontology, Chinese Academy of Sciences; Huangyangsi of Lingling, Hunan; Early Jurassic middle part (lower part?) of Guanyintan Formation.

Vittifoliolum multinerve **Zhou, 1984**
1984　Zhou Zhiyan, p. 50, pl. 32, figs. 1, 2; leaves and cuticles; Reg. No.: PB8944, PB8945; Holotype: PB8944 (pl. 32, fig. 1); Repository: Nanjing Institute of Geology and Palaeontology, Chinese Academy of Sciences; Huangyangsi of Lingling, Hunan; Early Jurassic middle part (lower part?) of Guanyintan Formation.

Genus *Wenshania* Zhu et Kenrick, 1999 (in English)

1999   Zhu Weiqing and Kenrick P, p. 112.

Type species: *Wenshania zhichangensis* Zhu et Kenrick, 1999

Taxonomic status: Zosterophyllophytes

Distribution and Age: Yunnan, China; Early Devonian.

Wenshania zhichangensis Zhu et Kenrick, 1999 (in English)

1999   Zhu Weiqing and Kenrick P, p. 112, pl. 1, figs. 1 — 6; text-figs. 1 — 3; plant and sporangia; No.: MPB-Y 885-1, MPB-Y 885-3a, MPB-Y 885-3b, MPB-Y 885-4; Holotype: MPB-Y 885-1 (pl. 1, fig. 1; text-fig. 1); Repository: National Museum of Plant History of China, Institute of Botany, Chinese Academy of Sciences; Wenshan, Yunnan; Early Devonian Posongchong Formation.

Genus *Wutubulaka* Wang, Fu, Xu et Hao, 2007 (in English)

2007   Wang Yi, Fu Qiang, Xu Honghe and Hao Shougang, in Wang Yi and others, p. 111.

Type species: *Wutubulaka multidichotoma* Wang, Fu, Xu et Hao, 2007

Taxonomic status: Plantae incertae sedis

Distribution and Age: Junggar Basin of Xinjiang, China; Late Silurian (Late Pridoli).

Wutubulaka multidichotoma Wang, Fu, Xu et Hao, 2007 (in English)

2007   Wang Yi, Fu Qiang, Xu Honghe and Hao Shougang, p. 113, figs. 1 — 4; plant with more than two orders of naked dichotomously branching axes; Holotype: PB20301 (figs. 1A, 1C); Paratypes: PB20302, PB20304 (figs. 4A — 4B, 4C); Repository: Nanjing Institute of Geology and Palaeontology, Chinese Academy of Sciences; Junggar Basin, Xinjiang; Late Silurian (Late Pridoli) middle part of Wutubulake Formation.

Genus *Wuxia* Berry, Wang et Cai, 2003 (in English)

2003   Berry C M, Wang Yi and Cai Chongyang, p. 268.

Type species: *Wuxia bistrobilata* Berry, Wang et Cai, 2003

Taxonomic status: Pteridopsida incertae sedis

Distribution and Age: Wuxi of Jiangsu, China; Late Devonian (Famennian).

Wuxia bistrobilata Berry, Wang et Cai, 2003 (in English)

2003   Berry C M, Wang Yi and Cai Chongyang, p. 268, figs. 2 — 6; lycopodiaceous plant; Holotype: PB18870 (fig. 3b); Paratypes: PB18862b, PB18864, PB18866 — PB18869, PB18871, PB18875 — PB18877, PB18879 (figs. 2, 3a, 3c — 3f, 5, 6); Repository: Nanjing Institute of Geology and Palaeontology, Chinese Academy of Sciences; Wuxi, Jiangsu; Late Devonian (Famennian) Wutong Formation.

Genus *Wuyunanthus* Wang, Li C, Li Z et Fu, 2001 (in English)
2001　Wang Yufei, Li Chengsen, Li Zhenyu and Fu Dezhi, p. 325.

Type species: *Wuyunanthus hexapetalus* Wang, Li C, Li Z et Fu, 2001

Taxonomic position: Celastraceae, Celastrales, Rosidae

Distribution and Age: Jiayin of Heilongjiang, Northeast China; Palaeocene.

Wuyunanthus hexapetalus Wang, Li C, Li Z et Fu, 2001 (in English)
2001　Wang Yufei, Li Chengsen, Li Zhenyu and Fu Dezhi, p. 325, figs. 1—3,5; flowers; No.: wy-92-101a, wy-92-101b; Holotypes: wy-92-101a, wy-92-101b (figs. 1, 2); Repository: National Museum of Plant History of China, Institute of Botany, Chinese Academy of Sciences; Wuyun Coal Mine of Jiayin, Heilongjiang; Palaeocene Wuyun Formation.

Genus *Xiajiajienia* Sun et Zheng, 2001 (in Chinese and English)
2001　Sun Ge and Zheng Shaolin, in Sun Ge and others, pp. 77, 187.

Type species: *Xiajiajienia mirabila* Sun et Zheng, 2001

Taxonomic status: Filicopsida

Distribution and Age: Liaoyuan of Jilin and Beipiao of western Liaoning, China; Middle — Late Jurassic.

Xiajiajienia mirabila Sun et Zheng, 2001 (in Chinese and English)
2001　Sun Ge and Zheng Shaolin, in Sun Ge and others, pp. 77, 187, pl. 10, figs. 3—6; pl. 39, figs. 1—10; pl. 56, fig. 7; frond; No.: PB19025 — PB190226, PB19028 — PB19032, ZY3015; Holotype: PB19025 (pl. 10, fig. 3); Repository: Nanjing Institute of Geology and Palaeontology, Chinese Academy of Sciences; Xiajiajie of Liaoyuan, Jilin; Middle Jurassic Xiajiajie Formation; Huangbanjigou in Shangyuan of Beipiao, western Liaoning; Late Jurassic Jianshangou Formation.

Genus *Xinganphyllum* Huang, 1977
1977　Huang Benhong, p. 60.

Type species: *Xinganphyllum aequale* Huang, 1977

Taxonomic status: Plantae incertae sedis

Distribution and Age: Tieli of Heilongjiang, China; Late Permian.

Xinganphyllum aequale Huang, 1977
1977　Huang Benhong, p. 60, pl. 6, figs. 1—2; pl. 7, figs. 1—3; text-fig. 20; leaves; Reg. No.: PFH0238, PFH0240, PFH0234, PFH0236, PFH0241; Repository: Shenyang Institute of Geology and Mineral Resources; Tieli, Heilongjiang; Late Permian Sanjiaoshan Formation.

The other species:
Xinganphyllum inaequale Huang,1977
1977　Huang Benhong,p. 61,pl. 27,fig. 2;text-fig. 21;leaves;Reg. No.:PFH0235; Repository:Shenyang Institute of Geology and Mineral Resources;Tieli,Heilongjiang; Late Permian Sanjiaoshan Formation.

Xinganphyllum sp.
1977　*Xinganphyllum* sp. ,Huang Benhong,p. 62,pl. 24,fig. 1;pl. 38,fig. 6;text-fig. 23; leaves;Tieli,Heilongjiang;Late Permian Sanjiaoshan Formation.

Genus *Xingxueanthus* Wang X et Wang S,2010 (in English)
2010　Wang Xin and Wang Shijun,p. 50.
Type species:*Xingxueanthus sinensis* Wang X et Wang S,2010
Taxonomic status:Angiosperms
Distribution and Age:Jinxi of western Liaoning,China;Middle Jurassic.

Xingxueanthus sinensis Wang X et Wang S,2010 (in English)
2010　Wang Xin and Wang Shijun,p. 50,figs. 2 — 5;female "inflorecence";Holotype:No. 8703a (fig. 2:a,d;fig. 3:a — d;fig. 4:e,h,i);Paratype:No. 8703b (fig. 2:b,c,e,g,i;fig. 3:e,f;fig. 4:a — d,f,g);Repository:Department of Palaeobotany,Chinese National Herbarium,Institute of Botany,Chinese Academy of Sciences;Sanjiaochengcun of Jinxi, western Liaoning (120°21′E,40°58′N);Middle Jurassic Haifanggou Formation.

Genus *Xingxueina* Sun et Dilcher,1997 (in Chinese and English)
1995a　Sun Ge and Dilcher D L,in Li Xingxue,p. 324. (Chinese) (nom. nud.)
1995b　Sun Ge and Dilcher D L,in Li Xingxue,p. 429. (English) (nom. nud.)
1996　Sun Ge and Dilcher D L,p. 396. (nom. nud.)
1997　Sun Ge and Dilcher D L,pp. 137,141.
Type species:*Xingxueina heilongjiangensis* Sun et Dilcher,1997
Taxonomic status:Dicotyledoneae
Distribution and Age:Jixi of Heilongjiang,China;Early Cretaceous.

Xingxueina heilongjiangensis Sun et Dilcher,1997 (in Chinese and English)
1995a　Sun Ge and Dilcher D L,in Li Xingxue,p. 324;text-fig. 9-2. 8;inflorescence and leaf; Jixi,Heilongjiang;Early Cretaceous Chengzihe Formation. (in Chinese) (nom. nud.)
1995b　Sun Ge and Dilcher D L,in Li Xingxue,p. 429;text-fig. 9-2. 8;inflorescence and leaf; Jixi,Heilongjiang;Early Cretaceous Chengzihe Formation. (in English) (nom. nud.)
1996　Sun Ge and Dilcher D L,pl. 2,figs. 1 — 6;text-fig. 1E;inflorescence and leaf;Jixi, Heilongjiang;Early Cretaceous Chengzihe Formation. (nom. nud.)
1997　Sun Ge and Dilcher D L,pp. 137,141,pl. 1,figs. 1 — 7;pl. 2,figs. 1 — 6;text-fig. 2;

inflorescence and leaf; Col. No.: WR47 — WR100; Reg. No.: SC10025, SC10026; Holotype: SC10026 (pl. 5, figs. 1B, 2; text-fig. 4G); Repository: Nanjing Institute of Geology and Palaeontology, Chinese Academy of Sciences; Jixi, Heilongjiang; Early Cretaceous Chengzihe Formation.

Genus *Xingxuephyllum* Sun et Dilcher, 2002 (in English)
2002 Sun Ge and Dilcher D L, p. 103.

Type species: *Xingxuephyllum jixiense* Sun et Dilcher, 2002

Taxonomic status: Dicotyledoneae

Distribution and Age: Jixi of Heilongjiang, China; Early Cretaceous.

Xingxuephyllum jixiense Sun et Dilcher, 2002 (in English)
2002 Sun Ge and Dilcher D L, p. 103, pl. 5, figs. 1B, 2; text-fig. 4G; leaves; No.: SC10025, SC10026; Holotype: SC10026 (pl. 5, figs. 1B, 2; text-fig. 4G); Jixi, Heilongjiang; Early Cretaceous Chengzihe Formation.

Genus *Xinjiangophyton* Sun, 1983
1983 Sun Zhehua, in Dou Yawei and others, p. 581.

Type species: *Xinjiangophyton spinosum* Sun, 1983

Taxonomic status: Lycopsida incertae sedis

Distribution and Age: Xinjiang, China; Middle Devonian.

Xinjiangophyton spinosum Sun, 1983
1983 Sun Zhehua, in Dou Yawei and others, p. 581, pl. 207, figs. 1 — 3; leaves; Col. No.: 63-7G-1-3341-b; Reg. No.: XPA167 — XPA169; Holotype: XPA167 (pl. 207, fig. 1); Qitai, Xinjiang; Middle Devonian.

Genus *Xinjiangopteris* Wu S Q et Zhou, 1986 (non Wu S Z, 1983)
[Notes: This generic name is a late homonym (homonymum junius) of *Xinjiangopteris* Wu S Z (Wu Xiangwu, 1993a, 1993b)]

1986 Wu Shunqing and Zhou Hanzhong, pp. 642, 645.

Type species: *Xinjiangopteris opposita* Wu S Q et Zhou, 1986

Taxonomic status: Filicopsida or Pteridospermopsida

Distribution and Age: Turpan Depression of Xinjiang, China; Early Jurassic.

Xinjiangopteris opposita Wu S Q et Zhou, 1986
[Notes: This species lately was referred as *Toksunopteris opposita* Wu S Q et Zhou (Wu Xiangwu, 1993a)]

1986 Wu Shunqing and Zhou Hanzhong, pp. 642, 645, pl. 5, figs. 1 — 8, 10, 10a; fronds; Col.

No.: K215 — K217, K219 — K223, K228, K229; Reg. No.: PB11780 — PB11786, PB11793, PB11794; Holotype: PB11785 (pl. 5, fig. 10); Repository: Nanjing Institute of Geology and Palaeontology, Chinese Academy of Sciences; Toksun of northwestern Turpan Depression, Xinjiang; Early Jurassic Badaowan Formation.

Genus *Xinjiangopteris* Wu S Z, 1983 (non Wu S Q et Zhou, 1986)

[Notes: This generic name *Xinjiangopteris* Wu S Q et Zhou, 1986 is a late homomum (homomum junius) of *Xinjiangopteris* Wu S Z, 1983 (Wu Xiangwu, 1993a, 1993b)]

1983　Wu Shaozu, in Dou Yawei and others, p. 607.

Type species: *Xinjiangopteris toksunensis* Wu S Z, 1983

Taxonomic status: Pteridospermopsida

Distribution and Age: Hejing of Xinjiang, China; Late Permian.

Xinjiangopteris toksunensis Wu S Z, 1983

1983　Wu Shaozu, in Dou Yawei and others, p. 607, pl. 223, figs. 1 — 6; leaves; Col. No.: 73KH1-6a; Reg. No.: XPB-032 — XPB-037; Syntypes: XPB-032 — XPB-037 (pl. 223, figs. 1 — 6); Hejing, Xinjiang; Late Permian.

Genus *Xinlongia* Yang, 1978

1978　Yang Xianhe, p. 516.

Type species: *Xinlongia pterophylloides* Yang, 1978

Taxonomic status: Bennettiales, Cycadopsida

Distribution and Age: Xinlong of Sichuan, China; Late Triassic.

Xinlongia pterophylloides Yang, 1978

1978　Yang Xianhe, p. 516, pl. 182, fig. 1; text-fig. 118; cycadophyte leaf; Reg. No.: Sp0116; Holotype: Sp0116 (pl. 182, fig. 1); Repository: Chengdu Institute of Geology and Mineral Resources; Xionglong of Xinlong, Sichuan; Late Triassic Lamayia Formation.

The other species:

Xinlongia hoheneggeri (Schenk) Yang, 1978

1869　*Podozamites hoheneggeri* Schenk, Schenk A, p. 9, pl. 11, figs. 3 — 6.

1978　Yang Xianhe, p. 516, pl. 178, fig. 7; cycadophyte leaf; Xujiahe (Hsuchiaho) of Guangyuan, Sichuan; Late Triassic Hsuchiaho Formation.

Genus *Xinlongophyllum* Yang, 1978

1978　Yang Xianhe, p. 505.

Type species: *Xinlongophyllum ctenopteroides* Yang, 1978

Taxonomic status: Pteridospermopsida

Distribution and Age: Xinlong of Sichuan, China; Late Triassic.

Xinlongophyllum ctenopteroides Yang, 1978

1978 Yang Xianhe, p. 505, pl. 182, fig. 2; cycadophyte leaf; Reg. No.: Sp0117; Holotype: Sp0117 (pl. 182, fig. 2); Repository: Chengdu Institute of Geology and Mineral Resources; Xionglong of Xinlong, Sichuan; Late Triassic Lamayia Formation.

The other species:

Xinlongophyllum multilineatum Yang, 1978

1978 Yang Xianhe, p. 506, pl. 182, figs. 3 — 4; cycadophyte leaf; Reg. No.: Sp0118, Sp0119; Syntype 1: Sp0118 (pl. 182, fig. 3); Syntype 2: Sp0119 (pl. 182, fig. 4); Repository: Chengdu Institute of Geology and Mineral Resources; Xionglong of Xinlong, Sichuan; Late Triassic Lamayia Formation.

Genus *Xitunia* Xue, 2009 (in English)

2009 Xue Jinzhuang, p. 505.

Type species: *Xitunia spinitheca* Xue, 2009

Taxonomic status: Zosterophyllophytes

Distribution and Age: Yunnan, China; Early Devonian.

Xitunia spinitheca Xue, 2009 (in English)

2009 Xue Jinzhuang, p. 505, pl. 1, figs. 1 — 5; text-figs. 1a, 1b; plant with sporangia; No.: PKU-XH200a, PKU-XH200b; Holotype: PKU-XH200a (pl. 1, fig. 1); Repository: Department of Geology, Peking University; Shengfeng of Qujing, Yunnan; Early Devonian Xitun Formation.

Genus *Yangzunyia* Yang, 2006 (in Chinese and English)

2006 Yang Guanxiu and others, pp. 79, 236.

Type species: *Yangzunyia henanensis* Yang, 2006

Taxonomic status: Protolepidodendrales, Lycophyta

Distribution and Age: Yuzhou of western Henan, China; Middle Permian.

Yangzunyia henanensis Yang, 2006 (in Chinese and English)

2006 Yang Guanxiu and others, pp. 79, 236, pl. 2, figs. 1 — 4; the stems with leaf cushions; Syntype 1: HEP0364 (pl. 2, fig. 3); Syntype 2: HEP0365 (pl. 2, fig. 4); Repository: China University of Geosciences, Beijing; Dafengkou of Yuzhou, western Henan; Middle Permian Xiaofengkou Formation.

Genus *Yanjiphyllum* Zhang, 1980

1980   Zhang Zhicheng, p. 338.

Type species: *Yanjiphyllum ellipticum* Zhang, 1980

Taxonomic status: Dicotyledoneae

Distribution and Age: Yanji of Jilin, China; Early Cretaceous.

Yanjiphyllum ellipticum Zhang, 1980

1980   Zhang Zhicheng, p. 338, pl. 192, figs. 7, 7a; leaf; Reg. No. : D631; Repository: Shenyang Institute of Geology and Mineral Resources; Dalazi of Yanji, Jilin; Early Cretaceous Dalazi Formation.

Genus *Yanliaoa* Pan, 1977

1977   P'an K, p. 70.

Type species: *Yanliaoa sinensis* Pan, 1977

Taxonomic status: Taxodiaceae, Coniferopsida

Distribution and Age: Jinxi of Liaoning, China; Middle — Late Jurassic.

Yanliaoa sinensis Pan, 1977

1977   P'an K, p. 70, pl. 5; twigs with cones; Reg. No. : L0064, L0027, L0034, L0040A; Repository: The Company of Geological Exploitation of Coal Fied, Liaoning; Jinxi, Liaoning; Middle — Late Jurassic.

Genus *Yimaia* Zhou et Zhang, 1988

1988a   Zhou Zhiyan and Zhang Bole, p. 217. (in Chinese)

1988b   Zhou Zhiyan and Zhang Bole, p. 1202. (in English)

Type species: *Yimaia recurva* Zhou et Zhang, 1988

Taxonomic status: Ginkgoales

Distribution and Age: Yima of Henan, China; Middle Jurassic.

Yimaia recurva Zhou et Zhang, 1988

1988a   Zhou Zhiyan and Zhang Bole, p. 217, fig. 3; fertile twigs; Reg. No. : PB14193; Holotype: PB14193 (fig. 3); Repository: Nanjing Institute of Geology and Palaeontology, Chinese Academy of Sciences; Yima, Henan; Middle Jurassic Yima Formation. (in Chinese)

1988b   Zhou Zhiyan and Zhang Bole, p. 1202, fig. 3; fertile twigs; Reg. No. : PB14193; Holotype: PB14193 (fig. 3); Repository: Nanjing Institute of Geology and Palaeontology, Chinese Academy of Sciences; Yima, Henan; Middle Jurassic Yima Formation. (in English)

Genus *Yixianophyllum* Zheng, Li N, Li Y, Zhang et Bian, 2005 (in English)
2005 Zheng Shaolin, Li Nan, Li Yong, Zhang Wu and Bian Xiongfei, p. 585.
Type species: *Yixianophyllum jinjiagouensie* Zheng, Li N, Li Y, Zhang et Bian, 2005
Taxonomic position: Cycadales
Distribution and Age: Yixian of Liaoning, China; Late Jurassic.

Yixianophyllum jinjiagouensie Zheng, Li N, Li Y, Zhang et Bian, 2005 (in English)
2004 *Taeniopteris* sp. (gen. et sp. nov.), Wang Wuli and others, p. 232, pl. 30, figs. 2 — 5; single leaf; Jinjiagou of Yixian, Liaoning; Late Jurassic Zhuanchengzi Bed of Yixian Formation.
2005 Zheng Shaolin, Li Nan, Li Yong, Zhang Wu and Bian Xiongfei, p. 585, pls. 1 — 2, figs. 2, 3A, 3B, 4A, 5J; single leaf and cuticles; Registration Number: JJG-7 — JJG-11; Holotype: JJG-7 (pl. 1, fig. 1); Paratypes: JJG-8 — JJG-10 (pl. 1, figs. 3, 5, 6); Repository: Shenyang Institute of Geology and Mineral Resources; Jinjiagou of Yixian, Liaoning; Late Jurassic lower part of Yixian Formation.

Genus *Yuania* Sze, 1953
1953 Sze H C, pp. 13, 18.
Type species: *Yuania striata* Sze, 1953
Taxonomic status: Noeggerathiales
Distribution and Age: Linyou of Shaanxi, China; Late Permian.

Yuania striata Sze, 1953
1953 Sze H C, pp. 13, 18, pl. 1, figs. 6 — 7a; text-fig. 1; fertile pinnae; Linyou, Shaanxi; Late Permian Shihchienfeng Series.

Genus *Yuguangia* Hao, Xue, Wang et Liu, 2007 (in English)
2007 Hao Shougang, Xue Jinzhuang, Wang Qi and Liu Zhenfeng, p. 1163.
Type species: *Yuguangia ordinata* Hao, Xue, Wang et Liu, 2007
Taxonomic status: Lycopodiales
Distribution and Age: Zhanyi of Yunnan, China; Middle Devonian (Late Givetian).

Yuguangia ordinata Hao, Xue, Wang et Liu, 2007 (in English)
2007 Hao Shougang, Xue Jinzhuang, Wang Qi and Liu Zhenfeng, p. 1163, figs. 2, 5 — 7; heterosporous, ligulate lycosid; Holotype: BUP. H-y07 (fig. 5d); Paratypes: BUP. H-y01, BUH-y1-t1-6, BUH-y1-11 (figs. 5c, 5e, 5g, 6), BUP. H-y02, BUP. H-y03, BUP. H-y04, BUP. H-y05, BUP. H-y06 (figs. 2a — 2c), SEM-y. t-01, 02, SEM-y. 1-03, 04 (figs. 2k — 2n), SEM-ypo-04, 06 (figs. 7d, 7g), LM-yspo-01, 02 (figs. 7e, 7h — 7j);

Repository: Department of Geology, Peking University; Yuguang of Zhanyi, Yunnan; Middle Devonian (Late Givetian) Haikou Formation.

Genus *Yungjenophyllum* Hsu et Chen,1974
1974   Hsu J and Chen Yeh,in Hsu J and others,p. 275.

Type species: *Yungjenophyllum grandifolium* Hsu et Chen,1974

Taxonomic status: Plantae incertae sedis

Distribution and Age: Yongren of Yunnan,China; Late Triassic.

Yungjenophyllum grandifolium Hsu et Chen,1974
1974   Hsu J and Chen Yeh,in Hsu J and others,p. 275,pl. 8,figs. 1 — 3; leaf; No. : No. 2883; Repository: Institute of Botany, Chinese Academy of Sciences; Baoding of Yongren, Yunnan; Late Triassic middle part of Daqiaodi Formation.

Genus *Yunia* Hao et Beck,1991
1991   Hao Shougang and Beck C B,p. 191.

Type species: *Yunia dichotoma* Hao et Beck,1991

Taxonomic status: Psilophytes incertae sedis

Distribution and Age: Wenshan of Yunnan,China; Early Devonian.

Yunia dichotoma Hao et Beck,1991
1991   Hao Shougang and Beck C B,p. 192,pl. 1,figs. 1 — 6; pl. 2,figs. 7 — 13; pl. 3,figs. 14 — 23; pl. 4,figs. 24 — 33; text-figs. 1 — 3; spiny axes (section) and sporangia; Holotype: BUHB-1101 (pl. 1, fig. 1); Paratypes: BUHB-1102 (pl. 1, fig. 2), BUHB-1103 (pl. 1, figs. 4,5), HS-11-1 (pl. 2, fig. 7), HB4-2,4 (pl. 2, figs. 8,9), HB1-2 (pl. 2, fig. 10), HB5-4,5,6 (pl. 3,figs. 14 — 16), HB6-2,4 (pl. 3,figs. 17,18), HS11-2 (pl. 3,fig. 19); Repository: Department of Geology, Peking University; Wenshan, Yunnan; Early Devonian Posongchong Formation.

Genus *Zeillerpteris* Koidzumi,1936
1936   Koidzumi G,p. 135.

Type species: *Zeillerpteris yunnanensis* Koidzumi,1936

Taxonomic status: Pteridospermopsida, Gigantopterides

Distribution and Age: Yunnan,China; Permian — Carboniferous.

Zeillerpteris yunnanensis Koidzumi,1936
1907   *Gigantopteris nicotinaefolia* Zeiller,Zeiller R,p. 480,pl. 14,figs. 15,15a.

1936   Koidzumi G,p. 135; Sine-si-kou, Yunnan; Permian — Carboniferous.

Genus *Zhengia* Sun et Dilcher, 2002 (in English)
1996   Sun Ge and Dilcher D L, pl. 1, fig. 15; pl. 2, figs. 7 — 9. (nom. nud.)
2002   Sun Ge and Dilcher D L, p. 103.
Type species: *Zhengia chinensis* Sun et Dilcher, 2002
Taxonomic status: Dicotyledonae
Distribution and Age: Jixi of Heilongjiang, China; Early Cretaceous.

Zhengia chinensis Sun et Dilcher, 2002 (in English)
1996   Sun Ge and Dilcher D L, pl. 1, fig. 15; pl. 2, figs. 7 — 9; leaves and cuticles; Chengzihe of Jixi, Heilongjiang; Early Cretaceous upper Chengzihe Formation. (nom. nud.)
2002   Sun Ge and Dilcher D L, p. 103, pl. 4, figs. 1 — 7; leaves and cuticles; Reg. No.: JS10004, SC10023, SD01996; Holotype: SC10023 (pl. 4, figs. 1, 3 — 6); Repository: Nanjing Institute of Geology and Palaeontology, Chinese Academy of Sciences; Chengzihe of Jixi, Heilongjiang; Early Cretaceous Chengzihe Formation.

Genus *Zhenglia* Hao, Wang D, Wang Q et Xue, 2006 (in English)
2006   Hao Shougang, Wang Deming, Wang Qi and Xue Jinzhuang, p. 13.
Type species: *Zhenglia radiata* Hao, Wang D, Wang Q et Xue, 2006
Taxonomic status: Lycopsida
Distribution and Age: Wenshan of Yunnan, China; Early Devonian.

Zhenglia radiata Hao, Wang D, Wang Q et Xue, 2006 (in English)
2006   Hao Shougang, Wang Deming, Wang Qi and Xue Jinzhuang, p. 13, pl. 1, figs. 1 — 12; text-fig. 2; herbaceous lycopsid; Holotype: PKU-HW-Ly. 06 (pl. 1, fig. 7; text-fig. 2d); Paratypes: PKU-HW-Ly. 01, 02, 04, 07, 08, 09 (pl. 1, figs. 1, 4, 6, 9, 11, 12); Wenshan, Yunnan; Early Devonian Posongchong Formation.

Genus *Zhongzhoucarus* Yang, 2006 (in Chinese and English)
2006   Yang Guanxiu and others, pp. 171, 327.
Type species: *Zhongzhoucarus deltatus* Yang, 2006
Taxonomic status: Gymnospermarum
Distribution and Age: Yuzhou of western Henan, China; Late Permian.

Zhongzhoucarus deltatus Yang, 2006 (in Chinese and English)
2006   Yang Guanxiu and others, pp. 171, 327, pl. 44, fig. 9; seed; Holotype: HEP0595 (pl. 44, fig. 9); Repository: China University of Geosciences, Beijing; Dafengkou of Yuzhou, western Henan; Late Permian Yungaishan Formation.

Genus *Zhongzhoupteris* Yang, 2006 (in Chinese and English)

2006　Yang Guanxiu and others, pp. 105, 258.

Type species: *Zhongzhoupteris cathaysicus* Yang, 2006

Taxonomic status: Osmundaceae?, Filicales

Distribution and Age: Yuzhou of western Henan, China; Middle Permian.

Zhongzhoupteris cathaysicus Yang, 2006 (in Chinese and English)

2006　Yang Guanxiu and others, pp. 105, 258, pl. 17, fig. 6; pl. 18; pl. 19, figs. 1, 2; vegetative fronds; Syntypes: HEP0260 (pl. 18, fig. 1), HEP0262 (pl. 19, fig. 1), HEP0263 (pl. 19, fig. 2), HEP0261 (pl. 17, fig. 6); Repository: China University of Geosciences, Beijing; Dajiancun of Yuzhou, western Henan; Middle Permian Shenhou Formation.

Genus *Zhouia* Zheng, Gao et Bo, 2008 (in Chinese and English)

2008　Zheng Shaolin, Gao Jiajun and Bo Xue, in Zheng Shaolin and others, pp. 331, 340.

Type species: *Zhouia beipiaoensis* Zheng, Gao et Bo, 2008

Taxonomic status: Monocots?

Distribution and Age: Beipiao of Liaoning, China; Early Cretaceous.

Zhouia beipiaoensis Zheng, Gao et Bo, 2008 (in Chinese and English)

2008　Zheng Shaolin, Gao Jiajun and Bo Xue, in Zheng Shaolin and others, pp. 332, 340; text-fig. 1. 1B; text-figs. 2. 1B, 3, 5; seed; Holotype: LBY2001 (text-fig. 1. 1B; text-figs. 2. 1B, 3, 5); Repository: Mr. Cai Shuren, Jinzhou, Liaoning; Huangbanjigou near Shangyuan of Beipiao, Liaoning; Early Cretaceous Jianshangou Bed of Yixian Formation.

Genus *Zhutheca* Liu, Li et Hilton, 2000 (in English)

2000　Liu Zhaohua, Li Chengsen and Hilton J, p. 150.

Type species: *Zhutheca densata* (Gu et Zhi) Liu, Li et Hilton, 2000

Taxonomic status: Marattaceae, Marattiales

Distribution and Age: Yunnan, Jiangsu, Guizhou and Xizang, China; Late Permian.

Zhutheca densata (Gu et Zhi) Liu, Li et Hilton, 2000 (in English)

1974　*Fascipteris (Ptychocarpus) densata* Gu et Zhi, Nanjing Institute of Geology and Palaeontology, Institute of Botany, Chinese Academy of Sciences, in *Palaeozoic Plants from China*, p. 100, pl. 69, figs. 8 — 14; text-figs. 85 — 86; fertile pinnules; Reg. No.: PB686, PB688, PB690; Syntypes: PB686 (pl. 69, fig. 8), PB688 (pl. 69, fig. 10), PB690 (pl. 69, fig. 12); Repository: Nanjing Institute of Geology and Palaeontology, Chinese Academy of Sciences; Jiangning, Jiangsu; Late Permian Longtan Formation; Panxian, Guizhou; Late Permian Xuanwei Formation; Qamdo, Xizang; Late Permian.

2000　Liu Zhaohua, Li Chengsen and Hilton J, p. 150, pl. 1, figs. 1 — 5; pl. 2, figs. 1 — 4; pl. 3, figs. 1 — 6; text-fig. 1; fertile pinnules; Holotypes: PB686, PB688, PB690; Repository: Nanjing Institute of Geology and Palaeontology, Chinese Academy of Sciences; No. : 9782, 9783, 9784; Paratypes: 9782, 9784 (pl. 1, figs. 1, 2, 3); Repository: Institute of Botany, Chinese Academy of Sciences; Xuanwei, Yunnan; Late Permian Xuanwei Formation.

APPENDIXES

Appendix 1　Index of Generic Names

[Arranged alphabetically, generic names and the page numbers (in English part / in Chinese part)]

A

Abropteris 华脉蕨属 ··· 175/1
Abrotopteris 华丽羊齿属 ·· 175/1
Acanthocladus 剌枝属 ··· 176/2
Acanthopteris 剌蕨属 ·· 176/2
Acerites 似槭树属 ·· 176/2
Aconititis 似乌头属 ·· 177/2
Acthephyllum 奇叶属 ·· 177/3
Actinophyllus Xiao,1985（non *Actinophyllum* Phillips,1848）辐射叶属 ············ 177/3
Aculeovinea 剌羊齿属 ··· 178/3
Adoketophyllum 奇异叶属 ·· 178/4
Aetheopteris 奇羊齿属 ··· 179/4
Aipteridium 准爱河羊齿属 ·· 179/4
Alloephedra 异麻黄属 ··· 179/5
Allophyton 奇异木属 ·· 180/5
Amdrupiopsis 拟安杜鲁普蕨属 ·· 180/5
Amentostrobus 花穗杉果属 ·· 180/6
Amphiephedra 疑麻黄属 ·· 181/6
Amplectosporangium 抱囊蕨属 ·· 181/6
Angustiphyllum 窄叶属 ··· 181/6
Annularites 似轮叶属 ··· 181 7
Archaefructus 古果属 ··· 182/7
Archimagnolia 始木兰属 ·· 182/8
Archimalus 始苹果属 ·· 183/8
Areolatophyllum 华网蕨属 ·· 183/8
Asiatifolium 亚洲叶属 ··· 183/8
Asiopteris 亚洲羊齿属 ··· 184/9
Astrocupulites 星壳斗属 ·· 184/9

B

Baiguophyllum 白果叶属 ·· 184/9

Batenburgia 荆棘果属	185/10
Beipiaoa 北票果属	185/10
Bennetdicotis 本内缘蕨属	186/11
Benxipteris 本溪羊齿属	186/11
Bicoemplectopteris 双网羊齿属	187/12
Bifariusotheca 双列囊蕨属	187/12
Bilobphyllum 两瓣叶属	187/12
Boseoxylon 鲍斯木属	12/188
Botrychites 似阴地蕨属	188/13
Bracteophyton 苞片蕨属	188/13

C

Callianthus 丽花属	189/13
Casuarinites 似木麻黄属	189/14
Catenalis 掌裂蕨属	189/14
Cathaiopteridium 中国羊齿属	190/14
Cathayanthus 华夏穗属	190/15
Cathaysiocycas 华夏苏铁属	191/15
Cathaysiodendron 华夏木属	191/15
Cathaysiophyllum 华夏叶属	192/16
Cathaysiopteridium 中华羊齿属	192/16
Cathaysiopteris 华夏羊齿属	192/17
Celathega 隐囊蕨属	193/17
Cervicornus 鹿角蕨属	193/17
Chamaedendron 纤木属	193/18
Chaneya 钱耐果属	194/18
Changwuia 长武蕨属	194/18
Changyanophyton 长阳木属	194/19
Chaoyangia 朝阳序属	195/19
Chengzihella 城子河叶属	195/19
Chiaohoella 小蛟河蕨属	196/20
Chilinia 吉林羽叶属	196/20
Ciliatopteris 细毛蕨属	197/20
Cladophlebidium 准枝脉蕨属	197/21
Cladotaeniopteris 枝带羊齿属	197/21
Clematites 似铁线莲叶属	197/21
Coenosophyton 普通蕨属	198/22
Cohaerensitheca 粘合囊蕨属	198/22
Conchophyllum 贝叶属	198/22
Cryptonoclea 隐羊齿属	199/23
Cycadeoidispermum 拟苏铁籽属	199/23
Cycadicotis 苏铁缘蕨属	199/23
Cycadolepophyllum 苏铁鳞叶属	200/24
Cycadostrobilus 铁花属	200/24

D

Daohugouthallus 道虎沟叶状体属 ········· 201/24
Datongophyllum 大同叶属 ············· 201/25
Decoroxylon 华美木属 ················ 202/25
Deltoispermum 正三角籽属 ············ 202/26
Demersatheca 扁囊蕨属 ··············· 202/26
Dengfengia 登封籽属 ················· 203/26
Denglongia 灯笼蕨属 ················· 203/26
Dentopteris 牙羊齿属 ················ 203/27
Dioonocarpus 苏铁籽属 ··············· 204/27
Discalis 盘囊蕨属 ··················· 204/27
Distichopteris 两列羊齿属 ············ 204/28
Distichotheca 缨囊属 ················ 205/28
Ditaxocladus 对枝柏属 ··············· 205/28
Dracopteris 龙蕨属 ·················· 205/29
Dukouphyllum 渡口叶属 ·············· 206/29
Dukouphyton 渡口痕木属 ············· 206/29

E

Eboraciopsis 拟爱博拉契蕨属 ··········· 206/30
Emplectopteris 织羊齿属 ·············· 207/30
Eoglyptostrobus 始水松属 ············· 207/30
Eogonocormus Deng, 1995 (non Deng, 1997) 始团扇蕨属 ··· 207/30
Eogonocormus Deng, 1997 (non Deng, 1995) 始团扇蕨属 ··· 208/31
Eogymnocarpium 始羽蕨属 ············ 208/31
Eolepidodendron 始鳞木属 ············ 209/31
Eophyllogonium 始叶羊齿属 ··········· 209/32
Eophyllophyton 始叶蕨属 ············· 210/32
Eragrosites 似画眉草属 ··············· 210/33
Eucommioites 似杜仲属 ·············· 210/33

F

Fascipteridium 准束羊齿属 ············ 211/33
Fascipteris 束羊齿属 ················· 211/34
Filicidicotis 羊齿缘蕨属 ·············· 212/35
Filiformorama 纤细蕨属 ·············· 213/35
Fimbriotheca 睫囊蕨属 ··············· 213/35
Foliosites 似茎状地衣属 ·············· 213/36
Fujianopteris 福建羊齿属 ············· 214/36

G

Gansuphyllite 甘肃芦木属 ·········· 215/37
Geminofoliolum 双生叶属 ·········· 215/37
Gigantonoclea 单网羊齿属 ·········· 215/37
Gigantonomia 带囊蕨属 ·········· 215/37
Gigantopteris 大羽羊齿属 ·········· 216/38
Gigantotheca 大囊蕨属 ·········· 216/38
Guangnania 广南蕨属 ·········· 217/39
Guangxiophyllum 广西叶属 ·········· 217/39
Guizhoua 黔囊属 ·········· 217/39
Guizhouoxylon 贵州木属 ·········· 218/40
Gumuia 古木蕨属 ·········· 218/40
Gymnogrammitites 似雨蕨属 ·········· 218/40

H

Hallea Mathews,1947－1948 (non Yang et Wu,2006) 哈勒角籽属 ·········· 219/40
Hallea Yang et Wu,2006 (non Mathews,1947－1948) 赫勒单网羊齿属 ·········· 219/41
Halleophyton 哈氏蕨属 ·········· 219/41
Hamatophyton 钩蕨属 ·········· 220/41
Hefengistrobus 和丰孢穗属 ·········· 220/42
Helicophyton 缠绕蕨属 ·········· 220/42
Henanophyllum 河南叶属 ·········· 221/42
Henanopteris 河南羊齿属 ·········· 221/43
Henanotheca 豫囊蕨属 ·········· 221/43
Hexaphyllum 六叶属 ·········· 222/43
Holozamites 全泽米属 ·········· 222/44
Hsiangchiphyllum 香溪叶属 ·········· 222/44
Hsuea 徐氏蕨属 ·········· 223/44
Huangia 汲清羊齿属 ·········· 223/45
Hubeiia 湖北蕨属 ·········· 223/45
Hubeiophyllum 湖北叶属 ·········· 224/45
Huia 先骕蕨属 ·········· 224/46
Hunanoequisetum 湖南木贼属 ·········· 224/46

I

Illicites 似八角属 ·········· 225/46

J

Jaenschea 耶氏蕨属 ·········· 225/47
Jiangxifolium 江西叶属 ·········· 225/47

Jiangxitheca 赣囊蕨属 ·· 226/47
Jingmenophyllum 荆门叶属 ··· 226/48
Jixia 鸡西叶属 ··· 227/48
Junggaria 准噶尔蕨属 ·· 227/48
Juradicotis 侏罗缘蕨属 ··· 227/49
Juramagnolia 侏罗木兰属 ·· 228/49
Jurastrobus 侏罗球果属 ·· 228/49

K

Kadsurrites 似南五味子属 ·· 228/50
Kaipingia 开平木属 ·· 229/50
Khitania 契丹穗属 ·· 229/50
Klukiopsis 似克鲁克蕨属 ·· 229/51
Koilosphenus 凹尖枝属 ·· 230/51
Kongshania 孔山羊齿属 ··· 230/51
Konnoa 今野羊齿属 ··· 230/52
Kuandiania 宽甸叶属 ··· 231/52

L

Leeites 李氏穗属 ··· 231/52
Lepingia 乐平苏铁属 ··· 231/53
Lhassoxylon 拉萨木属 ··· 232/53
Lianshanus 连山草属 ··· 232/53
Liaoningdicotis 辽宁缘蕨属 ·· 232/54
Liaoningocladus 辽宁枝属 ·· 233/54
Liaoningoxylon 辽宁木属 ·· 233/54
Liaoxia 辽西草属 ·· 234/55
Liella 李氏苏铁属 ··· 234/55
Lilites 似百合属 ··· 235/55
Lingxiangphyllum 灵乡叶属 ··· 235/56
Linophyllum 网叶属 ·· 235/56
Lioxylon 李氏木属 ·· 236/56
Liulinia 柳林果属 ··· 236/57
Lixotheca 李氏蕨属 ··· 236/57
Lobatannulariopsis 拟瓣轮叶属 ··· 237/57
Longjingia 龙井叶属 ··· 237/58
Longostachys 长穗属 ··· 237/58
Lopadiangium 碟囊属 ·· 58/238
Lophotheca 梳囊属 ·· 238/58
Lopinopteris 乐平蕨属 ··· 238/59
Loroderma 带状鳞穗属 ·· 239/59
Lorophyton 条形蕨属 ··· 239/59
Luereticopteris 吕蕨属 ··· 239/60

M

Macroglossopteris 大舌羊齿属 ··· 240/60
Manchurostachys 东北穗属 ··· 240/60
Manica 袖套杉属·· 240/61
　　Manica (*Chanlingia*) 袖套杉(长岭杉)亚属 ·· 240/61
　　Manica (*Manica*) 袖套杉(袖套杉)亚属 ·· 240/61
Mediocycas 中间苏铁属 ··· 242/62
Megalopteris Schenk,1883 (non Andrews E B,1875) 大叶羊齿属 ············ 242/62
Meia 梅氏叶属 ·· 243/63
Membranifolia 膜质叶属 ··· 243/63
Metacladophyton 异枝蕨属 ··· 244/64
Metalepidodendron 变态鳞木属 ·· 244/64
Metasequoia 水杉属·· 244/64
Metzgerites 似叉苔属 ·· 245/65
Minarodendron 塔状木属 ··· 245/65
Mirabopteris 奇异羊齿属 ··· 245/65
Mironeura 奇脉叶属 ··· 246/66
Mixophylum 间羽叶属 ··· 246/66
Mixopteris 间羽蕨属 ··· 247/66
Mnioites 似提灯藓属 ·· 247/66
Monogigantonoclea 单叶单网羊齿属 ··· 247/67
Monogigantopteris 单叶大羽羊齿属 ··· 248/68
Muricosperma 尖籽属 ··· 249/68
Myriophyllum 密叶属 ··· 249/68

N

Nanpiaophyllum 南票叶属 ··· 249/69
Nanzhangophyllum 南漳叶属 ··· 250/69
Neoannularia 新轮叶属 ··· 250/69
Neocordaites 新科达属 ·· 250/70
Neogigantopteridium 新准大羽羊齿属 ··· 251/70
Neostachya 新孢穗属 ··· 251/70
Neurophyllites 翅叶属 ··· 251/70
Ningxiaphyllum 宁夏叶属 ··· 252/71
Norinia 那琳壳斗属 ··· 252/71
Nudasporestrobus 裸囊穗属 ··· 252/71
Nystroemia 新常富籽属 ··· 253/72

O

Odontosorites 似齿囊蕨属 ·· 253/72
Orchidites 似兰属 ··· 253/72

Otofolium 耳叶属 ·· 254/73

P

Palaeoginkgoxylon 古银杏型木属 ·· 254/73
Palaeognetaleaana 古买麻藤属 ·· 255/73
Palaeoskapha 古舟藤属 ··· 255/74
Pania 潘氏果属 ·· 255/74
Pankuangia 潘广叶属 ··· 256/74
Papilionifolium 蝶叶属 ··· 256/75
Paracaytonia 副开通尼亚属 ··· 256/75
Paraconites 副球果属 ··· 257/75
Paradoxopteris Mi et Liu,1977 (non Hirmer,1927) 奇异羊齿属 ··· 257/75
Paradrepanozamites 副镰羽叶属 ·· 257/76
Parasphenophyllum 拟楔叶属 ··· 258/76
Parasphenopteris 拟楔羊齿属 ··· 258/76
Parastorgaardis 拟斯托加枝属 ·· 258/77
Parataxospermum 类紫杉籽属 ·· 259/77
Paratingia 拟齿叶属 ··· 259/77
Paratingiostachya 拟丁氏蕨穗属 ··· 259/78
Pavoniopteris 雅蕨属 ··· 259/78
Pectinangium 篦囊属 ··· 260/78
Perisemoxylon 雅观木属 ··· 260/79
Phoenicopsis 拟刺葵属 ··· 261/79
　　Phoenicopsis (*Stephenophyllum*) 拟刺葵(斯蒂芬叶)亚属 ··· 261/79
Phoroxylon 贼木属 ··· 262/80
Phylladendroid 叶茎属 ··· 262/81
Pinnagigantonoclea 羽叶单网羊齿属 ··· 263/81
Pinnagigantopteris 羽叶大羽羊齿属 ··· 264/82
Pinnatiramosus 羽枝属 ·· 265/83
Plagiozamiopsis 拟斜羽叶属 ··· 265/83
Polygatites 似远志属 ·· 266/83
Polygonites Wu S Q,1999 (non Saporta,1865) 似蓼属 ··· 266/84
Polypetalophyton 多瓣蕨属 ··· 267/84
Polythecophyton 多囊枝属 ·· 267/85
Primocycas 始苏铁属 ··· 267/85
Primoginkgo 始拟银杏属 ·· 268/85
Primozamia 始查米苏铁属 ··· 268/86
Prionophyllopteris 锯叶羊齿属 ·· 268/86
Procycas 原苏铁属 ··· 269/86
Progigantonoclea 原单网羊齿属 ··· 269/87
Progigantopteris 原大羽羊齿属 ·· 269/87
Proginkgoxylon 原始银杏型木属 ·· 270/87
Protoglyptostroboxylon 原始水松型木属 ·· 271/88
Protopteridophyton 原始蕨属 ·· 271/88

Protosciadopityoxylon 原始金松型木属 ······ 271/89
Pseudopolystichu 假耳蕨属 ······ 272/89
Pseudorhipidopsis 假拟扇叶属 ······ 272/89
Pseudotaeniopteris 假带羊齿属 ······ 272/90
Pseudotaxoxylon 白豆杉型木属 ······ 273/90
Pseudotsugxylon 黄杉型木属 ······ 273/90
Pseudoullmannia 假鳞杉属 ······ 273/90
Pteridiopsis 拟蕨属 ······ 274/91

Q

Qinlingopteris 秦岭羊齿属 ······ 274/91
Qionghaia 琼海叶属 ······ 274/92

R

Radiatifolium 辐叶属 ······ 275/92
Ramophyton 多枝蕨属 ······ 275/92
Ranunculophyllum 毛茛叶属 ······ 275/93
Rastropteris 耙羊齿属 ······ 276/93
Rehezamites 热河似查米亚属 ······ 276/93
Renifolium 肾叶属 ······ 277/94
Reteophlebis 网格蕨属 ······ 277/94
Reticalethopteris 网延羊齿属 ······ 277/94
Rhizoma 根状茎属 ······ 278/95
Rhizomopsis 拟根茎属 ······ 278/95
Rhomboidopteris 菱羊齿属 ······ 278/95
Riccardiopsis 拟片叶苔属 ······ 278/96
Rireticopteris 日蕨属 ······ 279/96
Rotafolia 轮叶蕨属 ······ 279/96

S

Sabinites 似圆柏属 ······ 280/97
Sagittopteris Zhang E et Xiao,1985 (non Zhang S et Xiao,1987) 箭羽羊齿属 ······ 280/97
Sagittopteris Zhang S et Xiao,1987 (non Zhang E et Xiao,1985) 箭羽羊齿属 ······ 280/97
Schizoneuropsis 拟裂鞘叶属 ······ 281/98
Sciadocillus 小伞属 ······ 281/98
Scoparia 帚羽叶属 ······ 281/98
Semenalatum 翅籽属 ······ 282/99
Setarites 似狗尾草属 ······ 282/99
Shangyuania 上园草属 ······ 282/99
Shanxicladus 山西枝属 ······ 283/100
Shanxioxylon 山西木属 ······ 283/100
Shenea 沈氏蕨属 ······ 284/100

Shenkuoia 沈括叶属 · 284/101
Shenzhouphyllum 神州叶属 · 284/101
Shenzhouspermum 神州籽属 · 285/102
Shenzhoutheca 神州聚囊属 · 285/102
Shuangnangostachya 双囊芦穗属 · 286/102
Shuichengella 水城蕨属 · 286/102
Siella 斯氏鞘叶属 · 286/103
Sinocarpus 中华古果属 · 287/103
Sinoctenis 中国篦羽叶属 · 287/103
Sinodicotis 中华缘蕨属 · 287/104
Sinopalaeospiroxylon 中国古螺纹木属 · 288/104
Sinophyllum 中国叶属 · 289/105
Sinozamites 中国似查米亚属 · 289/105
Siphonospermum 管子麻黄属 · 289/106
Solaranthus 太阳花属 · 290/106
Speirocarpites 似卷囊蕨属 · 290/106
Sphenobaieroanthus 楔叶拜拉花属 · 291/107
Sphenobaierocladus 楔叶拜拉枝属 · 291/107
Sphenopecopteris 楔栉羊齿属 · 292/107
Sphinxia Li, Hilton et Hemsley, 1997 (non Reid et Chandler, 1933) 斯芬克斯籽属 · 292/108
Sphinxiocarpon 仙籽属 · 292/108
Spinolepidodendron 刺鳞木属 · 293/108
Squamocarpus 鳞籽属 · 293/109
Squarmacarpus 翅鳞籽属 · 293/109
Stachybryolites 穗藓属 · 204/109
Stachyophyton 穗蕨属 · 294/110
Stalagma 垂饰杉属 · 294/110
Stephanofolium 金藤叶属 · 295/110
Strigillotheca 刷囊属 · 295/111
Suturovagina 缝鞘杉属 · 295/111
Symopteris 束脉蕨属 · 296/111
Szea 天石蕨属 · 296/112
Szecladia 斯氏松属 · 297/112
Szeioxylon 斯氏木属 · 297/112

T

Tachingia 大箐羽叶属 · 297/113
Taeniocladopsis 拟带枝属 · 297/113
Taipingchangella 太平场蕨属 · 298/113
Taiyuanitheca 太原蕨属 · 298/113
Tchiaohoella 蛟河羽叶属 · 298/114
Tenuisa 细轴始蕨属 · 299/114
Tetrafolia 四叶属 · 299/114
Tharrisia 哈瑞士叶属 · 299/115

Thaumatophyllum 奇异羽叶属	300/115
Thelypterites 似金星蕨属	201/116
Tianoxylon 田氏木属	301/116
Tianshanopteris 天山羊齿属	302/117
Tianshia 天石枝属	302/117
Tingia 丁氏羊齿属	302/117
Toksunopteris 托克逊蕨属	302/117
Tongchuanophyllum 铜川叶属	303/118
Tonglucarpus 桐庐籽属	303/118
Tongshania 钟囊属	304/119
Torreyocladus 榧型枝属	304/119
Tricoemplectopteris 三网羊齿属	304/119
Tricrananthus 三裂穗属	305/119
Trinerviopteris 三脉蕨属	305/120
Triqueteria 三棱果属	305/120
Tsaia 蔡氏蕨属	306/120
Tsiaohoella 蛟河蕉羽叶属	306/121

V

Vasovinea 导管羊齿属	306/121
Vittifoliolum 条叶属	307/121

W

Wenshania 文山蕨属	308/122
Wutubulaka 乌图布拉克蕨属	308/122
Wuxia 无锡蕨属	308/123
Wuyunanthus 乌云花属	309/123

X

Xiajiajienia 夏家街蕨属	309/123
Xinganphyllum 兴安叶属	309/124
Xingxueanthus 星学花属	310/124
Xingxueina 星学花序属	310/124
Xingxuephyllum 星学叶属	311/125
Xinjiangophyton 新疆木属	311/125
Xinjiangopteris Wu S Q et Zhou,1986 (non Wu S Z,1983) 新疆蕨属	311/125
Xinjiangopteris Wu S Z,1983 (non Wu S Q et Zhou,1986) 新疆蕨属	312/126
Xinlongia 新龙叶属	312/126
Xinlongophyllum 新龙羽叶属	312/126
Xitunia 西屯蕨属	313/127

Y

Yangzunyia 杨氏木属 ·· 313/127
Yanjiphyllum 延吉叶属 ·· 314/127
Yanliaoa 燕辽杉属 ··· 314/128
Yimaia 义马果属 ··· 314/128
Yixianophyllum 义县叶属 ·· 315/128
Yuania 卵叶属 ··· 315/129
Yuguangia 玉光蕨属 ··· 315/129
Yungjenophyllum 永仁叶属 ·· 316/129
Yunia 云蕨属 ·· 316/130

Z

Zeillerpteris 蔡耶羊齿属 ··· 316/130
Zhengia 郑氏叶属 ·· 317/130
Zhenglia 正理蕨属 ··· 317/131
Zhongzhoucarus 中州籽属 ··· 317/131
Zhongzhoupteris 中州蕨属 ··· 318/131
Zhouia 周氏籽属 ··· 318/131
Zhutheca 朱氏囊蕨属 ·· 318/132

Appendix 2　Index of Specific Names

[Arranged alphbetically, generic or specific names and the page numbers (in English part / in Chinese part)]

A

Abropteris 华脉蕨属 ··· 175/1
　　Abropteris virginiensis 弗吉尼亚华脉蕨 ··· 175/1
　　Abropteris yongrenensis 永仁华脉蕨 ·· 175/1
Abrotopteris 华丽羊齿属 ·· 175/1
　　Abrotopteris guizhouensis 贵州华丽羊齿 ··· 175/1
Acanthocladus 刺枝属 ·· 176/2
　　Acanthocladus xyloides 木质刺枝 ··· 176/2
Acanthopteris 刺蕨属 ··· 176/2
　　Acanthopteris gothani 高腾刺蕨 ··· 176/2
Acerites 似槭树属 ·· 176/2
　　Acerites sp. indet. 似槭树(sp. indet.) ··· 176/2
Aconititis 似乌头属 ·· 177/2
　　Aconititis sp. indet. 似乌头(sp. indet.) ··· 177/3
Acthephyllum 奇叶属 ··· 177/3
　　Acthephyllum kaixianense 开县奇叶 ·· 177/3
Actinophyllus Xiao,1985 (non *Actinophyllum* Phillips,1848) 辐射叶属 ········· 177/3
　　Actinophyllus cordaioides 科达状辐射叶 ··· 178/3
Aculeovinea 刺羊齿属 ·· 178/3
　　Aculeovinea yunguiensis 云贵刺羊齿 ·· 178/4
Adoketophyllum 奇异叶属 ··· 178/4
　　Adoketophyllum subverticillatum 亚轮生奇异叶 ··· 178/4
Aetheopteris 奇羊齿属 ·· 179/4
　　Aetheopteris rigida 坚直奇羊齿 ·· 179/4
Aipteridium 准爱河羊齿属 ··· 179/4
　　Aipteridium pinnatum 羽状准爱河羊齿 ·· 179/5
Alloephedra 异麻黄属 ·· 179/5
　　Alloephedra xingxuei 星学异麻黄 ··· 179/5
Allophyton 奇异木属 ··· 180/5
　　Allophyton dengqenensis 丁青奇异木 ·· 180/5
Amdrupiopsis 拟安杜鲁普蕨属 ·· 180/5
　　Amdrupiopsis sphenopteroides 楔羊齿型拟安杜鲁普蕨 ··························· 180/5
Amentostrobus 花穗杉果属 ··· 180/6
　　Amentostrobus sp. indet. 花穗杉果(sp. indet.) ·· 180/6
Amphiephedra 疑麻黄属 ··· 181/6
　　Amphiephedra rhamnoides 鼠李型疑麻黄 ··· 181/6
Amplectosporangium 抱囊蕨属 ··· 181/6
　　Amplectosporangium jiagyouense 江油抱囊蕨 ··· 181/6

Angustiphyllum 窄叶属 181/6
    Angustiphyllum yaobuense 腰埠窄叶 181/7
Annularites 似轮叶属 181/7
    Annularites ensilolius 剑瓣似轮叶 182/7
    Annularites lingulatus 舌形似轮叶 182/7
    Annularites heianensis 平安似轮叶 182/7
    Annularites sinensis 中国似轮叶 182/7
Archaefructus 古果属 182/7
    Archaefructus liaoningensis 辽宁古果 182/7
Archimagnolia 始木兰属 182/8
    Archimagnolia rostrato-stylosa 喙柱始木兰 182/8
Archimalus 始苹果属 183/8
    Archimalus calycina 大萼始苹果 183/8
Areolatophyllum 华网蕨属 183/8
    Areolatophyllum qinghaiense 青海华网蕨 183/8
Asiatifolium 亚洲叶属 183/8
    Asiatifolium elegans 雅致亚洲叶 183/9
Asiopteris 亚洲羊齿属 184/9
    Asiopteris huairenensis 怀仁亚洲羊齿 184/9
Astrocupulites 星壳斗属 184/9
    Astrocupulites acuminatus 渐尖星壳斗 184/9

B

Baiguophyllum 白果叶属 184/9
    Baiguophyllum lijianum 利剑白果叶 184/10
Batenburgia 荆棘果属 185/10
    Batenburgia sakmarica 萨克马尔荆棘果 185/10
Beipiaoa 北票果属 185/10
    Beipiaoa spinosa 强刺北票果 185/10
    Beipiaoa parva 小北票果 185/10
    Beipiaoa rotunda 圆形北票果 185/10
Bennetdicotis 本内缘蕨属 186/11
    Bennetdicotis sp. indet. 本内缘蕨(sp. indet.) 186/11
Benxipteris 本溪羊齿属 186/11
    Benxipteris acuta 尖叶本溪羊齿 186/11
    Benxipteris densinervis 密脉本溪羊齿 186/11
    Benxipteris partita 裂缺本溪羊齿 186/11
    Benxipteris polymorpha 多态本溪羊齿 187/11
Bicoemplectopteris 双网羊齿属 187/12
    Bicoemplectopteris hallei 赫勒双网羊齿 187/12
Bifariusotheca 双列囊蕨属 187/12
    Bifariusotheca qinglongensis 晴隆双列囊蕨 187/12
Bilobphyllum 两瓣叶属 187/12
    Bilobphyllum fengchengensis 丰城两瓣叶 187/12

Boseoxylon 鲍斯木属 ··· 12/188
  Boseoxylon andrewii 安德鲁斯鲍斯木 ··· 188/13
Botrychites 似阴地蕨属 ·· 188/13
  Botrychites reheensis 热河似阴地蕨 ··· 188/13
Bracteophyton 苞片蕨属 ·· 188/13
  Bracteophyton variatum 变异苞片蕨 ·· 188/13

C

Callianthus 丽花属 ·· 189/13
  Callianthus dilae 迪拉丽花 ··· 189/14
Casuarinites 似木麻黄属 ··· 189/14
  Casuarinites sp. indet. 似木麻黄(sp. indet.) ·· 189/14
Catenalis 掌裂蕨属 ·· 189/14
  Catenalis digitata 指状掌裂蕨 ·· 189/14
Cathaiopteridium 中国羊齿属 ·· 190/14
  Cathaiopteridium minutum 细小中国羊齿 ·· 190/14
Cathayanthus 华夏穗属 ··· 190/15
  Cathayanthus ramentrarus 少鳞华夏穗 ··· 190/15
  Cathayanthus sinensis 中国华夏穗 ··· 190/15
Cathaysiocycas 华夏苏铁属 ··· 191/15
  Cathaysiocycas rectanervis 直脉华夏苏铁 ··· 191/15
Cathaysiodendron 华夏木属 ·· 191/15
  Cathaysiodendron incertum 不定华夏木 ··· 191/16
  Cathaysiodendron chuseni 朱森华夏木 ··· 191/16
  Cathaysiodendron nanpiaoense 南票华夏木 ·· 192/16
Cathaysiophyllum 华夏叶属 ··· 192/16
  Cathaysiophyllum lobifolium 裂瓣华夏叶 ·· 192/16
Cathaysiopteridium 中华羊齿属 ··· 192/16
  Cathaysiopteridium fasciculatum 束脉中华羊齿 ··· 192/17
Cathaysiopteris 华夏羊齿属 ··· 192/17
  Cathaysiopteris whitei 怀特华夏羊齿 ·· 193/17
Celathega 隐囊蕨属 ··· 193/17
  Celathega beckii 贝氏隐囊蕨 ·· 193/17
Cervicornus 鹿角蕨属 ·· 193/17
  Cervicornus wenshanensis 文山鹿角蕨 ··· 193/17
Chamaedendron 纤木属 ·· 193/18
  Chamaedendron multisporangiatum 异囊纤木 ·· 193/18
Chaneya 钱耐果属 ··· 194/18
  Chaneya tenuis 细小钱耐果 ·· 194/18
  Chaneya kokangensis 科干钱耐果 ·· 194/18
Changwuia 长武蕨属 ··· 194/18
  Changwuia schweitzeri 施魏策尔长武蕨 ··· 194/18
Changyanophyton 长阳木属 ·· 194/19
  Changyanophyton hupeiense 湖北长阳木 ·· 195/19

Chaoyangia 朝阳序属 ·········· 195/19
    Chaoyangia liangii 梁氏朝阳序 ·········· 195/19
Chengzihella 城子河叶属 ·········· 195/19
    Chengzihella obovata 倒卵城子河叶 ·········· 195/19
Chiaohoella 小蛟河蕨属 ·········· 196/20
    Chiaohoella mirabilis 奇异小蛟河蕨 ·········· 196/20
    Chiahooella neozamioide 新查米叶型小蛟河蕨 ·········· 196/20
Chilinia 吉林羽叶属 ·········· 196/20
    Chilinia ctenioides 篦羽叶型吉林羽叶 ·········· 196/20
Ciliatopteris 细毛蕨属 ·········· 197/20
    Ciliatopteris pecotinata 栉齿细毛蕨 ·········· 197/21
Cladophlebidium 准枝脉蕨属 ·········· 197/21
    Cladophlebidium wongi 翁氏准枝脉蕨 ·········· 197/21
Cladotaeniopteris 枝带羊齿属 ·········· 197/21
    Cladotaeniopteris shaanxiensis 陕西枝带羊齿 ·········· 197/21
Clematites 似铁线莲叶属 ·········· 197/21
    Clematites lanceolatus 披针似铁线莲叶 ·········· 198/22
Coenosophyton 普通蕨属 ·········· 198/22
    Coenosophyton tristichus 三叉普通蕨 ·········· 198/22
Cohaerensitheca 粘合囊蕨属 ·········· 198/22
    Cohaerensitheca sahnii 沙尼粘合囊蕨 ·········· 198/22
Conchophyllum 贝叶属 ·········· 198/22
    Conchophyllum richthofenii 李氏贝叶 ·········· 199/22
Cryptonoclea 隐羊齿属 ·········· 199/23
    Cryptonoclea primitiva 原始隐羊齿 ·········· 199/23
Cycadeoidispermum 拟苏铁籽属 ·········· 199/23
    Cycadeoidispermum petiolatum 具柄拟苏铁籽 ·········· 199/23
Cycadicotis 苏铁缘蕨属 ·········· 199/23
    Cycadicotis nissonervis 蕉羽叶脉苏铁缘蕨 ·········· 200/23
    Cycadicotis sp. indet. 苏铁缘蕨(sp. indet.) ·········· 200/24
Cycadolepophyllum 苏铁鳞叶属 ·········· 200/24
    Cycadolepophyllum minor 较小苏铁鳞叶 ·········· 200/24
    Cycadolepophyllum aequale 等形苏铁鳞叶 ·········· 200/24
Cycadostrobilus 铁花属 ·········· 200/24
    Cycadostrobilus paleozoicus 古生铁花 ·········· 201/24

D

Daohugouthallus 道虎沟叶状体属 ·········· 201/24
    Daohugouthallus ciliiferus 细毛道虎沟叶状体 ·········· 201/25
Datongophyllum 大同叶属 ·········· 201/25
    Datongophyllum longipetiolatum 长柄大同叶 ·········· 201/25
    Datongophyllum sp. 大同叶(未定种) ·········· 201/25
Decoroxylon 华美木属 ·········· 202/25
    Decoroxylon chaoyangense 朝阳华美木 ·········· 202/25

Deltoispermum 正三角籽属 ·· 202/26
    Deltoispermum henanense 河南正三角籽 ························· 202/26
Demersatheca 扁囊蕨属 ·· 202/26
    Demersatheca contigua 紧贴扁囊蕨 ······························· 202/26
Dengfengia 登封籽属 ·· 203/26
    Dengfengia bifurcata 双翅登封籽 ·································· 203/26
Denglongia 灯笼蕨属 ··· 203/26
    Denglongia hubeiensis 湖北灯笼蕨 ································ 203/27
Dentopteris 牙羊齿属 ·· 203/27
    Dentopteris stenophylla 窄叶牙羊齿 ······························ 203/27
    Dentopteris platyphylla 宽叶牙羊齿 ······························ 204/27
Dioonocarpus 苏铁籽属 ·· 204/27
    Dioonocarpus ovatus 卵形苏铁籽 ································· 204/27
Discalis 盘囊蕨属 ··· 204/27
    Discalis longistipa 长柄盘囊蕨 ····································· 204/28
Distichopteris 两列羊齿属 ··· 204/28
    Distichopteris heteropinna 异常两列羊齿 ························ 205/28
Distichotheca 缨囊属 ·· 205/28
    Distichotheca crossothecoides 具边缨囊 ························ 205/28
Ditaxocladus 对枝柏属 ·· 205/28
    Ditaxocladus planiphyllus 扁叶对枝柏 ···························· 205/29
Dracopteris 龙蕨属 ··· 205/29
    Dracopteris liaoningensis 辽宁龙蕨 ······························ 206/29
Dukouphyllum 渡口叶属 ··· 206/29
    Dukouphyllum noeggerathioides 诺格拉齐蕨型渡口叶 ········ 206/29
Dukouphyton 渡口痕木属 ··· 206/29
    Dukouphyton minor 较小渡口痕木 ······························· 206/29

E

Eboraciopsis 拟爱博拉契蕨属 ··· 206/30
    Eboraciopsis trilobifolia 三裂叶拟爱博拉契蕨 ·················· 207/30
Emplectopteris 织羊齿属 ··· 207/30
    Emplectopteris trangularis 三角织羊齿 ·························· 207/30
Eoglyptostrobus 始水松属 ·· 207/30
    Eoglyptostrobus sabioides 清风藤型始水松 ····················· 207/30
Eogonocormus Deng, 1995 (non Deng, 1997) 始团扇蕨属 ······················· 207/30
    Eogonocormus cretaceum Deng, 1995 (non Deng, 1997) 白垩始团扇蕨 ····· 207/30
    Eogonocormus linearifolium 线形始团扇蕨 ······················ 207/31
Eogonocormus Deng, 1997 (non Deng, 1995) 始团扇蕨属 ······················· 208/31
    Eogonocormus cretaceum Deng, 1997 (non Deng, 1995) 白垩始团扇蕨 ····· 208/31
Eogymnocarpium 始羽蕨属 ··· 208/31
    Eogymnocarpium sinense 中国始羽蕨 ···························· 208/31
Eolepidodendron 始鳞木属 ·· 209/31
    Eolepidodendron jurongense 句容始鳞木 ························ 209/32

Eolepidodendron wusihense 无锡始鳞木 ·········· 209/32
Eolepidodendron cf. *wusihense* 无锡始鳞木(比较种) ·········· 209/32
Eolepidodendron sp. 始鳞木(未定种) ·········· 209/32
Eophyllogonium 始叶羊齿属 ·········· 209/32
Eophyllogonium cathayense 华夏始叶羊齿 ·········· 209/32
Eophyllophyton 始叶蕨属 ·········· 210/32
Eophyllophyton bellum 优美始叶蕨 ·········· 210/33
Eragrosites 似画眉草属 ·········· 210/33
Eragrosites changii 常氏似画眉草 ·········· 210/33
Eucommioites 似杜仲属 ·········· 210/33
Eucommioites orientalis 东方似杜仲 ·········· 211/33

F

Fascipteridium 准束羊齿属 ·········· 211/33
Fascipteridium ellipticum 椭圆准束羊齿 ·········· 211/34
Fascipteris 束羊齿属 ·········· 211/34
Fascipteris hallei 赫勒束羊齿 ·········· 211/34
Fascipteris recta 垂束羊齿 ·········· 211/34
Fascipteris sinensis 中国束羊齿 ·········· 212/34
Fascipteris (*Ptychocarpus*) *densata* 密囊束羊齿(皱囊蕨) ·········· 212/34
Fascipteris stena 狭束羊齿 ·········· 212/34
Filicidicotis 羊齿缘蕨属 ·········· 212/35
Filicidicotis sp. indet. 羊齿缘蕨(sp. indet.) ·········· 212/35
Filiformorama 纤细蕨属 ·········· 213/35
Filiformorama simplexa 简单纤细蕨 ·········· 213/35
Fimbriotheca 睫囊蕨属 ·········· 213/35
Fimbriotheca tomentosa 毛状睫囊蕨 ·········· 213/35
Foliosites 似茎状地衣属 ·········· 213/36
Foliosites formosus 美丽似茎状地衣 ·········· 213/36
Fujianopteris 福建羊齿属 ·········· 214/36
Fujianopteris fukianensis 闽福建羊齿 ·········· 214/36
Fujianopteris angustiangla 狭角福建羊齿 ·········· 214/36
Fujianopteris cladonervis 枝脉福建羊齿 ·········· 214/36
Fujianopteris intermedia 中间福建羊齿 ·········· 214/37

G

Gansuphyllite 甘肃芦木属 ·········· 215/37
Gansuphyllite multivervis 多脉甘肃芦木 ·········· 215/37
Geminofoliolum 双生叶属 ·········· 215/37
Geminofoliolum gracilis 纤细双生叶 ·········· 215/37
Gigantonoclea 单网羊齿属 ·········· 215/37
Gigantonoclea lagrelii 波缘单网羊齿 ·········· 215/37
Gigantonomia 带囊蕨属 ·········· 215/38

Gigantonomia (*Gigatonoclea*) *fukienensis* 福建带囊蕨(单网羊齿)	216/38
Gigantopteris 大羽羊齿属	216/38
Gigantopteris nicotianaefolia 烟叶大羽羊齿	216/38
Gigantotheca 大囊蕨属	216/38
Gigantotheca paradoxa 奇异大囊蕨	216/38
Guangnania 广南蕨属	217/39
Guangnania cuneata 楔形广南蕨	217/39
Guangxiophyllum 广西叶属	217/39
Guangxiophyllum shangsiense 上思广西叶	217/39
Guizhoua 黔囊属	217/39
Guizhoua gregalis 堆黔囊	217/39
Guizhouoxylon 贵州木属	218/40
Guizhouoxylon dahebianense 大河边贵州木	218/40
Gumuia 古木蕨属	218/40
Gumuia zyzzata 曲轴古木蕨	218/40
Gymnogrammitites 似雨蕨属	218/40
Gymnogrammitites ruffordioides 鲁福德似雨蕨	218/40

H

Hallea Mathews,1947 – 1948 (non Yang et Wu,2006) 哈勒角籽属	219/40
Hallea pekinensis 北京哈勒角籽	219/41
Hallea Yang et Wu,2006 (non Mathews,1947 – 1948) 赫勒单网羊齿属	219/41
Hallea dengfengensis 登封赫勒单网羊齿	219/41
Halleophyton 哈氏蕨属	219/41
Halleophyton zhichangense 纸厂哈氏蕨	219/41
Hamatophyton 钩蕨属	220/41
Hamatophyton verticillatum 轮生钩蕨	220/42
Hefengistrobus 和丰孢穗属	220/42
Hefengistrobus bifurcus 二歧和丰孢穗	220/42
Helicophyton 缠绕蕨属	220/42
Helicophyton dichotomum 二叉缠绕蕨	220/42
Henanophyllum 河南叶属	221/42
Henanophyllum palamifolium 掌河南叶	221/43
Henanopteris 河南羊齿属	221/43
Henanopteris lanceolatus 披针河南羊齿	221/43
Henanotheca 豫囊蕨属	221/43
Henanotheca (*Sphenopteris*) *ovata* 卵豫囊蕨(楔羊齿)	221/43
Hexaphyllum 六叶属	222/43
Hexaphyllum sinense 中国六叶	222/43
Holozamites 全泽米属	222/44
Holozamites hongtaoi 洪涛全查米亚	222/44
Hsiangchiphyllum 香溪叶属	222/44
Hsiangchiphyllum trinerve 三脉香溪叶	222/44
Hsuea 徐氏蕨属	223/44

Hsuea robusta 粗壮徐氏蕨 ·········· 223/44
Huangia 汲清羊齿属 ·········· 223/45
　　Huangia elliptica 椭圆汲清羊齿 ·········· 223/45
Hubeiia 湖北蕨属 ·········· 223/45
　　Hubeiia dicrofollia 叉叶湖北蕨 ·········· 223/45
Hubeiophyllum 湖北叶属 ·········· 224/45
　　Hubeiophyllum cuneifolium 楔形湖北叶 ·········· 224/45
　　Hubeiophyllum angustum 狭细湖北叶 ·········· 224/45
Huia 先骕蕨属 ·········· 224/46
　　Huia recurvata 回弯先骕蕨 ·········· 224/46
Hunanoequisetum 湖南木贼属 ·········· 224/46
　　Hunanoequisetum liuyangense 浏阳湖南木贼 ·········· 225/46

I

Illicites 似八角属 ·········· 225/46
　　Illicites sp. indet. 似八角(sp. indet.) ·········· 225/46

J

Jaenschea 耶氏蕨属 ·········· 225/47
　　Jaenschea sinensis 中国耶氏蕨 ·········· 225/47
Jiangxifolium 江西叶属 ·········· 225/47
　　Jiangxifolium mucronatum 短尖头江西叶 ·········· 225/47
　　Jiangxifolium denticulatum 细齿江西叶 ·········· 226/47
Jiangxitheca 赣囊蕨属 ·········· 226/47
　　Jiangxitheca xinanensis 新安赣囊蕨 ·········· 226/47
Jingmenophyllum 荆门叶属 ·········· 226/48
　　Jingmenophyllum xiheense 西河荆门叶 ·········· 226/48
Jixia 鸡西叶属 ·········· 226/48
　　Jixia pinnatipartita 羽裂鸡西叶 ·········· 227/48
Junggaria 准噶尔蕨属 ·········· 227/48
　　Junggaria spinosa 刺状准噶尔蕨 ·········· 227/48
Juradicotis 侏罗缘蕨属 ·········· 227/49
　　Juradicotis sp. indet. 侏罗缘蕨(sp. indet.) ·········· 227/49
Juramagnolia 侏罗木兰属 ·········· 228/49
　　Juramagnolia sp. indet. 侏罗木兰(sp. indet.) ·········· 228/49
Jurastrobus 侏罗球果属 ·········· 228/49
　　Jurastrobus chenii 陈氏侏罗球果 ·········· 228/49

K

Kadsurrites 似南五味子属 ·········· 228/50
　　Kadsurrites sp. indet. 似南五味子(sp. indet.) ·········· 228/50
Kaipingia 开平木属 ·········· 229/50

Kaipingia sinica 中国开平木 ·· 229/50
Khitania 契丹穗属 ·· 229/50
 Khitania columnispicata 柱状契丹穗 ································· 229/50
Klukiopsis 似克鲁克蕨属 ·· 229/51
 Klukiopsis jurassica 侏罗似克鲁克蕨 ································ 229/51
Koilosphenus 凹尖枝属 ··· 230/51
 Koilosphenus cuneifolius 楔裂凹尖枝 ······························· 230/51
 ？*Koilosphenus* sp.？凹尖枝（未定种） ··························· 230/51
Kongshania 孔山羊齿属 ·· 230/51
 Kongshania synangioides 类连生孔山羊齿 ······················ 230/51
Konnoa 今野羊齿属 ··· 230/52
 Konnoa koraiensis 高丽今野羊齿 ······································ 231/52
 Konnoa penchihuensis 本溪今野羊齿 ································ 231/52
Kuandiania 宽甸叶属 ·· 231/52
 Kuandiania crassicaulis 粗茎宽甸叶 ································· 231/52

L

Leeites 李氏穗属 ·· 231/52
 Leeites oblongifolis 椭圆李氏穗 ·· 231/52
Lepingia 乐平苏铁属 ·· 231/53
 Lepingia emarginata 缺顶乐平苏铁 ·································· 232/53
Lhassoxylon 拉萨木属 ··· 232/53
 Lhassoxylon aptianum 阿普特拉萨木 ································ 232/53
Lianshanus 连山草属 ·· 232/53
 Lianshanus sp. indet. 连山草（sp. indet.） ······················· 232/53
Liaoningdicotis 辽宁缘蕨属 ··· 232/54
 Liaoningdicotis sp. indet. 辽宁缘蕨（sp. indet.） ··········· 233/54
Liaoningocladus 辽宁枝属 ··· 233/54
 Liaoningocladus boii 薄氏辽宁枝 ····································· 233/54
Liaoningoxylon 辽宁木属 ··· 233/54
 Liaoningoxylon chaoyangehse 朝阳辽宁木 ······················ 233/54
Liaoxia 辽西草属 ·· 234/55
 Liaoxia chenii 陈氏辽西草 ·· 234/55
Liella 李氏苏铁属 ··· 234/55
 Liella mirabilis 奇异李氏苏铁 ··· 234/55
Lilites 似百合属 ··· 235/55
 Lilites reheensis 热河似百合 ··· 235/55
Lingxiangphyllum 灵乡叶属 ··· 235/56
 Lingxiangphyllum princeps 首要灵乡叶 ··························· 235/56
Linophyllum 网叶属 ·· 235/56
 Linophyllum xuanweiense 宣威网叶 ································· 235/56
Lioxylon 李氏木属 ·· 236/56
 Lioxylon liaoningense 辽宁李氏木 ···································· 236/56
Liulinia 柳林果属 ·· 236/57

Liulinia lacinulata 条裂柳林果 ··· 236/57
Lixotheca 李氏蕨属 ··· 236/57
　　Lixotheca (*Cladophlebis*) *permica* 二叠李氏蕨(枝脉蕨) ·· 237/57
Lobatannulariopsis 拟瓣轮叶属 ··· 237/57
　　Lobatannulariopsis yunnanensis 云南拟瓣轮叶 ·· 237/57
Longjingia 龙井叶属 ·· 237/58
　　Longjingia gracilifolia 细叶龙井叶 ··· 237/58
Longostachys 长穗属 ··· 237/58
　　Longostachys latisporophyllus 宽叶长穗 ·· 238/58
Lopadiangium 碟囊属 ·· 58/238
　　Lopadiangium acmodontum 齿缘碟囊 ··· 238/58
Lophotheca 梳囊属 ··· 238/58
　　Lophotheca panxianensis 盘县梳囊 ·· 238/59
Lopinopteris 乐平蕨属 ··· 238/59
　　Lopinopteris intercalata 插入乐平蕨 ··· 239/59
Loroderma 带状鳞穗属 ··· 239/59
　　Loroderma henania 河南带状鳞穗 ·· 239/59
Lorophyton 条形蕨属 ··· 239/59
　　Lorophyton goense 高氏条形蕨 ··· 239/60
Luereticopteris 吕蕨属 ··· 239/60
　　Luereticopteris megaphylla 大叶吕蕨 ··· 240/60

M

Macroglossopteris 大舌羊齿属 ·· 240/60
　　Macroglossopteris leeiana 李氏大舌羊齿 ··· 240/60
Manchurostachys 东北穗属 ··· 240/60
　　Manchurostachys manchuriensis 裂鞘叶东北穗 ·· 240/60
Manica 袖套杉属 ·· 240/61
　　Manica parceramosa 希枝袖套杉 ··· 240/61
　　Manica (*Chanlingia*) 袖套杉(长岭杉)亚属 ··· 241/61
　　　　Manica (*Chanlingia*) *tholistoma* 穹孔袖套杉(长岭杉) ··· 241/61
　　Manica (*Manica*) 袖套杉(袖套杉)亚属 ·· 241/61
　　　　Manica (*Manica*) *parceramosa* 希枝袖套杉(袖套杉) ·· 241/61
　　　　Manica (*Manica*) *dalatzensis* 大拉子袖套杉(袖套杉) ··· 241/61
　　　　Manica (*Manica*) *foveolata* 窝穴袖套杉(袖套杉) ·· 241/62
　　　　Manica (*Manica*) *papillosa* 乳突袖套杉(袖套杉) ·· 242/62
Mediocycas 中间苏铁属 ·· 242/62
　　Mediocycas kazuoensis 喀左中间苏铁 ··· 242/62
Megalopteris Schenk,1883 (non Andrews E B,1875) 大叶羊齿属 ·· 242/62
　　Megalopteris nicotianaefolia 烟叶大叶羊齿 ··· 243/63
Meia 梅氏叶属 ··· 243/63
　　Meia mingshanensis 鸣山梅氏叶 ·· 243/63
　　Meia magnifolia 大叶梅氏叶 ··· 243/63
Membranifolia 膜质叶属 ··· 243/63

Membranifolia admirabilis 奇异膜质叶 243/63
Metacladophyton 异枝蕨属 244/64
 Metacladophyton tetraxylum 四木质柱异枝蕨 244/64
Metalepidodendron 变态鳞木属 244/64
 Metalepidodendron sinensis 中国变态鳞木 244/64
 Metalepidodendron xiabanchengensis 下板城变态鳞木 244/64
Metasequoia 水杉属 244/64
 Metasequoia glyptostroboides 水松型水杉 244/64
 Metasequoia disticha 二列水杉 245/64
Metzgerites 似叉苔属 245/65
 Metzgerites yuxinanensis 蔚县似叉苔 245/65
Minarodendron 塔状木属 245/65
 Minarodendron cathaysiense 华夏塔状木 245/65
Mirabopteris 奇异羊齿属 245/65
 Mirabopteris hunjiangensis 浑江奇异羊齿 246/65
Mironeura 奇脉叶属 246/66
 Mironeura dakengensis 大坑奇脉叶 246/66
Mixophylum 间羽叶属 246/66
 Mixophylum simplex 简单间羽叶 246/66
Mixopteris 间羽蕨属 247/66
 Mixopteris intercalaris 插入间羽蕨 247/66
Mnioites 似提灯藓属 247/66
 Mnioites brachyphylloides 短叶杉型似提灯藓 247/67
Monogigantonoclea 单叶单网羊齿属 247/67
 Monogigantonoclea colocasifolia 芋叶单叶单网羊齿 247/67
 Monogigantonoclea rotundifolia 圆形单叶单网羊齿 248/67
 Monogigantonoclea latiovata 阔卵单叶单网羊齿 248/67
 Monogigantonoclea grandidenia 巨齿单叶单网羊齿 248/67
 Monogigantonoclea aceroides 似槭单叶单网羊齿 248/67
Monogigantopteris 单叶大羽羊齿属 248/68
 Monogigantopteris clathroreticulatus 格网单叶大羽羊齿 248/68
 Monogigantopteris densireticulatus 密网单叶大羽羊齿 248/68
Muricosperma 尖籽属 249/68
 Muricosperma guizhouensis 贵州尖籽 249/68
Myriophyllum 密叶属 249/68
 Myriophyllum shanxiense 山西密叶 249/68

N

Nanpiaophyllum 南票叶属 249/69
 Nanpiaophyllum cordatum 心形南票叶 249/69
Nanzhangophyllum 南漳叶属 250/69
 Nanzhangophyllum donggongense 东巩南漳叶 250/69
Neoannularia 新轮叶属 250/69
 Neoannularia shanxiensis 陕西新轮叶 250/69

Neoannularia chuandianensis 川滇新轮叶	250/69
Neocordaites 新科达属	250/70
Neocordaites lanceolatus 披针新科达	250/70
Neogigantopteridium 新准大羽羊齿属	251/70
Neogigantopteridium spiniferum 具刺新准大羽羊齿	251/70
Neostachya 新孢穗属	251/70
Neostachya shanxiensis 陕西新孢穗	251/70
Neurophyllites 翅叶属	251/70
Neurophyllites pecopteroides 栉状翅叶	251/71
Ningxiaphyllum 宁夏叶属	252/71
Ningxiaphyllum trilobatum 三裂宁夏叶	252/71
Norinia 那琳壳斗属	252/71
Norinia cucullata 僧帽状那琳壳斗	252/71
Nudasporestrobus 裸囊穗属	252/71
Nudasporestrobus ningxicus 宁夏裸囊穗	252/71
Nystroemia 新常富籽属	253/72
Nystroemia pectiniformis 篦形新常富籽	253/72

O

Odontosorites 似齿囊蕨属	253/72
Odontosorites heerianus 海尔似齿囊蕨	253/72
Orchidites 似兰属	253/72
Orchidites linearifolius 线叶似兰	253/72
Orchidites lancifolius 披针叶似兰	254/73
Otofolium 耳叶属	254/73
Otofolium polymorphum 多形耳叶	254/73
Otofolium ovatum 卵耳叶	254/73

P

Palaeoginkgoxylon 古银杏型木属	254/73
Palaeoginkgoxylon zhoui 周氏古银杏型木	254/73
Palaeognetaleaana 古买麻藤属	255/73
Palaeognetaleaana auspicia 吉祥古买麻藤	255/74
Palaeoskapha 古舟藤属	255/74
Palaeoskapha sichuanensis 四川古舟藤	255/74
Pania 潘氏果属	255/74
Pania cycadina 似苏铁潘氏果	255/74
Pankuangia 潘广叶属	256/74
Pankuangia haifanggouensis 海房沟潘广叶	256/74
Papilionifolium 蝶叶属	256/75
Papilionifolium hsui 徐氏蝶叶	256/75
Paracaytonia 副开通尼亚属	256/75
Paracaytonia hongtaoi 洪涛副开通尼亚	256/75

Paraconites 副球果属 ·· 257/75
 Paraconites longifolius 伸长副球果 ··· 257/75
Paradoxopteris Mi et Liu,1977（non Hirmer,1927）奇异羊齿属 ······················· 257/75
 Paradoxopteris hunjiangensis 浑江奇异羊齿 ·· 257/76
Paradrepanozamites 副镰羽叶属 ··· 257/76
 Paradrepanozamites dadaochangensis 大道场副镰羽叶 ······························ 257/76
Parasphenophyllum 拟楔叶属 ··· 258/76
 Parasphenophyllum shansiense 山西拟楔叶 ·· 258/76
Parasphenopteris 拟楔羊齿属 ·· 258/76
 Parasphenopteris orientalis 东方拟楔羊齿 ·· 258/77
Parastorgaardis 拟斯托加枝属 ·· 258/77
 Parastorgaardis mentoukouensis 门头沟拟斯托加枝 ··································· 258/77
Parataxospermum 类紫杉籽属 ·· 259/77
 Parataxospermum taiyuanesis 太原类紫杉籽 ·· 259/77
Paratingia 拟齿叶属 ·· 259/77
 Paratingia datongensis 大同拟齿叶 ·· 259/77
Paratingiostachya 拟丁氏蕨穗属 ··· 259/78
 Paratingiostachya cathaysiana 华夏拟丁氏蕨穗 ··· 259/78
Pavoniopteris 雅蕨属 ·· 259/78
 Pavoniopteris matonioides 马通蕨型雅蕨 ·· 260/78
Pectinangium 篦囊属 ··· 260/78
 Pectinangium lanceolatum 披针篦囊 ··· 260/78
Perisemoxylon 雅观木属 ··· 260/79
 Perisemoxylon bispirale 双螺纹雅观木 ·· 260/79
 Perisemoxylon sp. 雅观木（未定种）·· 260/79
Phoenicopsis 拟刺葵属 ··· 261/79
 Phoenicopsis angustifolia 狭叶拟刺葵 ·· 261/79
 Phoenicopsis (*Stephenophyllum*) 拟刺葵（斯蒂芬叶）亚属 ····························· 261/79
 Phoenicopsis (*Stephenophyllum*) *solmsi* 索尔姆斯拟刺葵（斯蒂芬叶）············· 261/79
 Phoenicopsis (*Stephenophyllum*) *decorata* 美形拟刺葵（斯蒂芬叶）················ 261/80
 Phoenicopsis (*Stephenophyllum*) *enissejensis* 厄尼塞捷拟刺葵（斯蒂芬叶）······ 261/80
 Phoenicopsis (*Stephenophyllum*) *mira* 特别拟刺葵（斯蒂芬叶）······················ 262/80
 Phoenicopsis (*Stephenophyllum*) *taschkessiensis* 塔什克斯拟刺葵（斯蒂芬叶） 262/80
Phoroxylon 贼木属 ·· 262/80
 Phoroxylon scalariforme 梯纹状贼木 ·· 262/80
Phylladendroid 叶茎属 ·· 262/81
 Phylladendroid jiangxiensis 江西叶茎 ··· 262/81
Pinnagigantonoclea 羽叶单网羊齿属 ·· 263/81
 Pinnagigantonoclea zelkovoides 似榉羽叶单网羊齿 ····································· 263/81
 Pinnagigantonoclea heteroeura 异常羽叶单网羊齿 ·· 263/81
 Pinnagigantonoclea mira 异脉羽叶单网羊齿 ··· 263/81
 Pinnagigantonoclea guizhouensis 贵州羽叶单网羊齿 ···································· 263/81
 Pinnagigantonoclea mucronata 尖头羽叶单网羊齿 ······································· 263/82
 Pinnagigantonoclea spatulata 匙羽叶单网羊齿 ·· 264/82
 Pinnagigantonoclea rosulata 莲座羽叶单网羊齿 ··· 264/82

Pinnagigantonoclea dryophylloides 似槲羽叶单网羊齿	264/82
Pinnagigantonoclea polymorpha 多形羽叶单网羊齿	264/82
Pinnagigantopteris 羽叶大羽羊齿属	264/82
Pinnagigantopteris nicotianaefolia 烟叶羽叶大羽羊齿	265/82
Pinnagigantopteris lanceolatus 披针羽叶大羽羊齿	265/83
Pinnagigantopteris oblongus 长圆羽叶大羽羊齿	265/83
Pinnatiramosus 羽枝属	265/83
Pinnatiramosus qianensis 黔羽枝	265/83
Plagiozamiopsis 拟斜羽叶属	265/83
Plagiozamiopsis podozamoides 苏铁杉型拟斜羽叶	266/83
Polygatites 似远志属	266/83
Polygatites sp. indet. 似远志(sp. indet.)	266/84
Polygonites Wu S Q, 1999 (non Saporta, 1865) 似蓼属	266/84
Polygonites polyclonus 多小枝似蓼	266/84
Polygonites planus 扁平似蓼	266/84
Polypetalophyton 多瓣蕨属	267/84
Polypetalophyton wufengensis 五峰多瓣蕨	267/85
Polythecophyton 多囊枝属	267/85
Polythecophyton demissum 下弯多囊枝	267/85
Primocycas 始苏铁属	267/85
Primocycas chinensis 中国始苏铁	267/85
Primocycas muscariformis 帚状始苏铁	268/85
Primoginkgo 始拟银杏属	268/85
Primoginkgo dissecta 深裂始拟银杏	268/86
Primozamia 始查米苏铁属	268/86
Primozamia sinensis 中国始查米苏铁	268/86
Prionophyllopteris 锯叶羊齿属	268/86
Prionophyllopteris spiniformis 多刺锯叶羊齿	268/86
Procycas 原苏铁属	269/86
Procycas densinervioides 密脉原苏铁	269/86
Progigantonoclea 原单网羊齿属	269/87
Progigantonoclea henanensis 河南原单网羊齿	269/87
Progigantopteris 原大羽羊齿属	269/87
Progigantopteris brevireticulatus 短网原大羽羊齿	269/87
Proginkgoxylon 原始银杏型木属	270/87
Proginkgoxylon benxiense 本溪原始银杏型木	270/87
Proginkgoxylon daqingshanense 大青山原始银杏型木	270/88
Protoglyptostroboxylon 原始水松型木属	271/88
Protoglyptostroboxylon giganteum 巨大原始水松型木	271/88
Protoglyptostroboxylon yimiense 伊敏原始水松型木	271/88
Protopteridophyton 原始蕨属	271/88
Protopteridophyton devonicum 泥盆纪原始蕨	271/88
Protosciadopityoxylon 原始金松型木属	271/89
Protosciadopityoxylon liaoningense 辽宁原始金松型木	272/89
Pseudopolystichu 假耳蕨属	272/89

Pseudopolystichu cretaceum 白垩假耳蕨 ········· 272/89
Pseudorhipidopsis 假拟扇叶属 ············ 272/89
    Pseudorhipidopsis brevicaulis 宽叶假拟扇叶 ········· 272/89
Pseudotaeniopteris 假带羊齿属 ············ 272/90
    Pseudotaeniopteris piscatorius 鱼形假带羊齿 ········· 272/90
Pseudotaxoxylon 白豆杉型木属 ············ 273/90
    Pseudotaxoxylon chinensis 中国白豆杉型木 ········· 90/273
Pseudotsugxylon 黄杉型木属 ············ 273/90
    Pseudotsugxylon pingzhangensis 平庄黄杉型木 ········· 273/90
Pseudoullmannia 假鳞杉属 ············ 273/90
    Pseudoullmannia frumentarioides 类麦假鳞杉 ········· 273/90
    Pseudoullmannia bronnioides 类纹假鳞杉 ········· 273/91
Pteridiopsis 拟蕨属 ············ 274/91
    Pteridiopsis didaoensis 滴道拟蕨 ········· 274/91
    Pteridiopsis tenera 柔弱拟蕨 ········· 274/91

Q

Qinlingopteris 秦岭羊齿属 ············ 274/91
    Qinlingopteris orientalis 东方秦岭羊齿 ········· 274/91
    Qinlingopteris sp. 秦岭羊齿(未定种) ········· 274/91
Qionghaia 琼海叶属 ············ 274/92
    Qionghaia carnosa 肉质琼海叶 ········· 275/92

R

Radiatifolium 辐叶属 ············ 275/92
    Radiatifolium magnusum 大辐叶 ········· 275/92
Ramophyton 多枝蕨属 ············ 275/92
    Ramophyton givetianum 吉维特多枝蕨 ········· 275/92
Ranunculophyllum 毛茛叶属 ············ 275/93
    Ranunculophyllum pinnatisctum 羽状全裂毛茛叶 ········· 276/93
Rastropteris 耙羊齿属 ············ 276/93
    Rastropteris pingquanensis 平泉耙羊齿 ········· 276/93
Rehezamites 热河似查米亚属 ············ 276/93
    Rehezamites anisolobus 不等裂热河似查米亚 ········· 276/93
    Rehezamites sp. 热河似查米亚(未定种) ········· 276/94
Renifolium 肾叶属 ············ 277/94
    Renifolium logipetiolatum 长柄肾叶 ········· 277/94
Reteophlebis 网格蕨属 ············ 277/94
    Reteophlebis simplex 单式网格蕨 ········· 277/94
Reticalethopteris 网延羊齿属 ············ 277/94
    Reticalethopteris yuani 袁氏网延羊齿 ········· 277/94
Rhizoma 根状茎属 ············ 278/95
    Rhizoma elliptica 椭圆形根状茎 ········· 278/95

Rhizomopsis 拟根茎属 ··· 278/95
    Rhizomopsis gemmifera 具芽拟根茎 ·· 278/95
Rhomboidopteris 菱羊齿属 ·· 278/95
    Rhomboidopteris yongwolensis 菱羊齿 ·· 278/95
Riccardiopsis 拟片叶苔属 ·· 278/96
    Riccardiopsis hsüi 徐氏拟片叶苔 ··· 279/96
Rireticopteris 日蕨属 ··· 279/96
    Rireticopteris microphylla 小叶日蕨 ··· 279/96
Rotafolia 轮叶蕨属 ··· 279/96
    Rotafolia songziensis 松滋轮叶蕨 ··· 279/96

S

Sabinites 似圆柏属 ··· 280/97
    Sabinites neimonglica 内蒙古似圆柏 ·· 280/97
    Sabinites gracilis 纤细似圆柏 ··· 280/97
Sagittopteris Zhang E et Xiao,1985（non Zhang S et Xiao,1987）箭羽羊齿属 ······· 280/97
    Sagittopteris belemnopteroides Zhang E et Xiao,1985（non Zhang S et Xiao,1987）
    戟形箭羽羊齿 ·· 280/97
Sagittopteris Zhang S et Xiao,1987（non Zhang E et Xiao,1985）箭羽羊齿属 ······· 280/97
    Sagittopteris belemnopteroides Zhang S et Xiao,1987（non Zhang E et Xiao,1985）
    戟形箭羽羊齿 ·· 281/98
Schizoneuropsis 拟裂鞘叶属 ·· 281/98
    Schizoneuropsis tokudae 德田拟裂鞘叶 ·· 281/98
Sciadocillus 小伞属 ··· 281/98
    Sciadocillus cuneifidus 楔裂小伞 ·· 281/98
Scoparia 帚羽叶属 ··· 281/98
    Scoparia plumaria 羽毛帚羽叶 ··· 282/98
Semenalatum 翅籽属 ·· 282/99
    Semenalatum paucum 珍贵翅籽 ··· 282/99
Setarites 似狗尾草属 ··· 282/99
    Setarites sp. indet. 似狗尾草（sp. indet.）··· 282/99
Shangyuania 上园草属 ·· 282/99
    Shangyuania caii 才氏上园草 ··· 283/99
Shanxicladus 山西枝属 ·· 283/100
    Shanxicladus pastulosus 疹形山西枝 ··· 283/100
Shanxioxylon 山西木属 ·· 283/100
    Shanxioxylon sinense 中国山西木 ··· 283/100
    Shanxioxylon taiyuanense 太原山西木 ·· 283/100
Shenea 沈氏蕨属 ·· 284/100
    Shenea hirschmeierii 希氏沈氏蕨 ··· 284/100
Shenkuoia 沈括叶属 ··· 284/101
    Shenkuoia caloneura 美脉沈叶 ·· 284/101
Shenzhouphyllum 神州叶属 ·· 284/101
    Shenzhouphyllum undulatum 波缘神州叶 ·· 284/101

Shenzhouphyllum rotundatum 圆形神州叶	285/101
Shenzhouphyllum spatulatum 匙形神州叶	285/101
Shenzhouspermum 神州籽属	285/102
Shenzhouspermum trichotomum 三歧神州籽	285/102
Shenzhoutheca 神州聚囊属	285/102
Shenzhoutheca aspergilliformis 刷状神州聚囊	285/102
Shuangnangostachya 双囊芦穗属	286/102
Shuangnangostachya gracilis 细小双囊芦穗	286/102
Shuichengella 水城蕨属	286/102
Shuichengella primitiva 原始水城蕨	286/103
Siella 斯氏鞘叶属	286/103
Siella leptocostata 细肋斯氏鞘叶	287/103
Sinocarpus 中华古果属	287/103
Sinocarpus decussatus 下延中华古果	287/103
Sinoctenis 中国篦羽叶属	287/103
Sinoctenis grabauiana 葛利普中国篦羽叶	287/104
Sinodicotis 中华缘蕨属	287/104
Sinodicotis sp. indet. 中华缘蕨(sp. indet.)	288/104
Sinopalaeospiroxylon 中国古螺纹木属	288/104
Sinopalaeospiroxylon baoligemiaoense 宝力格庙中国古螺纹木	288/104
Sinopalaeospiroxylon napiaoense 南票中国古螺纹木	288/105
Sinopalaeospiroxylon pingquanense 平泉中国古螺纹木	289/105
Sinophyllum 中国叶属	289/105
Sinophyllum suni 孙氏中国叶	289/105
Sinozamites 中国似查米亚属	289/105
Sinozamites leeiana 李氏中国似查米亚	289/105
Siphonospermum 管子麻黄属	289/106
Siphonospermum simplex 简单管子麻黄	290/106
Solaranthus 太阳花属	290/106
Solaranthus daohugouensis 道虎沟太阳花	290/106
Speirocarpites 似卷囊蕨属	290/106
Speirocarpites virginiensis 弗吉尼亚似卷囊蕨	290/106
Speirocarpites dukouensis 渡口似卷囊蕨	290/106
Speirocarpites rireticopteroides 日蕨型似卷囊蕨	291/107
Speirocarpites zhonguoensis 中国似卷囊蕨	291/107
Sphenobaieroanthus 楔叶拜拉花属	291/107
Sphenobaieroanthus sinensis 中国楔叶拜拉花	291/107
Sphenobaierocladus 楔叶拜拉枝属	291/107
Sphenobaierocladus sinensis 中国楔叶拜拉枝	291/107
Sphenopecopteris 楔栉羊齿属	292/107
Sphenopecopteris beaniata 豆子楔栉羊齿	292/108
Sphinxia Li,Hilton et Hemsley,1997 (non Reid et Chandler,1933) 斯芬克斯籽属	292/108
Sphinxia wuhania 武汉斯芬克斯籽	292/108
Sphinxiocarpon 仙籽属	292/108
Sphinxiocarpon wuhania 武汉斯芬克斯仙籽	292/108

Spinolepidodendron 刺鳞木属 ·· 293/108
    Spinolepidodendron hangzhouense 杭州刺鳞木 ································· 293/109
Squamocarpus 鳞籽属 ·· 293/109
    Squamocarpus papilioformis 蝶形鳞籽 ·· 293/109
Squarmacarpus 翅鳞籽属 ·· 293/109
    Squarmacarpus cuneiformis 楔形翅鳞籽 ·· 293/109
Stachybryolites 穗藓属 ·· 294/109
    Stachybryolites zhoui 周氏穗藓 ·· 294/109
Stachyophyton 穗蕨属 ·· 294/110
    Stachyophyton yunnanense 云南穗蕨 ·· 294/110
Stalagma 垂饰杉属 ·· 294/110
    Stalagma samara 翅籽垂饰杉 ·· 294/110
Stephanofolium 金藤叶属 ·· 295/110
    Stephanofolium ovatiphyllum 卵形金藤叶 ·· 295/110
Strigillotheca 刷囊属 ·· 295/111
    Strigillotheca fasciculata 束囊刷囊 ·· 295/111
Suturovagina 缝鞘杉属 ·· 295/111
    Suturovagina intermedia 过渡缝鞘杉 ·· 295/111
Symopteris 束脉蕨属 ·· 296/111
    Symopteris helvetica 瑞士束脉蕨 ·· 296/111
    Symopteris densinervis 密脉束脉蕨 ·· 296/111
    Symopteris zeilleri 蔡耶束脉蕨 ·· 296/112
Szea 天石蕨属 ·· 296/112
    Szea sinensis 中国天石蕨 ·· 296/112
Szecladia 斯氏松属 ·· 297/112
    Szecladia multinervia 多脉斯氏松 ·· 297/112
Szeioxylon 斯氏木属 ·· 297/112
    Szeioxylon xuzhouene 徐州斯氏木 ·· 297/112

T

Tachingia 大箐羽叶属 ·· 297/113
    Tachingia pinniformis 大箐羽叶 ·· 297/113
Taeniocladopsis 拟带枝属 ·· 297/113
    Taeniocladopsis rhizomoides 假根茎型拟带枝 ·· 298/113
Taipingchangella 太平场蕨属 ·· 298/113
    Taipingchangella zhongguoensis 中国太平场蕨 ······································ 298/113
Taiyuanitheca 太原蕨属 ·· 298/113
    Taiyuanitheca tetralinea 四线形太原蕨 ·· 298/114
Tchiaohoella 蛟河羽叶属 ·· 298/114
    Tchiaohoella mirabilis 奇异蛟河羽叶 ·· 299/114
    Tchiaohoella sp. 蛟河羽叶(未定种) ·· 299/114
Tenuisa 细轴始蕨属 ·· 299/114
    Tenuisa frasniana 弗拉细轴始蕨 ·· 299/114
Tetrafolia 四叶属 ·· 299/114

Tetrafolia changshaense 长沙四叶	299/114
Tharrisia 哈瑞士叶属	299/115
Tharrisia dinosaurensis 迪纳塞尔哈瑞士叶	299/115
Tharrisia lata 侧生哈瑞士叶	300/115
Tharrisia spectabilis 优美哈瑞士叶	300/115
Thaumatophyllum 奇异羽叶属	300/115
Thaumatophyllum ptilum 羽毛奇异羽叶	300/116
Thelypterites 似金星蕨属	301/116
Thelypterites sp. A 似金星蕨(未定种 A)	301/116
Thelypterites sp. B 似金星蕨(未定种 B)	301/116
Tianoxylon 田氏木属	301/116
Tianoxylon duanmutouense 段木头田氏木	301/116
Tianshanopteris 天山羊齿属	302/117
Tianshanopteris wensuensis 温宿天山羊齿	302/117
Tianshia 天石枝属	302/117
Tianshia patens 伸展天石枝	302/117
Tingia 丁氏羊齿属	302/117
Tingia carbonica 石炭丁氏羊齿	302/117
Toksunopteris 托克逊蕨属	302/117
Toksunopteris opposita 对生托克逊蕨	303/118
Tongchuanophyllum 铜川叶属	303/118
Tongchuanophyllum trigonus 三角形铜川叶	303/118
Tongchuanophyllum concinnum 优美铜川叶	303/118
Tongchuanophyllum shensiense 陕西铜川叶	303/118
Tonglucarpus 桐庐籽属	303/118
Tonglucarpus spectabilis 奇丽桐庐籽	304/118
Tongshania 钟囊属	304/119
Tongshania dentate 齿状钟囊	304/119
Torreyocladus 榧型枝属	304/119
Torreyocladus spectabilis 明显榧型枝	304/119
Tricoemplectopteris 三网羊齿属	304/119
Tricoemplectopteris taiyuanensis 太原三网羊齿	304/119
Tricrananthus 三裂穗属	305/119
Tricrananthus sagittatus 箭头状三裂穗	305/120
Tricrananthus lobatus 瓣状三裂穗	305/120
Trinerviopteris 三脉蕨属	305/120
Trinerviopteris cardiophylla 心叶三脉蕨	305/120
Triqueteria 三棱果属	305/120
Triqueteria sinensis 中国三棱果	306/120
Tsaia 蔡氏蕨属	306/120
Tsaia denticulata 细齿蔡氏蕨	306/121
Tsiaohoella 蛟河蕉羽叶属	306/121
Tsiaohoella mirabilis 奇异蛟河蕉羽叶	306/121
Tsiaohoella neozamioides 新似查米亚型蛟河蕉羽叶	306/121

V

Vasovinea 导管羊齿 ········· 306/121
    Vasovinea tianii 田氏导管羊齿 ········· 307/121
Vittifoliolum 条叶属 ········· 307/121
    Vittifoliolum segregatum 游离条叶 ········· 307/122
    Vittifoliolum segregatum forma *costatum* 游离条叶脊条型 ········· 307/122
    Vittifoliolum multinerve 多脉条叶 ········· 307/122

W

Wenshania 文山蕨属 ········· 308/122
    Wenshania zhichangensis 纸厂文山蕨 ········· 308/122
Wutubulaka 乌图布拉克蕨属 ········· 308/122
    Wutubulaka multidichotoma 多叉乌图布拉克蕨 ········· 308/122
Wuxia 无锡蕨属 ········· 308/123
    Wuxia bistrobilata 双穗无锡蕨 ········· 308/123
Wuyunanthus 乌云花属 ········· 309/123
    Wuyunanthus hexapetalus 六瓣乌云花 ········· 309/123

X

Xiajiajienia 夏家街蕨属 ········· 309/123
    Xiajiajienia mirabila 奇异夏家街蕨 ········· 309/123
Xinganphyllum 兴安叶属 ········· 309/124
    Xinganphyllum aequale 对称兴安叶 ········· 309/124
    Xinganphyllum inaequale 不对称兴安叶 ········· 310/124
    Xinganphyllum sp. 兴安叶(未定种) ········· 310/124
Xingxueanthus 星学花属 ········· 310/124
    Xingxueanthus sinensis 中国星学花 ········· 310/124
Xingxueina 星学花序属 ········· 310/124
    Xingxueina heilongjiangensis 黑龙江星学花序 ········· 310/125
Xingxuephyllum 星学叶属 ········· 311/125
    Xingxuephyllum jixiense 鸡西星学叶 ········· 311/125
Xinjiangophyton 新疆木属 ········· 311/125
    Xinjiangophyton spinosum 刺状新疆木 ········· 311/125
Xinjiangopteris Wu S Q et Zhou,1986 (non Wu S Z,1983) 新疆蕨属 ········· 311/125
    Xinjiangopteris opposita 对生新疆蕨 ········· 311/126
Xinjiangopteris Wu S Z,1983 (non Wu S Q et Zhou,1986) 新疆蕨属 ········· 312/126
    Xinjiangopteris toksunensis 托克逊新疆蕨 ········· 312/126
Xinlongia 新龙叶属 ········· 312/126
    Xinlongia pterophylloides 侧羽叶型新龙叶 ········· 312/126
    Xinlongia hoheneggeri 和恩格尔新龙叶 ········· 312/126
Xinlongophyllum 新龙羽叶属 ········· 312/126

Xinlongophyllum ctenopteroides 篦羽羊齿型新龙羽叶 ·················· 313/127
Xinlongophyllum multilineatum 多条纹新龙羽叶 ·················· 313/127
Xitunia 西屯蕨属 ·················· 313/127
Xitunia spinitheca 刺囊西屯蕨 ·················· 313/127

Y

Yangzunyia 杨氏木属 ·················· 313/127
Yangzunyia henanensis 河南杨氏木 ·················· 313/127
Yanjiphyllum 延吉叶属 ·················· 314/127
Yanjiphyllum ellipticum 椭圆延吉叶 ·················· 314/128
Yanliaoa 燕辽杉属 ·················· 314/128
Yanliaoa sinensis 中国燕辽杉 ·················· 314/128
Yimaia 义马果属 ·················· 314/128
Yimaia recurva 外弯义马果 ·················· 314/128
Yixianophyllum 义县叶属 ·················· 315/128
Yixianophyllum jinjiagouensie 金家沟义县叶 ·················· 315/128
Yuania 卵叶属 ·················· 315/129
Yuania striata 条纹卵叶 ·················· 315/129
Yuguangia 玉光蕨属 ·················· 315/129
Yuguangia ordinata 规则玉光蕨 ·················· 315/129
Yungjenophyllum 永仁叶属 ·················· 316/129
Yungjenophyllum grandifolium 大叶永仁叶 ·················· 316/129
Yunia 云蕨属 ·················· 316/130
Yunia dichotoma 二叉云蕨 ·················· 316/130

Z

Zeillerpteris 蔡耶羊齿属 ·················· 316/130
Zeillerpteris yunnanensis 云南蔡耶羊齿 ·················· 316/130
Zhengia 郑氏叶属 ·················· 317/130
Zhengia chinensis 中国郑氏叶 ·················· 317/130
Zhenglia 正理蕨属 ·················· 317/131
Zhenglia radiata 辐射正理蕨 ·················· 317/131
Zhongzhoucarus 中州籽属 ·················· 317/131
Zhongzhoucarus deltatus 三角中州籽 ·················· 317/131
Zhongzhoupteris 中州蕨属 ·················· 318/131
Zhongzhoupteris cathaysicus 尾羽中州蕨 ·················· 318/131
Zhouia 周氏籽属 ·················· 318/131
Zhouia beipiaoensis 北票周氏籽 ·················· 318/132
Zhutheca 朱氏囊蕨属 ·················· 318/132
Zhutheca densata 密囊朱氏囊蕨 ·················· 318/132

REFERENCES

Palaeozoic Plants from China Writing Group of Nanjing Institute of Geology and Palaeontology, Institute of Botany, Chinese Academy of Sciences (Gu et Zhi), 1974. Palaeozoic Plants from China. Beijing: Science Press: 1-226, pls. 1-130, text-figs. 1-142. (in Chinese)

Andrews E B, 1875. Descriptions of fossil plants from the Coal Measures of Ohio: Ohio Geol. Survey Rept., v. 2, Geology and Palaeontology, pt. 2: 415-426, pls. 46-53.

Asama K, 1959. Systematic study of so-called *Gigantopteris*. Science Reports of Tohoku University, Sendai, Japan, Series 2 (Geology), 31 (1): 1-72, pls. 1-20, text-figs. 1-6.

Asama K, 1970. Evolution and classification of Sphenophyllales in Cathaysia land. Bulletin of Natural Science Museum of Tokyo, Japan, 13 (2): 291-317, pls. 1-7.

Berry C M, Wang Yi (王怿), Cai Chongyang (蔡重阳), 2003. A lycopsid with novel reproductive structures from the Upper Devonian of Jiangsu, China. International Journal of Plant Sciences, 164: 263-273.

Bohlin B, 1971. Late Palaeozoic plants from Yuerhhung, Kansu, China // Reports from the scientific expedition to the north-western provinces of China under the leadership of Dr. Seven Hedin: the Sino-Sewedish expedition publication 51. Ⅳ. Palaeobotany. Stockholm: The Sven Hedin Foundation: Ⅰ, 1-150; Ⅱ, pls. 1-25, figs. 1-296.

Bose M N, Sah S G D, 1954. On *Sahnioxylon rajmahalense*, a new name for *Homoxylon rajmahalense* Sahni, and *S. andrewsii*, a new species of *Sahnioxylon* from Amrapara in the Rajmahal Hills, Behar. Palaeobotanist, 3: 1-8, pls. 1, 2.

Cao Zhengyao (曹正尧), Wu Shunqing (吴舜卿), Zhang Pingan (张平安), Li Jieru (李杰儒), 1997. Discovery of fossil monocotyledons from Yixian Formation, western Liaoning. Chinese Science Bulletin, 43 (3): 230-233, pls. 1, 2, figs. 1, 2. (in English)

Cao Zhengyao (曹正尧), Wu Shunqing (吴舜卿), Zhang Pingan (张平安), Li Jieru (李杰儒), 1998. Discovery of fossil monocotyledons from Yixian Formation, western Liaoning. Chinese Science Bulletin, 42 (16): 1764-1766, pls. 1, 2, figs. 1, 2. (in Chinese)

Cao Zhengyao (曹正尧), 1999. Early Cretaceous flora of Zhejiang. Palaeontologia Sinica, Whole Number 187, New Series A, 13: 1-174, pls. 1-40, text-figs. 1-35. (in Chinese and English)

Chang Chichen (张志诚), 1980. Subphyllum Angiospermae // Shenyang Institute of Geology and Mineral Resources ed. Paleontological atlas of Northeast China, Ⅱ. Mesozoic and Cenozoic. Beijing: Geological Publishing House: 308-342, pls. 192-210, text-figs. 208-211. (in Chinese with English title)

Chang Chichen (张志诚), 1981. Several Cretaceous angiospermous from Mudanjiang Basin, Heilongjiang. Bulletin of the Chinese Academy of Geological Sciences, Series V, 2 (1): 154-160, pls. 1, 2. (in Chinese with English summary)

Chen Gongxin (陈公信), 1984. Pteridophyta, Spermatophyta // Regional Geological Surveying Team of Hubei Province ed. The palaeontological atlas of Hubei Province. Wuhan: Hubei Science and Technology Press: 556-615, 797-812, pls. 216-270, figs. 117-133. (in Chinese with English title)

Chen Qishi (陈其奭), Zhu Deshou (朱德寿), 1994. Discovery and establishment for the new genus and species in Late Permian fossil of seed. Geology of Zhejiang, 10 (2): 5-8, pl. 1, text-figs. 1, 2. (in Chinese with English summary)

Chen Qishi (陈其奭), 1999. Fossil plants Lycopodiales from Late Devonian Xihu Formation and Wutong Formation in Xiaoshan and Changxing, Zhejiang. Geology of Zhejiang, 15 (2): 15-23, pls. 1-5, text-figs. 1-4. (in Chinese with English summary)

Chow Tseyen (周志炎), Tsao Chengyao (曹正尧), 1977. On eight species of conifers from the Cretaceous of East China with reference to their taxonomic position and phylogenetic relationship. Acta Palaeontologica Sinica, 16 (2): 165-181, pls. 1-5, text-figs. 1-6. (in Chinese with English summary)

Dawson J W, 1868. Acadian Geology. 2d ed. Edinburgh: Oliver and Boyd: 694.

Deng Shenghui (邓胜徽), Chen Fen (陈芬), 2001. The Early Cretaceous Filicopsida from Northeast China. Geological Publishing House: 1-249, pls. 1-123, text-figs. 1-41. (in Chinese with English summary)

Deng Shenghui (邓胜徽), Wang Shijun (王士俊), 1999. *Klukiopsis jurassica*: a new Jurassic schizaeaceous fern from China. Science in China, Series D, 29 (6): 551－557, fig. 1. (in Chinese)

Deng Shenghui (邓胜徽), Wang Shijun (王士俊), 2000. *Klukiopsis jurassica*: a new Jurassic schizaeaceous fern from China. Science in China, Series D, 43 (4): 356-363, fig. 1. (in English)

Deng Shenghui (邓胜徽), 1993. Four new species of Early Cretaceous ferns. Geoscience, 7 (3): 255-260, pl. 1, text-fig. 1. (in Chinese with English summary)

Deng Shenghui (邓胜徽), 1994. *Dracopteris liaoningensis* gen. et sp. nov.: a new Early Cretaceous fern from NE China. Geophytology, 24 (1): 13-22, pls. 1-4, text-figs. 1, 2.

Deng Shenghui (邓胜徽), 1995. Early Cretaceous flora of Huolinhe Basin, Inner Mongolia, Northeast China. Beijing: Geological Publishing House: 1-125, pls. 1-48, text-figs. 1-23. (in Chinese with English summary)

Deng Shenghui (邓胜徽), 1997. *Eogonocormus*: a new Early Cretaceous fern of Hymenophyllaceae from Northeast China. Australian Systematic Botany, 10 (1): 59-67, pls. 1-4, fig. 1.

Dilcher D L, Mei Meitang (梅美棠), Du Meili (杜美利), 1997. A new winged seed from the Permian of China. Review of Palaeobotany and Palynology, 98 (3-4): 247-256, pl. 1, text-figs. 1, 2.

Dilcher D L, Sun Ge (孙革), Ji Qiang (季强), Li Hongqi (李洪起), 2007. An early infructescence Hyrcantha decussata (comb. nov.) from the Yixian Formation in northeastern China. PNAS,104（22）:9370-9374.

Dou Yawei（窦亚伟）,Sun Zhehua（孙喆华）,Wu Shaozu（吴绍祖）,Gu Daoyuan（顾道源）,1983. Vegetable kingdom// Regional Geological Surveying Team, Institute of Geosciences of Xinjiang Bureau of Geology, Geological Surveying Department, Xinjiang Bureau of Petroleum eds. Palaeontological atlas of Northwest China, Uygur Autonomous Region of Xinjiang,2. Beijing:Geological Publishing House:561-614, pls. 189-226. (in Chinese)

Duan Shuying（段淑英）,Chen Ye（陈晔）,1982. Mesozoic fossil plants and coal formation of eastern Sichuan Basin// Compilatory Group of Continental Mesozoic Stratigraphy and Paleontology in Sichuan Basin ed. Continental Mesozoic stratigraphy and paleontology in Sichuan Basin of China, Part II (Paleontological professional papers). Chengdu:People's Publishing House of Sichuan:491-519,pls. 1-16. (in Chinese with English summary)

Duan Shuying（段淑英）,1987. The Jurassic flora of Zhai Tang, Western Hill of Beijing. Department of Geology, University of Stockholm, Department of Palaeonbotang, Swedish Museum of Natural History, Stockholm:1-95,pls. 1-22,text-figs. 1-17.

Duan Shuying（段淑英）,1997. The oldest angiosperm:a tricarpous female reproductive fossil from western Liaoning Province, NE China. Science in China, Series D,27 (6):519-524, figs. 1-4. (in Chinese)

Duan Shuying（段淑英）,1998. The oldest angiosperm:a tricarpous female reproductive fossil from western Liaoning Province, NE China. Science in China, Series D,41 (1):14-20,figs. 1-4. (in English)

Endo S,1939. Some new and interesting Miocene plants from Tyosen Korea. Pages 333-349 in Jubilee publication in the commemoration of Prof. H. Yabe's 60th birthday, Vol. 1.

Fairon-Demaret M,Li Chengsen（李承森）,1993. *Lorophyton goense* gen. et sp. nov. from the Lower Givetian of Belgium and a discussion of the Middle Devonian Cladoxylopsida. Review of Palaeobotany and Palynology,77:1-22.

Feng Shaonan（冯少南）,Chen Gongxin（陈公信）,Xi Yunhong（席运宏）,Zhang Caifan（张采繁）,1977b. Plants// Hupei Institute of Geological Sciences et al. eds. Fossil atlas of Middle-South China,II. Beijing:Geological Publishing House:622-674,pls. 230-253. (in Chinese)

Feng Shaonan（冯少南）,Ma Jie（马洁）,1991. Study on the genus *Hamatophyton*. Acta Botanica Sinica,33 (2):140-146,pls. 1, 2, text-figs. 1-5. (in Chinese with English summary)

Feng Shaonan（冯少南）,1984. Plant kingdom// Feng Shaonan（冯少南）,Xu Shouyong（许寿永）,Lin Jiaxing（林甲兴）,Yang Deli（杨德骊） eds. Biostratigraphy of the Yangtze Gorge area (3), Late Palaeozoic Era. Beijing:Geological Publishing House:293-305,pls. 46-49. (in Chinese with English summary)

Feng Zhuo（冯卓）,Wang Jun（王军）,Bek J,2008. *Nudasporestrobus ningxicus* gen. et sp. nov.,a novel sigillarian megasporangiate cone from the Bashkirian (Early Pennsylvanian) of Ningxia, northwestern China. Review of Palaeobotany and Palynology,149 (3－4):150－162.

Feng Zhuo（冯卓）,Wang Jun（王军）,Roessler R,2010. *Palaeoginkgoxylon zhoui*, a new ginkgophyte wood from the Guadalupian (Permian) of China and its evolutionary implications. Rev. Palaeobot. Palynol. ,162:146-158.

Florin R,1936. Die fossilen Ginkgophyten von Franz-Joseph-Land nebst Erörterungen ueber vermeintliche Cordaaitales mesozoischen Alters,Ⅰ. Spezieller Teil. Palaeontographica, ABT. B,Band 81:71-173.

Fontaine W M,1883. Contributions to the knowledge of the older Mesozoic flora of Virgina. U S Geol. Surv. ,6:1-144,pls. 1-54.

Fontaine W M,1889. The Potomac or younger Mesozoic flora. Monogr. U S Geol. Surv. ,15:1-377,pls. 1-80.

Fucini Alberto, 1936. Problematica verrucana tavole iconografiche delle vestigia vegetali, animali, fisiche e meccaniche del Wealdiano dei Monti Pisani, Part Ⅰ: Palaeontographia italica,app. 1:126,pls. 76.

Galtier J,Wang Shijun（王士俊）,Li Chengsen（李承森）,Hilton,2001. A new genus of filicalean fern from the Lower Permian of China. Botanical Journal of the Linnean Society, 137:429-442.

Gao Zhifeng（高志峰）,Thomas B A,1991. An enigmatic cone from the Lower Permian of Taiyuan,China. Review of Palaeobotany and Palynology,68（3-4）:197-201, pl. 1, text-fig. 1.

Gao Zhifeng（高志峰）,Thomas B A,1993. A new fern from the Lower Permian of China and its bearing on the evolution of the marattialeans. Palaeontology,36（1）:81-89,text-figs. 1-4.

Geng Baoyin（耿宝印）,Hilton J,1999. New coniferophyte ovulate structures from the Early Permian of China. Botanical Journal of the Linnean Society,129:115-138,figs. 1-22.

Geng Baoyin（耿宝印）,1983. *Stachyophyton* gen. nov. discovers from Lower Devonian of Yunnan and its significance. Acta Botanica Sinica,25（6）:574-579,pls. 1,2,text-fig. 1.（in Chinese with English summary）

Geng Baoyin（耿宝印）, 1985. *Huia recurvata*: a new plant from Lower Devonian of southeastern Yunnan,China. Acta Botanica Sinica,27（4）:419-426,pls. 1,2.（in Chinese with English summary）

Geng Baoyin（耿宝印）,1986. Anatomy and morphology of *Pinnatiramosus*,a new plant from the Middle Silurian（Wenlockian）of China. Acta Botanica Sinica,28（6）:664-670,pls. 1-6. （in Chinese with English summary）

Geng Baoyin（耿宝印）, 1992a. Studies on Early Devonian flora of Sichuan. Acta Phytotaxonomica Sinica,30（3）:197-211,pls. 1-8.（in Chinese with English summary）

Geng Baoyin（耿宝印）,1992b. *Amplectosporangium*: a new genus of plant from the Lower Devonian of Sichuan,China. Acta Botanica Sinica,34（6）:450-455,pl. 1,text-fig. 1.（in Chinese with English summary）

Gothan W,Sze H C（斯行健）,1933. Ueber die palaeozoische Flora der Provinz Kiangsu. Memoirs of National Research Institute Geology,Chinese Academy of Sciences,13:1-40, pls. 1-4.

Guo Shuangxing（郭双兴）,Sha Jingeng（沙金庚）,Bian Lizeng（边力曾）,Qiu Yinlong（仇寅龙）,2009. Male spike strobiles with *Gnetum* affinity from the Early Cretaceous in western Liaoning,Northeast China. Journal of Systematics and Evolution,47（2）:93-102.

Guo Shuangxing（郭双兴）,Sun Zhehua（孙喆华）,Li Haomin（李浩敏）,Dou Yawei（窦亚伟）,

1984. Paleocene megafossil flora from Altai of Xinjiang. Bulletin of Nanjing Institute of Geology and Palaeontology,Chinese Academy of Sciences,8:119-146,pls. 1-8. (in Chinese with English summary)

Guo Shuangxing（郭双兴）,Wu Xiangwu（吴向午）,2000. *Ephedrites* from latest Jurassic Yixian Formation in western Liaoning, Northeast China. Acta Palaeontologica Sinica,39(1):81-91,pls. 1,2. (in Chinese and English)

Guo Shuangxing（郭双兴）,2000. New material of the Late Cretaceous flora from Hunchun of Jilin,Northeast China. Acta Palaeontologica Sinica,39（Supplement）:226-250,pls. 1-8. (in English with Chinese summary)

Halle T G,1925. *Tingia*,a new genus of fossil plants from the Permian of China (preliminary note). Bulletin of Geological Survey of China,7:1-12,pls. 1,2.

Halle T G,1927. Palaeozoic plants from central Shansi. Palaeontologia Sinica,Series A,2(1): 1-316,pls. 1-64.

Halle T G,1936. On *Drepanophycus*,*Protolepidodendron* and *Protopteridium*,with notes on the Palaeozoic flora of Yunnan. Palaeontologia Sinica,Series A,1(4):1-38,pls. 1-5.

Hao Shougang（郝守刚）,Beck C B,1991a. *Catenalis digitata* gen. et sp. nov.,a plant from the Lower Devonian (Siegenian) of Yunnan,China. Canadian Journal of Botany,69(4):873-882,figs. 1-31.

Hao Shougang（郝守刚）,Beck C B,1991b. *Yunia dichotoma*,a Lower Devonian plant from Yunnan,China. Review of Palaeobotany and Palynology,68:181-195,pls. 1-4,text-figs. 1-3.

Hao Shougang（郝守刚）,Gensel P G,Wang Deming（王德明）,2001. *Polythecophyton Demissum*,Gen. Et Sp. Nov.,a New Plant from the Lower Devonian (Pragian) of Yunnan,China and Its Phytogeographic Significance. Review of Palaeobotany and Palynology,116:55-71.

Hao Shougang（郝守刚）,Gensel P G,1995. A new genus and species,*Celatheca beckii* from the Siegenian (Early Devonian) of southeastern Yunnan,China. International Journal of Plant Sciences,156:896-909,figs. 1-35.

Hao Shougang（郝守刚）,Wang Deming（王德明）,Wang Qi（王祺）,Xue Jinzhuang（薛进庄）,2006. A New Lycopsid,*Zhenglia Radiata* Gen. Et Sp. Nov.,from the Lower Devonian Posongchong Formation of Southeastern Yunnan,China,and Its Evolutionary Significance. Acta Geologica Sinica,80:11-19.

Hao Shougang（郝守刚）,Xue Jinzhuang（薛进庄）,Wang Qi（王祺）,Liu Zhenfeng（刘振锋）,2007. *Yuguangia ordinata* gen. et sp. nov.,A new Lycopsid from the Middle Devonian (Late Givetan) of Yunnan,China,and its Phylogenectic Implications. International Journal of Plant Sciences,168(8):1161-1175.

Hao Shougang（郝守刚）,1988. A new Lower Devonian genus from Yunnan,with notes on the origin of leaf. Acta Botanica Sinica,30(4):441-448,pls. 1-3,text-figs. 1,2. (in Chinese with English summary)

Hao Shougang（郝守刚）,1989a. A new zosterophyll from the Lower Devonian (Siegenian) of Yunnan,China. Review of Palaeobotany and Palynology,57:155-171,pls. 1-4,figs. 1-7.

Hao Shougang（郝守刚）,1989b. *Gumuia zyzzata*:a new plant from the Lower Devonian of

Yunnan,China. Acta Botanica Sinica,31（12）:954-961,pls. 1,2,text-figs. 1-3.（in Chinese with English summary）

Harris Thomas Maxwell,1931. The fossil flora of Scoresby Sound,East Greenland:Part 1, Cryptogams（exclusive of Lycopodiales）. Medd. om Gronland,85（2）:1-102,pls. 1-18.

Harris Thomas Maxwell,1932. The fossil flora of Scoresby Sound,East Greenland:Part 2, Description of seed plants incertae sedis together with a discussion of certain cycadophyte cuticles. Medd. om Gronland,85（3）:1-112,pls. 1-9.

He Dechang（何德长）,Zhang Xiuyi（张秀仪）,1993. Some species of coal-forming plants in the seams of the Middle Jurassic in Yima,Henan Province and Ordos Basin. Geoscience,7（3）: 261-265,pls. 1-4.（in Chinese with English summary）

He Dechang（何德长）,1995. The coal-forming plants of Late Mesozoic in Da Hinggan Mountains. Beijing:China Coal Industry Publishing House:1-35,pls. 1-16.（in Chinese and English）

He Xilin（何锡麟）,Liang Dunshi（梁敦士）,Shen Shuzhong（沈树忠）,1996. Research on the Permian flora from Jiangxi Province,China. Xuzhou:China Mining and Technology University Publishing House:1-201,pls. 1-98.（in Chinese with English summary）

He Yuanliang（何元良）,Wu Xiuyuan（吴秀元）,Wu Xiangwu（吴向午）,Li Pejuan（李佩娟）, Li Haomin（李浩敏）,Guo Shuangxing（郭双兴）,1979. Plants // Nanjing Institute of Geology and Palaeontology,Chinese Academy of Sciences,Qinghai Institute of Geological Sciences eds. Fossil atlas of Northwest China Qinghai volume,Ⅱ. Beijing:Geological Publishing House:129-167,pls. 50-82.（in Chinese）

Heer O,1876. Beitrage zur fossilen Flora Spitzbergens,in Flora fossilis arctica,Band 4,Heft 1. Kgl. Svenska vetenskapsakad. Handlingar,14:1-141,pls. 1-32.

Hilton J,Geng Baoyin（耿宝印）,1998. *Batenburgia sakmanica* Hilton et Geng,gen. et sp. nov.,a new genus of conifer from the Lower Permian of China. Review of Palaeobotany and Palynology,103（3-4）:263-287,pls. 1-6,text-figs. 1-4.

Hilton J,Geng Baoyin（耿宝印）,Kenrick P,2003. A novel Late Devonian（Frasnian）woody cladoxylopsid from China. International Journal of Plant Sciences,164:793-805.

Hilton J,Li Chengsen（李承森）,2000. Novel branching structures from the Lower Devonian and a note of caution. Acta Palaeobotanica,40（1）:9-15,pl. 1,fig. 1.

Hsu J（徐仁）,Chu C N（朱家楠）,Chen Yeh（陈晔）,Tuan Shuying（段淑英）,Hu Yufan（胡雨帆）,Chu W C（朱为庆）,1974. New genera and species of Late Traissic plants from Yungjen,Yunnan. Ⅰ. Acta Botanica Sinica,16（3）:266-278,pls. 1-8,text-figs. 1-5.（in Chinese with English summary）

Hsu J（徐仁）,Chu C N（朱家楠）,Chen Yeh（陈晔）,Tuan Shuying（段淑英）,Hu Yufan（胡雨帆）,Chu W C（朱为庆）,1979. Late Triassic Baoding flora,SW Sichuan,China. Beijing: Science Press:1-130,pls. 1-75,text-figs. 1-18.（in Chinese）

Hsu J（徐仁）,Chu C N（朱家楠）,Chen Yeh（陈晔）,Hu Yufan（胡雨帆）,Tuan Shuying（段淑英）,1975. New genera and species of the Late Triassic plants from Yungjen,Yunnan. Ⅱ. Acta Botanica Sinica,17（1）:70-76,pls. 1-6,text-figs. 1,2.（in Chinese with English summary）

Hsu J（徐仁）,1952. Fossil plants from the Kuangshangchang Coal Series of north-esatern

Yunnan, China. Palaeobotanist, 1:245-262, pls. 1-6.

Hu Hsenhsu (胡先骕), Cheng Wanchun (郑万钧), 1948. On the new family Metasequoiaceae and *Metasequoia glyptostroboides*, a living species of the genus *Metasequoia* found in Szechuan and Hupeh. Bull Fan Mem Inst Boil:153-161.

Hu Yufan (胡雨帆), 1984. Fossil plants from the original "Huairen Group" in Meiyukou, Datong, Shanxi, and correction of their age. Geological Review, 30 (6):569-574, fig. 1. (in Chinese with English summary)

Huang Benhong (黄本宏), 1977. Permian flora from the southeastern part of the Xiao Hinggan Ling (Lesser Khingan Mt.), NE China. Beijing:Geological Publishing House:1-79, pls. 1-43. (in Chinese)

Huang Lianmeng (黄联盟), Huang Yuning (黄玉宁), Mei Meitang (梅美棠), Li Shengsheng (李生盛), 1989. The Early Permian coal-bearing strata and flora from southwestern Fujian Province, South China. Beijing:China Coal Industry Publishing House:1-101, pls. 1-43. (in Chinese with English summary)

Huang Qisheng (黄其胜), 1983. The Early Jurassic Xiangshan flora from the Yangzi River Valley in Anhui Province of eastern China. Earth Science — Journal of Wuhan College of Geology, (2):25-36, pls. 2-4. (in Chinese with English summary)

Huang Qisheng (黄其胜), 1992. Plants // Yin Hongfu et al. eds. The Triassic of Qinling Mountains and nieghboring areas. Wuhan:Press of China University of Geosciences:77-85, 174-180, pls. 16-20. (in Chinese with English title)

Huang Zhigao (黄枝高), Zhou Huiqin (周惠琴), 1980. Fossil plants // Mesozoic stratigraphy and palaeontology from the basin of Shaanxi, Gansu and Ningxia (Ⅰ). Beijing:Geological Publishing House:43-104, pls. 1-60. (in Chinese)

Jacques M B Frédérrie, Guo Shuangxing (郭双兴), 2007. *Palaeoskapha sichuanensis* gen. et sp. nov. (Menispermaceae) Eocene Relu Formation in western Sichuan, West China. Journal of Systematics and Evolution, 45 (4):576-582. (in Chinese with English summary)

Kawasaki S, Kon'no E, 1932. The flora of the Heian System, Part 3. Bulletin on Geological Survey of Chosen, 6 (3):31-43, 61-279, pls. 100-104.

Kawasaki S, 1927-1931. The flora of the Heian System. Bulletin on Geological Survey of Chosen (Korea), 6 (1-2):1-30, pls. 1-99.

Kawasaki S, 1934. The flora of the Heian System, Part 2. Bulletin on the Geological Survey of Chosen, 6 (4):47-311, pls. 105-110.

Kimura T, Ohana T, Zhao Liming (赵立明), Geng Baoyin (耿宝印), 1994. *Pankuangia haifanggouensis* gen. et sp. nov., a fossil plant with unknown affinity from the Middle Jurassic Haifanggou Formation, Western Liaoning, Northeast China. Bulletin of Kitakyushu Museum of Natural History, 13:255-261, figs. 1-8.

Kobayashi T, Yosida T, 1944. *Odontosorites* from North Manchuria. Japanese Journal of Geology and Geography, 19 (1-4):255-273, pl. 28, text-figs. 1, 2.

Koidzumi G, 1934. On the *Gigantopteris* flora. Acta Phytotaxonomica et Geobotanica, Japan, 3 (2):112-113. (in Japanese with English summary)

Koidzumi G, 1936. On the *Gigantopteris* flora. Acta Phytotaxonomica et Geobotanica, Japan, 5 (2):130-139. (in Japanese with English summary)

Kon'no E, 1960. *Schizoneura manchuriensis* Kon'no and its fructification (*Manchurostachys* n. gen.) from the *Gigantopteris nicotianaefolia*-bearing formation in Penchihu Coal-field, northeastern China. Science Reports of Tohoku University, Sendai, Japan, 2nd Series (Geology), Special Volume, (4): 163-188, pls. 16-20, text-figs. 1-4.

Krasser F, 1901. Die von W. A. Obrutschew in China und Centralasien 1893-1894: geasmmelten fossilien Pflanzen. Denkschriften der Könglische Akadedmie der Wissenschaften, Wien, Mathematik-Naturkunde Classe, 70: 139-154, pls. 1-4.

Lee H H (李星学), Wang S (王水), 1956. Note on a new Permian species of *Cladophlebis* from China. Acta Palaeontologica Sinica, 4 (3): 345-353, pls. 1-3, text-fig. 1. (in Chinese and English)

Lee H H (李星学), 1963. Fossil plants of the Yuehmenkou Series, North China. Palaeontologia Sinica, Whole Number 148, New Series A, 6: 1-185, pls. 1-45. (in Chinese and English)

Lee P C (李佩娟), Tsao Chengyao (曹正尧), Wu Shunching (吴舜卿), 1976. Mesozoic plants from Yunnan // Nanjing Institute of Geology and Palaeontology, Chinese Academy of Sciences ed. Mesozoic plants from Yunnan, Ⅰ. Beijing: Science Press: 87-150, pls. 1-47, text-figs. 1-3. (in Chinese)

Leng Q (冷琴), Frii E M, 2003. *Sinocarpus decussatus* gen. et sp. nov., a new angiosperm with basally syncarpous fruits from the Yixian Formation of Northeast China. Plant Systematics and Evolution, 241 (1-2): 77-88.

Lesquereux L, 1883. Contributions to the fossil flora of the Western Territories. Ⅲ. The Cretaceous and Tertiary floras. Rep US Geol Surv Territ 8: 1-283.

Li Chengsen (李承森), Cui Jinzhong (崔金钟), 1995. Atlas of fossil plant an atomy in China. Beijing: Science Press: 1-132, pls. 1-117.

Li Chengsen (李承森), Edwards D, 1992. A new genus of early land plants with novel strobilar construction from the Lower Devonian Posongchong Formation, Yunnan Province, China. Palaeontology, 35 (2): 257-272, pls. 1-4, text-figs. 1-3.

Li Chengsen (李承森), Edwards D, 1996. *Demersatheca* Li et Edwards, gen. nov., a new genus of early land plants from the Lower Devonian, Yunnan Province, China. Review of Palaeobotany and Palynology, 93: 77-88, pls. 1-4, text-figs. 1, 2.

Li Chengsen (李承森), Edwards D, 1997. A new microphyllous plant from the Lower Devonian of Yunnan Province, China. American Journal of Botany, 84 (10): 1441-1448, figs. 1-29.

Li Chengsen (李承森), Hilton J, Hemsley A R, 1997. Frasnian (Upper Devonian) evidence for the multiple origin of seed-like structures. Botanical Journal of the Linnean Society, 123: 133-146, figs. 1-24.

Li Chengsen (李承森), Hsu J (徐仁), 1987. Studies on a new Devonian plant *Protopteridophyton devonicum* assigned to primitive fern from South China. Palaeontographica, B, 207 (1-6): 111-131, pls. 1-16, text-figs. 1-5.

Li Chengsen (李承森), Hueber F M, 2000. *Cervicornus*, a Siegenian (Early Devonian) plant with forked leaves from Yunnan Province, China. Review of Palaeobotany and Palynology, 109 (2): 113-119, pl. 1, text-fig. 1.

Li Chengsen (李承森), 1982. *Hsua robusta*, a new land plant from the Lower Devonian of Yunnan, China. Acta Phytotaxonomica Sinica, 20 (3): 331-342, pls. 3-10, text-figs. 1, 2. (in

Chinese with English summary)

Li Chengsen (李承森), 1990. *Minarodendron cathaysiense* (gen. et comb. nov.), a lycopod from the late Middle Devonian of Yunnan, China. Palaeontographica, B, 220 (5-6): 97-117, pls. 1-11, text-figs. 1-8.

Li Hanmin (李汉民), Lan Shanxian (蓝善先), Li Xingxue (李星学), Cai Chongyang (蔡重阳), Wu Xiuyuan (吴秀元), Mo Zhuangguan (莫壮观), Chen Qishi (陈其奭), Wang Guoping (王国平), 1982. Plants// Nanjing Institute of Geology and Mineral Resources ed. Palaeontological atlas of East China (2), Late Palaeozoic. Beijing: Geological Publishing House: 336-378, pls. 129-157, text-fig. 92. (in Chinese)

Li Hongqi (李洪起), Taylor D W, 1998. *Aculeovinea yunguiensis* gen. et sp. nov. (Gigantopteridales), a new taxon of gigantopterid stem from the Upper Permian of Guizhou Province, China. International Journal of Plant Sciences, 159 (6): 1023-1033, text-figs. 1-5.

Li Hongqi (李洪起), Taylor D W, 1999. Vessel-bearing stems of *Vasovinea tianii* gen. et sp. nov. (Gigantopteridales) from the Upper Permian of Guizhou Province, China. American Journal of Botany, 86 (11): 1563-1575, figs. 1-32.

Li Jieru (李杰儒), 1983. Middle Jurassic flora from Houfulongshan region of Jingxi, Liaoning. Bulletin of Geological Society of Liaoning Province, China, (1): 15-29, pls. 1-4. (in Chinese with English summary)

Li Nan (李楠), Fu Xiaoping (傅晓平), Zhang Wu (张武), Zheng Shaolin (郑少林), Cao Yu (曹雨), 2005. A new genus of Cycadalean plants from the Early Triassic of western Liaoning, China: Mediocycas Gen. Nov. and its evolutional Significance. Acta Palaeontologica Sinica, 44 (3): 423-434. (in Chinese with English summary)

Li Peijuan (李佩娟), He Yuanliang (何元良), 1986. Late Triassic plants from Mt. Burhan Budai, Qinghai// Qinghai Institute of Geological Sciences, Nanjing Institute of Geology and Palaeontology, Chinese Academy of Sciences eds. Carboniferous and Triassic strata and fossils from the southern slope of Mt. Burhan Budai, Qinghai, China. Hefei: Anhui Science and Technology Publishing House: 275-293, pls. 1-10. (in Chinese with English summary)

Li Peijuan (李佩娟), He Yuanliang (何元良), Wu Xiangwu (吴向午), Mei Shengwu (梅盛吴), Li Bingyou (李炳有), 1988. Early and Middle Jurassic strata and their floras from northeastern border of Qaidam Basin, Qinghai. Nanjing: Nanjing University Press: 1-231, pls. 1-140, text-figs. 1-24. (in Chinese with English summary)

Li Xingxue (李星学), 1995a. Fossil floras of China through the geological ages. Guangzhou: Guangdong Science and Technology Press: 1-542, pls. 1-144. (Chinese Edition)

Li Xingxue (李星学), 1995b. Fossil floras of China through the geological ages. Guangzhou: Guangdong Science and Technology Press: 1-695, pls. 1-144. (English Edition)

Li Xingxue (李星学), Cai Chongyang (蔡重阳), 1977. Early Devonian *Zosterophyllum*-remains from Southwest China. Acta Palaeontologica Sinica, 16 (1): 12-34, pls. 1-5. (in Chinese with English summary)

Li Xingxue (李星学), Cai Chongyang (蔡重阳), 1978. A type-section of Lower Devonian strata in SW China with brief notes on the succession and correlation of its plant assemblages. Acta Geologica Sinica, 52 (1): 1-12, pls. 1, 2. (in Chinese with English summary)

Li Xingxue (李星学), Shen Guanglong (沈光隆), Wu Xiuyuan (吴秀元), 1993. *Reticalethopteris*, a new genus of Carboniferous plants with restudy on *Palaeoweichselia yuani* Sze. Acta Palaeontologica Sinica, 32 (5): 540-549, pls. 1-4. (in Chinese with English summary)

Li Xingxue (李星学), Yao Zhaoqi (姚兆奇), 1983. Current studies of gigantopterids. Palaeotologia Cathayana, 1: 319-326, text-fig. 1.

Li Xingxue (李星学), Ye Meina (叶美娜), Zhou Zhiyan (周志炎), 1986. Late Early Cretaceous flora from Shansong, Jiaohe, Jilin Province, Northeast China. Palaeontologia Cathayana, 3: 1-53, pls. 1-45, text-figs. 1-12.

Li Xingxue (李星学), Ye Meina (叶美娜), 1980. Middle-late Early Cretacous floras from Jilin, EN China. Paper for the 1st Conf. IOP London & Reading, 1980. Nanjing Institute Geology Palaeontology Chinese Academy of Sciences, Nanjing: 1-13, pls. 1-5.

Li Zhongming (李中明), 1983. *Palaeosmunda* emended and two new species. Acta Phytotaxonomica Sinica, 21 (2): 153-160, pls. 9-14, text-fig. 1. (in Chinese with English summary)

Li Zhongming (李中明), 1993a. The genus *Shuichengella* gen. nov. and systematic classification of the order Osmundales. Review of Palaeobotany and Palynology, 77: 51-63, pls. 1-4.

Li Zhongming (李中明), 1993b. Studies on *Parataxospermum taiyuanensis* gen. et sp. nov. from coal balls. Review of Palaeobotany and Palynology, 77: 65-74, pls. 1-4, text-fig. 1.

Li Zhongming (李中明), 1993c. The recent status and future on reproductive biology of fossil plants. Chinese Bulletin of Botany, 10 (3): 28-30. (in Chinese with English title)

Liu Lujun (刘陆军), Yao Zhaoqi (姚兆奇), 2002. *Lepingia*: A new genus of probable cycadalean affinity with taeniopterid lamina from the Permian of South China. International Journal of Plant Science, 163 (1): 175-183.

Liu Lujun (刘陆军), Yao Zhaoqi (姚兆奇), 2004. *Fujianopteris*: A new genus of Gigatopterid plants from South China. Acta Palaeontologica Sinica, 43 (4): 472-488, pls. 1-3. (in Chinese with English summary)

Liu Lujun (刘陆军), Yao Zhaoqi (姚兆奇), 2006. *Cohaerensitheca*: A new Palaeozoic marattialean genus for compression-impression plant fossils from China. Palaeoworld, 15: 68-76.

Liu Zhaohua (刘照华), Li Chengsen (李承森), Hilton J, 2000. *Zhutheca* Liu, Li et Hilton gen. nov., the fertile pinnules of *Fascipteris densata* Gu et Zhi and their significance in marattialean evolution. Review of Palaeobotany and Palynology, 109 (2): 149-160, pls. 1-3, text-fig. 1.

Liu Zijin (刘子进), 1982. Vegetable kingdom // Xi'an Institute of Geology and Mineral Resources ed. Paleontological atlas of Northwest China, Shaanxi, Gansu Ningxia volume, Part Ⅲ. Mesozoic and Cenozoic. Beijing: Geological Publishing House: 116-139, pls. 56-75. (in Chinese with English title)

Ma Jie (马洁), Du Xianming (杜贤铭), 1989. A new fossil Ginkgo, *Primoginkgo dissecta* gen. et sp. nov. from Shihezi (Shihhotse) Formation of Dongshan in Taiyuan. Memoirs of Beijing Natural History Museum, 3 (43): 1-4, pls. 1, 2. (in Chinese with English summary)

Mathews G B,1947-1948. On some fructifications from the Shuantsuang Series in the Western Hill of Peking. Bulletin of National History Peking,16 (3-4):239-241.

Mei Meitang(梅美棠),Dilcher D L,Wan Zhihui(万志辉),1992. A new seed-bearing leaf from the Permian of China. Palaeobotanist,41:98-109,pls. 1-5,text-fig. 1.

Meng Fansong(孟繁松),1981. Fossil plants of the Lingxiang Group of southeastern Hubei and their implications. Bulletin of the Yichang Institute of Geology and Mineral Resources, Chinese Academy of Geological Sciences, 1981 (special issue of stratigraphy and paleontology):98-105,pls. 1,2,fig. 1. (in Chinese with English summary)

Meng Fansong(孟繁松),1983. New materials of fossil plants from the Jiuligang Formation of Jingmen-Dangyang Basin, W. Hubei. Professional Papers of Stratigraphy and Palaeontology,10:223-238. (in Chinese with English summary)

Meng Fansong(孟繁松),1992. New genus and species of fossil plants from Jiuligang Formation in W. Hubei. Acta Palaeontologica Sinica,31(6):703-707,pls. 1-3. (in Chinese with English summary)

Mi Jiarong(米家榕),Sun Chunlin(孙春林),Sun Yuewu(孙跃武),Cui Shangsen(崔尚森), Ai Yongliang(艾永亮),1996. Early-Middle Jurassic phytoecology and coal-accumulating environments in northern Hebei and western Liaoning. Beijing:Geological Publishing House:1-169,pls. 1-39,text-figs. 1-20. (in Chinese with English summary)

Mi Jiarong(米家榕),Zhang Chuanbo(张川波),Sun Chunlin(孙春林),Luo Guichang(罗桂昌),Sun Yuewu(孙跃武) et al., 1993. Late Triassic stratigraphy, palaeontology and paleogeography of the northern part of the Circum Pacific Belt, China. Beijing: Science Press:1-219,pls. 1-66,text-figs. 1-47. (in Chinese with English title)

Miki S,1964. Mesozoic flora of *Lycoptera* Bed in South Manchuria. Bulletin of Mukogawa Women's University,(12):13-22. (in Japanese with English summary)

Newberry J S,1865 (1867). Description of fossil plants from the Chinese coal-bearing rocks// Pumpelly R ed. Geological researches in China, Mongolia and Japan during the years 1862-1865. Smithsonian Contributions to Knowledge,Washington,15 (202):119-123,pl. 9.

Ngo C K(敖振宽),1956. Preliminary notes on the Rhaetic flora from Siaoping Coal Series of Kwangtung. Journal of Central-South Institute of Mining and Metallurgy,(1):18-32,pls. 1-7,text-figs. 1-4. (in Chinese)

Ngo C K(敖振宽),1963. On a new species of ?*Sphenophyllum changshaense* from the Lower Carboniferous of Changsha, Hunan. Acta Palaeontologica Sinica, 11 (4):610, pl. 1. (in Chinese)

Obrhel Jiri,1966. *Protopteridium hostinense* Krejei und Bemerkungen zu den uebrigin Arten der gattung *Protopteridium*. Casopis pro Mineralogii a Geologii,11 (4):441-443.

P'an C H(潘钟祥),1936-1937. Notes on Kawasaki and Konno's *Rhipidopsis brevicaulis* and *Rh. baieroides* of Korea with description of similar form from Yuhsien, Honan. Bulletin of Geological Society of China,16:261-280,pls. 1-5.

P'an C H(潘钟祥),1936. Older Mesozoic plants from North Shensi. Palaeontologia Sinica, Series A,4 (2):1-49,pls. 1-15.

Pan Guang(潘广),1983. Notes on the Jurassic precursors of angiosperms from Yan-Liao region of North China and the origin of angiosperms. A Monthly Journal of Science (Kexue

Tongbao),28 (24):1520. (in Chinese)

Pan Guang (潘广),1984. Notes on the Jurassic precursors of angiosperms from Yan-Liao region of North China and the origin of angiosperms. A Monthly Journal of Science (Kexue Tongbao),29 (7):958-959. (in English)

Pan K (潘广),1977. A Jurassic conifer *Yanliaoa sinensis* gen. et sp. nov. from Yanliao region. Acta Phytotaxonomica Sinica,15 (1):69-71,pl. 1. (in Chinese with English summary)

Phillips John, Salter John William,1848. Paleontological appendix: Great Britain Geol. Survey Mem,2 (1):331-386,pls. 1-30.

Prakash U,Du Naizheng (杜乃正),1995. Fossil woods from the Miocene sediments of China with remarks on environmental implications of Miocene floras of the region//Pant D D ed. Global environment and diversification of plants through geological time. Allahabad: South Asian Publishers:341-360.

Reid E M,Chandler M E J,1933. The London Clay Flora. British Museum (Natural History), London:1-561,figs. 1-17,pls. 1-33.

Ren Shouqin (任守勤), Chen Fen (陈芬),1989. Fossil plants from Early Cretaceous Damoguaihe Formation in Wujiu Coal Basin, Hailar, Inner Mongolia. Acta Palaeontologica Sinica,28 (5):634-641,pls. 1-3,text-figs. 1,2. (in Chinese with English summary)

Rydin C,Friis E M,2010. A new Early Cretaceous relative of Gnetales: *Siphonospermum simplex* gen. et sp. Nov. from Yixian Formation of Northeast China. BMC Evolutionary Biology,10 (183):1-6.

Sahni B,1932. *Homoxylon rajmahalense*,gen. et sp. nov.,a fossil angiospermous wood,devoid of vessels, from the Rajmahal Hills, Behar. India Geol. Survey Mem. 2, Paleontologia Indica,New Ser. ,20:1-19,pls. 1,2.

Saport G,1865. Etudes sur la vegetation du sud-est de la Francea l'epoque tertiaire. Annales Sci. Nat. ,Botanique,Ser. 5,4:5-264,pls. 1-13.

Schenk A,1869. Beitraege zue Flora vorwelt. Palaeontographica,19:1-34,pls. 1-7.

Schenk A,1883. Pflanzliche Versteinerungen. Pflanzen der Steinkohleformation//Richthofen F (Von). China,Ⅳ. Berlin:211-244,taf. 30-45,49,fig. 1.

Schweitzer H J,Cai Chongyang (蔡重阳),1987. Beiträge zur Mitteldevon-Flora Sudchinas. Palaeontographica,B,207 (1-6):1-109,taf. 1-20.

Schweitzer H J,Li Chengsen (李承森),1996. *Chamaedendron* nov. gen. ,eine multisporangiate Lycophyte aus dem Frasnium Sudchinas. Palaeontographica,B,238 (1-2):45-69,taf. 1-4, abb. 1-13.

Seward A C,1919. Fossil Plants,vol. Ⅳ,Ginkgoales,Coniferales,Gnetales. Cambrige: Cambrige University Press.

Seyfullah L J, Hilton J, Liang Mingmei (梁明媚), Wang Shijun (王士俊),2010. Resolving the systematics and phylogenetic position of isolated ovules: a case study on a new genus from the Permian of China. Botanical Journal of the Linnean Society,164:84-108.

Si Xingjian (Sze Hsingchien) (斯行健),1989. Late Palaeozoic plants from the Qingshuihe region of Inner Mongolia and the Hequ district of northwestern Shanxi. Palaeontologia Sinica,Whole Number 176,New Series A,11:1-268,pls. 1-93.

Steere William C,1946. Cenozoic and Mesozoic bryophytes of North America, in Symposium on paleobotanical taxonomy. Am. Midland Naturalist, 36: 298-324, pls. 1, 2.

Stockmans F, Mathieu F F, 1939. La flore paleozoique du bassin houiller de Kaiping (Chine). Bulletin du Musée Royal d'Histoire Naturelle de Belgique, Bruxelles: 49-165, pls. 1-34.

Stockmans F, Mathieu F F, 1941. Contribution a l'etude de la flore jurassique de la Chine septentrionale. Bulletin du Musee Royal d'Histoire Naturelle de Belgique, Bruxelles: 33-67, pls. 1-7.

Stockmans F, Mathieu F F, 1957. La flore paléozoïque du bassin houiller de Kaiping (Chine) (deuxiéme partie). Assocation pour Etude de la Palaeontologie et de la Stratigraphie Houilleres, Bruxelles, Publication 32: 1-89, pls. 1-15.

Sun Ge (孙革), Dilcher D L, Zheng Shaolin (郑少林), Zhou Zhekun (周浙昆), 1998. In research of the first flower: a Jurassic angiosperm, *Archaefructus*, from Northeast China. Science, 282 (5394): 1692-1695, figs. 1, 2.

Sun Ge (孙革), Dilcher D L, 1996. Early angiosperms from Lower Cretaceous of Jixi, China and their significance for study of the earliest occurrence of angiosperms in the world. Palaeobotanist, 45: 393-399, pls. 1, 2, text-figs. 1, 2.

Sun Ge (孙革), Dilcher D L, 1997. Discovery of the oldest known angiosperm inflorescences in the world from Lower Cretaceous of Jixi, China. Acta Palaeontologica Sinica, 36 (2): 135-142, pls. 1, 2, text-figs. 1, 2. (in Chinese with English summary)

Sun Ge (孙革), Dilcher D L, 2002. Early angiosperms from the Lower Cretacous of Jixi, eastern Heilongjiang, China. Review of Palaeobotany and Palynology, 121 (2): 91-112, pls. 1-6, figs. 1-4. (in English)

Sun Ge (孙革), Guo Shuangxing (郭双兴), Zheng Shaolin (郑少林), Piao Taiyuan (朴泰元), Sun Xuekun (孙学坤), 1992. First discovery of the earliest angiospermous megafossils in the world. Science in China, Series B, 35 (5): 543-548, pls. 1, 2. (in Chinese)

Sun Ge (孙革), Guo Shuangxing (郭双兴), Zheng Shaolin (郑少林), Piao Taiyuan (朴泰元), Sun Xuekun (孙学坤), 1993. First discovery of the earliest angiospermous megafossils in the world. Science in China, Series B, 36 (2): 249-256, pls. 1, 2. (in English)

Sun Ge (孙革), Zheng Shaolin (郑少林), Dilcher D L, Wang Yongdong (王永栋), Mei Shengwu (梅盛吴), 2001. Early Angiosperms and their Associated Plants from western Liaoning, China. Shanghai: Shanghai Scientific and Technological Education Publishing House: 1-227, pls. 1-75. (in Chinese and English)

Sun Ge (孙革), Zheng Shaolin (郑少林), MEI Shengwu (梅盛吴), 2000. Discovery of *Liaoningocladus* gen. nov. from the lower part of Yixian Formation (Upper Jurassic) in western Liaoning, China. Acta Palaeontologica Sinica, 39 (Supplement): 200-208, pls. 1-4, text-fig. 1. (in English with Chinese summary)

Sun Keqin (孙克勤), Deng Shenghui (邓胜徽), 2006. *Parasphenopteris* Sun et Deng, a new genus from the lower Permian of Wuda, Nei Mongol, China. Acta Phytotaxonomica Sinica, 44 (2): 161-164.

Sun Keqin (孙克勤), Deng Shenghui (邓胜徽), Cui Jinzhong (崔金钟), Shang Ping (商平), 1999. Discovery of *Paratingia* and *Paratingiostachya* from the Shanxi Formation of the early Early Permian in the Wuda area of Inner Mongolia. Acta Botanica Sinica, 41 (9): 1024-1026, pls. 1, 2. (in Chinese with English summary)

Surveying Group of Department of Geological Exploration of Changchun Institute of Geology, Regional Geological Surveying Team of Geological Bureau of Kirin Province, 102 Surveying Team of Coal Geology Exploration Company of Kirin Province (长春地质学院地勘系、吉林省地质局区测大队、吉林省煤田地质勘探公司 102 队调查队), 1977. Late Triassic stratigraphy and plants of Hunkiang, Kirin. Journal of Changchun College of Geology, (3): 2-12, pls. 1-4, text-fig. 1. (in Chinese)

Sze H C (斯行健), Lee H H (李星学), 1945. Palaeozoic plants from Ninghsia. Bulletin of Geological Society of China, 25 (1-4): 227-260, pls. 1-3, text-fig. 1.

Sze H C (斯行健), Lee H H (李星学), 1952. Jurassic plants from Szechuan. Palaeontologia Sinica, Whole Number 135, New Series A, (3): 1-38, pls. 1-9, text-figs. 1-5. (in Chinese and English)

Sze H C (斯行健), 1931. Beiträge zur liasischen Flora von China. Memoirs of National Research Institute of Geology, Chinese Academy of Sciences, 12: 1-85, pls. 1-10.

Sze H C (斯行健), 1933. On the occurrence of a new species of *Palaeoweichselia* in Kansu. Memoirs of National Research Institute of Geology, Chinese Academy of Sciences, 13: 59-64, pls. 6, 7.

Sze H C (斯行健), 1936. Über die altkarbonische Flora der Prov. Kiangsu mit besonderer Berucksichtigung des Alters des Wutung Quartzites. Bulletin of Geological Society of China, 15 (2): 135-164, pls. 1-6.

Sze H C (斯行健), 1942. *Cycadolepis corrugata* Zeiller und *Pterophyllum aequale* Brongniart. Bulletin of Geological Society of China, 22 (3-4): 189-194, pl. 1.

Sze H C (斯行健), 1943. On the occurrence of *Sublepidodendron*, a lepidodendroid plant from Wutong Formation. Bulletin of Geological Society of China, 23 (1-2): 61-68, pl. 1.

Sze H C (斯行健), 1945. *Plagiozamiopsis podozamioides*, n. g. et sp., a new Permian plant from the *Gigantopteris* Coal Series. Science Record, 1 (3-4): 511-517, pl. 1.

Sze H C (斯行健), 1949. Die mesozoische Flora aus der Hsiangchi Kohlen Serie in Westhupeh. Palaeontologia Sinica, Whole Number 133, New Series A, 2: 1-71, pls. 1-15.

Sze H C (斯行健), 1951a. Über einen problematischen Fossilrest aus der Wealdenformation der suedlichen Mandschurei. Science Record, 4 (1): 81-83, pl. 1.

Sze H C (斯行健), 1951b. Petrified wood from northern Manchuria. Science Record, 4 (4): 443-457, pls. 1-7, text-figs. 1-3. (in English with Chinese summary)

Sze H C (斯行健), 1952a. Upper Devonian plants from China. Palaeontologia Sinica, Whole Number 136, New Series A, 4: 1-30, pls. 1-8.

Sze H C (斯行健), 1952b. Upper Devonian plants from China. Acta Scientia Sinica, 1 (2): 166-192, pls. 1-6, text-figs. 1, 2.

Sze H C (斯行健), 1953. Notes on some fossil remains from the Shihchienfeng Series in northwestern Shensi. Acta Palaeontologica Sinica, 1 (1): 11-22, pl. 1. (in Chinese with English summary)

Sze H C (斯行健), 1956a. Older Mesozoic plants from the Yenchang Formation, northern Shensi. Palaeontologia Sinica, Whole Number 139, New Series A, 5: 1-217, pls. 1-56, text-fig. 1. (in Chinese and English)

Sze H C (斯行健), 1956b (1). On some specimens of *Lepidodendropsis hirmeri* Lutz from the

Sze H C (斯行健), 1956b (2). On some specimens of *Lepidodendropsis hirmeri* Lutz from the Wutung Series of Kiangsu. Scientia Sinica, 5 (1): 137-143, pl. 1.

Sze H C (斯行健), 1958. On a Westphalian flora of the Tzushan Coal Series in Loping district, northeastern Kiangsi. Acta Palaeontologica Sinica, 6 (4): 375-388, pls. 1-3. (in Chinese with English summary)

Sze H C (斯行健), 1959. On a Westphalian flora of the Tzushan Coal Series in Loping district, northeastern Kiangsi. Scientia Sinica, 8 (3): 314-322, pls. 1-3.

Tan Lin (谭琳), Zhu Jianan (朱家楠), 1982. Palaeobotany // Bureau of Geology and Mineral Resources of Nei Monggol Autonomous Region ed. The Mesozoic stratigraphy and paleontology of Guyang Coal-bearing Basin, Nei Monggol Autonomous Region, China. Beijing: Geological Publishing House: 137-160, pls. 33-41. (in Chinese with English title)

Tao Junrong (陶君容), Xiong Xianzheng (熊宪政), 1986. The latest Cretaceous flora of Heilongjiang Province and the floristic relationship between East Asia and North America. Acta Phytotaxonomica Sinica, 24 (1): 1-15, pls. 1-16, fig. 1; 24 (2): 121-135. (in Chinese with English summary)

Tao Junrong (陶君容), Yang Jiaju (杨家驹), Wang Yufei (王宇飞), 1994. Miocene wood fossils and paleoclimate in Inner Mongolia. Acta Botanica Yunnanica, 16 (2): 111-116, pls. 1, 2. (in Chinese with English summary)

Tao Junrong (陶君容), Yang Yong (杨永), 2003. *Alloephedra xingxuei* gen. et sp. nov., an Early Cretacous member of Epherdaceae from Dalazi Formation in Yanji Basin, Jilin Province of China. Acta Palaeontologica Sinica, 42 (2): 208-213, pls. 1, 2. (in Chinese with English summary)

Tao Junrong (陶君容), Zhang Chuanbo (张川波), 1990. Early Cretaceous angiosperms of the Yanji Basin, Jilin Province. Acta Botanica Sinica, 32 (3): 220-229, pls. 1, 2, fig. 1. (in Chinese with English summary)

Tao Junrong (陶君容), Zhang Chuanbo (张川波), 1992. Two angiosperm reproductive organs from the Early Cretaceous of China. Acta Phytotaxonomica Sinica, 30 (5): 423-426, pl. 1. (in Chinese with English summary)

Tao Junrong (陶君容), 1992. Study on the fossil flowers of angiosperm. Acta Botanica Sinica, 34 (3): 240-242, pl. 1. (in Chinese with English summary)

Tian Baolin (田宝霖), Li Hongqi (李洪起), 1992. A new special petrified stem, *Guizhouoxylon dahebianense* gen. et sp. nov., from Upper Permian in Shuicheng district, Guizhou, China. Acta Palaeontologica Sinica, 31 (3): 336-345, pls. 1-4, text-fig. 1. (in Chinese with English summary)

Tian Baolin (田宝霖), Wang Shijun (王士俊), 1987. On cordaitean stems in coal balls from Taiyuan Formation in Xishan, Taiyuan, Shanxi. Acta Palaeontologica Sinica, 26 (2): 196-204, pls. 1-3, text-figs. 1-4. (in Chinese with English summary)

Tokunaga S, 1951. Preliminary Note on the Geology of Heijo Coal Field. Jour. Geol. Soc. Tokyo., Vol. XVIII, No. 257. (in Japanese)

Vozenin-Serra C, Pons D, 1990. Interets phylogenetique et paleoelogigue des structures ligneuses homoxyles decouvertes dams le Cretace inferieur du Tibetmeridional. Palaeontographica, B, 216 (1-4): 107-127, pls. 1-6, tex-figs. 1-3.

Wang Deming（王德明）, Hao Shougang（郝守刚）, Wang Qi（王祺）, 2005. *Rotafolia songziensis* gen. et comb. nov., a Sphenopsid from the Late Devonian of Hubei. Botanical Journal of the Linnean Society, 148: 21-37.

Wang Deming（王德明）, Hao Shougang（郝守刚）, 2002. *Guangnania cuneata* Gen. et sp. Nov. from the Lower Devonian of Yunnan Pronince, China. Review of Palaeobotany and Palynology, 122: 13-27, pls. 1-3, text-figs. 1-3.

Wang Deming（王德明）, Hao Shougang（郝守刚）, 2004. *Bracteophyton variatum* Gen. et Sp. Nov., An Early plant from the Xujiachong Formation of Yunnan, China. International Journal of Plant Sciences, 165 (2): 337-345, figs. 1-5.

Wang Deming（王德明）, 2007. *Tenuisa frasniana* gen. et sp. nov., a Plant of Euphyllophyte Affinity from the Late Devonian of China. International Journal of Plant Sciences, 168 (9): 1341-1349.

Wang Deming（王德明）, 2008. A new iridopteridalean plant from the Middle Devonian of northwest China. International Journal of Plant Sciences, 169 (8): 1100-1115, figs. 1-13.

Wang Qi（王祺）, Xue Jinzhuang（薛进庄）, Prestianni C, 2007. Sphinxiocarpon, a New Name for Sphinxia Li, Hilton & Hemsley, 1997 — Not Reid & Chandler, 1933. Lethaia, 40 (4): 393.

Wang Qingzhi（王庆之）, 1993. Flora from Lower Shihhotse Formation, Lingshan, Hebei, North China. Acta Palaeontologica Sinica, 32 (2): 218-226, pls. 1-4, fig. 1. (in Chinese with English summary)

Wang Shijun（王士俊）, Jiang Yaofa（姜尧发）, Qin Yong（秦勇）, 1994. A new petrified wood from Taiyuan Formation, Xuzhou Coal Field, Jiangsu Province. Acta Botanica Sinica, 36 (Supplement): 194-198, pls. 1, 2. (in Chinese with English summary)

Wang Shijun（王士俊）, Tian Baolin（田宝霖）, Galtier J, 2003. Cordaitalean seed plants from the early Permian of North China. Ⅰ. Delimitation and reconstruction of the *Shanxioxylon sinense* plants. International Journal of Plant Sciences, 164 (1): 89-112.

Wang Shijun（王士俊）, Tian Baolin（田宝霖）, 1991. On male cordaitean reproductive organs in coal balls from Taiyuan Formation, Xishan Coal-field, Taiyuan. Acta Palaeontologica Sinica, 30 (6): 743-749, pls. 1-3, text-figs. 1-4. (in Chinese with English summary)

Wang Shijun（王士俊）, Tian Baolin（田宝霖）, 1993. On female cordaitean reproductive organ in coal balls from Taiyuan Formation, Xishan Coal-field, Taiyuan, Shanxi. Acta Palaeontologica Sinica, 32 (6): 760-764, pls. 1-3. (in Chinese with English summary)

Wang Wuli（王五力）, Zhang Hong（张宏）, Zhang Lijun（张立君）, Zheng Shaolin（郑少林）, Yang Fanglin（杨芳林）, Li Zhitong（李之彤）, Zheng Yuejuan（郑月娟）, Ding Qiuhong（丁秋红）, 2004. Stand sections of Tuchaengzi Stage and Yixian Stage and their stratigraphy, palaeontology and tectonic-volcanic actions. Beijing: Geological Publishing House: 1-514, pls. 1-37. (in Chinese with English summar)

Wang Xifu（王喜富）, 1977. On the new genera of *Annularia*-like plants from the Upper Triassic in Sichuan-Shaanxi area. Acta Palaeontologica Sinica, 16 (2): 185-190, pls. 1, 2, text-fig. 1. (in Chinese with English summary)

Wang Xifu(王喜富),1984. A supplement of Mesozoic plants from Hebei//Tianjin Institute of Geology and Mineral Rescurces ed. Palaeontological atlas of North China,Ⅱ. Mesozoic. Beijing:Geological Publishing House:297-302,pls. 174-178. (in Chinese)

Wang Xin(王鑫),Krings M,Taylor T N,2010. A thalloid organism with possible lichen affinity from the Jurassic of northeastern China. Review of Palaeobotany and Palynology,162(4):591-598.

Wang Xin(王鑫),Li Nan(李楠),Cui Jinzhong(崔金钟),2006. *Jurastrobus chenii* gen. et sp. nov.,a cycadlean pollen cone connected with vegetative parts from Inner Mongolia,China. Progress in Natural Science,16(Special Issue):213-221,figs. 1-6.

Wang Xin(王鑫),Li Nan(李楠),Wang Yongdong(王永栋),Zheng Shaolin(郑少林),2009a. The discovery of whole-plant fossil cycad from the Upper Triassic in waestern Liaoning and its significance. Chinese Science Bulletin,54(13):1937-1939,fig. 1. (in Chinese)

Wang Xin(王鑫),Li Nan(李楠),Wang Yongdong(王永栋),Zheng Shaolin(郑少林),2009b. The discovery of whole-plant fossil cycad from the Upper Triassic in waestern Liaoning and its significance. Chinese Science Bulletin,54(17):3116-3119,fig. 1. (in English)

Wang Xin(王鑫),Wang Shijun(王士俊),2010. *Xingxueanthus*:An enigmatic Jurassc seed plant and its implications for the Origin of Angiospermy. Acta Geologica Sinica,84(1):47-55,figs. 1-6. (English Edition)

Wang Xin(王鑫),Zheng Shaolin(郑少林),2009. The Eariest normal flower from Liaoning Province,China. Journal of Integrative Plant Biology,51(8):800-811,figs. 1-5.

Wang Xin(王鑫),2010. Axial nature of the cupule-bearing organ in Caytoniales. Journal Systematics Evolution,48(3):207-214,figs. 1-3.

Wang Yi(王怿),Berry C M,2001. A new plant from the Xichong Formation (Middle Devonian),South China. Review of Palaeobotany and Palynology,116(1):73-85.

Wang Yi(王怿),Hao Shougang(郝守刚),Cai Chongyang(蔡重阳),Xu Honghe(徐洪河),2006. A diminutive plant from the Late Silurian of Xinjiang,China. Alcheringa,30:23-31.

Wang Yi(王怿),Fu Qiang(傅强),Xu Honghe(徐洪河),Hao Shougang(郝守刚),2007. A new Late Silurian plant with complex branching from Xinjiang,China. Alcheringa,31:111-120.

Wang Yi(王怿),Xu Honghe(徐洪河),2002. A new fossil plant from the earliest Carboniferous of China. International Journal of Plant Science,163:475-483.

Wang Yi(王怿),Xu Honghe(徐洪河),2003. Studies on a new Earliest Carboniferous plant: *Coenosophyton tristichus* gen. et sp. nov. from China. International Journal of Plant Science,164:77-87.

Wang Yi(王怿),2000. *Kongshania* gen. nov.,a new plant from the Wutong Formation (Upper Devonian) of Jiangning County,Jiangsu,China. Acta Palaeontologica Sinica,39(Supplement):42-56,pls. 1-4,text-figs. 1-6. (in Chinese with English summary)

Wang Yongdong(王永栋),Li Nan(李楠),Yang Xiaoju(杨小菊),Zhang Wu(张武),2006. Jurassic Fossil Woods // Shenzhen Urban Managerment Bureau,Shenzhen Fairy Lake Botanical Garden,Shenyang Institute of Geology and Mineral Resources,Ministry of Land and Resources eds. Fossil Woods China. Beijing:China Forestry Publishing House:120-174. (in Chinese)

Wang Yongdong(王永栋), Li Nan(李楠), Yang Xiaoju(杨小菊), Zhang Wu(张武), 2008. Jurassic Fossil Woods // Shenyang Institute of Geology and Mineral Resources, Ministry of Land and Resources, Shenzhen Urban Managerment Bureau, Shenzhen Fairy Lake Botanical Garden, Nanjing Institute of Geology and Palaeontology, Chinese Academy of Sciences eds. Fossil Woods China. Beijing: China Forestry Publishing House: 120-174. (in English)

Wang Yufei(王宇飞), Li Chengsen(李承森), Li Zhenyu(李振宇), Fu Dezhi(傅德志), 2001. *Wuyunanthus* gen. nov., a flower of Celastraceae from the Palaeocene of north-east China. Botanical Journal of the Linnean Society, 136: 323-327, figs. 1-6.

Wang Yufei(王宇飞), Manchester R, 2000. *Chaneya*, a new genus of winged fruit from the Tertiary of North America and eastern Asia. International Journal of Plant Sciences (Botanical Gazette), 161 (1): 167-178, figs. 1-7.

Wang Zhong (王忠), Geng Baoyin (耿宝印), 1997. A new Middle Devonian plant: *Metacladophyton tetraxylum* gen. et sp. nov. Palaeontographica, B, 243 (4-6): 85-102, pls. 1-11, text-figs. 1-7.

Wang Ziqiang(王自强), Wang Lixin(王立新), 1986. Late Permian fossil plants from the lower part of the Shiqianfeng (Shihchienfeng) Group in North China. Bulletin of the Tianjin Institute of Geology and Mineral Resources, Chinese Academy of Geological Sciences, 15: 1-80, pls. 1-40, text-figs. 1-24. (in Chinese with English summary)

Wang Ziqiang(王自强), Wang Lixin(王立新), 1990a. Late Early Triassic fossil plants from upper part of the Shiqianfeng Group in North China. Shanxi Geology, 5 (2): 97-154, pls. 1-26, figs. 1-7. (in Chinese with English summary)

Wang Ziqiang(王自强), Wang Lixin(王立新), 1990b. A new plant assemblage from the bottom of the mid-Triassic Ermaying Formation. Shanxi Geology, 5 (4): 303-315, pls. 1-10, figs. 1-5. (in Chinese with English summary)

Wang Ziqiang(王自强), 1984. Plant kingdom // Tianjin Institute of Geology and Mineral Resources ed. Palaeontological atlas of North China, II. Mesozoic. Beijing: Geological Publishing House: 223-296, 367-384, pls. 108-174. (in Chinese with English title)

Wang Ziqiang(王自强), 1986. *Liulinia lacinulata*, a new male cone of cycads from latest Permian in Shanxi. Acta Palaeontologica Sinica, 25 (6): 610-616, pls. 1-3, text-fig. 1. (in Chinese with English summary)

Wang Ziqiang(王自强), 2004. A new Permian Gnetalean cone as fossil evidence for supporting current molecular phylogeny. Annals of Botany, 94: 281-288, figs. 1-6.

White David, 1912. The characters of the fossil plant Gigantopteris Schenk and its occurrence in North America. U S National Museum, Vol. 41, No. 1873: 493-516, pls. 43-49.

Wu Shunqing(吴舜卿), Zhou Hanzhong(周汉忠), 1986. Early Liassic plants from East Tianshan Mountains. Acta Palaeontologica Sinica, 25 (6): 636-647, pls. 1-6. (in Chinese with English summary)

Wu Shunqing(吴舜卿), 1999. A preliminary study of the Jehol flora from western Liaoning. Palaeoworld, 11: 7-57, pls. 1-20. (in Chinese with English)

Wu Xiangwu(吴向午), Li Baoxian(厉宝贤), 1992. A study of some Bryophytes from Middle Jurassic Qiaoerjian Formation in Yuxian district of Huber, China. Acta Palaeontologica

Sinica,31（3）:257-279,pls. 1-6,text-figs. 1-8. (in Chinese with English summary)

Wu Xiangwu（吴向午）,Wu Xiuyuan（吴秀元）,Wang Yongdong（王永栋）,2000. Two new forms of Bryiidae（Musci）from the Jurrasic of Junggar Basin in Xinjiang,China. Acta Palaeontologica Sinica,39（Supplement）:167-175,pls. 1-3. (in English with Chinese summary)

Wu Xiangwu（吴向午）,1982. Fossil plants from the Upper Triassic Tumaingela Formation in Amdo-Baqen area,northern Xizang// The Comprehensive Scientific Expedition Team to the Qinghai-Xizang Plateau,Chinese Academy of Sciences ed. Palaeontology of Xizang,V. Beijing:Science Press:45-62,pls. 1-9. (in Chinese with English summary)

Wu Xiangwu（吴向午）,1993a. Record of generic names of Mesozoic megafossil plants from China（1865-1990）. Nanjing:Nanjing University Press:1-250. (in Chinese with English summary)

Wu Xiangwu（吴向午）,1993b. Index of generic names founded on Mesozoic-Cenozoic specimens from China in 1865-1990. Acta Palaeontologica Sinica,32（4）:495-524. (in Chinese with English summary)

Wu Xiangwu（吴向午）,2006. Record of Mesozoic-Cenozoic Megafossil plant Generic names founded on Chinese specimens（1991-2000）. Acta Palaeontologica Sinica,45（1）:114-140. (in Chinese and English)

Wu Xiangwu（吴向午）,2013. Record of Mesozoic and Cenozoic Megafossil plant Generic names established for Chinese specimens（2001-2010）. Acta Palaeontologica Sinica,52（1）:121-140. (in Chinese and English)

Wu Xiuyuan（吴秀元）,Wang Jun（王军）,2004. Two new fan-veined species from the Carboniferous of Qinling Mt. , Shaanxi. Acta Palaeontologica Sinica,43（4）:489-499,pls. 1,2. (in Chinese with English summary)

Wu Xiuyuan（吴秀元）,Zhao Xiuhu（赵修祜）,1981. Fossil plants from the Kaolishan Formation（Lower Carboniferous）in Jurong, southern Jiangsu. Acta Palaeontologica Sinica,20（1）:50-59,pls. 1-3. (in Chinese with English summary)

Xiao Suzhen（肖素珍）,Zhang Enpeng（张恩鹏）,1985. Plant kingdom// Tianjin Institute of Geology and Mineral Resources ed. Palaeontological atlas of North China（1）Palaeozoic. Beijing:Geological Publishing House:530-586,pls. 164-205. (in Chinese)

Xu Honghe（徐洪河）,Wang Yi（王泽）,2002. A new Lycopsid cone from the Upper Devonian of Western Junggar Basin,Xinjiang,China. Acta Palaeontologica Sinica,41（2）:251-458,pls. 1,2. (in Chinese with English summary)

Xue Jinzhuang（薛进庄）,Hao Shougang（郝守刚）,Wang Deming（王德明）and Liu Zhenfeng（刘振峰）,2005. A new lycopsid from the Upper Devonian of China. International Journal of Plant Sciences,166（3）:519-531.

Xue Jinzhuang（薛进庄）,Hao Shougang（郝守刚）,2008. *Denglongia hubeiensis* gen. et sp. nov. ,a New Plant Attributed to Cladoxylopsida from the Upper Devonian（Frasnian）of South China. International Journal of Plant Sciences,169（9）:1314-1331,figs. 1-14.

Xue Jinzhuang（薛进庄）,2009. Two Zosterophyll Plants from the Lower Devonian（Lochkovian）Xitun Formation of Nortneastern Yunnan,China. Acta Geologica Sinica,83（3）:504-512,pl. 1,figs. 1,2.

Yabe H, Ôishi S, 1937-1938. Notes on some fossil plants from Fukien Province, China. Science Reports of Tohoku Imperial University, Sendai, Japan, Series 2 (Geology), 19 (2): 222-234, pl. 32 (1), text-figs. 1-12.

Yabe H, Shimakura M, 1940a. *Schizoneuropsis tokudae* gen. et sp. nov., a Palaeozoic plant from China. Japanese Journal of Geology and Geography, 17 (3-4): 177, 178, pl. 15.

Yabe H, Shimakura M, 1940b. *Distichopteris heteropinna* gen. et sp. nov., a Palaeozoic plant from China. Japanese Journal of Geology and Geography, 17 (3-4): 179, 180, pl. 16.

Yang Guanxiu (杨关秀) and others, 2006. The Permian Cathaysian Flora in Western Henan Province, China: Yuzhou Flora. Beijing: Geological Publishing House: 1-361, pls. 1-76. (in Chinese with English summary)

Yang Guanxiu (杨关秀), Chen Fen (陈芬), 1979. Palaeobotany // Hou Hungfei et al. eds. The coal-bearing strata and fossils of Late Permian from Guangdong. Beijing: Geological Publishing House: 104-139, pls. 16-47. (in Chinese)

Yang Guanxiu (杨关秀), 1987a. Plant fossil assemblage, stratigraphic division and palaeoclimatic analysis of Permian coal measures in Yuxian // Yang Qi ed. Depositional environments and coal-forming characteristics of Late Palaeozoic coal measures in Yuxian, Henan Province. Beijing: Geological Publishing House: 11-54, pls. 2-17. (in Chinese with English summary)

Yang Guanxiu (杨关秀), 1987b. The evolution of the Permian gigantopterids in Yuxian County, western Henan and its geological significance. Geoscience, 1 (2): 173-195, pls. 1-3, text-figs. 1-10. (in Chinese with English summary)

Yang Guanxiu (杨关秀), 1990. A new genus of slender axis cycadophytes: *Cathaysiocycas* and its evolutionary significance. Geoscience, 4 (3): 38-43, pl. 1, figs. 1, 2. (in Chinese with English summary)

Yang Xianhe (杨贤河), 1978. The vegetable kingdom (Mesozoic) // Chengdu Institute of Geology and Mineral Resources (The Southwest China Institute of Geological Science) ed. Atlas of fossils of Southwest China Sichuan volume, Part II: Carboniferous to Mesozoic. Beijing: Geological Publishing House: 469-536, pl. 156-190. (in Chinese with English title)

Yang Xianhe (杨贤河), 1986. *Sphenobaierocladus*: a new ginhgophytes genus (Sphenobaieraceae n. fam.) and its affinites. Bulletin of the Chengdu Institute of Geology and Mineral Resources, Chinese Academy of Geological Sciences, 7: 49-60, pl. 1, figs. 1, 2. (in Chinese with English summary)

Yang Xuelin (杨学林), Lih Baoxian (厉宝贤), Li Wenben (黎文本), Chow Tseyen (周志炎), Wen Shixuan (文世宣), Chen Peichi (陈丕基), Yeh Meina (叶美娜), 1978. Younger Mesozoic continental strata of the Jiaohe Basin, Jilin. Acta Stratigraphica Sinica, 2 (2): 131-145, pls. 1-3, text-figs. 1-3. (in Chinese)

Yao Zhaoqi (姚兆奇), Liu Lujun (刘陆军), Rothwell G W, Mapes G, 2000. *Szecladia* new genus, a Late Permian conifer with multiveined leaves from South China. Journal of Palaeontology, 74 (3): 527-531, text-figs. 1-5.

Yao Zhaoqi (姚兆奇), Liu Lujun (刘陆军), Zhang Shi (张士), 1993. *Lixotheca*: a new generic name for Permian cladphleboid fern. Acta Palaeontologica Sinica, 32 (5): 525-539, pls. 1-4, text-figs. 1-3. (in Chinese with English summary)

Yao Zhaoqi（姚兆奇）,Taylor T N,1988. On a new gleicheniaceous fern from the Permian of South China. Review of Palaeobotany and Palynology,54（1-2）:121-134,pls. 1-4,text-figs. 1-3.

Ye Meina（叶美娜）,Liu Xingyi（刘兴义）,Huang Guoqing（黄国清）,Chen Lixian（陈立贤）,Peng Shijiang（彭时江）,Xu Aifu（许爱福）,Zhang Bixing（张必兴）,1986. Late Triassic and Early-Middle Jurassic fossil plants from northeastern Sichuan. Hefei:Anhui Science and Technology Publishing House:1-141,pls. 1-56. (in Chinese with English summary)

Ye Meina（叶美娜）,1981. On the preparation methods of fossil cuticle // Palaeontological Society of China ed. Selected papers of the 12th Annual Conference of the Palaeontogical Society of China. Beijing:Science Press:170-179,pls. 1,2. (in Chinese with English title)

Yokoyama M,1889. Jurassic plants from Kaga and Echizen. Tokyo Univ. Coll. Sci. Jour. ,v. 3,pt. 1:1-66,pls. 1-14.

Zeiller R,1907. Note sur quelques empreintes vegetales des gîtes de charbon du Yunnan meridional. Annales des Mines de Paris,10（10）:472-503.

Zeng Yong（曾勇）,Shen Shuzhong（沈树忠）,Fan Bingheng（范炳恒）,1995. Flora from the coal-bearing strata of Yima Formation in western Henan. Nanchang:Jiangxi Science and Technology Publishing House:1-92,pls. 1-30,figs. 1-9. (in Chinese with English summary)

Zhang Caifan（张采繁）,1986. Early Jurassic flora from eastern Hunan. Professional Papers of Stratigraphy and Palaeontology,14:185-206,pls. 1-6,figs. 1-10. (in Chinese with English summary)

Zhang Hong（张泓）,1987. Plant fossils // Institute of Geology and Exploration,China Coal Research Institute,Ministry of Coal Industry,Provincial Coal Exploration Corporation of Shanxi eds. Sedimentary environment of the coal-bearing strata in Pinglu-Shuoxian Mining Area,China. Xi'an:Shaanxi People's Education Press:195-205,pls. 1-21,text-fig. 129. (in Chinese with English summary)

Zhang Shanzhen（张善桢）,Mo Zhuangguan（莫壮观）,1981. On the occurrence of cycadophytes with slender growth habit in the Permian of China. Geological Society of America,Special Paper 187:237-246,pls. 1-3.

Zhang Shanzhen（张善桢）,Mo Zhuangguan（莫壮观）,1985. New forms of seea-bearing fronds from the Cathaysia flora in Henan,China // Thomas D,Pfefferkorn H eds. Papers for the 9th International Congress of Carboniferous Stratigraphy and Geology,5. Urbana:South Illinois University Press:173-178,pls. 1-3.

Zhang Shanzhen（张善桢）,Xiao Suzhen（肖素珍）,1987. Discovery of *Sagittopteris belemnopteroides* gen. et sp. nov. from Shanxi. Acta Palaeontologica Sinica,26（2）:181-186,pls. 1,2,text-fig. 1. (in Chinese with English summary)

Zhang Wu（张武）,Fu Xiaoping（傅小平）,Zheng Shaolin（郑少林）,Li Nan（李楠）,2006. Triassc Fossil Woods // Shenzhen Urban Managerment Bureau,Shenzhen Fairy Lake Botanical Garden,Shenyang Institute of Geology and Mineral Resources,Ministry of Land and Resources eds. Fossil Woods China. Beijing:China Forestry Publishing House:101-119. (in Chinese)

Zhang Wu（张武）,Fu Xiaoping（傅小平）,Zheng Shaolin（郑少林）,Li Nan（李楠）,2008.

Triassc Fossil Woods // Shenyang Institute of Geology and Mineral Resources, Ministry of Land and Resources, Shenzhen Urban Managerment Bureau, Shenzhen Fairy Lake Botanical Garden, Nanjing Institute of Geology and Palaeontology, Chinese Academy of Sciences eds. Fossil Woods China. Beijing: China Forestry Publishing House: 101-119. (in English)

Zhang Wu(张武), Wang Yongdong(王永栋), Saiki Ken'ichi, Li Nan(李楠), Zheng Shaolin (郑少林), 2006. A structurall preserved cycad-like stem, *Lioxylon liaoningense* gen. Et sp. nov., from Middle Jurassic in Western Liaoning, China. Progress in Natural Science, 16 (Special Issue): 236-248.

Zhang Wu(张武), Wang Yongdong(王永栋), Zheng Shaolin(郑少林), Yang Xiaoju(杨小菊), Li Yong(李勇), Fu Xiaoping(傅小平), Li Nan(李楠), 2007. Taxonomic investigations no permineralized conifer woods from the Late Paleozoic Angaran deposits of northeastern Inner Mongolia, China and their palaeoclimatic significance. Review of Palaeobotany and Palynology, 144: 261-285, figs. 1-12.

Zhang Wu(张武), Zhang Zhicheng(张志诚), Zheng Shaolin(郑少林), 1980. Phyllum Pteridophyta, subphyllum Gymnospermae // Shenyang Institute of Geology and Mineral Resources ed. Paleontological atlas of Northeast China, II. Mesozoic and Cenozoic. Beijing: Geological Publishing House: 222-308, pls. 112-191, text-figs. 156-206. (in Chinese with English title)

Zhang Wu(张武), Zheng Shaolin(郑少林), 1984. New fossil plants from the Laohugou Formation (Upper Triassic) in the Jinlingsi-Yangshan Basin, western Liaoning. Acta Palaeontologica Sinica, 23(3): 382-393, pls. 1-3. (in Chinese with English summary)

Zhang Wu(张武), Zheng Shaolin(郑少林), Ding Qiuhong(丁秋红), 1999. A new genus (*Protosciadopityoxylon* gen. nov.) of Early Cretaceous fossil wood from Liaoning, China. Acta Botanica Sinica, 41(2): 1312-1316, pls. 1, 2. (in Chinese with English summary)

Zhang Wu(张武), Zheng Shaolin(郑少林), Li Yong(李勇), Li Nan(李楠), 2006. Cabniferous-Permian Fossil Woods // Shenzhen Urban Managerment Bureau, Shenzhen Fairy Lake Botanical Garden, Shenyang Institute of Geology and Mineral Resources, Ministry of Land and Resources eds. Fossil Woods China. Beijing: China Forestry Publishing House: 33-119. (in Chinese)

Zhang Wu(张武), Zheng Shaolin(郑少林), Li Yong(李勇), Li Nan(李楠), 2008. Cabniferous-Permian Fossil Woods // Shenyang Institute of Geology and Mineral Resources, Ministry of Land and Resources, Shenzhen Urban Managerment Bureau, Shenzhen Fairy Lake Botanical Garden, Nanjing Institute of Geology and Palaeontology, Chinese Academy of Sciences eds. Fossil Woods China. Beijing: China Forestry Publishing House: 33-119. (in English)

Zhao Xiuhu(赵修祜), Mo Zhuangguan(莫壮观), Zhang Shanzhen(张善桢), Yao Zhaoqi(姚兆奇), 1980. Late Permian flora from W. Guizhou and E. Yunnan // Nanjing Institute Geology and Palaeontology, Chinese Academy of Sciences ed. Stratigraphy and palaeontology of Upper Permian coal measures W. Guizhou and E. Yunnan. Beijing: Science Press: 70-99, pls. 1-23. (in Chinese)

Zhao Xiuhu(赵修祜), Wu Xiuyuan(吴秀元), Gu Qichang(顾其昌), 1986. Late Devonian flora from southern Ningxia. Acta Palaeontologica Sinica,25(5):544-559,pls. 1-5,text-figs. 1-9. (in Chinese with English summary)

Zheng Shaolin(郑少林),Gao Jiajun(高家俊),Bo Xue(薄学),2008. Two new Taxa of Monocotyledoous Angiosperm from Lower Cretaceous Yixian Formation of Beipiao, Lianing. Acta Palaeontologica Sinica,47(3):326-440. (in Chinese and English)

Zheng Shaolin(郑少林),Li Nan(李楠),Li Yong(李勇),Zhang Wu(张武),Bian Xiongfei(边雄飞),2005. A new genus of fossil Cycads *Yixianophyllum* gen. nov. from the Late Jurassic Yixian Formation, western Liaoning, China. Acta Geologica Sinica,79(5):582-592,pls. 1,2. (English Edition)

Zheng Shaolin(郑少林),Li Yong(李勇),Wang Yongdong(王永栋),Zhang Wu(张武),Yang Xiaoju(杨小菊),Li Nan(李楠),2005. Jurassic fossil wood of Sahnioxylon from wester Liaoning, China and special references to its systematic affinity. Global Geology,24(3):209-216,pls. 1,2.

Zheng Shaolin(郑少林),Wang Xin(王鑫),2010. An undercover Angiosperm from the Jurassic of China. Acta Geologica Sinica,84(4):895-902,figs. 1-4. (English Edition)

Zheng Shaolin(郑少林),Zhang Wu(张武),1983. A new genus of Pteridiaceae from Late Jurassic East Heilongjiang Province. Acta Botanica Sinica,25(4):380-384,pls. 1,2. (in Chinese with English summary)

Zheng Shaolin(郑少林),Zhang Wu(张武),1986. New discovery of Early Triassic fossil plants from western Liaoning Province. Bulletin of the Shenyang Institute of Geology and Mineral Resources, Chinese Academy of Geological Sciences,14:173-184,pls. 1-4,figs. 1-3. (in Chinese with English summary)

Zheng Shaolin(郑少林),Zhang Wu(张武),2000. Late Palaeozoic ginkgoalean woods from Northern China. Acta Palaeont Sin,39(Supplement):119-126.

Zhou Tongshun(周统顺),1978. On the Mesozoic coal-bearing strata and fossil plants from Fujian Province. Professional Papers of Stratigraphy and Palaeontology,4:88-134,pls. 15-30,text-figs. 1-5. (in Chinese)

Zhou Xianding(周贤定),1988. *Jiangxifolium*, a new genus of fossil plants from Anyuan Formation in Jiangxi. Acta Palaeontologica Sinica,27(1):125-128,pl. 1,text-fig. 1. (in Chinese with English summary)

Zhou Zhiyan(周志炎),Li Baoxian(厉宝贤),1979. A preliminary study of the Early Triassic plants from the Qionghai district, Hainan Island. Acta Palaeontologica Sinica,18(5):444-462,pls. 1,2,text-figs. 1,2. (in Chinese with English summary)

Zhou Zhiyan(周志炎),Wu Xiangwu(吴向午),Chief Compilers,2002. Chinese Bibliography of Palaeobotany (Megafossils)(1865-2000). Hefei: University of Science and Technology of China Press:1-231 (in Chinese),1-307 (in English).

Zhou Zhiyan(周志炎),Wu Xiangwu(吴向午),Zhang Bole(章伯乐),2001. Tharrisia, a new fossil leaf organ genus, with description of three Jurassic species from China. Review of Palaeobotany and Palynology,120:92-105.

Zhou Zhiyan(周志炎),Zhang Bole(章伯乐),1988a. Two new ginkgolaean female reproductive organs from the Middle Jurassic of Henan Province. Science Bulletin (Kexue Tongbao),33(3):216-217,text-fig. 1.

Zhou Zhiyan (周志炎), Zhang Bole (章伯乐), 1988b. Two new ginkgolaean female reproductive organs from the Middle Jurassic of Henan Province. Science Bulletin (Kexue Tongbao), 33 (4): 1201-1203, text-fig. 1.

Zhou Zhiyan (周志炎), Zhang Bole (章伯乐), 1989. A Middle Jurassic *Ginkgo* with ovule-bearing organs from Henan, China. Palaeontographica, B, 211: 113-133, pls. 1-8, text-figs. 1-7.

Zhou Zhiyan (周志炎), 1983. *Stalagma samara*, a new podocarpaceous conifer with monocolpate pollen from the Upper Triassic of Hunan, China. Palaeontographica, B, 185: 56-78, pls. 1-12, text-figs. 1-7.

Zhou Zhiyan (周志炎), 1984. Early Liassic Plants from southeastern Hunan, China. Palaeontologia Sinica, Whole Number 165, New Series A, 7: 1-91, pls. 1-34, text-figs. 1-14. (in Chinese with English summary)

Zhu Jianan (朱家楠), Chen Gongxin (陈公信), 1981. *Fimbriotheca tomentosa* Zhu et Chen: a new genus and species from Permian of China and its systematic position. Acta Botanica Sinica, 23 (6): 487-491, pl. 1. (in Chinese with English summary)

Zhu Jianan (朱家楠), Du Xianming (杜贤铭), 1981. A new cycad: *Primocycas chinensis* gen. et sp. nov. discovers from the Lower Permian in Shanxi, China and its significance. Acta Botanica Sinica, 23 (5): 401-404, pls. 1, 2, text-figs. 1-4. (in Chinese with English summary)

Zhu Jianan (朱家楠), Hu Yufan (胡雨帆), Du Xianming (杜贤铭), 1982. Supplementary study of Upper Shihezi (Shihhotse) plants from Taiyuan, Shanxi Province. Acta Botanica Sinica, 24 (1): 77-84, pls. 1-3, figs. 1-4. (in Chinese with English summary)

Zhu Jianan (朱家楠), Hu Yufan (胡雨帆), Feng Shaonan (冯少南), 1983. On occurrence and significance about the fossil plants from the Yuntaiguan Formation located between Hunan and Hubei. Acta Botanica Sinica, 25 (1): 75-81, pl. 1. (in Chinese with English summary)

Zhu Jianan (朱家楠), Zhang Xiusheng (张秀生), Ma Jie (马洁), 1994. A new genus and species: *Cycadostrobilus paleozoicus* Zhu of Cycadaceae from the Permian of China. Acta Phytotaxonomica Sinica, 32 (4): 340-344, pl. 1. (in Chinese with English summary and the Latin description)

Zhu Jianan (朱家楠), Zhang Xiusheng (张秀生), 1995. *Trinerviopteris cardiophylla* (Zhu et Geng) Zhu gen. et comb. nov. and the classification of gigantopterids. Acta Botanica Sinica, 37 (4): 314-320, pls. 1, 2, text-figs. 1-4. (in Chinese with English summary)

Zhu Weiqing (朱为庆), Kenrick P, 1999. A *Zosterophyllum*-like plant from the Lower Devonian of Yunnan Province, China. Review of Palaeobotany and Palynology, 105 (1-2): 111-118, pl. 1, text-figs. 1-4.

Zodrow E L, Gao Zhifeng (高志峰), 1991. *Leeites oblongifolis* nov. gen. et sp. (sphenophyllaean, Carboniferous), Sydney Coalfield, Nova Scotia, Canada. Palaeontographica, B, 223: 61-80, pls. 1-8, text-figs. 1-10.